Undergraduate Texts in Mathematics

L. R. Foulds

Optimization Techniques
An Introduction

With 72 Illustrations

Springer-Verlag
New York   Heidelberg   Berlin

L. R. Foulds
Department of Economics
University of Canterbury
Christchurch 1
New Zealand

AMS Classification: 49-01, 90-01

Library of Congress Cataloging in Publication Data

Foulds, L. R., 1948–
 Optimization techniques.

 Bibliography: p.
 Includes index.
 1. Mathematical optimization. 2. Programming
(Mathematics) I. Title.
QA402.5.F68   519   81-5642
                AACR2

9 8 7 6 5 4 3 2 1

ISBN-13:978-1-4613-9460-0   e-ISBN-13:978-1-4613-9458-7
DOI: 10.1007/978-1-4613-9458-7

This book is dedicated to the memory of my father

Richard Seddon Foulds

Contents

Preface

Optimization is the process by which the optimal solution to a problem, or optimum, is produced. The word *optimum* has come from the Latin word *optimus*, meaning best. And since the beginning of his existence Man has strived for that which is best. There has been a host of contributions, from Archimedes to the present day, scattered across many disciplines. Many of the earlier ideas, although interesting from a theoretical point of view, were originally of little practical use, as they involved a daunting amount of computational effort. Now modern computers perform calculations, whose time was once estimated in man-years, in the figurative blink of an eye. Thus it has been worthwhile to resurrect many of these earlier methods. The advent of the computer has helped bring about the unification of optimization theory into a rapidly growing branch of applied mathematics. The major objective of this book is to provide an introduction to the main optimization techniques which are at present in use. It has been written for final year undergraduates or first year graduates studying mathematics, engineering, business, or the physical or social sciences. The book does not assume much mathematical knowledge. It has an appendix containing the necessary linear algebra and basic calculus, making it virtually self-contained.

This text evolved out of the experience of teaching the material to finishing undergraduates and beginning graduates. A feature of the book is that it adopts the sound pedagogical principle that an instructor should proceed from the known to the unknown. Hence many of the ideas in the earlier chapters are introduced by means of a concrete numerical example to which the student can readily relate. This is followed by generalization to the underlying theory. The courses on which the book is based usually have a significant number of students of Business and Engineering. The interests of these people have been taken into account in the development of the courses and hence in the writing of this book. Hence many of its arguments are intuitive rather than rigorous. Indeed plausibility and clarity have been given precedence before rigour for the sake of itself.

Chapter 1 contains a brief historical account and introduces the basic terminology and concepts common to all the theory of optimization. Chapters 2 and 3 are concerned with linear programming and complications of the basic model. Chapter 2 on the simplex method, duality, and sensitivity analysis can be covered in an undergraduate course. However some of the topics in Chapter 3 such as considerations of efficiency and parametric programming, may be best left to graduate level. Chapter 4 deals with only the basic strategies of integer linear programming. It is of course dependent on Chapter 2. It does contain a number of formulations of applications of integer programming. Some of this material has never appeared before in book form. Chapter 5 is on network analysis and contains a section on using networks to analyze some practical problems.

Chapter 6 introduces dynamic programming. It is beyond the scope of this book to provide a detailed account of this vast topic. Hence techniques suitable for only deterministic, serial systems are presented. The interested reader is referred to the extensive literature. Chapter 7 serves as an introduction to Chapter 8, which is on nonlinear programming. It presents some of the classical techniques: Jacobian and Lagrangian methods together with the Kuhn–Tucker conditions. The ideas in this chapter are used in devising the more computationally efficient strategies of Chapter 8.

This text contains enough material for one semester at the undergraduate level and one more at the graduate level. The first course could contain Chapters 1, 2, the first half of Chapter 3, and parts of Chapter 4 and Chapter 5. The remainder can be covered in the second course. A plan outlining this follows.

The book contains a large number of exercises. Students are strongly encouraged to attempt them. One cannot come to grips with the concepts by solely looking at the work of others. Mathematics is not a spectator sport!

The author is grateful for this opportunity to express his thanks for the support of his employers, the University of Canterbury, which he enjoyed while finishing this book. He is also thankful for the faith and encouragement of his wife, Maureen, without which it would never have been written. He is also grateful to a number of friends including David Robinson, Hans Daellenbach, Michael Carter, Ian Coope and Susan Byrne, who read parts of the manuscript and made valuable suggestions. A vote of thanks should also go to his student, Trevor Kearney, who read the entire manuscript and discovered an embarrassing number of errors.

Christchurch, New Zealand   L. R. Foulds
November 1980

Plan of the Book

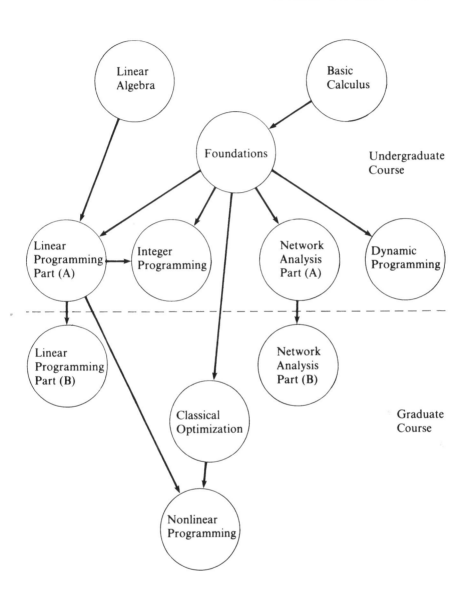

Linear Algebra

Basic Calculus

Foundations

Undergraduate Course

Linear Programming Part (A)

Integer Programming

Network Analysis Part (A)

Dynamic Programming

Linear Programming Part (B)

Network Analysis Part (B)

Classical Optimization

Graduate Course

Nonlinear Programming

Chapter 1

Introduction

1.1 Motivation for Studying Optimization

There exist an enormous variety of activities in the everyday world which can usefully be described as systems, from actual physical systems such as chemical processing plants to theoretical entities such as economic models. The efficient operation of these systems often requires an attempt at the optimization of various indices which measure the performance of the system. Sometimes these indices are quantified and represented as algebraic variables. Then values for these variables must be found which maximize the gain or profit of the system and minimize the waste or loss. The variables are assumed to be dependent upon a number of factors. Some of these factors are often under the control, or partial control, of the analyst responsible for the performance of the system.

The process of attempting to manage the limited resources of a system can usually be divided into six phases: (i) mathematical analysis of the system; (ii) construction of a mathematical model which reflects the important aspects of the system; (iii) validation of the model; (iv) manipulation of the model to produce a satisfactory, if not optimal, solution to the model; (v) implementation of the solution selected; and (vi) the introduction of a strategy which monitors the performance of the system after implementation. It is with the fourth phase, the manipulation of the model, that the theory of optimization is concerned. The other phases are very important in the management of any system and will probably require greater total effort than the optimization phase. However, in the presentation of optimization theory here it will be assumed that the other phases have been, or will be, carried out. Because the theory of optimization provides this link in the chain of systems management it is an important body of mathematical knowledge.

1.2 The Scope of Optimization

One of the most important tools of optimization is *linear programming*. A linear programming problem is specified by a linear, multivariable function which is to be optimized (maximized or minimized) subject to a number of linear constraints. The mathematician G. B. Dantzig (1963) developed an algorithm called the *simplex method* to solve problems of this type. The original simplex method has been modified into an efficient algorithm to solve large linear programming problems by computer. Problems from a wide variety of fields of human endeavor can be formulated and solved by means of linear programming. Resource allocation problems in government planning, network analysis for urban and regional planning, production planning problems in industry, and the management of transportation distribution systems are just a few. Thus linear programming is one of the successes of modern optimization theory.

Integer programming is concerned with the solution of optimization problems in which at least some of the variables must assume only integer values. In this book only integer programming problems in which all terms are linear will be covered. This subtopic is often called *integer linear programming*. However, because little is known about how to solve nonlinear integer programming problems, the word linear will be assumed here for all terms. Many problems of a combinatorial nature can be formulated in terms of integer programming. Practical examples include facility location, job sequencing in production lines, assembly line balancing, matching problems, inventory control, and machine replacement. One of the important methods for solving these problems, due to R. E. Gomory (1958), is based in part on the simplex method mentioned earlier. Another approach is of a combinatorial nature and involves reducing the original problem to smaller, hopefully easier, problems and partitioning the set of possible solutions into smaller subsets which can be analyzed more easily. This approach is called *branch and bound* or *branch and backtrack*. Two of the important contributions to this approach have been by Balas (1965) and Dakin (1965). Although a number of improvements have been made to all these methods, there does not exist as yet a relatively efficient method for solving realistically-sized integer programming problems.

Another class of problems involves the *management of a network*. Problems in traffic flow, communications, the distribution of goods, and project scheduling are often of this type. Many of these problems can be solved by the methods mentioned previously—linear or integer programming. However because these problems usually have a special structure, more efficient specialized techniques have been developed for their solution. Outstanding contributions have been made in this field by Ford and Fulkerson (1962). They developed the *labelling method* for maximizing the flow of a commodity through a network and the *out-of-kilter method* for minimizing the cost of transporting a given quantity of a commodity through a network. These ideas can be combined with those of integer programming to analyze a whole host of practical network problems.

Some problems can be decomposed into parts, the decision processes of which are then optimized. In some instances it is possible to attain the optimum for the original problem solely by discovering how to optimize these constituent parts. This decomposition process is very powerful, as it allows one to solve a series of smaller, easier problems rather than one large, intractable problem. Systems for which this approach will yield a valid optimum are called *serial multistage systems*. One of the best known techniques to attack such problems was named *dynamic programming* by the mathematician who developed it, R. E. Bellman (1957). Serial multistage systems are characterized by a process which is performed in stages, such as manufacturing processes. Rather than attempting to optimize some performance measure by looking at the problem as a whole, dynamic programming optimizes one stage at a time to produce an optimal set of decisions for the whole process. Problems from all sorts of areas, such as capital budgeting, machine reliability, and network analysis, can be viewed as serial multistage systems. Thus dynamic programming has wide applicability.

In the formulation of many optimization problems the assumption of linearity cannot be made, as it was in the case of linear programming. There do not exist general procedures for nonlinear problems. A large number of specialized algorithms have been developed to treat special cases. Many of these procedures are based on the mathematical theory concerned with analysing the structure of such problems. This theory is usually termed *classical optimization*. One of the outstanding modern contributions to this theory has been made by Kuhn and Tucker (1951) who developed what are known as the Kuhn–Tucker conditions.

The collection of techniques developed from this theory is called *nonlinear programming*. Despite the fact that many nonlinear programming problems are very difficult to solve, there are a number of practical problems which can be formulated nonlinearly and solved by existing methods. These include the design of such entities as electrical transformers, chemical processes, vapour condensors, microwave matching networks, gallium–arsenic light sources, digital filters, and also problems concerning maximum likelihood estimation and optimal parts replacement.

1.3 Optimization as a Branch of Mathematics

It can be seen from the previous section that the theory of optimization is mathematical in nature. Typically it involves the maximization or minimization of a function (sometimes unknown) which represents the performance of some system. This is carried out by the finding of values for those variables (which are both quantifiable and controllable) which cause the function to yield an optimal value. A knowledge of linear algebra and differential multivariable calculus is required in order to understand how the algorithms operate. A sound knowledge of analysis is necessary for an understanding of the theory.

Some of the problems of optimization theory can be solved by the classical techniques of advanced calculus—such as Jacobian methods and the use of Lagrange multipliers. However, most optimization problems do not satisfy the conditions necessary for solution in this manner. Of the remaining problems many, although amenable to the classical techniques, are solved more efficiently by methods designed for the purpose. Throughout recorded mathematical history a collection of such techniques has been built up. Some have been forgotten and reinvented, others received little attention until modern-day computers made them feasible.

The bulk of the material of the subject is of recent origin because many of the problems, such as traffic flow, are only now of concern and also because of the large numbers of people now available to analyze such problems. When the material is catalogued into a meaningful whole the result is a new branch of applied mathematics.

1.4 The History of Optimization

One of the first recorded instances of optimization theory concerns the finding of a geometric curve of given length which will, together with a straight line, enclose the largest possible area. Archimedes conjectured correctly that the optimal curve is a semicircle. Some of the early results are in the form of principles which attempt to describe and explain natural phenomena. One of the earliest examples was presented approximately 100 years after Archimedes' conjecture. It was formulated by Heron of Alexandria in C. 100 B.C., who postulated that light always travels by the shortest path. It was not until 1657 that Fermat correctly generalized this postulate by stating that light always travels by the path which incurs least time rather than least distance.

The fundamental problem of another branch of optimization is concerned with the choosing of a function that minimizes certain functionals. (A functional is a special type of function whose domain is a set of real-valued functions.) Two problems of this nature were known at the time of Newton. The first involves finding a curve such that the solid of revolution created by rotating the curve about a line through its endpoints causes the minimum resistance when this solid is moved through the air at constant velocity. The second problem is called the *brachistochrone*. In this problem two points in space are given. One wishes to find the shape of a curve joining the two points, such that a frictionless bead travelling on the curve from one point to the other will cover the journey in least time. This problem was posed as a competiton by John Bernoulli in 1696. The problem was successfully solved by Bernoulli himself, de l'Hôpital, Leibniz, and Newton (who took less than a day!). Problems such as these led Euler to develop the ideas involved into a systematic discipline which he called the *calculus of variations* in 1766. Also at the time of Euler many laws of mechanics were first formulated in terms of principles of optimality (examples are the least action principle of Maupertuis, the principle of least restraint of Gauss, and Lagrange's kinetic principle). Lagrange and Gauss both made other contributions. In 1760 Lagrange invented a method for solving optimization problems that had equality constraints using his *Lagrange multipliers.* Lagrange transformations are, among other uses, employed to examine the behaviour of a function in the neighbourhood of a suspected optimum. And Gauss, who made contributions to many fields, developed the method of *least squares curve fitting* which is of interest to those working in optimization as well as statistics.

In 1834 W. R. Hamilton developed a set of functions called Hamiltonians which were used in the statement of a principle of optimality that unified what was known of optics and mechanics at that time. In 1875 J. W. Gibbs presented a further principle of optimality concerned with the equilibrium of a thermodynamical system. Between that time and the present there have been increasing numbers of contributions each year. Among the most outstanding recent achievements, the works of Dantzig and of Bellman have already been mentioned. Another is the work of Pontryagin (1962) and others, who developed the *maximum principle* which is used to solve problems in the theory of optimal control.

1.5 Basic Concepts of Optimization

This section introduces some of the basic concepts of optimization. Each concept is illustrated by means of the following example.

The problem is to:

$$\text{Maximize:} \quad x_0 = f(X) = f(x_1, x_2) \tag{1.1}$$

$$\text{subject to:} \quad h_1(X) \leq 0 \tag{1.2}$$

$$x_1 \geq 0 \tag{1.3}$$

$$x_2 \geq 0. \tag{1.4}$$

This is a typical problem in the theory of optimization—the maximization (or minimization) of a real-valued function of a number of real variables (sometimes just a single variable) subject to a number of constraints (sometimes the number is zero). The special case of functionals, where the domain of the function is a set of functions, will be dealt with under the section on the calculus of variations in Chapter 7.

The function f is called the *objective function*. The set of constraints, in this case a set of inequalities, is called the *constraint set*. The problem is to find real values for x_1 and x_2, satisfying (1.2), (1.3) and (1.4), which when inserted in (1.1) will cause $f(x_1, x_2)$ to take on a value no less than that for any other such x_1, x_2 pair. Hence x_1 and x_2 are called *independent variables*.

Three objective function contours are present in Figure 1.1. The objective function has the same value at all points on each line, so that the contours can be likened to isobar lines on a weather map. Thus it is not hard to see

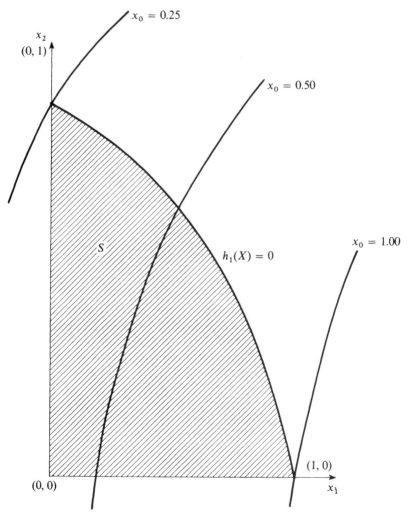

Figure 1.1. Objective function contours and the feasible region for an optimization problem.

that the solution to the problem is:

$$X^* = (x_1^*, x_2^*) = (1, 0).$$

This means that

$$f(X^*) \geq f(X) \quad \text{for all } X \in S. \tag{1.5}$$

When a solution $X^* \in S$ satisfies (1.5) it is called the *optimal solution*, and in this case the *maximal solution*. If the symbol in (1.5) were "$\leq$", X^* would be called the *minimal solution*. Also, $f(X^*)$ is called the *optimum* and is written x_0^*.

On looking at Figure 1.1 it can be seen that greater values for f could be obtained by choosing certain x_1, x_2 outside S. Any ordered pair of real numbers is called a *solution* to the problem and the corresponding value of f is called the *value* of the solution. A solution X such that

$$X \in S$$

is called a *feasible solution*.

Let us examine which x_1, x_2 pairs are likely candidates to achieve this maximization. In Figure 1.1 the set of points which satisfy this constraint set has been shaded. The set is defined as S:

$$S = \{(x_1, x_2) : h(x_1, x_2) \leq 0, x_1 \geq 0, x_2 \geq 0\}.$$

Such a set S for an optimization problem is often a connected region and is called the *feasible region*.

Many optimization problems do not have unique optimal solutions. For instance, suppose a fourth constraint

$$h_2(x_1, x_2) \leq 0 \tag{1.6}$$

is added to the problem. The feasible region is shown in Figure 1.2. In this case one of the boundaries of S coincides with an objective function contour. Thus all points on that boundary represent maximum solutions.

However, if it exists the optimal value is always unique.

As another example of a problem which does not have an optimal solution, suppose (1.2) is replaced by:

$$h_1(X) < 0. \tag{1.7}$$

On examining Figure 1.2, it becomes apparent that (1.7) does not hold for $X^* = (1, 0)$, hence $X^* \notin S$. In fact, there is no solution which will satisfy (1.5), as points successively closer to (but a positive distance away from) $(1, 0)$ correspond to successively larger x_0 values. To recognize this situation we called $f(X')$ an *upper bound for f under S* if

$$f(X') \geq f(X) \quad \text{for all } X \in S. \tag{1.8}$$

Also $f(X')$ is called a *least upper bound* or *supremum for f under S* if $f(X')$ is an upper bound for f under S and

$$f(X') \leq f(X) \quad \text{for all upper bounds } f(X) \text{ for } f \text{ under } S. \tag{1.9}$$

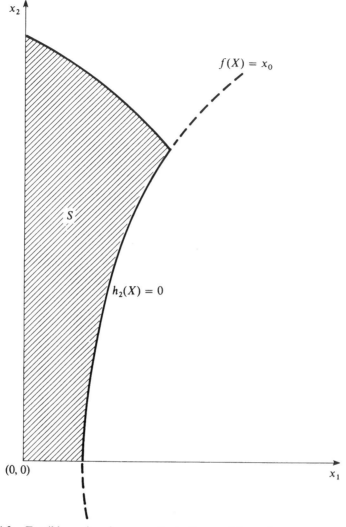

Figure 1.2. Feasible region for an optimization problem where one constraint is identical with an objective function contour.

Most of the preceding ideas have been concerned with maximization. Of course many optimization problems have the aim of minimization and each of the above concepts has a minimization counterpart. The sense of the inequalities in (1.7), (1.8), and (1.9) need to be reversed for minimization. The counterparts of the terms are:

minimum	maximum
lower bound	upper bound
greatest lower bound	least upper bound
infimum	supremum

Throughout the remainder of book we shall deal mainly with maximization problems only, because of the following theorem.

Theorem 1.1. *If* X^* *is the optimal solution to problem* P1:

$$\text{Maximize:} \quad f(X),$$
$$\text{subject to:} \quad g_j(X) = 0, \qquad j = 1, 2, \ldots, m$$
$$\qquad\qquad\quad h_j(X) \leq 0, \qquad j = 1, 2, \ldots, k$$

then X^* *is the optimal solution to problem* P2:

$$\text{Minimize:} \quad -f(X),$$
$$\text{subject to:} \quad g_j(X) = 0, \qquad j = 1, 2, \ldots, m$$
$$\qquad\qquad\quad h_j(X) \leq 0, \qquad j = 1, 2, \ldots, k.$$

PROOF. Because X^* is the optimal solution for P1, it is a feasible solution for P1, hence

$$g_j(X^*) = 0, \qquad j = 1, 2, \ldots, m$$
$$h_j(X^*) \leq 0, \qquad j = 1, 2, \ldots, k.$$

Hence X^* is a feasible solution for P2.

Also,

$$f(X^*) \geq f(X) \quad \text{for all } X \in S$$

where

$$S = \{X : g_j(X) = 0, j = 1, 2, \ldots, m; h_j(X) \leq 0, j = 1, 2, \ldots, k\}.$$

Hence

$$-f(X^*) \leq -f(X) \quad \text{for all } X \in S.$$

Hence X^* is optimal for P2. $\qquad\qquad\qquad\qquad\qquad\qquad\qquad \square$

This result allows us to solve any minimization problem by multiplying its objective function by -1 and solving a maximization problem under the same constraints. Of course we could have just as easily proven another theorem concerning the conversion of any maximization problem into an equivalent minimization problem.

Chapter 2

Linear Programming

2.1 Introduction

This present chapter is concerned with a most important area of optimization, in which the objective function and all the constraints are linear. Problems in which this is not the case fall in the nonlinear programming category and will be covered in Chapters 7 and 8.

There are a large number of real problems that can be either formulated as linear programming (L.P.) problems or formulated as models which can be successfully approximated by linear programming. Relatively small problems can readily be solved by hand, as will be explained later in the chapter. Large problems can be solved by very efficient computer programs. The mathematical structure of L.P. allows important questions to be answered concerning the sensitivity of the optimum to data changes. L.P. is also used as a subroutine in the solving of more complex problems in nonlinear and integer programming.

This chapter will begin by introducing the basic ideas of L.P. with a simple example and then generalize. A very efficient method for solving L.P. problems, the simplex method, will be developed and it will be shown how the method deals with the different types of complications that can arise. Next the idea of a dual problem is introduced with a view to analyzing the behaviour of the optimal L.P. solution when the problem is changed. This probing is called postoptimal analysis. Algorithms for special L.P. problems will also be looked at.

2.2 A Simple L.P. Problem

A coal mining company producing both lignite and anthracite finds itself in the happy state of being able to sell all the coal it can process. The present profit is \$4.00 and \$3.00 (in hundreds of dollars) for a ton of lignite and anthracite, respectively. However, because of various restrictions the cutting machine at the coal face, the screens, and the washing plant can be operated for no more than 12, 10, and 8 hours per day, respectively. It requires 3, 3, and 4 hours for the cutting machine, the screens, and the washing plant, respectively, to process one ton of lignite. It requires 4, 3, and 2 hours for the cutting machine, the screens, and the washing plant, respectively, to process one ton of anthracite. The problem is to decide how many tons of each type of coal will be produced so as to maximize daily profits.

In order to solve this problem we need to express it in mathematical terms. Toward this end the decision (independent) variables are defined as follows. Let

x_1 = the daily production of lignite in tons,
x_2 = the daily production of anthracite in tons,
x_0 = the profit gained by producing x_1 and x_2 tons of lignite and anthracite, respectively.

If x_1 tons of lignite are produced each day, and the profit per ton is \$4.00 then the daily profit for lignite is

$$\$4x_1.$$

Similarly, if x_2 tons of anthracite are produced each day with a profit of \$3.00 per ton, then the daily profit is

$$\$3x_2.$$

Thus for a daily production schedule of x_1 and x_2 tons of lignite and anthracite, the total daily profit, in dollars, is:

$$4x_1 + 3x_2 \, (=x_0).$$

It is this expression whose value we must maximize.

We can formulate similar expressions for the constraints of time on the various machines. For instance, consider the cutting operation. If x_1 tons of lignite are produced each day and each ton of lignite requires 3 hours' cutting time, then the total cutting time required to produce those x_1 tons of lignite is

$$3x_1 \text{ hours.}$$

Similarly, if x_2 tons of anthracite are produced each day with each ton taking 4 hours to cut, the total cutting time required to produce those x_1 tons of anthracite is

$$4x_2 \text{ hours.}$$

Thus the total cutting time for x_1 tons of lignite and x_2 tons of anthracite is

$$3x_1 + 4x_2.$$

But only 12 hours' cutting time are available each day. Hence we have the constraint:

$$3x_1 + 4x_2 \leq 12.$$

We can formulate similar constraints for the screening and washing times. This has been done below. The problem can now be stated mathematically:

$$\text{Maximize:} \quad 4x_1 + 3x_2 = x_0 \tag{2.1}$$

$$\text{subject to:} \quad 3x_1 + 4x_2 \leq 12 \tag{2.2}$$

$$3x_1 + 3x_2 \leq 10 \tag{2.3}$$

$$4x_1 + 2x_2 \leq 8 \tag{2.4}$$

$$x_1 \qquad \geq 0 \tag{2.5}$$

$$x_2 \geq 0. \tag{2.6}$$

The above expressions are now explained:

(2.1): The objective is to maximize daily profit.
(2.2): A maximum of 12 hours cutting time is available each day.
(2.3): A maximum of 10 hours screening time is available each day.
(2.4): A maximum of 8 hours washing time is available each day.
(2.5), (2.6): A nonnegative amount of each type of coal must be produced.

Because only two independent variables are present it is possible to solve the problem graphically. This can be achieved by first plotting the constraints (2.2)–(2.6) in two-dimensional space. The origin can be used to test which half-plane created by each constraint contains feasible points. The feasible region is shown in Figure 2.1. It can be seen that constraint (2.3) is *redundant*, in the sense that it does not define part of the boundary of the feasible region. The arrow on constraint (2.3) denotes the feasible half-plane defined by the constraint. The problem now becomes that of selecting the point in the feasible region which corresponds to the maximum objective function value—the optimum. This point is found by setting the objective function equal to a number of values and plotting the resulting lines. Clearly, the maximum value corresponds to point $(\frac{4}{5}, \frac{12}{5})$. Thus the optimal solution is

$$x_1^* = \tfrac{4}{5} \quad \text{and} \quad x_2^* = \tfrac{12}{5},$$

with value $10\frac{2}{5}$. Hence the best profit the company can hope to make is \$1,040 by producing 0.8 tons of lignite and 2.4 tons of anthracite per day.

When more than two independent variables are present, linear programs are solved by analytic methods, as it is difficult to draw in three dimensions and impossible in higher dimensions. The next section introduces the general problem.

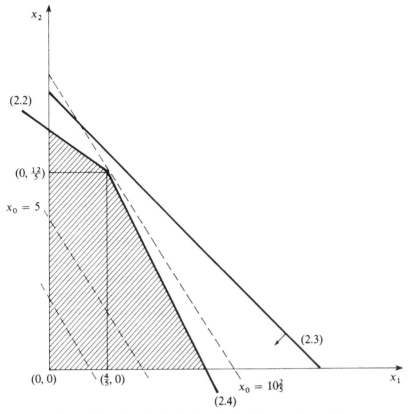

Figure 2.1.   Graphical solution to the L.P. example problem.

2.3 The General L.P. Problem

The problem of (2.1)–(2.6) can be generalized as follows:

$$\text{Maximize:} \quad c_1 x_1 + c_2 x_2 + \cdots + c_n x_n = x_0$$
$$\text{subject to:} \quad a_{11} x_1 + a_{12} x_2 + \cdots + a_{1n} x_n \le b_1$$
$$a_{21} x_1 + a_{22} x_2 + \cdots + a_{2n} x_n \le b_2$$
$$\vdots$$
$$a_{m1} x_1 + a_{m2} x_2 + \cdots + a_{mn} x_n \le b_m$$
$$x_i \ge 0, \quad i = 1, 2, \ldots, n.$$

Of course this problem can be stated in matrix form:

$$\text{Maximize:} \quad C^T X$$
$$\text{subject to:} \quad AX \le B,$$
$$X \ge 0,$$

where

$$C = (c_1, c_2, \ldots, c_n)^T$$
$$X = (x_1, x_2, \ldots, x_n)^T$$
$$A = (a_{ij})_{m \times n}$$
$$B = (b_1, b_2, \ldots, b_m)^T$$
$$O = (0)_{1 \times n}.$$

Here $(x_1, x_2, \ldots, x_n)^T$ represents the transpose of $(x_1, x_2, \ldots, x_n)$. The general minimizing linear program has an analogous form:

$$\text{Minimize:} \quad C^T X$$
$$\text{subject to:} \quad AX \geq B$$
$$X \geq 0.$$

We are now in a position to discover some basic features of the general linear programming problem.

1. The objective function and all the constraints are linear functions of the independent variables. This assumption has some important implications. It means that both the contribution of the level of each activity represented by its decision variable value (for the objective function) and the drain on resources of each activity (for the constraints) are directly proportional to the level of the activity. That is, for example, doubling the amount of a product produced will double both the profit gained by the product and the amount of each resource used on the product. It also means that both the total contribution to the objective and the total drain on each resource of all activities is, in each case, the sum of those of the individual activities.

2. The independent variables are all nonnegative. Nearly all problems which come from real situations have this property. In the few cases where this is not so, no great hardship need occur. A method for replacing variables unrestricted in sign by nonnegative ones will be explained later in this section.

3. The independent variables are all continuous. This feature does restrict the application of linear programming. It does not make sense to advocate the allocation of a noninteger number of ships to a task, as this would be indiscrete in more ways than one! When the variables concerned have relatively large values at the optimum they can often be rounded to the nearest feasible combination of integral values to yield a satisfactory solution. When this is not true, specialized artillery, collectively called integer linear programming, must be brought into service. Some of the shots that can be fired are examined in Chapter 4.

4. Each constraint involves either a "$\leq$" or a "$\geq$" sign. In many problems, one or more constraints contain an equality sign. A method for replacing

such equations by inequalities will be explained later. In the previous chapter we found that a problem with a strict inequality constraint (involving either a " $<$ " or a " $>$ " sign) does not necessarily have an optimal solution. This is also true for linear programming. Most problems from real situations do not contain strict inequality constraints, and methods for solving L.P. problems do not allow for strict inequalities. Thus we shall confine our attention to problems in which all the inequalities are nonstrict, i.e., are of the " $\leq$ " or " $\geq$ " type, not the " $<$ " or " $>$ " type.

Although all L.P. problems possess all four features outlined above, it is obvious that there can be many variations. The problem could be one of maximization or minimization, it may contain variables unrestricted in sign, and it may contain a mixture of constraint signs. Rather than devise a method for each class of problems, a method will be presented which will solve the problems of one common class. The method is completely general, as it will be shown that any L.P. problem can be made a member of the class by a series of simple steps. L.P.'s belonging to the class of interest are said to be in *standard form*.

An L.P. is in standard form if it can be expressed as:

$$\text{Maximize:} \quad C^T X \tag{2.7}$$

$$\text{subject to:} \quad AX = B \tag{2.8}$$

$$X \geq 0, \tag{2.9}$$

where

$$B \geq 0.$$

Thus the features of a problem in standard form are

1. The objective function is to be maximized.
2. All constraints except the nonnegativity conditions are strict equations.
3. The independent variables are all nonnegative.
4. The constant to the right of each equality sign in each constraint is nonnegative.

The steps that transform any L.P. into standard form are as follows.

1. A minimizing problem can be transformed into a maximizing problem by replacing the objective function by a new function in which the signs of the objective function coefficients have all been changed. (See Section 1.5).
2. Each variable unrestricted in sign can be replaced by an expression representing the difference between two new nonnegative variables. For example, if x_i is unrestricted, it is replaced by

$$x_j - x_k,$$

where x_j and x_k are new variables. Two new constraints,

$$x_j \geq 0$$
$$x_k \geq 0,$$

are added to the problem.

3. Each negative right-hand-side constraint constant can be made positive by multiplying the entire equation or inequality by minus one.
4. Each inequality constraint can be made an equation by adding a non-negative variable to a "$\leq$" constraint, or subtracting a nonnegative variable from a "$\geq$" constraint. For example, consider the constraint

$$3x_1 + 4x_2 \leq 6.$$

This becomes

$$3x_1 + 4x_2 + x_i = 6,$$

and a new constraint is added:

$$x_i \geq 0.$$

Similarly, a constraint of the form

$$5x_3 - 9x_4 \geq 18$$

becomes

$$5x_3 - 9x_4 - x_j = 18,$$

with the additional constraint:

$$x_j \geq 0.$$

Note that as all decision variables must be nonnegative the new variables which force equality must be added for "$\leq$" constraints and subtracted for "$\geq$" constraints. The new variables added to the constraints are called *slack variables*. The original variables are called *structural variables*.

The problem of Section 2.2 has the following standard form:

PROBLEM 2.1

Maximize:	$4x_1 + 3x_2$	$= x_0$	(2.10)
subject to:	$3x_1 + 4x_2 + x_3$	$= 12$	(2.11)
	$3x_1 + 3x_2 \qquad + x_4$	$= 10$	(2.12)
	$4x_1 + 2x_2 \qquad\qquad + x_5 = 8$		(2.13)
	$x_i \geq 0, \qquad i = 1, 2, \ldots, 5.$		

Now that the problem is in a form suitable to be attacked, we can consider ways to find its solution. It is apparent that realistically-sized problems will present quite a challenge and thus trial-and-error methods would be futile. Before unveiling the algorithm, some mathematical preliminaries are presented which are essential to the understanding of the method.

2.4 The Basic Concepts of Linear Programming

Consider the L.P. problem (2.7)–(2.9). Suppose that the problem has n variables and m constraints:

$$X = (x_1, x_2, \ldots, x_n)^T$$

and

$$B = (b_1, b_2, \ldots, b_m)^T.$$

A solution X is *feasible* if it satisfies (2.8) and (2.9). Let us now consider (2.8):

$$AX = B.$$

This represents a system of m equations in n unknowns.

If

$$m > n,$$

some of the constraints are redundant.

If

$$m = n,$$

and A is nonsingular (see Section 9.1.5), a unique solution can be found:

$$X = A^{-1}B.$$

If

$$m < n,$$

$n - m$ of the variables can be set equal to zero. This corresponds to the formation of an $m \times m$ submatrix $\bar{A}$ of A.

As an example of this last possibility, consider Problem 2.1, where

$$m = 3 \quad \text{and} \quad n = 5.$$

Here

$$A = \begin{pmatrix} 3 & 4 & 1 & 0 & 0 \\ 3 & 3 & 0 & 1 & 0 \\ 4 & 2 & 0 & 0 & 1 \end{pmatrix}.$$

By setting

$$x_4 = 0 \quad \text{and} \quad x_5 = 0,$$

we obtain

$$\bar{A} = \begin{pmatrix} 3 & 4 & 1 \\ 3 & 3 & 0 \\ 4 & 2 & 0 \end{pmatrix}.$$

Provided $\bar{A}$ is nonsingular, the values of the remaining variables can be found, as there are now m equations in m unknowns. Such a solution is called a *basic solution* and the m variables are called *basic variables*.

If this basic solution, which must satisfy (2.8), also satisfies (2.9) it is called a *basic feasible solution*. A basic feasible solution is called *degenerate* if at least one of the basic variables has a zero value.

A subset, S of R^n is said to be *convex* if the line segment joining any two points of S is also in S. That is, S is *convex* $\Leftrightarrow \alpha X_1 + (1 - \alpha)X_2 \in S$, for all $X_1, X_2 \in S, 0 \le \alpha \le 1$. Using this definition we can form some idea of what a convex set is like in two dimensions. In Figure 2.2, sets D and E are convex, sets F and G are not.

It is not difficult to show that the set S of all feasible solutions to a L.P. problem in standard form is convex. If the set is nonempty it must be examined in order to identify which of its points corresponds to the optimum. A point X of a convex set, S is said to be an *extreme point* of S if x cannot be expressed as:

$$X = \alpha X_1 + (1 - \alpha)X_2, \quad \text{for some } \alpha, 0 < \alpha < 1; X_1 \ne X_2; X_1, X_2 \in S.$$

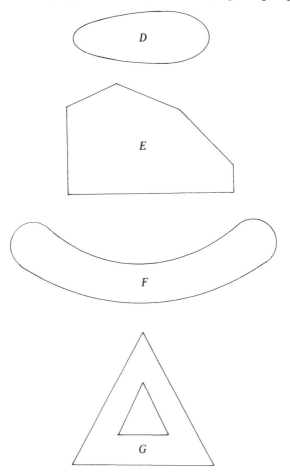

Figure 2.2.   Convex and nonconvex sets.

Suppose that the convex set of feasible solutions to an L.P. problem is denoted by S and S is bounded. Then an optimal solution to the problem corresponds to an extreme point of S. This fact considerably reduces the effort required to examine S for an optimal solution. We need examine only the extreme points of S to find an optimum. The next section introduces the method which takes advantage of this fact.

2.5 The Simplex Algorithm

2.5.1 Background

In the previous section it was noted that the optimal solution to the L.P. problem corresponds to an extreme point of the feasible region of the problem. Each extreme point can be determined by a basic solution. Now by (2.9) all the variables have to be non-negative in a feasible solution. Thus it is necessary to examine only the basic feasible solutions, rather than all basic solutions, in order to find the optimum. This amounts to examining only those extreme points for which all variables are non-negative. The algorithm is a process by which successive basic feasible solutions are identified and in which each has an objective function value which is greater than the preceding solution. Each basic feasible solution in this series is obtained from the previous one (after the first has been selected) by replacing one of the basic variables by a nonbasic variable. This is attained by setting one of the basic variables equal to zero and calculating the values of the other basic variables and the new variable (which is now part of the basis) which satisfy (2.8). This replacement of one variable by another is carried out with the following criterion in mind. The new variable that is becoming part of the basis (the entering variable) is selected so as to improve the objective function value. This happens if the nonbasic variable with the largest per unit increase is selected (as long as the solution is not degenerate). The variable to leave the basis is selected so as to guarantee that feasibility has been preserved. This procedure is repeated until no improvement in objective function value can be made. When this happens the optimal solution has found.

Consider once again Problem (2.1). Suppose we choose an initial basis of (x_3, x_4, x_5). The nonbasic variables are then x_1 and x_2, which are set equal to zero. The submatrix $\bar{A}$ corresponding to this basis is the identity matrix I and is of course nonsingular. Hence we can solve for the basic variables:

$$x_3 = 12$$
$$x_4 = 10$$
$$x_5 = 8.$$

As all these basic variables are nonnegative, we have found a basic feasible solution. The next step is to find a new basic feasible (b.f.) solution with an improved (larger) value. Recall that when a new b.f. solution is created exactly one new variable replaces one existing variable in the basis. Which variable should be brought into the basis in the present problem? On looking at (2.10) it can be seen that x_1 has the largest gain per unit (4) in the objective function. Hence it seems wise to prefer x_1 to x_2. In some cases this criterion will not always yield the greatest improvement; however, it has been shown that other criteria usually require more overall computation to find the optimum. Now that x_1 has been chosen to enter, which of x_3, x_4, or x_5 should leave the basis? Two factors must be considered:

1. We wish to allow x_1 to assume as large a value as possible in order to make the objective function take on the largest possible value.
2. The new basic solution must be feasible: all variables must be nonnegative.

How much can we increase x_1 and still satisfy factor 2? Suppose we write the constraints of (2.10) as functions of x_1:

$$x_1 = 4 - \tfrac{4}{3}x_2 - \tfrac{1}{3}x_3$$
$$x_1 = \tfrac{10}{3} - x_2 - \tfrac{1}{3}x_4$$
$$x_1 = 2 - \tfrac{1}{2}x_2 - \tfrac{1}{4}x_5.$$

Now, as

$$x_2 = 0,$$

these equations reduce to

$$x_1 = 4 - \tfrac{1}{3}x_3 \qquad\qquad (2.14)$$
$$x_1 = \tfrac{10}{3} - \tfrac{1}{3}x_4 \qquad\qquad (2.15)$$
$$x_1 = 2 - \tfrac{1}{4}x_5. \qquad\qquad (2.16)$$

Consider in turn the removal of one of x_3 or x_4 or x_5 from the basis. That is, set x_3, or x_4 or x_5 equal to zero one at a time. Here x_1 will take on the following values:

$$x_3 = 0 \Rightarrow x_1 = 4$$
$$x_4 = 0 \Rightarrow x_1 = \tfrac{10}{3}$$
$$x_5 = 0 \Rightarrow x_1 = 2.$$

Now it can be seen from (2.15) and (2.16) that

$$x_3 = 0 \Rightarrow x_4 < 0,\ x_5 < 0,$$
$$x_4 = 0 \Rightarrow x_5 < 0.$$

Hence setting either of x_3 or x_4 equal to zero will cause the new basis to be infeasible. Therefore, the leaving variable should be x_5, and the new basis is (x_1, x_3, x_4). It should be noted that the leaving variable belongs to the equation which has the minimum positive constant out of (2.14), (2.15), and (2.16). This is no coincidence, and will always occur.

Now that the new basis has been chosen, the values of its variables can be found. We have:

$$\bar{A}(x_1, x_3, x_4)^T = (12, 10, 8)^T,$$

where

$$\bar{A} = \begin{pmatrix} 3 & 1 & 0 \\ 3 & 0 & 1 \\ 4 & 0 & 0 \end{pmatrix}.$$

Thus

$$x_1 = 2$$
$$x_2 = 6$$

and

$$x_4 = 4.$$

The corresponding objective function value is 8.

What has been performed here is basically one iteration of the simplex method. In order to perform the iterations of the simplex algorithm it is convenient to set out the problem in a tableau. How this is done is discussed in the next section.

2.5.2 Canonical Form

As was mentioned previously, the calculations of the simplex method are most easily performed when the problem is set out in a tableau. We shall assume that all the inequalities of the problem are of the "$\leq$" type, with a nonnegative right-hand-side (r.h.s.) constant. Thus in converting the problem into standard form a slack variable is *added* to each inequality to transform it into an equation. Other cases shall be considered in Section 2.5.4. Problem 2.1 is of the required form and will be used for illustrative purposes.

Refer to Table 2.1. Each column of the tableau corresponds to a variable, except the last column, which corresponds to the r.h.s. of each standard form equation. For consistency, the objective function equation must be put in the same form as the constraint equations. In Problem 2.1, (2.10)

Table 2.1

Constraint	Variables						r.h.s.
	x_0	x_1	x_2	x_3	x_4	x_5	
(2.11)	0	3	4	1	0	0	12
(2.12)	0	3	3	0	1	0	10
(2.13)	0	4	2	0	0	1	8
(2.10)	1	-4	-3	0	0	0	0

can be expressed as:

$$x_0 - 4x_1 - 3x_2 = 0.$$

The x_0 column is usually not included in the tableau. Each row of the tableau corresponds to a constraint equation, the last row corresponding to the objective function. The r.h.s. entry of the objective function row equals the value of the objective function for the current basis.

We must now select an initial basis for the problem and calculate the values of the basic variables. An initial basic feasible solution can always be found by letting all the slack variables only be basic. Each basic variable then has a value equal to the r.h.s. constant of its equation. The value of this solution is zero, as all basic variables have a zero objective function coefficient. It can be seen from Table 2.1 that the coefficients in A corresponding to the basis form an identity matrix. As the simplex method is applied to the elements of the tableau, their values will be manipulated. However, at the end of each iteration, the coefficients of the current basis will form an identity matrix (within a permutation of rows) and the objective function coefficients of basis variables will be zero. A tableau which possesses this property is said to be in *canonical form*.

2.5.3 The Algorithm

Before discussing the steps of the algorithm it is necessary to make a digression into the area of matrix manipulation. It has been noted that the columns in the simplex tableau corresponding to the basic variables form an identity matrix (within a permutation of rows). When another iteration is performed (if necessary), one of the basic variables is replaced by a nonbasic variable. This new basis must have coefficients in the tableau which form an identity matrix. How is the tableau to be transformed so as to create this new identity matrix?

As an example, consider Table 2.1. It was decided that x_1 should replace x_5 in the basis. Thus the x_1 column should be manipulated until it looks like the present x_5 column. It can be shown (Hu (1969)) that Gauss–Jordan elimination can achieve this without altering the set of feasible solutions to the problem. For convenience, Table 2.1 is reproduced in Table 2.2 with extraneous matter omitted and the objective function row labelled x_0 rather than (2.10).

Table 2.2

Constraints	x_1	x_2	x_3	x_4	x_5	r.h.s.
(2.11)	3	4	1	0	0	12
(2.12)	3	3	0	1	0	10
(2.13)	④	2	0	0	1	8
x_0	-4	-3	0	0	0	0

The entry which lies at the intersection of the entering variable column and the row containing the unit element of the leaving variable is called the pivot element. It is circled in Table 2.2. The first step is to divide each element in the pivot row (the row containing the pivot element) by the pivot element. This produces Table 2.3. We have now produced a unit element in the correct position in the x_1 column.

Table 2.3

Constraints	x_1	x_2	x_3	x_4	x_5	r.h.s.
(2.11)	3	4	1	0	0	12
(2.12)	3	3	0	1	0	10
(2.13)	1	$\frac{1}{2}$	0	0	$\frac{1}{4}$	2
x_0	-4	-3	0	0	0	0

Next, each row other than the pivot row has an amount subtracted from it, element by element. The amount subtracted from each element is equal to the present entry of the corresponding pivot row element multiplied by a constant. That constant is equal to the entry in the row concerned which lies in the pivot column—the column containing the pivot element (the entering variable column.)

For example, let us subtract from the first row of Table 2.3 element by element. The constant to be subtracted is the entry in row (2.11) in the x_1 column: 3. Thus row (2.11) becomes:

$$3 - 3(1) \qquad 4 - 3(\tfrac{1}{2}) \qquad 1 - 3(0) \qquad 0 - 3(0) \qquad 0 - 3(\tfrac{1}{4}) \qquad 12 - 3(2)$$

This produces Table 2.4.

Table 2.4

Constraints	x_1	x_2	x_3	x_4	x_5	r.h.s.
(2.11)	0	$\frac{5}{2}$	1	0	$-\frac{3}{4}$	6
(2.12)	3	3	0	1	0	10
(2.13)	1	$\frac{1}{2}$	0	0	$\frac{1}{4}$	2
x_0	-4	-3	0	0	0	0

We have now produced a zero element in the first entry of the x_1 column. Performing the same operation for each other row (other than the pivot row) produces Table 2.5. The new basis (x_1, x_2, x_4) now has coefficients which form an identity matrix, (within a permutation of rows).

The simplex method can now be outlined.

1. Transform the problem into standard form.
2. Set up the initial simplex tableau.

Table 2.5

Constraints	x_1	x_2	x_3	x_4	x_5	r.h.s.
(2.11)	0	$\frac{5}{2}$	1	0	$-\frac{3}{4}$	6
(2.12)	0	$\frac{3}{2}$	0	1	$-\frac{3}{4}$	4
(2.13)	1	$\frac{1}{2}$	0	0	$\frac{1}{4}$	2
x_0	0	-1	0	0	1	8

3. Identify the negative entry which is largest in magnitude among all entries corresponding to nonbasic variables in the objective function row. Ties may be settled arbitrarily. (If all such entries are nonnegative, go to step 10). Suppose the entry in column i is identified.
4. Identify all nonnegative elements in column i.
5. For each element identified in step 4, form a ratio of the r.h.s. constant for the row of the element to the element itself.
6. Choose the minimum such ratio and identify to which row it belongs, say row j. Ties may be settled arbitrarily.
7. Identify the basic variable which has a unit entry in row j, say x_k.
8. Replace variable x_k by variable x_i in the basis using Gauss–Jordan elimination.
9. Go to step 3.
10. The optimal solution has been found. Each basic variable is set equal to the entry in the r.h.s. column corresponding to the row in which the variable has a unit entry. All other variables are set equal to zero. The optimal solution value is equal to the entry at the intersection of the x_0 row and the r.h.s. column.

Problem 2.1 will now be solved by the simplex method. Refer to Table 2.6. The initial basis is (x_3, x_4, x_5), with values

$$x_3 = 12$$
$$x_4 = 10$$
$$x_5 = 8$$

and

$$x_0 = 0.$$

Table 2.6

Constraints	x_1	x_2	x_3	x_4	x_5	r.h.s.	Ratio
(2.11)	3	4	1	0	0	12	$\frac{12}{3}$
(2.12)	3	3	0	1	0	10	$\frac{10}{3}$
(2.13)	④	2	0	0	1	8	$\frac{8}{4}$
x_0	-4	-3	0	0	0	0	

Table 2.7

Constraints	x_1	x_2	x_3	x_4	x_5	r.h.s.	Ratio
(2.11)	0	$(\frac{5}{2})$	1	0	$-\frac{3}{4}$	6	$\frac{12}{5}$
(2.12)	0	$\frac{3}{2}$	0	1	$-\frac{3}{4}$	4	$\frac{8}{3}$
(2.13)	1	$\frac{1}{2}$	0	0	$\frac{1}{4}$	2	$\frac{4}{1}$
x_0	0	-1	0	0	1	8	

The entering variable is x_1, as it has the smallest x_0-row (objective-function-row) coefficient. The leaving variable is x_5, as it has a unit element in the row corresponding to the minimum ratio $(\frac{8}{4})$. The pivot element has been circled. Gauss–Jordan elimination produces Table 2.7.

The new basis is (x_1, x_3, x_4), with values

$$x_1 = 2$$
$$x_3 = 6$$
$$x_4 = 4$$

and

$$x_0 = 8.$$

The entering variable is x_2, as it has the smallest x_0-row coefficient (-1). The leaving variable is x_3, as it has a unit element in the row corresponding to the minimum ratio $(\frac{12}{5})$. The pivot element has been circled. Gauss–Jordan elimination produces Table 2.8.

As there are no more negative entries in the x_0 row, the optimal solution has been found. It can be read off from the tableau, the basic variables being equal to the r.h.s. values of the rows in which their column entry is a unit element. Thus

$$x_1^* = \frac{4}{5}$$
$$x_2^* = \frac{12}{5}$$
$$x_4^* = \frac{2}{5}.$$

All other variables are zero. The optimum is

$$x_0^* = \frac{52}{5}.$$

Table 2.8

Constraints	x_1	x_2	x_3	x_4	x_5	r.h.s.
(2.11)	0	1	$\frac{2}{5}$	0	$-\frac{3}{10}$	$\frac{12}{5}$
(2.12)	0	0	$-\frac{3}{5}$	1	$-\frac{3}{10}$	$\frac{2}{5}$
(2.13)	1	0	$-\frac{1}{5}$	0	$\frac{2}{5}$	$\frac{4}{5}$
x_0	0	0	$\frac{2}{5}$	0	$\frac{7}{10}$	$\frac{52}{5}$

The slack variables in constraints (2.11) and (2.13) are zero at the optimal solution. This means that the amount of resource available in each of these constraints (cutting and washing time, respectively) is to be fully used. There are 12 and 8 hours' cutting and washing time available per day, respectively, and all this is going to be used in the optimal solution. Such constraints are called *binding* constraints. The slack variable of constraint (2.12) is positive. This means that not all of the available screening time of 10 hours is to be used. The amount unused per day is equal to the optimal value of the slack variable, $\frac{2}{5}$ hour. A constraint such as (2.12) is called a *slack* constraint.

The algorithm presented above is designed to solve maximization problems only. A minimization problem can be converted into a maximization problem by maximizing the negative of its objective function. However, the algorithm can instead be easily modified to solve such problems directly. At the beginning of each iteration in which a minimization problem is being solved, the x_0-row element that is the minimum of all negative elements is identified. The column of this element becomes the pivot column. The iteration then proceeds as before. When all elements in the x_0 row are nonnegative the optimum has been found.

2.5.4 Artificial Variables

Until now it has been assumed that all constraints in the linear programming problem were of the "≤" type. This allowed slack variables to be *added* to (rather than subtracted from) each inequality to transform it to an equation. The positive unit coefficients of these slack variables meant that an identity submatrix was present in A. Thus the collection of slack variables conveniently formed an initial basis which represented a basic feasible solution. Hence the simplex algorithm could be easily initiated using this easily found basis.

With constraints of the "=" or "≥" type the procedures differ: no slack variable need be introduced in the former case and the slack variable is *subtracted* in the latter, so each equation does not necessarily contain a unique element with a positive unit element as coefficient. Therefore an identity submatrix of A is not necessarily present. As many problems contain constraints of these types, we must develop a systematic method for creating an initial feasible basis so that the simplex algorithm can be used.

2.5.4.1 *The Big M Method*

The problem is first transformed into standard form. Next a new variable is added to the left-hand side of each constraint equation which was of the "=" or "≥" type. The collection of these variables together with the slack variables in the equations that were of the "≤" type form the initial feasible basis. As with all other variables these new variables are constrained to be non-negative. Any feasible solution must contain these new variables all at the zero level, for any positive new variable causes its constraint to be violated. In order to ensure that all the new variables are forced to zero in any feasible solution, each is included in the objective function. The coefficient of each new variable in the objective function is assigned a relatively large negative (positive) value for a maximization (minimization) problem. These coefficients are usually represented by the symbol M. Thus this technique is sometimes called the *big M method*. The new variables introduced have no physical interpretation and are called *artificial* variables.

To illustrate the method, suppose an additional constraint is added to Problem 2.1. Because of contractual commitments at least one ton of coal must be produced and the buyers are not concerned about the ratio of lignite to anthracite. The new constraint is

$$x_1 + x_2 \geq 1, \tag{2.17}$$

which, on the introduction of the slack variable x_6, becomes

$$x_1 + x_2 - x_6 = 1.$$

When the artificial variable x_7 is introduced we have

$$x_1 + x_2 - x_6 + x_7 = 1,$$

and the new objective function is

$$x_0 = 4x_1 + 3x_2 - Mx_7.$$

In mathematical form the new problem is

PROBLEM 2.2

$$\begin{array}{lllll}
\text{Maximize:} & 4x_1 + 3x_2 & & -Mx_7 = x_0 \\
\text{subject to:} & 3x_1 + 4x_2 + x_3 & & = 12 \\
& 3x_1 + 3x_2 & + x_4 & = 10 \\
& 4x_1 + 2x_2 & + x_5 & = 8 \\
& x_1 + x_2 & -x_6 + x_7 = 1 \\
& x_j \geq 0, & i = 1, 2, \ldots, 7.
\end{array}$$

The feasible region for this problem is shown in Figure 2.3. The optimal solution remains unchanged because the optimal solution of the previous problem is still a solution to the new problem, whose feasible set is a subset of the original feasible set. The initial tableau for the problem is displayed in Table 2.9.

The initial basis is (x_3, x_4, x_5, x_7). However, because the objective function coefficient of the basic variable x_7 is nonzero, the tableau is not yet in canonical form. Gauss–Jordan elimination is used to remedy this by replacing the x_0 row by the sum of the x_0 row and $-M$ times (2.17). This creates

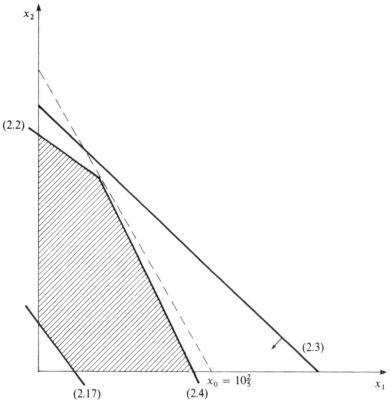

Figure 2.3.   The graphical solution to the expanded example problem.

Table 2.9

Constraints	x_1	x_2	x_3	x_4	x_5	x_6	x_7	r.h.s.
(2.11)	3	4	1	0	0	0	0	12
(2.12)	3	3	0	1	0	0	0	10
(2.13)	4	2	0	0	1	0	0	8
(2.17)	1	1	0	0	0	-1	1	1
x_0	-4	-3	0	0	0	0	M	0

Table 2.10. The simplex iterations required to reach the optimal solution are displayed in Tables 2.11–2.13. The optimal solution is

$$x_1^* = \tfrac{4}{5}, \qquad x_2^* = \tfrac{12}{5}$$
$$x_4^* = \tfrac{2}{5}, \qquad x_6^* = \tfrac{11}{5}$$
$$x_3^*, x_5^*, x_7^* = 0$$
$$x_0^* = \tfrac{52}{5}.$$

Table 2.10

Constraints	x_1	x_2	x_3	x_4	x_5	x_6	x_7	r.h.s.	Ratio
(2.11)	3	4	1	0	0	0	0	12	$\frac{12}{3}$
(2.12)	3	3	0	1	0	0	0	10	$\frac{10}{3}$
(2.13)	4	2	0	0	1	0	0	8	$\frac{8}{4}$
(2.17)	①	1	0	0	0	-1	1	1	$\frac{1}{1}$
x_0	$-M-4$	$-M-3$	0	0	0	M	0	$-M$	

Table 2.11

Constraints	x_1	x_2	x_3	x_4	x_5	x_6	x_7	r.h.s.	Ratio
(2.11)	0	1	1	0	0	3	-3	9	$\frac{9}{3}$
(2.12)	0	0	0	1	0	3	-3	7	$\frac{7}{3}$
(2.13)	0	-2	0	0	1	④	-4	4	$\frac{4}{4}$
(2.17)	1	1	0	0	0	-1	1	1	
x_0	0	1	0	0	0	-4	$(M+4)$	4	

Table 2.12

Constraints	x_1	x_2	x_3	x_4	x_5	x_6	x_7	r.h.s.	Ratio
(2.11)	0	⑤⁄₂ $\left(\frac{5}{2}\right)$	1	0	$-\frac{3}{4}$	0	0	6	$\frac{12}{5}$
(2.12)	0	$\frac{3}{2}$	0	1	$-\frac{3}{4}$	0	0	4	$\frac{8}{3}$
(2.13)	0	$-\frac{1}{2}$	0	0	$\frac{1}{4}$	1	-1	1	—
(2.17)	1	$\frac{1}{2}$	0	0	$\frac{1}{4}$	0	0	2	$\frac{4}{1}$
x_0	0	-1	0	0	1	0	M	8	

Table 2.13

Constraints	x_1	x_2	x_3	x_4	x_5	x_6	x_7	r.h.s.
(2.11)	0	1	$\frac{2}{5}$	0	$-\frac{3}{10}$	0	0	$\frac{12}{5}$
(2.12)	0	0	$-\frac{3}{5}$	1	$-\frac{3}{10}$	0	0	$\frac{2}{5}$
(2.13)	0	0	$\frac{1}{5}$	0	$\frac{1}{10}$	1	-1	$\frac{11}{5}$
(2.17)	1	0	$-\frac{1}{5}$	0	$\frac{2}{5}$	0	0	$\frac{4}{5}$
x_0	0	0	$\frac{2}{5}$	0	$\frac{7}{10}$	0	M	$\frac{52}{5}$

2.5.4.2 *The Two-Phase Method*

There exists another method for finding an initial feasible solution to an L.P. problem with "=" or "$\geq$" constraints. It is called the *two-phase method*. Phase I of the method begins by introducing slack and artificial variables as before. The objective function is then replaced by the sum of the artificial variables. In terms of the present example, the new objective function is $x_0' = x_7$. This creates a new problem in which this new objective function is to be *minimized* subject to the original constraints.

When this minimization has taken place, the optimal solution value is analyzed. A value greater than zero indicates that the original problem does not have a feasible solution. A value of zero corresponds to a solution which is basic and feasible for the original problem as all the artificial variables have value zero. In this case the original objective function is substituted in the x_0 row of the final tableau, and this basic feasible solution without the artificial variables is used as a starting solution for further iterations of the simplex method. This is phase II.

The two-phase method is usually preferred to the big M method as it does not involve the problem of roundoff error that occurs in using the large value assigned to M. It will now be illustrated by employing Problem 2.2.

PHASE I

$$\text{Minimize:} \qquad\qquad\qquad\qquad\qquad\qquad\qquad\qquad x_7 = x_0'$$

$$\text{Subject to:} \quad 3x_1 + 4x_2 + x_3 \qquad\qquad\qquad\qquad = 12$$
$$3x_1 + 3x_2 + \qquad x_4 \qquad\qquad\qquad = 10$$
$$4x_1 + 2x_2 \qquad\qquad\quad + x_5 \qquad\qquad = 8$$
$$x_1 + x_2 \qquad\qquad\qquad\qquad - x_6 + x_7 = 1$$
$$x_i \geq 0, \qquad i = 1, 2, \ldots, 7.$$

Table 2.14 shows the initial tableau for phase I. Note that the x_0-row coefficient of x_7 is $+1$ rather than -1 as the objective has been changed to one of maximization. Transforming the problem to canonical form, we

Table 2.14

Constraints	x_1	x_2	x_3	x_4	x_5	x_6	x_7	r.h.s.
(2.11)	3	4	1	0	0	0	0	12
(2.12)	3	3	0	1	0	0	0	10
(2.13)	4	2	0	0	1	0	0	8
(2.17)	1	1	0	0	0	-1	1	1
x_0'	0	0	0	0	0	0	1	0

Table 2.15

Constraints	x_1	x_2	x_3	x_4	x_5	x_6	x_7	r.h.s.	Ratio
(2.11)	3	4	1	0	0	0	0	12	$\frac{12}{3}$
(2.12)	3	3	0	1	0	0	0	10	$\frac{10}{3}$
(2.13)	4	2	0	0	1	0	0	8	$\frac{8}{4}$
(2.17)	①	1	0	0	0	-1	1	1	$\frac{1}{1}$
x_0'	-1	-1	0	0	0	1	0	-1	

Table 2.16

Constraints	x_1	x_2	x_3	x_4	x_5	x_6	x_7	r.h.s.
(2.11)	0	1	1	0	0	3	-3	9
(2.12)	0	0	0	1	0	3	-3	7
(2.13)	0	-2	0	0	1	4	-4	4
(2.17)	1	1	0	0	0	-1	1	1
x_0'	0	0	0	0	0	0	1	0

obtain Tables 2.15 and 2.16. It is clear from Table 2.16 that phase I is now complete, as the objective function has value zero. (Note that the objective can never attain an optimal negative value as it is the sum of a set of variables all constrained to be nonnegative.) The solution in Table 2.16 represents a basic feasible solution to the original problem.

PHASE II. The original objective function is substituted, neglecting the artificial variable x_7. This gives Table 2.17, which is expressed in canonical form as Table 2.18. Subsequent iterations are shown in Tables 2.19 and 2.20.

Table 2.20 displays the same optimal solution as that found by the big M method in Table 2.13. It can be seen that the iterations in phase II are identical to those of the big M method. This is no coincidence, and will always happen.

Table 2.17

Constraints	x_1	x_2	x_3	x_4	x_5	x_6	r.h.s.
(2.11)	0	1	1	0	0	3	9
(2.12)	0	0	0	1	0	3	7
(2.13)	0	-2	0	0	1	4	4
(2.17)	1	1	0	0	0	-1	1
x_0	-4	-3	0	0	0	0	0

Table 2.18

Constraints	x_1	x_2	x_3	x_4	x_5	x_6	r.h.s.	Ratio
(2.11)	0	1	1	0	0	3	9	$\frac{9}{3}$
(2.12)	0	0	0	1	0	3	7	$\frac{7}{3}$
(2.13)	0	-2	0	0	1	④	4	$\frac{4}{4}$
(2.17)	1	1	0	0	0	-1	1	—
x_0	0	1	0	0	0	-4	4	

Table 2.19

Constraints	x_1	x_2	x_3	x_4	x_5	x_6	r.h.s.	Ratio
(2.11)	0	$\frac{5}{2}$	1	0	$-\frac{3}{4}$	0	6	$\frac{12}{5}$
(2.12)	0	$\frac{3}{2}$	0	1	$-\frac{3}{4}$	0	4	$\frac{8}{3}$
(2.13)	0	$-\frac{1}{2}$	0	0	$\frac{1}{4}$	1	1	—
(2.17)	1	$\frac{1}{2}$	0	0	$\frac{1}{4}$	0	2	$\frac{4}{1}$
x_0	0	-1	0	0	1	0	8	

Table 2.20

Constraints	x_1	x_2	x_3	x_4	x_5	x_6	r.h.s.
(2.11)	0	1	$\frac{2}{5}$	0	$-\frac{3}{10}$	0	$\frac{12}{5}$
(2.12)	0	0	$-\frac{3}{5}$	1	$-\frac{3}{10}$	0	$\frac{2}{5}$
(2.13)	0	0	$\frac{1}{5}$	0	$\frac{1}{10}$	1	$\frac{11}{5}$
(2.17)	1	0	$-\frac{1}{5}$	0	$\frac{2}{5}$	0	$\frac{4}{5}$
x_0	0	0	$\frac{2}{5}$	0	$\frac{7}{10}$	0	$\frac{52}{5}$

2.5.5 Multiple Optimal Solutions

Suppose that in order to compete with other companies in the sale of lignite, the firm must reduce its price per ton. The profit is now \$3 per ton. In order to compensate, the profit on anthracite is raised to \$4/ton. Although the feasible region of the problem remains unchanged, as given in Figure 2.1, the new objective function is:

$$x_0 = 3x_1 + 4x_2.$$

The problem is solved graphically in Figure 2.4. When the objective function is drawn at the optimal level, it coincides with constraint line (2.2). This means that all points on the line from point $(0, 3)$ to $(\frac{4}{5}, \frac{12}{5})$ represent optimal solutions. This situation can be stated as follows:

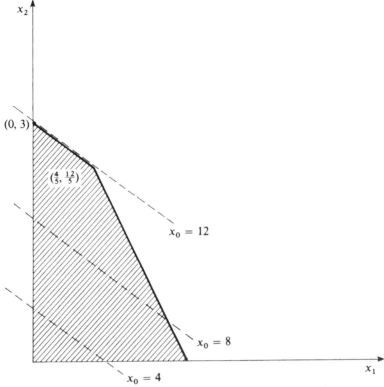

Figure 2.4.   An L.P. problem with multiple optimal solutions.

$$3x_1^* + 4x_2^* = 12$$
$$0 \leq x_1^* \leq \tfrac{4}{5}$$
$$\tfrac{12}{5} \leq x_2^* \leq 3$$
$$x_0^* = 12.$$

Note that for the multiple optimal solutions to be present the objective function line, plane, or hyperplane (in two, three, or more dimensions, respectively) must be parallel to that of a binding constraint. When this occurs there is always an infinite number of optimal solutions (except when the solution is degenerate,—see Section 2.5.6).

The problem is now solved using the simplex method (see Tables 2.21 and 2.22). Table 2.22 yields the following optimal solution:

$$x_2^* = 3$$
$$x_4^* = 1$$
$$x_5^* = 2$$
$$x_1^*, x_3^* = 0$$
$$x_0^* = 12.$$

Table 2.21

Constraints	Variable					r.h.s.	Ratio
	x_1	x_2	x_3	x_4	x_5		
(2.11)	3	④	1	0	0	12	$\frac{12}{4}$
(2.12)	3	3	0	1	0	10	$\frac{10}{3}$
(2.13)	4	2	0	0	1	8	$\frac{8}{2}$
x_0	-3	-4	0	0	0	0	

Table 2.22

Constraints	x_1	x_2	x_3	x_4	x_5	r.h.s.	Ratio
(2.11)	$\frac{3}{4}$	1	$\frac{1}{4}$	0	0	3	$\frac{12}{3}$
(2.12)	$\frac{3}{4}$	0	$-\frac{3}{4}$	1	0	1	$\frac{4}{3}$
(2.13)	$\frac{5}{2}$	0	$-\frac{1}{2}$	0	1	2	$\frac{4}{5}$
x_0	0	0	1	0	0	12	

However, the nonbasic variable x_1 has a zero x_0-row coefficient, indicating that the objective function value would remain unchanged if x_1 was brought into the basis. This is carried out in Table 2.23, this tableau yields the optimal solution:

$$x_1^* = \tfrac{4}{5}$$
$$x_2^* = \tfrac{12}{5}$$
$$x_4^* = \tfrac{2}{5}$$
$$x_3^*, x_5^* = 0$$

and

$$x_0^* = 12.$$

Of course the x_0-row value of x_5 is zero, indicating that x_5 could replace x_1 in the basis at no change in objective function value. This would produce Table. 2.22.

The significance of this example is that we have discovered two basic optimal solutions. It is straightforward to prove that if more than one basic

Table 2.23

Constraint	x_1	x_2	x_3	x_4	x_5	r.h.s.
(2.11)	0	1	$\frac{2}{5}$	0	$-\frac{3}{10}$	$\frac{12}{5}$
(2.12)	0	0	$-\frac{3}{5}$	1	$-\frac{3}{10}$	$\frac{2}{5}$
(2.13)	1	0	$-\frac{1}{5}$	0	$\frac{2}{5}$	$\frac{4}{5}$
x_0	0	0	1	0	0	12

feasible solution is optimal, then any linear combination of those points is also optimal (see, for example, Gass (1969)). As we have seen from Figure 2.4, any point on the line segment joining the two basic optimal solutions is optimal. Multiple optimal solutions are present if nonbasic variables have zero entries in the x_0-row of the simplex tableau which displays an optimal solution.

2.5.6 Degeneracy

Suppose that the management of the mining company would like to reduce the number of hours of screening time available each day. They reason that, as it is not all being used in the present optimal plan, why not reduce it? Exactly $9\frac{3}{5}$ hours are used daily, so this becomes the amount available. Mathematically the new problem is the same as Problem 2.1, except that constraint (2.12) is replaced by

$$3x_1 + 3x_2 + x_4 = 9\tfrac{3}{5}. \qquad (2.18)$$

This problem is solved graphically in Figure 2.5. Notice that constraint (2.18) coincides with exactly one point of the feasible region—the optimal

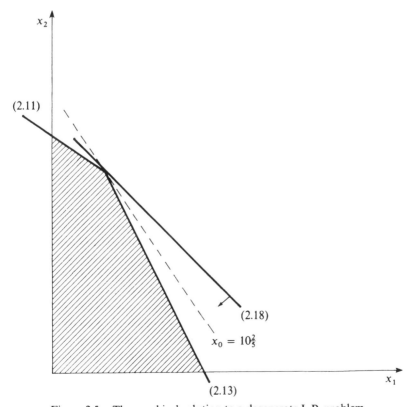

Figure 2.5.   The graphical solution to a degenerate L.P. problem.

Table 2.24

Constraints	x_1	x_2	x_3	x_4	x_5	r.h.s.	Ratio
(2.11)	3	4	1	0	0	12	$\frac{12}{3}$
(2.18)	3	3	0	1	0	$9\frac{3}{5}$	$\frac{16}{5}$
(2.13)	④	2	0	0	1	8	$\frac{8}{4}$
x_0	-4	-3	0	0	0	0	

Table 2.25

Constraints	x_1	x_2	x_3	x_4	x_5	r.h.s.	Ratio
(2.11)	0	$\left(\frac{5}{2}\right)$	1	0	$-\frac{3}{4}$	6	$\frac{12}{5}$
(2.18)	0	$\frac{3}{2}$	0	1	$-\frac{3}{4}$	$3\frac{3}{5}$	$\frac{12}{5}$
(2.13)	1	$\frac{1}{2}$	0	0	$\frac{1}{4}$	2	$\frac{4}{1}$
x_0	0	-1	0	0	1	8	

Table 2.26

Constraints	x_1	x_2	x_3	x_4	x_5	r.h.s.
(2.11)	0	1	$\frac{2}{5}$	0	$-\frac{3}{10}$	$\frac{12}{5}$
(2.18)	0	0	$-\frac{3}{5}$	1	$-\frac{3}{10}$	0
(2.13)	1	0	$-\frac{1}{5}$	0	$\frac{2}{5}$	$\frac{4}{5}$
x_0	0	0	$\frac{2}{5}$	0	$\frac{7}{10}$	$\frac{52}{5}$

point. This problem is solved by the simplex method in Tables 2.24–2.26. It can be seen from the tableau of Table 2.25 that x_2 should enter the basis. However, on forming the ratios to decide which variable leaves the basis, a tie occurs. Whenever this happens the next iteration will produce one or more basic variables with value zero. Such basic feasible solutions are called *degenerate* solutions. As it happens, we have reached the optimum in the same tableau as the first instance of degeneracy, so no problems occur. However, if Table 2.26 did not display the optimum, complications might have arisen. These are best explained by means of another example.

Suppose that a new screening plant is built and it now takes 4 hours to process one ton of lignite and 1 hour to process one ton of anthracite. There are 8 hours' screening time available per day. This means that the problem is the same as Problem 2.1 except that constraint (2.12) is replaced by

$$4x_1 + x_2 + x_4 = 8. \qquad (2.19)$$

The problem is solved graphically in Figure 2.6. It is solved in by the simplex method in Tables 2.27–2.29.

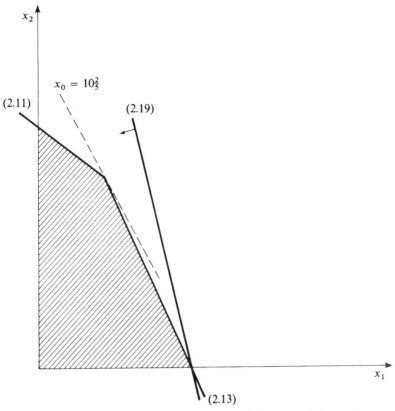

Figure 2.6.   The graphical solution to a second degenerate L.P. problem.

Table 2.27

Constraints	x_1	x_2	x_3	x_4	x_5	r.h.s.	Ratio
(2.11)	3	4	1	0	0	12	$\frac{12}{3}$
(2.19)	4	1	0	1	0	8	$\frac{8}{4}$
(2.13)	④	2	0	0	1	8	$\frac{8}{4}$
x_0	-4	-3	0	0	0	0	

Table 2.28

Constraints	x_1	x_2	x_3	x_4	x_5	r.h.s.	Ratio
(2.11)	0	$\left(\frac{5}{2}\right)$	1	0	$-\frac{3}{4}$	6	$\frac{12}{5}$
(2.19)	0	-1	0	1	-1	0	—
(2.13)	1	$\frac{1}{2}$	0	0	$\frac{1}{4}$	2	$\frac{4}{1}$
x_0	0	-1	0	0	1	8	

Table 2.29

Constraints	x_1	x_2	x_3	x_4	x_5	r.h.s.
(2.11)	0	1	$\frac{2}{5}$	0	$-\frac{3}{10}$	$\frac{12}{5}$
(2.19)	0	0	$\frac{2}{5}$	1	$-\frac{13}{10}$	$\frac{12}{5}$
(2.13)	1	0	$-\frac{1}{5}$	0	$\frac{2}{5}$	$\frac{4}{5}$
x_0	0	0	$\frac{2}{5}$	0	$\frac{7}{5}$	$\frac{52}{5}$

We can see from Table 2.28 that one of the basic feasible solutions produced by the simplex method was degenerate, as the variable x_4 has zero value. However, there is no degeneracy in the tableau of the next iteration. This is because the entering-variable (x_2) coefficient is negative (-1) in (2.19). Thus no ratio is formed, hence the dash in the ratio column. What would have happened if that x_2 coefficient had been positive and the appropriate ratio was formed? This would have caused the ratio to be zero. Thus that x_2 coefficient would become the pivot element. Then the next basic feasible solution would also be degenerate. Also there would be no improvement in the value of the objective function.

But the simplex algorithm assumes that each new basic feasible solution value is an improvement over the preceding one. When this does not happen, there is a danger that eventually a previous basis will reappear, and an endless series of iterations will be performed, with no improvement in the objective function value. And the optimal solution would never be found. This unhappy phenomonen is termed *cycling*.

Degeneracy occurs often in realistic large-scale problems. However, there do not appear to be any reported cases of cycling of the simplex technique in solving realistic problems. Because of this most computer codes do not contain measures to prevent cycling. This appears to be quite safe, because the accumulation of rounding errors will usually prevent any basic variable from assuming a value of exactly zero. There are a number of theoretical techniques which do prevent cycling (see, for example, Gass (1969)).

In the previous paragraph we asked the question, what would happen if the x_2 coefficient in the (2.19) row of Table 2.28 had been positive. This will come about if constraint (2.19) is replaced by

$$4x_1 + 2\tfrac{1}{2}x_2 + x_4 = 8 \qquad (2.20)$$

in Table 2.20. We are solving the following problem:

$$
\begin{aligned}
\text{Maximize:} \quad & 4x_1 + 3x_2 & = x_0 \\
\text{subject to:} \quad & 3x_1 + 4x_2 + x_3 & = 12 \\
& 4x_1 + 2\tfrac{1}{2}x_2 + x_4 & = 8 \\
& 4x_1 + 2x_2 + x_5 & = 8 \\
& x_i \geq 0, \quad i = 1, 2, \ldots, 5.
\end{aligned}
$$

$$\text{Maximize:} \quad 3x_1 + 3x_2 \qquad\qquad = x_0$$

$$\text{subject to:} \quad 3x_1 + 4x_2 + x_3 \qquad\qquad = 12 \qquad (2.11)$$

$$3x_1 + 3x_2 \quad\; + x_4 \qquad\; = 9\tfrac{3}{5} \qquad (2.18)$$

$$4x_1 + 2x_2 \qquad\qquad + x_5 = 8 \qquad (2.13)$$

$$x_1, x_2, \ldots, x_5 \geq 0.$$

This problem is solved graphically in Figure 2.8 and analytically in the following Tables 2.34–2.36. The optimal solution is

$$x_1^* = \tfrac{4}{5}$$
$$x_2^* = \tfrac{12}{5}$$
$$x_3^* = x_4^* = x_5^* = 0$$
$$x_0^* = 9\tfrac{3}{5}.$$

Consider Table 2.36. As there are no nonbasic x_0-row coefficients with zero value, there are no alternative optimal solutions. Yet the objective

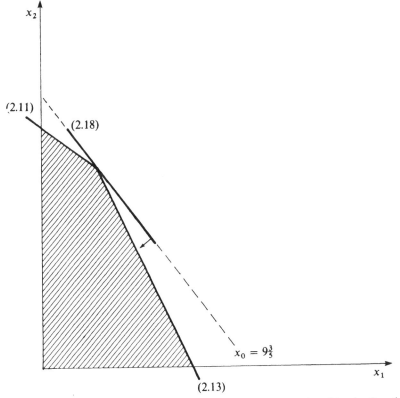

Figure 2.8. The optimal solution to an L.P. problem in which the objective function is parallel to a binding redundant constraint.

Table 2.34

Constraints	x_1	x_2	x_3	x_4	x_5	r.h.s.	Ratio
(2.11)	3	4	1	0	0	12	$\frac{12}{3}$
(2.18)	3	3	0	1	0	$9\frac{3}{5}$	$\frac{16}{5}$
(2.13)	④	2	0	0	1	8	$\frac{8}{4}$
x_0	-3	-3	0	0	0	0	

Table 2.35

Constraints	x_1	x_2	x_3	x_4	x_5	r.h.s.	Ratio
(2.11)	0	$\textcircled{\frac{5}{2}}$	1	0	$-\frac{3}{4}$	6	$\frac{12}{5}$
(2.18)	0	$\frac{3}{2}$	0	1	$-\frac{3}{4}$	$3\frac{3}{5}$	$\frac{12}{5}$
(2.13)	1	$\frac{1}{2}$	0	0	$\frac{1}{4}$	2	$\frac{4}{1}$
x_0	0	$-\frac{3}{2}$	0	0	$\frac{3}{4}$	6	

Table 2.36

Constraints	x_1	x_2	x_3	x_4	x_5	r.h.s.
(2.11)	0	1	$\frac{2}{5}$	0	$-\frac{3}{10}$	$\frac{12}{5}$
(2.18)	0	0	$-\frac{3}{5}$	1	$-\frac{3}{10}$	0
(2.13)	1	0	$-\frac{1}{5}$	0	$\frac{2}{5}$	$\frac{4}{5}$
x_0	0	0	$\frac{3}{5}$	0	$\frac{3}{10}$	$9\frac{3}{5}$

function is parallel to the binding constraint (2.18). (Constraint (2.18) is binding because its slack variable, x_4, has zero optimal value.) This is possible because the constraint is redundant (although binding), as shown in Figure 2.8. This creates a degenerate optimal solution.

2.5.7 Nonexistent Feasible Solutions

Suppose now that the company management, heartened by the efficiency of the L.P. approach, demands a plan that guarantees that at least 4 tons of coal are produced each day. The coal produced no longer need be screened. This means that the problem is the same as Problem 2.1 except that constraint (2.12) is to be replaced by a constraint representing the new guarantee. Mathematically, this guarantee can be expressed as

$$x_1 + x_2 \geq 4.$$

When we introduce a slack variable (x_6) and an artificial variable (x_4), it becomes

$$x_1 + x_2 + x_4 - x_6 = 4. \tag{2.21}$$

So the problem is the following:

$$\text{Maximize:} \quad 4x_1 + 3x_2 \quad\quad - Mx_4 \quad\quad\quad = x_0$$

$$\text{subject to:} \quad 3x_1 + 4x_2 + x_3 \quad\quad\quad\quad\quad = 12 \quad (2.11)$$

$$x_1 + x_2 \quad\quad + x_4 \quad - x_6 = 4 \quad (2.21)$$

$$4x_1 + 2x_2 \quad\quad\quad\quad + x_5 \quad\quad = 8 \quad (2.13)$$

$$x_i \geq 0, \quad i = 1, 2, \ldots, 6.$$

When this problem is expressed graphically, as in Figure 2.9, it can be seen that there does not exist a point which will satisfy all constraints simultaneously. Hence the problem does not have a feasible solution. We need a strategy for detecting this situation in the simplex method. Towards this end

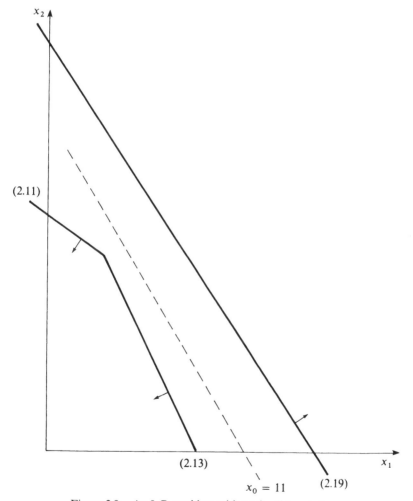

Figure 2.9. An L.P. problem with no feasible solution.

the present problem shall be "solved" by the simplex algorithm. Table 2.37 shows the initial tableau. The first step is to transform the problem into canonical form, as in Table 2.38. Tables 2.39 and 2.40 complete the process.

Table 2.40 displays the "optimal" solution. However, it will be noticed that the artificial variable x_4 has a *positive* value ($\frac{4}{5}$) in this solution. Whenever this occurs in the final simplex tableau, it can be concluded that the

Table 2.37

Constraints	x_1	x_2	x_3	x_4	x_5	x_6	r.h.s.
(2.11)	3	4	1	0	0	0	12
(2.21)	1	1	0	1	0	-1	4
(2.13)	4	2	0	0	1	0	8
x_0	-4	-3	0	M	0	0	0

Table 2.38

Constraints	x_1	x_2	x_3	x_4	x_5	x_6	r.h.s.	Ratio
(2.11)	3	4	1	0	0	0	12	$\frac{12}{3}$
(2.21)	1	1	0	1	0	-1	4	$\frac{4}{1}$
(2.13)	④	2	0	0	1	0	8	$\frac{8}{4}$
x_0	$-(M+4)$	$-(M+3)$	0	0	0	M	$-4M$	

Table 2.39

Constraints	x_1	x_2	x_3	x_4	x_5	x_6	r.h.s.	Ratio
(2.11)	0	$\frac{5}{2}$	1	0	$-\frac{3}{4}$	0	6	$\frac{12}{5}$
(2.21)	0	$\frac{1}{2}$	0	1	$-\frac{1}{4}$	-1	2	$\frac{4}{1}$
(2.13)	1	$\frac{1}{2}$	0	0	$\frac{1}{4}$	0	2	$\frac{4}{1}$
x_0	0	$-(M/2+1)$	0	0	$M/4+1$	M	$-2M+8$	

Table 2.40

Constraints	x_1	x_2	x_3	x_4	x_5	x_6	r.h.s.
(2.11)	0	1	$\frac{2}{5}$	0	$-\frac{3}{10}$	0	$\frac{12}{5}$
(2.21)	0	0	$-\frac{1}{5}$	1	$-\frac{1}{10}$	-1	$\frac{4}{5}$
(2.13)	1	0	$-\frac{1}{5}$	0	$\frac{2}{5}$	0	$\frac{4}{5}$
x_0	0	0	$\frac{1}{5}(M+2)$	0	$\frac{1}{10}(M+7)$	M	$\frac{1}{5}(52-4M)$

problem has no feasible solution. If the two-phase method had been used, we would have obtained a positive x'_0 at the end of the first phase, indicating no feasible solution.

2.5.8 Unboundedness

Consider the following L.P. problem.

$$\begin{aligned}
\text{Maximize:} \quad & x_1 + 2x_2 = x_0 \\
\text{subject to:} \quad -4x_1 + & x_2 \le 2 \\
x_1 + & x_2 \ge 3 \\
x_1 + & 2x_2 \ge 4 \\
x_1 - & x_2 \le 2 \\
& x_1 \ge 0 \\
& x_2 \ge 0.
\end{aligned}$$

On looking at the graphical solution to the problem in Figure 2.10 it can be seen that the feasible region is unbounded. Because of the slope of the objective function (dashed line), the x_0 line can be moved parallel to itself an arbitrary distance from the origin and still coincide with feasible points. Therefore this problem does not have a bounded optimal solution value.

We shall now attempt to apply the simplex method to the problem. Transforming the problem into standard form yields:

$$\begin{aligned}
\text{Maximize:} \quad & x_1 + 2x_2 && - Mx_5 && - Mx_7 && = x_0 \\
\text{subject to:} \quad -4x_1 + & x_2 + x_3 && && && = 2 && (2.22) \\
x_1 + & x_2 && - x_4 + x_5 && && = 3 && (2.23) \\
x_1 + & 2x_2 && && - x_6 + x_7 && = 4 && (2.24) \\
x_1 - & x_2 && && && + x_8 = 2 && (2.25)
\end{aligned}$$

$$x_i \ge 0, \quad i = 1, 2, \ldots, 8.$$

The simplex tableaux are given in Tables 2.41–2.45.

From Table 2.45 it can be seen that x_4 should enter the basis next. Which variable should leave the basis in order to ensure feasibility? All of the x_4 constraint coefficients are now negative. So x_4 can be bought into the basis at an arbitrarily large positive level. This will cause the objective function to assume an arbitrarily large value. Thus there is no bounded optimal solution to the problem. This situation can always be detected in the simplex algorithm by the presence of a negative x_0-row variable coefficient with column entries all nonpositive.

Sometimes a problem will have an unbounded feasible region but still have a bounded optimum. This is illustrated by the dotted line in Figure

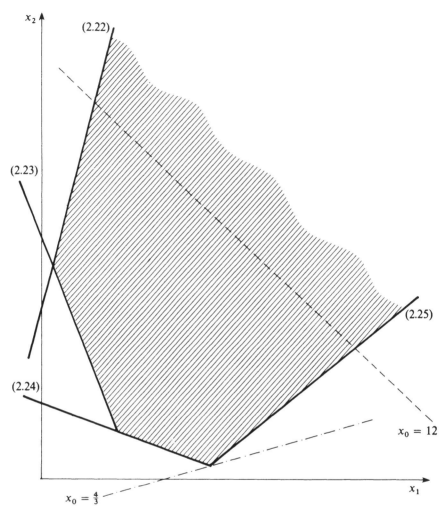

Figure 2.10.   L.P. problems with an unbounded feasible region.

Table 2.41

Constraints	x_1	x_2	x_3	x_4	x_5	x_6	x_7	x_8	r.h.s.
(2.22)	-4	1	1	0	0	0	0	0	2
(2.23)	1	1	0	-1	1	0	0	0	3
(2.24)	1	2	0	0	0	-1	1	0	4
(2.25)	1	-1	0	0	0	0	0	1	2
x_0	-1	-2	0	0	M	0	M	0	0

Table 2.42

Constraints	x_1	x_2	x_3	x_4	x_5	x_6	x_7	x_8	r.h.s.	Ratio
(2.22)	-4	①	1	0	0	0	0	0	2	$\frac{2}{1}$
(2.23)	1	1	0	-1	1	0	0	0	3	$\frac{3}{1}$
(2.24)	1	2	0	0	0	-1	1	0	4	$\frac{4}{2}$
(2.25)	1	-1	0	0	0	0	0	1	2	—
x_0	$-(1+2M)$	$-(3M+2)$	0	M	0	M	0	0	$-7M$	

Table 2.43

Constraints	x_1	x_2	x_3	x_4	x_5	x_6	x_7	x_8	r.h.s.	Ratio
(2.22)	-4	1	1	0	0	0	0	0	2	—
(2.23)	5	0	-1	-1	1	0	0	0	1	$\frac{1}{5}$
(2.24)	⑨	0	-2	0	0	-1	1	0	0	$\frac{0}{9}$
(2.25)	-3	0	1	0	0	0	0	1	4	—
x_0	$-(14M+9)$	0	$(2+3M)$	M	0	M	0	0	$4-M$	

Table 2.44

Constraints	x_1	x_2	x_3	x_4	x_5	x_6	x_7	x_8	r.h.s.	Ratio
(2.22)	0	1	$-\frac{1}{9}$	0	0	$-\frac{4}{9}$	$\frac{4}{9}$	0	2	—
(2.23)	0	0	$\frac{1}{9}$	-1	1	⑤⁄⑨	$-\frac{5}{9}$	0	1	$\frac{9}{5}$
(2.24)	1	0	$-\frac{2}{9}$	0	0	$-\frac{1}{9}$	$\frac{1}{9}$	0	0	—
(2.25)	0	0	$\frac{1}{3}$	0	0	$-\frac{1}{3}$	$\frac{1}{3}$	1	4	—
x_0	0	0	$-\dfrac{M}{9}$	M	0	$-\left(\dfrac{5M+9}{9}\right)$	$\left(\dfrac{14M+9}{9}\right)$	0	$4-M$	

Table 2.45

Constraints	x_1	x_2	x_3	x_4	x_5	x_6	x_7	x_8	r.h.s.
(2.22)	0	1	$\frac{1}{5}$	$-\frac{4}{5}$	$\frac{4}{5}$	0	0	0	$\frac{14}{5}$
(2.23)	0	0	$\frac{1}{5}$	$-\frac{9}{5}$	$\frac{9}{5}$	1	-1	0	$\frac{9}{5}$
(2.24)	1	0	-1	$-\frac{1}{5}$	$\frac{1}{5}$	0	0	0	$\frac{1}{5}$
(2.25)	0	0	$\frac{2}{5}$	$-\frac{3}{5}$	$\frac{3}{5}$	0	0	1	$\frac{23}{5}$
x_0	0	0	$\frac{1}{5}$	$-\frac{9}{5}$	$M+\frac{9}{5}$	0	M	0	$\frac{29}{5}$

2.10, where the problem is the same as the previous one, except the objective is now to maximize:

$$x_1 - 2x_2 = x_0.$$

This problem is solved by the simplex method in Tables 2.46–2.52. The optimal solution is

$$x_1^* = \tfrac{8}{3}$$
$$x_2^* = \tfrac{2}{3}$$
$$x_3^* = 12$$
$$x_4^* = \tfrac{1}{3}$$
$$x_5^* = x_6^* = x_7^* = x_8^* = 0$$
$$x_0^* = \tfrac{4}{3}.$$

Table 2.46

Constraints	x_1	x_2	x_3	x_4	x_5	x_6	x_7	x_8	r.h.s.
(2.22)	-4	1	1	0	0	0	0	0	2
(2.23)	1	1	0	-1	1	0	0	0	3
(2.24)	1	2	0	0	0	-1	1	0	4
(2.25)	1	-1	0	0	0	0	0	1	2
x_0	-1	2	0	0	M	0	M	0	0

Table 2.47

Constraints	x_1	x_2	x_3	x_4	x_5	x_6	x_7	x_8	r.h.s.	Ratio
(2.22)	-4	①	1	0	0	0	0	0	2	$\tfrac{2}{1}$
(2.23)	1	1	0	-1	1	0	0	0	3	$\tfrac{3}{1}$
(2.24)	1	2	0	0	0	-1	1	0	4	$\tfrac{4}{2}$
(2.25)	1	-1	0	0	0	0	0	1	2	—
x_0	$-(1+2M)$	$2-3M$	0	M	0	M	0	0	$-7M$	

Table 2.48

Constraints	x_1	x_2	x_3	x_4	x_5	x_6	x_7	x_8	r.h.s.	Ratio
(2.22)	-4	1	1	0	0	0	0	0	2	—
(2.23)	5	0	-1	-1	1	0	0	0	1	$\tfrac{1}{5}$
(2.24)	⑨	0	-2	0	0	-1	1	0	0	$\tfrac{0}{9}$
(2.25)	-3	0	1	0	0	0	0	1	4	—
x_0	$7-14M$	0	$3M-2$	M	0	M	0	0	$-(M+4)$	

Table 2.49

Constraints	x_1	x_2	x_3	x_4	x_5	x_6	x_7	x_8	r.h.s.	Ratio
(2.22)	0	1	$\frac{1}{9}$	0	0	$-\frac{4}{9}$	$\frac{4}{9}$	0	2	—
(2.23)	0	0	$\frac{1}{9}$	-1	1	$\frac{5}{9}$	$-\frac{5}{9}$	0	1	$\frac{9}{5}$
(2.24)	1	0	$-\frac{2}{9}$	0	0	$-\frac{1}{9}$	$\frac{1}{9}$	0	0	—
(2.25)	0	0	$\frac{1}{3}$	0	0	$-\frac{1}{3}$	$\frac{1}{3}$	1	4	—
x_0	0	0	$-\left(\dfrac{M+4}{9}\right)$	M	0	$\dfrac{7-5M}{9}$	$\dfrac{14M-7}{9}$	0	$-(M+4)$	

Table 2.50

Constraints	x_1	x_2	x_3	x_4	x_5	x_6	x_7	x_8	r.h.s.	Ratio
(2.22)	0	1	$\frac{1}{5}$	$-\frac{4}{5}$	$\frac{4}{5}$	0	0	0	$\frac{14}{5}$	$\frac{14}{1}$
(2.23)	0	0	$\left(\frac{1}{5}\right)$	$-\frac{9}{5}$	$\frac{9}{5}$	1	-1	0	$\frac{9}{5}$	$\frac{9}{1}$
(2.24)	1	0	$-\frac{1}{5}$	$-\frac{1}{5}$	$\frac{1}{5}$	0	0	0	$\frac{1}{5}$	—
(2.25)	0	0	$\frac{2}{5}$	$-\frac{3}{5}$	$\frac{3}{5}$	0	0	1	$\frac{23}{5}$	$\frac{23}{2}$
x_0	0	0	$-\frac{3}{5}$	$\frac{7}{5}$	$\left(\dfrac{5M-7}{5}\right)$	0	M	0	$-\frac{27}{5}$	

Table 2.51

Constraints	x_1	x_2	x_3	x_4	x_5	x_6	x_7	x_8	r.h.s.	Ratio
(2.22)	0	1	0	1	-1	-1	1	0	1	$\frac{1}{1}$
(2.23)	0	0	1	-9	9	5	-5	0	9	—
(2.24)	1	0	0	-2	2	1	-1	0	2	—
(2.25)	0	0	0	(3)	-3	-2	2	1	1	$\frac{1}{3}$
x_0	0	0	0	-4	$(4+M)$	3	$M-3$	0	0	

Table 2.52

Constraints	x_1	x_2	x_3	x_4	x_5	x_6	x_7	x_8	r.h.s.
(2.22)	0	1	0	0	0	$-\frac{1}{3}$	$\frac{1}{3}$	$-\frac{1}{3}$	$\frac{2}{3}$
(2.23)	0	0	1	0	0	-1	1	3	12
(2.24)	1	0	0	0	0	$-\frac{1}{3}$	$\frac{1}{3}$	$\frac{2}{3}$	$\frac{8}{3}$
(2.25)	0	0	0	1	-1	$-\frac{2}{3}$	$\frac{2}{3}$	$\frac{1}{3}$	$\frac{1}{3}$
x_0	0	0	0	0	M	$\frac{1}{3}$	$\dfrac{3M-1}{3}$	$\frac{4}{3}$	$\frac{4}{3}$

2.6 Duality and Postoptimal Analysis

Duality is an important concept and we now present some of the reasons for this importance. In the previous section it became obvious that the more constraints an L.P. problem had, the longer it took to solve. Experience with efficient computer codes has shown that computational time is more sensitive to the number of constraints than to the number of variables. In order to solve a relatively large problem it would therefore be convenient to reduce its number of constraints. This can often be done by constructing a new L.P. problem from the given problem, where the new problem has fewer constraints. This new problem is then solved more easily than the original one. The information obtained in the final simplex tableau can be used to deduce the optimal solution to the original problem. The new problem constructed for this purpose is called the *dual* problem to the original problem. The original problem is called the *primal*.

After an L.P. problem has been solved one would often like to know the sensitivity of the solution to changes in the objective function, constraint coefficients and the r.h.s. constants and to the addition of new variables and constraints. Duality can be used to answer such questions.

2.6.1 Duality

2.6.1.1 *The Relationship Between the Primal and the Dual*

Consider once again the initial L.P. problem outlined in Section 2.2. Suppose that a corporation is considering hiring the equipment of the mining company. The corporation is uncertain about the hourly hireage rates it should offer the company for the three types of implements. During negotiations the mining company reveals that its profits per ton of lignite and anthracite are \$4 and \$3, respectively. The company states that it will not accept hireage rates which amount to less revenue than these present figures. For the purpose of fixing acceptable rates the following variables are defined. Let

$y_1 =$ the hourly hireage rate of the cutting machine,
$y_2 =$ the hourly hireage rate of the screens,
$y_3 =$ the hourly hireage rate of the washing plant.

Recall that it requires 3,3, and 4 hours for the cutting machine, the screens, and the washing plant, respectively, to process 1 ton of lignite. The revenue of the company from hiring out the machines for the corporation to process one ton of lignite is then

$$3y_1 + 3y_2 + 4y_3.$$

Because the company requires a revenue no less than its present profit, this revenue must be such that

$$3y_1 + 3y_2 + 4y_3 \geq 4.$$

By analogy, the constraint for anthracite is

$$4y_1 + 3y_2 + 2y_3 \geq 3.$$

The corporation obviously wishes to minimize the total daily hireage cost it has to pay. Recall that the cutting machine, the screens, and the washing plant can be operated for no more than 12, 10, and 8 hours per day, respectively. The objective of the corporation is to minimize

$$12y_1 + 10y_2 + 8y_3.$$

Of course all hireage costs have to be nonnegative. The corporation is then faced with the following L.P. problem:

$$
\begin{aligned}
\text{Minimize:} \quad & 12y_1 + 10y_2 + 8y_3 = y_0 \\
\text{subject to:} \quad & 3y_1 + 3y_2 + 4y_3 \geq 4 \\
& 4y_1 + 3y_2 + 2y_3 \geq 3 \\
& y_i \geq 0, \quad i = 1, 2, 3.
\end{aligned}
$$

Let us now compare this problem with the original L.P. problem, which is reproduced here for convenience:

$$
\begin{aligned}
\text{Maximize:} \quad & 4x_1 + 3x_2 = x_0 \\
\text{subject to:} \quad & 3x_1 + 4x_2 \leq 12 \\
& 3x_1 + 3x_2 \leq 10 \\
& 4x_1 + 2x_2 \leq 8 \\
& x_i \geq 0, \quad i = 1, 2.
\end{aligned}
$$

A moment's comparison shows that both problems have the same set of constants, but in different positions. In particular, each "row" of the hireage problem contains the same coefficients as one "column" of the original problem. When two L.P. problems have the special relationship displayed here, the original problem is called the *primal* and the new problem is called the *dual*. We shall now formalize this relationship by showing how the dual is constructed from the primal.

1. Replace each primal equality constraint by a "$\leq$" constraint and a "$\geq$" constraint. For example, replace

$$3x_1 + 4x_2 + 5x_3 = 6,$$

by

$$3x_1 + 4x_2 + 5x_3 \geq 6,$$

and

$$3x_1 + 4x_2 + 5x_3 \leq 6.$$

2. If the primal is a maximization (minimization) problem, multiply all "$\geq$" ("$\leq$") constraints by (-1). This ensures all constraints are of the "$\leq$" type for maximization and of the "$\geq$" type for minimization.
3. Define a unique nonnegative dual variable for each primal constraint.

4. Define each dual objective function coefficient to be equal to the r.h.s. constant of the primal constraint of the variable.
5. If the primal objective is maximization, define the dual objective to be minimization, and vice versa.
6. Define the dual r.h.s. constraint constants to be the primal objective function coefficients.
7. If the primal objective is maximization (minimization), define the dual constraint inequalities to be of the "$\geq$" ("$\leq$") type.
8. Define the dual constraint coefficient matrix, A (defined in Section 2.3) to be the transpose of primal constraint coefficient matrix.

These steps can be summed up in mathematical form. The primal:

$$\text{Maximize:} \quad c_1 x_1 + c_2 x_2 + \cdots + c_n x_n = x_0$$
$$\text{subject to:} \quad a_{11} x_1 + a_{12} x_2 + \cdots + a_{1n} x_n \leq b_1$$
$$a_{21} x_1 + a_{22} x_2 + \cdots + a_{2n} x_n \leq b_2$$
$$\vdots$$
$$a_{m1} x_1 + a_{m2} x_2 + \cdots + a_{mn} x_n \leq b_m$$
$$x_i \geq 0, \quad i = 1, 2, \ldots, n,$$

has dual:

$$\text{Minimize:} \quad b_1 y_1 + b_2 y_2 + \cdots + b_m y_m = y_0$$
$$\text{subject to:} \quad a_{11} y_1 + a_{21} y_2 + \cdots + a_{m1} y_m \geq c_1$$
$$a_{12} y_1 + a_{22} y_2 + \cdots + a_{m2} y_m \geq c_2$$
$$\vdots$$
$$a_{1n} y_1 + a_{2n} y_2 + \cdots + a_{mn} y_m \geq c_n$$
$$y_i \geq 0, \quad i = 1, 2, \ldots, m.$$

This can also be expressed in matrix form. The primal:

$$\text{Maximize:} \quad C^T X$$
$$\text{subject to:} \quad AX \leq B \qquad\qquad (2.26a)$$
$$X \geq 0$$

has dual:

$$\text{Minimize:} \quad B^T Y,$$
$$\text{subject to:} \quad A^T Y \geq C \qquad\qquad (2.26b)$$
$$Y \geq 0,$$

where

C is $n \times 1$,
X is $n \times 1$,
A is $m \times n$,
B is $m \times 1$, and
Y is $m \times 1$.

Suppose the dual of (2.26b) is constructed:

$$\text{Maximize:} \quad C^T X$$
$$\text{subject to:} \quad (A^T)^T X \leq (B^T)^T$$
$$X \geq 0.$$

Because the transpose of the transpose of a matrix (vector) is the matrix (vector), we have proven:

Theorem 2.1. *The dual of the dual of a primal L.P. problem is the primal L.P. problem itself.*

2.6.1.2 The Optimal Solution to the Dual

The dual problem introduced in the last section will now be solved by the two-phase method. In standard form the problem is as follows:

PROBLEM 2.3

$$\text{Maximize:} \quad -12y_1 - 10y_2 - 8y_3 \qquad\qquad\qquad = y_0 \qquad (2.27)$$
$$\text{subject to:} \quad 3y_1 + 3y_2 + 4y_3 - y_4 + y_5 \qquad\qquad = 4 \qquad (2.28)$$
$$4y_1 + 3y_2 + 2y_3 \qquad\quad - y_6 + y_7 = 3 \qquad (2.29)$$
$$y_i \geq 0, \qquad i = 1, 2, \ldots, 7.$$

Phase I, with $y_0' = y_5 + y_7$, is shown in Tables 2.53–2.56.

Phase II, with columns corresponding to artificial variables removed, is shown in Tables 2.57 and 2.58.

Table 2.53

Constraints	y_1	y_2	y_3	y_4	y_5	y_6	y_7	r.h.s.
(2.28)	3	3	4	−1	1	0	0	4
(2.29)	4	3	2	0	0	−1	1	3
y_0'	0	0	0	0	1	0	1	0

Table 2.54

Constraints	y_1	y_2	y_3	y_4	y_5	y_6	y_7	r.h.s.	Ratio
(2.28)	3	3	4	−1	1	0	0	4	$\frac{4}{3}$
(2.29)	④	3	2	0	0	−1	1	3	$\frac{3}{4}$
y_0'	−7	−6	−6	1	0	1	0	−7	

Table 2.55

Constraints	y_1	y_2	y_3	y_4	y_5	y_6	y_7	r.h.s.	Ratio
(2.28)	0	$\frac{3}{4}$	$\left(\frac{5}{2}\right)$	-1	1	$\frac{3}{4}$	$-\frac{3}{4}$	$\frac{7}{4}$	$\frac{7}{10}$
(2.29)	1	$\frac{3}{4}$	$\frac{1}{2}$	0	0	$-\frac{1}{4}$	$\frac{1}{4}$	$\frac{3}{4}$	$\frac{3}{2}$
y'_0	0	$-\frac{3}{4}$	$-\frac{5}{2}$	1	0	$-\frac{3}{4}$	$\frac{7}{4}$	$-\frac{7}{4}$	

Table 2.56

Constraints	y_1	y_2	y_3	y_4	y_5	y_6	y_7	r.h.s.
(2.28)	0	$\frac{3}{10}$	1	$-\frac{2}{5}$	$\frac{2}{5}$	$\frac{3}{10}$	$-\frac{3}{10}$	$\frac{7}{10}$
(2.29)	1	$\frac{3}{5}$	0	$\frac{1}{5}$	$-\frac{1}{5}$	$-\frac{2}{5}$	$\frac{2}{5}$	$\frac{2}{5}$
y'_0	0	0	0	0	1	0	1	0

Table 2.57

Constraints	y_1	y_2	y_3	y_4	y_6	r.h.s.
(2.28)	0	$\frac{3}{10}$	1	$-\frac{2}{5}$	$\frac{3}{10}$	$\frac{7}{10}$
(2.29)	1	$\frac{3}{5}$	0	$\frac{1}{5}$	$-\frac{2}{5}$	$\frac{2}{5}$
y_0	12	10	8	0	0	0

Table 2.58

Constraints	y_1	y_2	y_3	y_4	y_6	r.h.s.
(2.28)	0	$\frac{3}{10}$	1	$-\frac{2}{5}$	$\frac{3}{10}$	$\frac{7}{10}$
(2.29)	1	$\frac{3}{5}$	0	$\frac{1}{5}$	$-\frac{2}{5}$	$\frac{2}{5}$
y_0	0	$\frac{2}{5}$	0	$\frac{4}{5}$	$\frac{12}{5}$	$-\frac{52}{5}$

The solution to the original minimization problem is:

$$y_1^* = \tfrac{2}{5}$$
$$y_3^* = \tfrac{7}{10}$$
$$y_i^* = 0, \quad \text{otherwise}$$
$$y_0^* = \tfrac{52}{5}.$$

2.6.1.3 Properties of the Primal–Dual Relationship

Compare Table 2.58, the optimum tableau for the dual, with Table 2.8, the optimum tableau for the primal, which is reproduced here for convenience. As the primal and the dual had the same set of constants in their mathematical formulation, it is not surprising to find some similarities in their optimal

Table 2.8

Constraints	x_1	x_2	x_3	x_4	x_5	r.h.s.
(2.11)	0	1	$\frac{2}{5}$	0	$-\frac{3}{10}$	$\frac{12}{5}$
(2.12)	0	0	$-\frac{3}{5}$	1	$-\frac{3}{10}$	$\frac{2}{5}$
(2.13)	1	0	$-\frac{1}{5}$	0	$\frac{2}{5}$	$\frac{4}{5}$
x_0	0	0	$\frac{2}{5}$	0	$\frac{7}{10}$	$\frac{52}{5}$

tableaux. These similarities are:

1. The value of optimal solutions of the primal and dual are equal.
2. The optimal value of each slack variable in one problem is equal to the objective function coefficient of the structural variable of the corresponding equation in the other.
3. (a) Whenever a primal structural variable has a positive optimal value, the corresponding dual slack variable has zero optimal value.
   (b) Whenever a primal slack variable has positive optimal value, the corresponding dual structural variable has zero optimal value.

Result 3 is an example of what is known as the *complementary slackness theorem*. Indeed, these results are true for any primal–dual pair of L.P. problems which have finite optimal solutions. We now go on to prove some general results concerning duality for the pair of problems defined by (2.26), with a view to proving the complementary slackness theorem in general.

Theorem 2.2. *If X and Y are feasible solutions for* (2.26a) *and* (2.26b), *respectively, then the value of Y is no less than the value of X. That is,*

$$C^T X \leq B^T Y.$$

PROOF. As X is feasible,

$$AX \leq B.$$

As Y is feasible,

$$Y \geq 0.$$

Therefore

$$Y^T AX \leq Y^T B.$$

As Y is feasible,

$$A^T Y \geq C.$$

As X is feasible,

$$X \geq 0.$$

Therefore

$$X^T A^T Y \geq X^T C.$$

But

$$X^T A^T Y = (AX)^T Y$$
$$= Y^T AX.$$

Therefore

$$C^T X = X^T C \leq X^T A^T Y = Y^T A X \leq Y^T B = B^T Y. \qquad \square$$

Theorem 2.3. *If X^* and Y^* are feasible solutions for (2.26a) and (2.26b) such that $C^T X^* = B^T Y^*$, then X^* and Y^* are optimal solutions for (2.26a) and (2.26b), respectively.*

PROOF. By Theorem 2.2,

$$C^T X \leq B^T Y^*, \quad \text{for any feasible } X.$$

But, by assumption,

$$C^T X^* = B^T Y^*.$$

Therefore

$$C^T X \leq C^T X^*, \quad \text{for any feasible } X.$$

Thus X^* is optimal for (2.26a). Similarly,

$$C^T X^* \leq B^T Y, \quad \text{for any feasible } Y.$$

And, by assumption,

$$C^T X^* = B^T Y^*.$$

Therefore

$$B^T Y^* \leq B^T Y, \quad \text{for any feasible } Y.$$

Thus Y^* is optimal for (2.26b). $\qquad \square$

We can make a number of inferences from these results. Firstly, the value of any feasible primal solution is a lower bound on the value of any feasible dual solution. Conversely, the value of any feasible dual solution is an upper bound on the value of any feasible primal solution. The reader should verify that these observations are true for the numerical example. Secondly, if the primal has an unbounded optimal solution value, the dual cannot have any feasible solutions.

The converse to Theorem 2.3 is also true:

Theorem 2.4. *If X^* and Y^* are optimal solutions for (2.26a) and (2.26b) respectively then*

$$C^T X^* = B^T Y^*.$$

For a proof of this theorem, see the book by David Gale (1960).

Theorem 2.5 (Complementary Slackness). *Feasible solutions X^* and Y^* are optimal for (2.26a) and (2.26b), respectively if and only if*

$$(X^*)^T [A^T Y^* - C^T] + (Y^*)^T [B - A X^*] = 0.$$

PROOF. Let U and V be the set of slack variables for (2.26a) and (2.26b), respectively, with respect to X^* and Y^*, i.e.,

$$AX^* + U = B$$
$$A^T Y^* - V = C$$
$$U, V \geq 0.$$

Premultiplying the first equation by $(Y^*)^T$, we obtain

$$(Y^*)^T AX^* + (Y^*)^T U = (Y^*)^T B;$$

premultiplying the second by $(X^*)^T$, we obtain

$$(X^*)^T A^T Y^* - (X^*)^T V = (X^*)^T C.$$

As

$$(X^*)^T A^T Y^* = (Y^*)^T AX^*,$$

we can eliminate this common expression from these two equations, to obtain

$$(Y^*)^T B - (Y^*)^T U = (X^*)^T C = (X^*)^T V. \tag{2.30}$$

In view of the way the slack variables have been introduced, we have

$$U = B - AX^*$$
$$V = A^T Y^* - C.$$

Thus, in order to prove the theorem we must show that X^* and Y^* are optimal for (2.26a) and (2.26b) if and only if

$$(X^*)^T V + (Y^*)^T U = 0. \tag{2.31}$$

($\Rightarrow$) If X^* and Y^* are assumed optimal for (2.26a) and (2.26b), respectively, then, by Theorem 2.4,

$$C^T X^* = B^T Y^*.$$

Thus (2.30) reduces to (2.31).

($\Leftarrow$) Assuming (2.31) holds, (2.30) reduces to

$$C^T X^* = B^T Y^*.$$

Thus, by Theorem 2.3, X^* and Y^* are optimal solutions for (2.26a) and (2.26b), respectively. $\qquad \square$

Let us examine (2.31) more closely in order to discover why Theorem 2.5 is named the complementary slackness theorem. Because X^*, Y^*, U, and V are all nonnegative we have

$$(X^*)^T V \geq 0,$$

and

$$(Y^*)^T U \geq 0.$$

Therefore

$$x_i^* v_i \geq 0, \qquad i = 1, 2, \ldots, n, \quad \text{and}$$
$$y_j^* u_j \geq 0, \qquad j = 1, 2, \ldots, m.$$

But by (2.31) we can conclude that

$$x_i^* v_i = 0, \qquad i = 1, 2, \ldots, n, \quad \text{and}$$
$$y_j^* u_j = 0, \qquad j = 1, 2, \ldots, m.$$

Thus we can conclude that the results 1, 2, and 3 hold for any pair of primal–dual L.P. problems with finite optimal solution values.

It will now be shown what happens to the dual when the primal either does not have a feasible solution or else has an unbounded optimum. Recall the problem of Section 2.5.7. The dual of that problem is

$$\text{Minimize:} \qquad 12y_1 - 4y_2 + 8y_3 = y_0 \qquad (2.32)$$

$$\text{subject to:} \qquad 3y_1 - y_2 + 4y_3 \geq 4 \qquad (2.33)$$

$$4y_1 - y_2 + 2y_3 \geq 3 \qquad (2.34)$$

$$y_1, y_2, y_3 \geq 0.$$

An attempt will now be made to solve this problem by the two-phase method. Phase I is shown in Tables 2.59–2.62. Here,

$$y_0' = y_5 + y_7.$$

Phase II is shown in Tables 2.63 and 2.64. This problem has an unbounded optimum because y_2 can be introduced to the basis at an arbitrarily high level, causing an arbitrarily large objective function value. It is true in general that *when a primal L.P. problem has no feasible solution the dual has either an unbounded optimum or no feasible solution.*

Table 2.59

Constraints	y_1	y_2	y_3	y_4	y_5	y_6	y_7	r.h.s.
(2.33)	3	-1	4	-1	1	0	0	4
(2.34)	4	-1	2	0	0	-1	1	3
y_0'	0	0	0	0	1	0	1	0

Table 2.60

Constraints	y_1	y_2	y_3	y_4	y_5	y_6	y_7	r.h.s.	Ratio
(2.33)	3	-1	4	-1	1	0	0	4	$\frac{4}{3}$
(2.34)	④	-1	2	0	0	-1	1	3	$\frac{3}{4}$
y_0'	-7	2	-6	1	0	1	0	-7	

Table 2.61

Constraints	y_1	y_2	y_3	y_4	y_5	y_6	y_7	r.h.s.	Ratio
(2.33)	0	$-\frac{1}{4}$	$\left(\frac{5}{2}\right)$	-1	1	$\frac{3}{4}$	$-\frac{3}{4}$	$\frac{7}{4}$	$\frac{7}{10}$
(2.34)	1	$-\frac{1}{4}$	$\frac{1}{2}$	0	0	$-\frac{1}{4}$	$\frac{1}{4}$	$\frac{3}{4}$	$\frac{3}{2}$
y_0'	0	$\frac{1}{4}$	$-\frac{5}{2}$	1	0	$-\frac{3}{4}$	$\frac{7}{4}$	$-\frac{7}{4}$	

Table 2.62

Constraints	y_1	y_2	y_3	y_4	y_5	y_6	y_7	r.h.s.
(2.33)	0	$-\frac{1}{10}$	1	$-\frac{2}{5}$	$\frac{2}{5}$	$\frac{3}{10}$	$-\frac{3}{10}$	$\frac{7}{10}$
(2.34)	1	$-\frac{1}{5}$	0	$\frac{1}{5}$	$-\frac{1}{5}$	$-\frac{2}{5}$	$\frac{2}{5}$	$\frac{2}{5}$
y_0'	0	0	0	0	1	0	1	0

Table 2.63

Constraints	y_1	y_2	y_3	y	y_5	y_6	y_7	r.h.s.
(2.33)	0	$-\frac{1}{10}$	1	$-\frac{2}{5}$	$\frac{2}{5}$	$\frac{3}{10}$	$-\frac{3}{10}$	$\frac{7}{10}$
(2.34)	1	$-\frac{1}{5}$	0	$\frac{1}{5}$	$-\frac{1}{5}$	$-\frac{2}{5}$	$\frac{2}{5}$	$\frac{2}{5}$
y_0	12	-4	8	0	0	0	0	0

Table 2.64

Constraints	y_1	y_2	y_3	y_4	y_5	y_6	y_7	r.h.s.
(2.33)	0	$-\frac{1}{10}$	1	$-\frac{2}{5}$	$\frac{2}{5}$	$\frac{3}{10}$	$-\frac{3}{10}$	$\frac{7}{10}$
(2.34)	1	$-\frac{1}{5}$	0	$\frac{1}{5}$	$-\frac{1}{5}$	$-\frac{2}{5}$	$\frac{2}{5}$	$\frac{2}{5}$
y_0	0	$-\frac{4}{5}$	0	$\frac{4}{5}$	$-\frac{4}{5}$	$-\frac{12}{5}$	$-\frac{12}{5}$	$-\frac{52}{5}$

Considering problem (2.32)–(2.34) as the primal, we have an example of the following statement which is true in general: *When a primal L.P. problem has an unbounded optimum the dual has no feasible solution.* This result is a corollary to Theorem 2.3.

2.6.2 Postoptimal Analysis

When the optimal solution to a linear program is analyzed to answer questions concerning changes in its formulation, the study is called *postoptimal analysis*. What changes can be made to an L.P. problem? Of the variety that can be studied, the following will be considered:

1. Changes in the coefficients of the objective function.
2. Changes in the r.h.s. constants of the constraints.
3. Changes in the l.h.s. coefficients of the constraints.
4. The introduction of new variables.
5. The introduction of new constraints.

Obviously, when the original L.P. is changed, the new problem could be solved from scratch. If the changes are minor, however, it seems a shame to ignore the valuable information gained in solving the original problem. The following sections show how the optimal solution to a modified problem can be found using duality and the solution to the original L.P.

2.6.2.1 Changes in the Objective Function Coefficients

When changes are made to the objective function only, the optimal solution is still feasible, as the feasible region is unaltered.

Consider once again problem 2.1. The optimal simplex tableau for this problem was presented in Table 2.8, which is reproduced here for convenience.

Table 2.8

Constraints	x_1	x_2	x_3	x_4	x_5	r.h.s.
(2.11)	0	1	$\frac{2}{5}$	0	$-\frac{3}{10}$	$\frac{12}{5}$
(2.12)	0	0	$-\frac{3}{5}$	1	$-\frac{3}{10}$	$\frac{2}{5}$
(2.13)	1	0	$-\frac{1}{5}$	0	$\frac{2}{5}$	$\frac{4}{5}$
x_0	0	0	$\frac{2}{5}$	0	$\frac{7}{10}$	$\frac{52}{5}$

(i) *Changes to Basic Variable Coefficients.* Suppose that the objective function coefficient c_2 of x_2 is going to be changed. What is the range from its present value of 3 for which the present solution will remain optimal? Suppose c_2 is changed from 3 to $3 + q$. The initial simplex tableau for the problem then is as shown in Table 2.65.

Table 2.65

Constraints	x_1	x_2	x_3	x_4	x_5	r.h.s.
(2.11)	3	4	1	0	0	12
(2.12)	3	3	0	1	0	10
(2.13)	4	2	0	0	1	8
x_0	-4	$-(3 + q)$	0	0	0	0

Table 2.66

Constraints	x_1	x_2	x_3	x_4	x_5	r.h.s.
(2.11)	0	1	$\frac{2}{5}$	0	$-\frac{3}{10}$	$\frac{12}{5}$
(2.12)	0	0	$-\frac{3}{5}$	1	$-\frac{3}{10}$	$\frac{2}{5}$
(2.13)	1	0	$-\frac{1}{5}$	0	$\frac{2}{5}$	$\frac{4}{5}$
x_0	0	$-q$	$\frac{2}{5}$	0	$\frac{7}{10}$	$\frac{52}{5}$

Table 2.67

Constraints	x_1	x_2	x_3	x_4	x_5	r.h.s.
(2.11)	0	1	$\frac{2}{5}$	0	$-\frac{3}{10}$	$\frac{12}{5}$
(2.12)	0	0	$-\frac{3}{5}$	1	$-\frac{3}{10}$	$\frac{2}{5}$
(2.13)	1	0	$-\frac{1}{5}$	0	$\frac{2}{5}$	$\frac{4}{5}$
x_0	0	0	$\frac{2}{5}+\frac{2}{5}q$	0	$\frac{7}{10}-\frac{3}{10}q$	$\frac{52}{5}+\frac{12}{5}q$.

It is easily verified that the tableau corresponding to table 2.8 is that shown in Table 2.66. In order for the present basis to remain optimal, x_2 must still be basic. Therefore, the x_2 value in the x_0 row must have zero value. This is achieved in Table 2.67 by adding q times (2.11) to the x_0 row.

For the present basis to remain optimal, all x_0-row values must be nonnegative. Thus,

$$\frac{2}{5} + \frac{2}{5}q \geq 0$$
$$\frac{7}{10} - \frac{3}{10}q \geq 0.$$

Therefore,

$$-1 \leq q \leq \frac{7}{3}.$$

Hence the range for c_2 is $(3 - 1, 3 + \frac{7}{3})$, with a corresponding optimum range of $(8, 16)$. This is illustrated in Figure 2.11.

This approach can be generalized. If the objective function coefficient c_i of a basic variable x_i is replaced by $(c_i + q)$, it is of interest to know whether the original optimal solution is still optimal or not. On considering the mechanics of the simplex method, it is clear that if the same iterations were repeated on the new problem the only change in the optimal tableau is that the x_i coefficient in x_0 is reduced by q. Hence this coefficient is $(-q)$, as it was originally zero as x_i was basic.

For the present basis to remain optimal, x_i must remain basic. That is, the x_i coefficient in the x_0 row must be zero. This is achieved by adding q times the equation containing x_i as its basic variable to the x_0 row. The tableau is now in canonical form. For the present basis to remain optimal, all the x_0-row coefficients must be nonnegative. Conditions on q can be deduced to

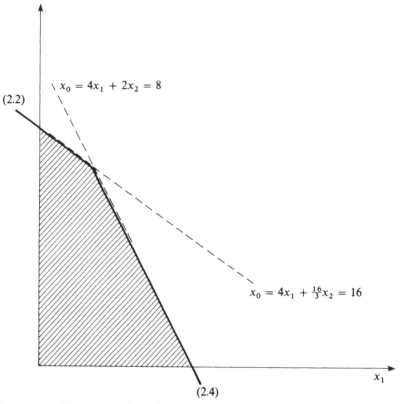

Figure 2.11.   The graphical solution to an L.P. when an objective function coefficient ranges.

achieve this. If a particular value of q is given, it can be deduced whether the present basis is optimal. If it is not, further simplex iterations can be carried out in the normal way.

(*ii*) *Changes to Nonbasic Variable Coefficients.* The situation is even simpler for a change of $+q$ to the objective function coefficient c_j of a variable x_j that turns out to be nonbasic in the optimal solution. The coefficient of x_j in the x_0 row is still reduced by q. However, in the present case there is no need for the coefficient to become zero (as x_j is nonbasic). Hence it simply remains to check whether the coefficient is nonnegative for the present basis to remain optimal. Thus once more conditions on q can be deduced.

2.6.2.2 *Changes in the r.h.s. Constants of the Constraints*

Suppose that a r.h.s. constant of an L.P. problem is altered. Is the current optimal solution still feasible? If it is still feasible it will still be optimal, as the x_0-row coefficients are unchanged.

For example, consider Problem 2.1. The optimal simplex tableau for this problem is given in Table 2.8.

(i) *Change in r.h.s. Constant Whose Slack Variable Is Basic.* Suppose that the r.h.s. constant of constraint (2.12) is changed from 10 to $(10 + r)$. For what values of r will the present solution remain feasible and hence optimal?

Recall that (2.12) was:

$$3x_1 + 3x_2 + x_4 = 10.$$

It now becomes:

$$3x_1 + 3x_2 + 1 \cdot x_4 = 10 + 1 \cdot r.$$

Note that the columns corresponding to x_4 and r are identical in the initial tableau of the new problem, i.e.,

$$(0, 1, 0, 0)^T.$$

Hence they will remain equal in any subsequent simplex tableau. But, as x_4 is basic in Table 2.8, its column of coefficients is unchanged. Hence when the same sequence of iterations that produced Table 2.8 is performed on the new problem the only place in which r will appear is the r.h.s. of (2.12). This new constant becomes $(\frac{2}{5} + r)$. For this solution to remain feasible, all the r.h.s. constants must be nonnegative, i.e.,

$$\tfrac{2}{5} + r \geq 0$$

or

$$r \geq -\tfrac{2}{5}.$$

Thus, as long as

$$b_2 \geq 10 - \tfrac{2}{5},$$

i.e.,

$$b_2 \geq 9\tfrac{3}{5},$$

the current solution will remain feasible and optimal.

Let us now generalize the above considerations. Suppose it is decided to increase the r.h.s. constant of constraint i from b_i to $b_i + r$, and the slack variable of the constraint is basic at the optimum. Then the only possible change in the new optimal tableau will occur in the final b_i entry. This entry could be negative, indicating that the present solution may be infeasible. However, if it is feasible it will still be optimal. Now if the final b_i entry was $\bar{b}_i$, the present solution will be optimal if

$$\bar{b}_i + r \geq 0,$$

i.e.,

$$r \geq -\bar{b}_i.$$

It may be that a specific value of r has been given that forces this inequality to be violated, and hence for the present solution to be infeasible. One then may ask what is the new optimal solution and its value? The negative r.h.s. entry $(\bar{b}_i + r)$ for constraint i is removed to attain feasibility. This is achieved by replacing the basic variable associated with this constraint by a nonbasic variable. How this is done forms the kernel to the *dual simplex method*, which is explained in the next chapter.

(*ii*) *Change in r.h.s. Constant Whose Slack Variable Is Nonbasic*. Suppose now that the r.h.s. constant of constraint (2.11) is changed from 12 to (12 + r). Once again we ask for what values of r will the present solution remain feasible and hence optimal? Equation (2.11) was

$$3x_1 + 4x_2 + x_3 = 12.$$

It now becomes

$$3x_1 + 4x_2 + 1 \cdot x_3 = 12 + 1 \cdot r.$$

As with the previous case, the columns in any simplex tableau corresponding to x_3 and r are identical, and will remain identical in any subsequent simplex tableau. But now x_3 is nonbasic, and hence its column in the optimal tableau is substantially changed. Hence the r.h.s. column in the tableau found by performing the same iterations to the new problem is:

$$(\tfrac{12}{5} + \tfrac{2}{5}r, \tfrac{2}{5} - \tfrac{3}{5}r, \tfrac{4}{5} - \tfrac{1}{5}r, \tfrac{52}{5} + \tfrac{2}{5}r)^T.$$

The first three entries must be nonnegative to preserve feasibility (and optimality):

$$\tfrac{12}{5} + \tfrac{2}{5}r \geq 0$$
$$\tfrac{2}{5} - \tfrac{3}{5}r \geq 0$$
$$\tfrac{4}{5} - \tfrac{1}{5}r \geq 0.$$

That is,

$$-6 \leq r \leq \tfrac{2}{3},$$

with a corresponding solution value

$$\tfrac{52}{5} + \tfrac{2}{5}r.$$

This can be generalized quite naturally. Suppose that the r.h.s. constant, b_i of constraint i is changed to $(b_i + r)$, where the starting basic variable in constraint i is x_j. That is, constraint i is changed from

$$a_{i1}x_1 + a_{i2}x_2 + \cdots + a_{in}x_n + x_j = b_i$$

to

$$a_{i1}x_1 + a_{i2}x_2 + \cdots + a_{in}x_n + 1 \cdot x_j = b_i + 1 \cdot r.$$

Assuming that x_j is nonbasic in the optimal tableau, let its coefficients in this tableau be given in the vector:

$$(\bar{a}_{1j}, \bar{a}_{2j}, \ldots, \bar{a}_{mj})^T$$

and let the final r.h.s. coefficients be given in the vector:

$$(\bar{b}_1, \bar{b}_2, \ldots, \bar{b}_m)^T.$$

Then the current solution will still be feasible (and optimal) if

$$\bar{b}_1 + \bar{a}_{1j}r \geq 0$$
$$\bar{b}_2 + \bar{a}_{2j}r \geq 0$$
$$\vdots$$
$$\bar{b}_m + \bar{a}_{mj}r \geq 0$$

all hold.

Those inequalities can be used to deduce a range in which the present solution remains optimal. Provided r remains within this range, this yields optimal solution values

$$x_0^* + a_{0j}r,$$

where x_0^* is the present optimal solution value and a_{0j} is the x_0-row coefficient of x_j in the optimal tableau.

2.6.2.3 Changes in the l.h.s. Coefficients of the Constraints

(i) Changes to Nonbasic Variable Coefficients. Consider changing a l.h.s. coefficient a_{ij} to $\bar{a}_{ij}$ when its associated variable x_j is nonbasic in the optimal tableau. Suppose that the same sequence of simplex iterations are carried out on the new problem. How will the new optimal tableau differ from the original optimal tableau? The only differences that can possibly occur are in the x_j column. However, we have assumed that x_j is nonbasic. Thus

$$x_j^* = 0.$$

Therefore changes in the x_j coefficients in the constraints have no effect, and the original solution will still be obtained and so must still be feasible. It remains to settle the question of its optimality. The new x_0-row coefficient of x_j can be obtained as follows.

Let the starting basic variable from constraint i be $x_{\alpha(i)}$. This variable does not appear in any original equation other than constraint i, where it has coefficient 1. So it is possible to deduce what multiple of a_{ij} was added to $(-c_j)$, the coefficient of x_j in the x_0 row. Indeed, if $a_{0\alpha(i)}^*$ is the final coefficient of x_i, then exactly $a_{0\alpha(i)}^*$ times equation i must have been somehow added to equation 0. Let the current x_0-row coefficient of x_j be a_{0j}^*. Then the new x-row coefficient of x_j should become

$$a_{0j}^* + a_{0\alpha(i)}^*(\bar{a}_{ij} - a_{ij}).$$

If this value is nonnegative, the present solution is still optimal. If the value is negative, further simplex iterations must be performed, beginning with x_j entering the basis. In order to decide which variable leaves the basis it is necessary to update the rest of the x_j column. It can be shown by an argument similar to that for a_{0j}^* that the coefficient of x_j in row k ($k = 1, 2, \ldots, m$) should be changed from a_{kj}^* to

$$a_{kj}^* + a_{k\alpha(i)}^*(\bar{a}_{ij} - a_{ij}).$$

It is also possible to decide the question of whether the change to a_{ij} affects the optimality of the current solution by analyzing the dual problem. The only change in the dual problem is that the jth constraint

$$a_{1j}y_1 + a_{2j}y_2 + \cdots + a_{ij}y_i + \cdots + a_{mj}y_m \geq c_j$$

becomes

$$a_{1j}y_1 + a_{2j}y_2 + \cdots + \bar{a}_{ij}y_i + \cdots + a_{mj}y_m \geq c_j.$$

It is possible to deduce the optimal y_i values, y_i^*, either by having solved the dual originally or from the x_0-row coefficients in the optimal primal tableau. Hence one can substitute in these y_i^* values and check whether this new constraint is satisfied. If it is the solution is still optimal. If it is not, and the optimal dual tableau is available, the optimal solution to the new problem can be obtained by using the *dual simplex method*. This method will be explained in Chapter 3.

(ii) *Changes to Basic Variable Coefficients.* Consider once again Problem 2.1. Suppose that a_{32} is changed from 2 to 3 in (2.13). Suppose now that the same sequence of simplex iterations is performed on the new problem as that which produced Table 2.8. This will produce a tableau which differs from Table 2.8 only in the x_2 column. The new x_2 column values can be calculated by the method outlined in the previous section. This new tableau is shown in Table 2.68. Here the condition of canonical form is destroyed, as x_2 is supposed to be a basic variable with column

$$(1, 0, 0, 0)^T.$$

This condition is restored by row manipulation in Table 2.69.

Table 2.68

Constraints	x_1	x_2	x_3	x_4	x_5	r.h.s.
(2.11)	0	$\frac{7}{10}$	$\frac{2}{5}$	0	$-\frac{3}{10}$	$\frac{12}{5}$
(2.12)	0	$-\frac{3}{10}$	$-\frac{3}{5}$	1	$-\frac{3}{10}$	$\frac{2}{5}$
(2.13)	1	$\frac{2}{5}$	$-\frac{1}{5}$	0	$\frac{2}{5}$	$\frac{4}{5}$
x_0	0	$\frac{7}{10}$	$\frac{2}{5}$	0	$\frac{7}{10}$	$\frac{52}{5}$

Table 2.69

Constraints	x_1	x_2	x_3	x_4	x_5	r.h.s.
(2.11)	0	1	$\frac{4}{7}$	0	$-\frac{3}{7}$	$\frac{24}{7}$
(2.12)	0	0	$-\frac{3}{7}$	1	$-\frac{3}{7}$	$\frac{10}{7}$
(2.13)	1	0	$-\frac{3}{7}$	0	$\frac{4}{7}$	$-\frac{4}{7}$
x_0	0	0	0	0	1	8

It can be seen that this solution is infeasible, as

$$x_1^* = -\tfrac{4}{7}.$$

The dual simplex method (detailed in Chapter 3) can be used to transform this tableau into an optimal one.

However, if the condition for optimality (all x_0-row coefficients non-negative) has not been satisfied the situation may be somewhat gloomy. In this case one can select an earlier tableau where the condition is satisfied and use the dual simplex method from there. If there are no such suitable earlier tableaux, little of the computation can be saved and it is necessary to solve the new problem from scratch.

If Table 2.69 had displayed a feasible but suboptimal solution, further simplex iterations would have been needed to produce optimality. If it had displayed a feasible solution satisfying the condition for optimality, nothing more would have been necessary

2.6.2.4 The Introduction of New Variables

Suppose that a new variable, x_6 is added to Problem 2.1 as follows:

$$
\begin{aligned}
\text{Maximize:} \quad & 4x_1 + 3x_2 && + 5x_6 \\
\text{subject to:} \quad & 3x_1 + 4x_2 + x_3 && + 2x_6 = 12 \\
& 3x_1 + 3x_2 + \quad x_4 && + 3x_6 = 10 \\
& 4x_1 + 2x_2 && + x_5 + 4x_6 = 8 \\
& x_i \geq 0, \quad i = 1, 2, \ldots, 6.
\end{aligned}
$$

The original optimal solution given in Table 2.8 can be considered a solution to this new problem with x_6 nonbasic, i.e.,

$$x_6^* = 0.$$

Hence it must still be a feasible solution. One must now decide whether it is optimal or not. The new dual problem can be used to make this decision. It will be identical to the original dual except that a new constraint,

$$2y_1 + 3y_2 + 4y_3 \geq 5, \tag{2.35}$$

based upon x_6 must be added to Problem 2.2. So the original dual solution given in Table 2.64 remains feasible if and only if it satisfies (2.35). And feasibility of the original dual solution implies optimality of the original primal solution (with $x_6^* = 0$) for the new primal problem. However the optimal dual solution given in Table 2.64 unfortunately does not satisfy (2.35). So more primal simplex iterations are necessary to produce optimality.

Before they can be carried out it is necessary to calculate the coefficients of x_6 in the tableau produced when the iterations that produced Table 2.8 are applied to the new problem. These coefficients can be found by considering x_6 as an *original* variable with constraint and objective function coefficients equal to zero. Then the introduction of x_6 corresponds to a change in the value of these coefficients from zero to their present values. How to perform calculations based on these changes was explained in Section 2.6.2.3. As each x_6 coefficient was assumed to be zero, it would remain zero when the simplex iterations are performed on the original problem. Hence $a_{k6}^* = 0, k = 0, 1, \ldots, 4$. For the purposes of the calculations it is assumed that the changes occur one at a time. This produces Table 2.70. Now further simplex iterations can be carried out with first x_6 entering the basis.

Table 2.70

Constraint	x_1	x_2	x_3	x_4	x_5	x_6	r.h.s.
(2.11)	0	1	$\frac{2}{5}$	0	$-\frac{3}{10}$	$[\frac{2}{5}(2) + 0(3) - \frac{3}{10}(4)]$	$\frac{12}{5}$
(2.12)	0	0	$-\frac{3}{5}$	1	$-\frac{3}{10}$	$[-\frac{3}{5}(2) + 1(3) - \frac{3}{10}(4)]$	$\frac{2}{5}$
(2.13)	1	0	$-\frac{1}{5}$	0	$\frac{2}{5}$	$[-\frac{1}{5}(2) + 0(3) + \frac{2}{5}(4)]$	$\frac{4}{5}$
x_0	0	0	$\frac{2}{5}$	0	$\frac{7}{10}$	$[-5 + \frac{2}{5}(2) + 0(3) + \frac{7}{10}(4)]$	$\frac{52}{5}$

2.6.2.5 *The Introduction of a New Constraint*

Suppose that a new constraint,

$$x_1 + x_2 \leq 3, \tag{2.36}$$

is added to Problem 2.1. Is the solution in Table 2.8 still feasible? When a further constraint is added to an L.P. problem, a new optimal solution cannot improve on the original one. So if the original optimal solution is still feasible, it is still optimal. However this is not true with regard to (2.36) and the solution in Table 2.8. Hence a new slack variable, x_6 is added to (2.36) to produce

$$x_1 + x_2 + x_6 = 3, \tag{2.37}$$

which is added to Table 2.8 to give Table 2.71.

Table 2.71

Constraints	x_1	x_2	x_3	x_4	x_5	x_6	r.h.s.
(2.11)	0	1	$\frac{2}{5}$	0	$-\frac{3}{10}$	0	$\frac{12}{5}$
(2.12)	0	0	$-\frac{3}{5}$	1	$-\frac{3}{10}$	0	$\frac{2}{5}$
(2.13)	1	0	$-\frac{1}{5}$	0	$\frac{2}{5}$	0	$\frac{4}{5}$
(2.37)	1	1	0	0	0	1	3
x_0	0	0	$\frac{2}{5}$	0	$\frac{7}{10}$	0	$\frac{52}{5}$

Table 2.72

Constraints	x_1	x_2	x_3	x_4	x_5	x_6	r.h.s.
(2.11)	0	1	$\frac{2}{5}$	0	$-\frac{3}{10}$	0	$\frac{12}{5}$
(2.12)	0	0	$-\frac{3}{5}$	1	$-\frac{3}{10}$	0	$\frac{2}{5}$
(2.13)	1	0	$-\frac{1}{5}$	0	$\frac{2}{5}$	0	$\frac{4}{5}$
(2.37)	0	0	$-\frac{1}{5}$	0	$-\frac{1}{10}$	1	$-\frac{1}{5}$
x_0	0	0	$\frac{2}{5}$	0	$\frac{7}{10}$	0	$\frac{52}{5}$

When this Table 2.71 is reduced to canonical form by subtracting (2.11) and (2.13) from (2.37) it is seen that the resulting solution is infeasible, as

$$x_6 = -\tfrac{1}{5},$$

even though the condition for optimality is satisfied. This is shown in Table 2.72. This situation can be remedied to produce optimality using the dual simplex method of Chapter 3.

2.7 Special Linear Programs

2.7.1 The Transportation Problem

The transportation problem is a special type of linear program. Because of its structure it can be solved more efficiently by a modification of the simplex technique than by the simplex technique itself. Consider a supply system comprising three factories which must supply the needs for a single commodity of three warehouses. The unit cost of shipping one item from each factory to each warehouse is known. The production capacity of each factory is limited to a known amount. Each warehouse must receive a minimum number of units of the commodity. The problem is to find the minimum cost supply schedule which satisfies the production and demand constraints. Figure 2.12 shows a typical supply system in diagrammatic form, the numbers associated with the arrows representing unit shipping costs. The supply schedule to be found consists of a list which describes how much of the commodity should be shipped from each factory to each warehouse. For this purpose, define x_{ij} to be the number of units shipped from factory i to warehouse j.

Consider factory 1 with capacity 20. Factory 1 cannot supply more than 20 units in total to warehouses 1, 2, and 3. Hence

$$x_{11} + x_{12} + x_{13} \leq 20.$$

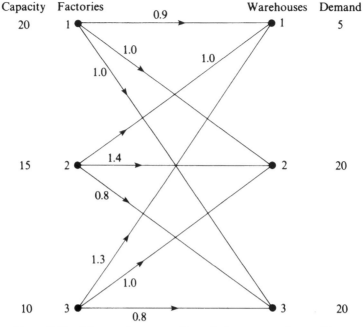

Figure 2.12.   The supply system of a typical transportation problem.

The production constraints for factories 2 and 3 are, respectively,

$$x_{21} + x_{22} + x_{23} \leq 15$$

and

$$x_{31} + x_{32} + x_{33} \leq 10.$$

Consider warehouse 1 with a demand of 5. Warehouse 1 must receive at least five units in total from factories 1, 2, and 3, hence

$$x_{11} + x_{21} + x_{31} \geq 5.$$

The demand constraints for warehouses 2 and 3 are, respectively,

$$x_{12} + x_{22} + x_{32} \geq 20$$

and

$$x_{13} + x_{23} + x_{33} \geq 20.$$

Of course, all quantities shipped must be nonnegative; thus,

$$x_{ij} \geq 0, \qquad i = 1, 2, 3$$
$$j = 1, 2, 3.$$

The objective is to find a supply schedule with minimum cost. The total cost is the sum of all costs from all factories to all warehouses. This cost x_0

can be expressed as

$$x_0 = 0.9x_{11} + 1.0x_{12} + 1.0x_{13} + 1.0x_{21} + 1.4x_{22}$$
$$+ 0.8x_{23} + 1.3x_{31} + 1.0x_{32} + 0.8x_{33}.$$

The problem can now be summarized in linear programming form as follows.

Minimize: $x_0 = 0.9x_{11} + 1.0x_{12} + 1.0x_{13} + 1.0x_{21} + 1.4x_{22}$
$$+ 0.8x_{23} + 1.3x_{31} + 1.0x_{32} + 0.8x_{33}$$

subject to: $x_{11} + x_{12} + x_{13} \leq 20$

 $x_{21} + x_{22} + x_{23} \leq 15$

 $x_{31} + x_{32} + x_{33} \leq 10$

 $x_{11} + x_{21} + x_{31} \geq 5$

 $x_{12} + x_{22} + x_{32} \geq 20$

 $x_{13} + x_{23} + x_{33} \geq 20$

$$x_{ij} \geq 0, \quad i = 1, 2, 3,$$
$$j = 1, 2, 3.$$

The problem can be generalized as follows. Let

$m = $ the number of factories;

$n = $ the number of warehouses;

$a_i = $ the number of units available at factory i, $i = 1, 2, \ldots, m$;

$b_j = $ the number of units required by warehouse j, $j = 1, 2, \ldots, n$;

$c_{ij} = $ the unit transportation cost from factory i to warehouse j.

Then the problem is to

Minimize: $$x_0 = \sum_{i=1}^{m} \sum_{j=1}^{n} c_{ij}x_{ij} \tag{2.38}$$

subject to: $$\sum_{i=1}^{m} x_{ij} \geq b_j, \quad j = 1, 2, \ldots, n \tag{2.39}$$

$$\sum_{j=1}^{n} x_{ij} \leq a_i, \quad i = 1, 2, \ldots, m \tag{2.40}$$

$$x_{ij} \geq 0, \quad i = 1, 2, \ldots, m,$$
$$j = 1, 2, \ldots, n.$$

Problems which belong to this class of L.P. problems are called *transportation problems*. However, many of the problems of this class do not involve the transporting of a commodity between sources and destinations. In the particular problem studied here, total supply is equal to total demand.

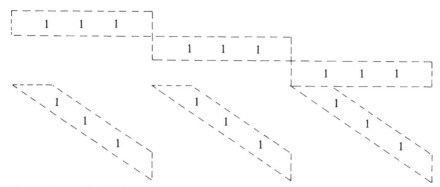

Figure 2.13. The distinctive pattern of the unit constraint coefficients in the transportation problem.

Hence in any feasible solution each factory will be required to ship its entire supply and each warehouse will receive exactly its demand. Therefore, all constraints will be binding in any feasible solution. The algorithm for the solution of the transportation problem, shortly to be explained, assumes that supply and demand is balanced in this way. Of course, there may exist well formulated problems in which "supply" exceeds "demand" or vice versa, as the problems may have nothing to do with the transportation of a commodity. In this case a fictitious "warehouse" or "factory" is introduced, whichever is required. Its "capacity" or "demand" is defined so as to balance total supply with total demand. All unit transportation costs to or from this fictitious location are defined to be zero. Then the value of the optimal solution to this balanced problem will equal that of the original problem.

It was mentioned earlier that, because of its structure, the transportation problem could be solved efficiently by a modified simplex procedure. This structure is.

1. All l.h.s. constraint coefficients are either zero or one.
2. All l.h.s. unit coefficients are always positioned in a distinctive pattern in the initial simplex tableau representing the problem (ignoring slack variables). This is shown in Figure 2.13.
3. All r.h.s. constraint constants are integers.

This structure implies a very important result, that the optimal values of the decision variables will be integer.

In solving problems by hand using the simplex method it was convenient to display each iteration in a tableau. This is also done in the transportation problem, except a different type of tableau is used. The general tableau is given in Table 2.73.

The tableau for the example problem is given in Table 2.73a. The value of each decision variable is written in each cell. A feasible solution to the problem is displayed in Table 2.74. Methods by which an initial feasible solution can be identified are outlined in the next section.

Table 2.73

			Warehouses			Supply
	1	2	$\ldots$	j	n	
1	c_{11}	c_{12}		c_{1j}	c_{1n}	a_1
2	c_{21}	c_{22}		c_{2j}	c_{2n}	a_2
i	c_{i1}	c_{i2}		c_{ij}	c_{in}	a_i
m	c_{m1}	c_{m2}		c_{mj}	c_{mn}	a_m
Demand	b_1	b_2		b_j	b_n	

Factories (row label at left)

Table 2.73a

	1	2	3	
1	0.9	1	1	20
2	1	1.4	0.8	15
3	1.3	1	0.8	10
	5	20	20	

Table 2.74

	1	2	3	
1	0.9 ⑤	1 ⑮	1	20
2	1	1.4 ⑤	0.8 ⑩	15
3	1.3	1	0.8 ⑩	10
	5	20	20	

2.7.1.1 *The Identification of an Initial Feasible Solution*

2.7.1.1.1 *The Northwest Corner Method.* This method starts by allocating as much as possible to the cell in the northwest corner of the tableau of the problem, cell $(1,1)$ (row 1, column 1). In the example problem, the maximum that can be allocated is five units, as the demand of warehouse 1 is five. This satisfies the demand of warehouse 1 and leaves factory 1 with 15 units left. As warehouse 1 is satisfied, column 1 is removed from consideration. Then cell $(1,2)$ becomes the new northwest corner. As much as possible is allocated to this cell. The maximum that can be allocated is 15, all that remains in factory 1. Warehouse 2 now has its demand reduced to 5, as it has just received 15 units from factory 1. Row 1 is dropped from consideration as it has now expended all its resources. This means that cell $(2,2)$ becomes the new northwest corner. This procedure continues until all demand is met. Table 2.74 shows the feasible solution thus obtained.

2.7.1.1.2 *The Least Cost Method.* Although the northwest corner method is easy to implement and always produces a feasible solution, it takes no account of the relative unit transportation costs. It is quite likely that the solution thus produced will be far from optimal. The methods of this section and the next usually produce less costly initial solutions. The least cost method starts by allocating the largest possible amount to the cell in the tableau with the least unit cost. In the example problem, this amounts to allocating to either cell $(2,3)$ or cell $(3,3)$. Suppose cell $(2,3)$ is chosen arbitrarily and 15 units are assigned to it. This procedure will always satisfy a row or column which is removed from consideration. In this case row 2 is removed. The demand of warehouse 3 is reduced to 5, as it has been allocated 15 by factory 2. (The cell with the next smallest unit cost is identified and the maximum is allocated to it. This means 5 units are allocated to cell $(3,3)$. This procedure continues until all demand is met. Table 2.75 shows the feasible solution thus obtained.

Table 2.75

	1	2	3	
1	0.9 ⑤	1 ⑮	1	20
2	1	1.4	0.8 ⑮	15
3	1.3	1 ⑤	0.8 ⑤	10
	5	20	20	

2.7.1.1.3 *The Vogel Approximation Method.* The Vogel approximation method often produces initial solutions which are even better than those of the least cost method. However, the price of this attractiveness is considerably more computation than the previous two methods. The approach is similar to that of the Hungarian method for the assignment problem, discussed later in this chapter, and also to that used in solving the travelling salesman problem by branch and bound enumeration, discussed in Chapter 4.

The variation of the Vogel approximation method described here begins by first *reducing* the matrix of unit costs. This reduction is achieved by subtracting the minimum quantity in each row from all elements in that row. This results in the following unit costs in the current example in Table 2.76. The costs are further reduced by carrying out this procedure on the columns of the new cost matrix. This produces Table 2.77.

A penalty is then calculated for each cell which currently has zero unit cost. Each cell penalty represents the unit cost incurred if a positive allocation is not made to that cell. Each cell penalty is found by adding together the second smallest costs of the row and column of the cell. These second

Table 2.76

	1	2	3	
1	0	0.1	0.1	(-0.9)
2	0.2	0.6	0	(-0.8)
3	0.5	0.2	0	(-0.8)

Table 2.77

	1	2	3
1	0	0	0.1
2	0.2	0.5	0
3	0.5	0.1	0
	(0)	(-0.1)	(0)

smallest costs for each row and column are shown alongside each row and column for the example problem in Table 2.78. The penalties are shown in the top right-hand corner of each appropriate cell.

Table 2.78

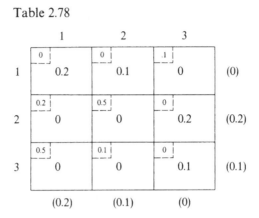

The cell with the largest penalty is identified. The maximum amount possible is then allocated to this cell. Ties are settled arbitrarily. In the example, either cell $(1, 1)$ or cell $(2, 3)$ could be chosen, each with a penalty of 0.2. Cell $(1, 1)$ will be arbitrarily chosen, and 5 units are allocated to it. This procedure will always satisfy a row or column (or both), which is then removed from further consideration. This removal may necessitate a further reduction in the cost matrix and a recalculation of some penalties. This results in Table 2.79. This process is repeated until all demand is met. The final allocation is given in Table 2.80.

A comparison of the three techniques shows that the northwest corner method produced an initial solution with value 42.5, the least cost method and the Vogel approximation method produced the same solution with value 40.5. It will be shown that this latter solution is optimal.

Table 2.79

5	0 0.2	0.1	(0.1)
	0.5	0 0.5	(0.5)
	0.1	0 0.1	(0.1)
	(0.1)	(0)	

Table 2.80

	1	2	3
1	0.9 ⑤	1 ⑮	1
2	1	1.4	8 ⑮
3	1.3	⑤	0.8 ⑤

2.7.1.2 The Stepping Stone Algorithm

Once an initial feasible solution has been found by one of the three preceding methods, it is desired to transform it into the optimal solution. This is achieved by the stepping stone algorithm. Consider the initial feasible solution found by the northwest corner method given in Table 2.74. To determine whether this solution is optimal or not it is necessary to ask, *for each cell individually*, if the allocation of one unit to that cell would reduce the total cost. This is done only for those cells which presently have no units assigned to them.

For example, cell $(1, 3)$ has nothing assigned to it. Would the total cost be reduced if at least one unit was assigned to that cell? Assume that one unit is assigned, i.e.,

$$x_{13} = 1.$$

This means row 1 and column 3 are unbalanced—the sum of their assignments do not add up to the appropriate capacity and demand. To balance row 1, one unit is subtracted from cell $(1, 2)$ so that now

$$x_{12} = 14.$$

Now column 2 is unbalanced. To correct this, one unit is added to cell 2, 2, so that now

$$x_{22} = 6.$$

Now row 2 is unbalanced. To correct this, one unit is subtracted from cell $(2, 3)$:

$$x_{23} = 9.$$

This also balances row 3.

What we have done is to trace out a *circuit of cells*, the only empty one being the cell under scrutiny. This circuit is shown in Table 2.81. Is this solution an improvement over the initial solution? The solution is displayed in Table 2.82.

Table 2.81

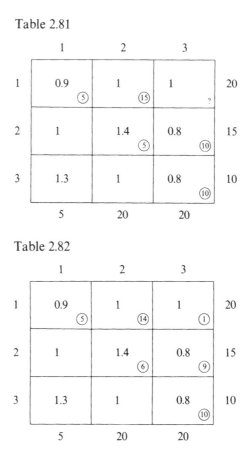

	1	2	3	
1	0.9 ⑤	1 ⑮	1 ?	20
2	1	1.4 ⑤	0.8 ⑩	15
3	1.3	1	0.8 ⑩	10
	5	20	20	

Table 2.82

	1	2	3	
1	0.9 ⑤	1 ⑭	1 ①	20
2	1	1.4 ⑥	0.8 ⑨	15
3	1.3	1	0.8 ⑩	10
	5	20	20	

The difference between the solutions in Tables 2.81 and 2.82 is

$+\$1.0$	for the unit shipped from factory 1 to warehouse 3	
$-\$1.0$	for the unit less from factory 1 to warehouse 2	
$+\$1.4$	for the unit shipped from factory 2 to warehouse 2	
$-\$0.8$	for the unit less from factory 2 to warehouse 3	
$\overline{\$0.6}$		

Thus an allocation of one unit to cell $(1,3)$ causes an increase of $\$0.6$. Hence such an allocation is not worthwhile. We can evaluate the worth of all other empty cells in a similar manner; that is, for each empty cell we can form a circuit of cells, the only empty cell in the circuit being the cell in question. The reader should verify that the changes in x_0 for a unit allocation to cells $(2,1)$, $(3,1)$, and $(3,2)$ is $-\$0.3$, $\$0.0$, and $-\$0.4$, respectively. The circuit of cells for this last proposed allocation is shown in Table 2.83.

As there is a decrease in x_0 of $\$0.4$ for each unit allocated to cell $(3,2)$, we wish to allocate the maximum possible amount to $(3,2)$. The cells which are going to have their allocations reduced are $(2,2)$ and $(3,3)$. The minimum

Table 2.83

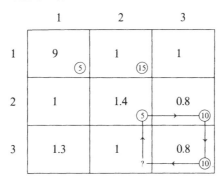

allocation among these is 5 units in $(2, 2)$. Hence the maximum allocation we can make to $(3, 2)$ is 5: any more and $(2, 2)$ would have a negative allocation, which would be infeasible. The new assignment is shown in Table 2.84. The total cost is decreased by

$$\$(5 \times (0.4)) = \$2.0.$$

Thus the new total cost is

$$\$(42.5 - 2.0) = \$40.5,$$

which can be verified by direct computation. We have now completed one iteration of the stepping stone method. All empty cells in Table 2.84 are examined in the same way. Of course we know from the previous iteration that a unit allocation to $(2, 2)$ will produce an increase in x_0 of \$0.4. Indeed an allocation to any empty cell in Table 2.84 will effect an increase in x_0. The circuit of cells for each empty cell is:

$(1, 3)$:	$\langle (1, 3), (3, 3), (3, 2), (1, 2), (1, 3) \rangle$	$(+\$0.2)$
$(2, 1)$:	$\langle (2, 1), (1, 1), (1, 2), (3, 2), (3, 3), (2, 3), (2,1) \rangle$	$(+\$0.1)$
$(2, 2)$:	$\langle (2, 2), (2, 3) \ (3, 3), (3, 2), (2, 2) \rangle$	$(+\$0.4)$
$(3, 1)$:	$\langle (3, 1), (1, 1), (1, 2), (3, 2), (3, 1) \rangle$	$(+\$0.4).$

Table 2.84

	1	2	3
1	0.9 ⑤	1 ⑮	1
2	1	1.4	0.8 ⑮
3	1.3	1 ⑤	0.8 ⑤

Note that the circuit for $(2, 1)$ crosses over itself, but this need not cause any alarm.

This means that we have arrived at the optimal solution. The shipping schedule is: ship

  5 units from factory 1 to warehouse 1,
  15 units from factory 1 to warehouse 2,
  15 units from factory 2 to warehouse 3,
  5 units from factory 3 to warehouse 2,
  5 units from factory 3 to warehouse 3,

for a total cost of \$40.5. This solution is the same as that obtained by the Vogel approximation method and the least cost method.

It has been stated that total supply must equal total demand for the problem to be in a form suitable for the stepping stone algorithm. This means that one of the constraints (2.39), (2.40) can be expressed in terms of the others and is redundant. Hence the problem possesses in effect $(m + n - 1)$ constraints. Thus any basic feasible solution should contain $(m + n - 1)$ basic variables. It may occur that a solution contains less than $(m + n - 1)$ basic (positive) variables. Such a solution is degenerate.

It is not possible to analyze all the empty cells of the tableau of the degenerate solution to find an improvement. This problem can be overcome by declaring basic as many cells as necessary to bring the number in the basis up to $(m + n - 1)$. This is achieved by allocating a very small positive real number ε to these cells. These allocations are made to cells in rows or columns where there is only one basic cell in order to enable circuits of cells to be created for all empty cells. These ε's are then removed when the optimal solution has been found.

2.7.1.3 *Dantzig's Method*

The stepping stone method will guarantee to find the minimal solution for any well formulated transportation problem in a finite number of steps. However, its implementation becomes very laborious on all but the smallest problems. For realistically sized problems the following simpler method due to Dantzig is recommended. Like the stepping stone method it evaluates each empty cell—in order to see whether it would be profitable to make a positive assignment to it. This evaluation is based on the theory of duality of Section 2.6.1. To be more specific, values are calculated for variables in the dual of the transportation problem regarded as an L.P.

Unlike the stepping stone method, Dantzig's method does not create a circuit of cells in order to evaluate the worth of an empty cell. Instead it calculates values for the dual variables; these enable one to determine which empty cell should be filled. It then creates *one* circuit of cells in order to determine how much should be allocated and which cell leaves the basis. As only one circuit is created at each iteration, this method is far simpler than the preceeding one.

We now explain how the method works by using it to solve our example problem. Consider once again the solution obtained by the northwest corner method, given in Table 2.74. We associate multipliers u_i with each row i and v_j with each column j. For each *basic* cell (i, j) set

$$u_i + v_j = c_{ij},$$

the unit transportation cost, and

$$u_1 = 0.$$

The values of all u_i's and v_j's can be then be calculated as the c_{ij}'s are known constants. From Table 2.74 we have

$$
\begin{aligned}
0 + v_1 &= 0.9, &\quad (u_1 = 0) \\
0 + v_2 &= 1, &\quad (u_1 = 0) \\
u_2 + v_2 &= 1.4 \\
u_2 + v_3 &= 0.8 \\
u_3 + v_3 &= 0.8,
\end{aligned}
$$

which can be solved to yield

$$u_1 = 0, \quad v_1 = 0.9, \quad v_2 = 1, \quad u_2 = 0.4, \quad v_3 = 0.4, \quad u_3 = 0.4.$$

Having determined values for what will be seen to be dual variables, we now calculate the change in x_0 for a unit allocation to each nonbasic cell (k, l):

$$\bar{c}_{kl} = c_{kl} - u_k - v_l.$$

The $\bar{c}_{kl}$'s will have the same values as those determined by the stepping stone method. For our example:

$$
\begin{aligned}
\bar{c}_{13} &= c_{13} - u_1 - v_3 = 1 - 0.4 = 0.6 \\
\bar{c}_{21} &= c_{21} - u_2 - v_1 = 1 - 0.4 - 0.9 = -0.3 \\
\bar{c}_{31} &= c_{31} - u_3 - v_1 = 1.3 - 0.4 - 0.9 = 0 \\
\bar{c}_{32} &= c_{32} - u_3 - v_2 = 1 - 0.4 - 1 = -0.4.
\end{aligned}
$$

Thus, as with the stepping stone method, we have discovered that the maximum amount possible should be allocated to cell $(3, 2)$. This allocation is made as in the previous method. We effect the change of basis, producing Table 2.84 from Table 2.83. The multipliers for Table 2.84 are now calculated:

$$
\begin{aligned}
0 + v_1 &= 0.9, &\quad (u_1 = 0) \\
0 + v_2 &= 1 &\quad (u_1 = 0) \\
u_2 + v_3 &= 0.8 \\
u_3 + v_2 &= 1 \\
u_3 + v_3 &= 0.8.
\end{aligned}
$$

These are solved to yield:

$$u_1 = 0, \quad v_1 = 0.9, \quad v_2 = 1, \quad u_3 = 0, \quad v_3 = 0.8, \quad u_2 = 0.$$

We can now calculate the change in x_0 for a unit allocation to each nonbasic cell:

$$\bar{c}_{13} = 1 - 0 - 0.8 = 0.2$$
$$\bar{c}_{21} = 1 - 0 - 0.9 = 0.1$$
$$\bar{c}_{22} = 1.3 - 0 - 0.9 = 0.4$$
$$\bar{c}_{31} = 1.3 - 0 - 0.9 = 0.4.$$

These values are identical with those obtained by the simplex method and are all nonnegative. As the optimal solution has been found, as displayed in Table 2.84, the method is terminated.

In order to explain why the method works, let us take the dual of the capacitated version of (2.38), (2.39), (2.40). That is, we assume equality in the constraints and the problem becomes

$$\text{Minimize:} \qquad x_0 = \sum_{i=1}^{m} \sum_{j=1}^{n} c_{ij} x_{ij}$$

$$\text{subject to:} \qquad \sum_{i=1}^{m} x_{ij} = b_j, \qquad j = 1, 2, \ldots, n$$

$$\sum_{j=1}^{n} x_{ij} = a_i, \qquad i = 1, 2, \ldots, m$$

$$x_{ij} \geq 0.$$

The reader unfamiliar with L.P. duality should refer to Section 2.6.1. In taking the dual, suppose we associate a dual variable v_j with each of the first n constraints and a dual variable, u_i with each of the next m constraints. The dual problem is:

$$\text{Maximize:} \qquad \sum_{j=1}^{n} b_j v_j + \sum_{i=1}^{m} a_i u_i$$

$$\text{subject to:} \qquad v_j + u_i \leq c_{ij}, \qquad i = 1, 2, \ldots, m$$
$$j = 1, 2, \ldots, n.$$

Note that the u_i's and v_j's are not restricted to nonnegative values, as they arise from equality constraints. The special nature of the inequality constraints in the dual arises because of the structure of the primal constraint matrix, as illustrated in Figure 2.13.

Suppose we are solving the transportation problem as a regular L.P. using the simplex method. We would wish to calculate the x_0-row coefficients $\bar{c}_{ij}$ at each iteration in order to test for optimality and, if the test is negative, decide which variable enters the basis. Now, according to property 3(a) of complementary slackness (see Section 2.6.1.3),

$$x_{ij} > 0 \Rightarrow v_j + u_i = c_{ij}$$

for every basic variable x_{ij}. This creates $(m + n - 1)$ equations in $(m + n)$

unknowns, which can be solved by assigning an arbitrary value to one of the unknowns. Traditionally u_1 is set to zero.

Now the value of each slack variable in the dual constraint

$$v_j + u_i \leq c_{ij}$$

is

$$c_{ij} - u_i - v_j.$$

Thus from property 2 of Section 2.6.1.3 we have

$$\bar{c}_{ij} = c_{ij} - u_i - v_j.$$

Thus in order to determine which variable x_{ij} should enter the basis (if optimality has not yet been reached) we must simply select the x_{ij} which has the most negative value of $c_{ij} - u_i - v_j$. Note that the most negative (rather than most positive) is selected, as our original objective is one of minimization.

2.7.2 The Assignment Problem

The assignment problem is a special type of transportation problem. Because of its structure it can be solved more efficiently by a special algorithm than by the stepping stone algorithm. Consider a collection of n workers and n machines. Each worker must be assigned one and only one machine. Each worker has been rated on each machine and a standardized time for him to complete a standard task is known. The problem is to make an assignment of workers to machines so as to minimize the total amount of standardized time of the assignment.

For the purposes of describing an assignment, define

$$x_{ij} = \begin{cases} 1, & \text{if worker } i \text{ is assigned to machine } j, \\ 0, & \text{otherwise} \end{cases}$$

$n =$ the number of workers and the number of machines

$c_{ij} =$ the standardized time of worker i on machine j, assumed to be non-negative.

Then the problem is to

$$\text{Minimize:} \quad x_0 = \sum_{i=1}^{n} \sum_{j=1}^{n} c_{ij} x_{ij} \tag{2.41}$$

$$\text{subject to:} \quad \sum_{i=1}^{n} x_{ij} = 1, \qquad j = 1, 2, \ldots, n \tag{2.42}$$

$$\sum_{j=1}^{n} x_{ij} = 1, \qquad i = 1, 2, \ldots, n \tag{2.43}$$

$$x_{ij} = 0, 1, \qquad \begin{aligned} i &= 1, 2, \ldots, n \\ j &= 1, 2, \ldots, n. \end{aligned}$$

It can be seen on comparison with (2.38), (2.39), (2.40) that this formulation is indeed a special case of the transportation problem the workers representing factories and the machines representing warehouses. Here each "factory" and each "warehouse" has a capacity and demand of one unit. The problem can be represented by a tableau like Table 2.73, shown in Table 2.85. The problem could be solved using the techniques developed for the transportation problem. However let us examine the problem a little more deeply in order to discover a more efficient method which exploits the special structure of this assignment problem.

Table 2.85

The matrix of standardized times

$$C = \begin{pmatrix} c_{11} & c_{12} & \cdots & c_{1i} & \cdots & c_{1n} \\ c_{21} & c_{22} & \cdots & c_{2i} & \cdots & c_{2n} \\ \vdots & \vdots & & \vdots & & \vdots \\ c_{i1} & c_{i2} & \cdots & c_{ii} & \cdots & c_{in} \\ \vdots & \vdots & & \vdots & & \vdots \\ c_{n1} & c_{n2} & & c_{ni} & & c_{nn} \end{pmatrix}$$

holds the key to the problem. Because there is a one-to-one assignment of workers to machines, our problem reduces to finding a set S of n entries of C with the properties that (i) exactly one entry of S appears in each row of C and (ii) exactly one entry of S appears in each column of C. Then among all sets S of n entries of C we require the one with the least sum. In order to make use of this fact, consider the following numerical example:

$$n = 5$$

and

$$C = \begin{pmatrix} 5 & ① & 5 & 1 & 3 \\ 4 & 6 & ① & 8 & 6 \\ ① & 3 & 2 & 2 & 3 \\ 5 & 6 & 5 & ① & 6 \\ 2 & 1 & 5 & 3 & ① \end{pmatrix}.$$

In this case it is possible to identify a minimal S quite easily. Because every entry in C is at least one, any set S with sum five must be minimal. Such a set has been circled and it has, of course, exactly one entry in each row and in each column, thus obeying properties (i) and (ii).

Suppose now that we subtract one unit from each entry in C to obtain C':

$$C' = \begin{pmatrix} 4 & ⓪ & 4 & 0 & 2 \\ 3 & 5 & ⓪ & 7 & 3 \\ ⓪ & 2 & 1 & 1 & 2 \\ 4 & 5 & 4 & ⓪ & 5 \\ 1 & 0 & 4 & 2 & ⓪ \end{pmatrix}.$$

Because the relative value of the entries remain unchanged, the minimal solution remains the same.

These observations hold true for any matrix C, and furthermore, because all entries are assumed to be nonnegative, once a set S of all zero entries has been identified it must be minimal. The *Hungarian method*, due to the Hungarian mathematician Konig has this as its aim. The method progressively reduces the entries in a manner similar to our step from C to C' until a set S of zeros can be identified.

The method is made up of three parts:

1. C is reduced by subtracting the least entry in column i from every element in column j, for each column i, $i = 1, 2, \ldots, n$. Then if any row has all positive entries, the same operation is applied to it.
2. A check is made to see whether a set S of all zeros can be found in the matrix. If so, S represents a minimal solution and the method is terminated. If not, step (c) is applied.
3. As a minimal S cannot yet be identified, the zeros in C are redistributed and possibly some new zeros are created. How this is carried out will be explained shortly. Then the check of step 2 is performed again. This cycle of steps 2 and 3 is repeated until a minimal S is found.

A few comments about these steps will now be made. We need to show that if C' is the matrix obtained from C by step (1), then the set of minimal sets S for C and C' are identical. Suppose that α_i and β_j, positive real numbers, are subtracted from the ith row and jth column of C, respectively, for each row i and each column j. Then, if c_{ij} and c'_{ij} are the $i - j$ elements of C and C', respectively,

$$c'_{ij} = c_{ij} - \alpha_i - \beta_j.$$

Further, if x'_0 is the objective function associated with the new assignment problem represented by C', then

$$x'_0 = \sum_{i=1}^{n} \sum_{j=1}^{n} c'_{ij} x_{ij}$$

$$= \sum_{i=1}^{n} \sum_{j=1}^{n} (c_{ij} - \alpha_i - \beta_j) x_{ij}$$

$$= \sum_{i=1}^{n} \sum_{j=1}^{n} c_{ij} x_{ij} - \sum_{i=1}^{n} \sum_{j=1}^{n} \alpha_i x_{ij} - \sum_{i=1}^{n} \sum_{j=1}^{n} \beta_j x_{ij}$$

$$= x_0 - \sum_{i=1}^{n} \left(\alpha_i \sum_{j=1}^{n} x_{ij} \right) - \sum_{j=1}^{n} \beta_j \left(\sum_{i=1}^{n} x_{ij} \right).$$

By (2.42) and (2.34),

$$x'_0 = x_0 - \sum_{i=1}^{n} \alpha_i - \sum_{j=1}^{n} \beta_j.$$

Thus x'_0 and x_0 differ only by the total amount subtracted, which is a constant. Therefore they have identical minimal sets.

We need to have an efficient way of performing step (2). That is, we need to be able to pronounce whether or not a set S of zeros exists, and if it does, which entries belong to it. A moment's reflection reveals that any such S has the property that its zeros in C' can be transformed into a leading diagonal of zeros by an interchange of rows. For example the C' of our numerical example:

$$\begin{pmatrix} ⓪ & 2 & 1 & 1 & 2 \\ 3 & 5 & ⓪ & 7 & 3 \\ 4 & ⓪ & 4 & 0 & 2 \\ 4 & 5 & 4 & ⓪ & 5 \\ 1 & 0 & 4 & 2 & ⓪ \end{pmatrix}$$

upon the interchange of rows 2 and 3 becomes:

$$\begin{pmatrix} ⓪ & 2 & 1 & 1 & 2 \\ 4 & ⓪ & 4 & 0 & 2 \\ 3 & 5 & ⓪ & 7 & 3 \\ 4 & 5 & 4 & ⓪ & 5 \\ 1 & 0 & 4 & 2 & ⓪ \end{pmatrix}.$$

One needs exactly n (in this case 5) straight lines in order to cross out all the circled zeros: no smaller number of straight lines will suffice:

$$
\begin{pmatrix}
\cancel{\textcircled{0}} & 2 & 1 & 1 & 2 \\
4 & \textcircled{0} & 4 & 0 & 2 \\
3 & 5 & \textcircled{0} & 7 & 3 \\
4 & 5 & 4 & \textcircled{0} & 5 \\
1 & 0 & 4 & 2 & \textcircled{0}
\end{pmatrix}
$$

Because the interchange of rows does not affect this minimum number of crossing lines, we have discovered a simple test to determine whether or not a minimal S can be found:

If the minimum number of lines necessary to cross out all the zeros equals n, a minimal S can be identified. If the minimum number of lines is strictly less than n, a minimal S is not yet at hand.

How can we be sure we are using the smallest possible number of crossing lines? The following rules of thumb are most helpful in this regard: (a) Identify a row (column) with exactly one uncrossed zero. Draw a vertical (horizontal) line through this zero. (b) If all rows or columns with zeros have at least two uncrossed zeros, choose the row or column with the least, identify one of the zeros and proceed as in (a). Ties are settled arbitrarily.

In order to make these rules clear, we illustrate them on the following example:

$$
\begin{matrix}
0 & 0 & 4 & 4 & 9 \\
6 & 0 & 0 & 5 & 0 \\
8 & 0 & 8 & 0 & 8 \\
3 & 0 & 1 & 8 & 0 \\
4 & 2 & 0 & 0 & 8
\end{matrix}
$$

The first column has exactly one zero, so according to (a) we cross out the first row:

$$
\begin{matrix}
\cancel{0} & \cancel{0} & \cancel{4} & \cancel{4} & \cancel{9} \\
6 & 0 & 0 & 5 & 0 \\
8 & 0 & 8 & 0 & 8 \\
3 & 0 & 1 & 8 & 0 \\
4 & 2 & 0 & 0 & 5
\end{matrix}
$$

We must now use (b). We arbitrarily choose c_{32} and cross out the second column:

$$
\begin{array}{ccccc}
\cancel{0} & \cancel{0} & 4 & 4 & 9 \\
6 & 0 & 0 & 5 & 0 \\
8 & 0 & 8 & 0 & 8 \\
3 & 0 & 1 & 8 & 0 \\
4 & 2 & 0 & 0 & 5
\end{array}
$$

Now row 4 has one uncrossed zero, thus we cross out column 5. Proceeding in this way we produce:

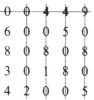

This requires five lines and thus contains a minimal S. Identifying such an S is usually not difficult if one begins by looking for rows or columns with exactly one zero.

Now we come to step (3). Suppose that strictly less than n lines are needed to cross out all the lines in C'. We know then that a minimal S cannot be found directly. In order to transform C' we make use of *König's theorem*:

If the elements of a matrix are divided into two classes by property R, then the minimum number of lines that contain all the elements with the property R is equal to the maximum number of elements with the property R, with no two on the same line.

Applying this to C', where R is the property of being zero, we now present a way to transform C'. We wish to change at least one of the uncrossed (and hence positive) numbers to become zero. This is brought about by (i) subtracting the minimum uncrossed entry from all uncrossed entries; and (ii) adding this same number to each doubly crossed entry (an entry with both horizontal and vertical lines passing through it). All lines are then removed and step (3) is completed.

Table 2.86

		Machines				
		1	2	3	4	5
	1	6	5	9	4	6
	2	3	8	3	9	5
Workers	3	2	7	6	5	6
	4	5	9	8	3	8
	5	1	3	7	4	2

2.7.2.1 A Numerical Example

The Hungarian method will now be used to solve the problem whose C matrix is given in Table 2.86. Here $n = 5$. Following step 1 we subtract a quantity from each column of the (c_{ij}) matrix. This amount is equal to the minimum quantity in that column. Thus the initial (c_{ij}) matrix becomes:

$$
\begin{array}{ccccc}
5 & 2 & 6 & 1 & 4 \\
2 & 5 & 0 & 6 & 3 \\
1 & 4 & 3 & 2 & 4 \\
4 & 6 & 5 & 0 & 6 \\
0 & 0 & 4 & 1 & 0 \\
(-1) & (-3) & (-3) & (-3) & (-2)
\end{array}
$$

The next step is to carry out the same operation for each row. The matrix becomes

$$
\begin{array}{ccccccc}
4 & 1 & 5 & 0 & 3 & (-1) \\
2 & 5 & 0 & 6 & 3 & (0) \\
0 & 3 & 2 & 1 & 3 & (-1) \\
4 & 6 & 5 & 0 & 6 & (0) \\
0 & 0 & 4 & 1 & 0 & (0)
\end{array}
$$

Following step 2, the minimum number of lines passing through all the zero elements are drawn:

$$
\begin{array}{ccccc}
4 & 1 & 5 & 0 & 3 \\
2 & 5 & 0 & 6 & 3 \\
0 & 3 & 2 & 1 & 3 \\
4 & 6 & 5 & 0 & 6 \\
0 & 0 & 4 & 1 & 0
\end{array}
$$

If the minimum number of lines had been 5, an optimum solution could have been found by inspection. This involves selecting five zero elements— one such element in each row and each column. This selection will be illustrated shortly. Such is not the case in the present problem, where only four lines are necessary. This means that an implementation of step 3 is required. The minimum uncrossed number is selected. It is subtracted from all uncrossed numbers:

$$
\begin{array}{ccccc}
4 & 0 & 4 & 0 & 2 \\
2 & 5 & 0 & 6 & 3 \\
0 & 2 & 1 & 1 & 2 & (-1) \\
4 & 5 & 4 & 0 & 5 \\
0 & 0 & 4 & 1 & 0
\end{array}
$$

This same number (1 in this case) is added to all numbers with two lines passing through them:

$$
\begin{array}{ccccc}
4 & 0 & 4 & 0 & 2 \\
3 & 5 & 0 & 7 & 3 \\
0 & 2 & 1 & 1 & 2 \quad (+1) \\
4 & 5 & 4 & 0 & 5 \\
1 & 0 & 4 & 2 & 0
\end{array}
$$

The minimum number of lines are again drawn through all the zero elements:

$$
\begin{array}{ccccc}
4 & 0 & 4 & 0 & 2 \\
3 & 5 & 0 & 7 & 3 \\
0 & 2 & 1 & 1 & 2 \\
4 & 5 & 4 & 0 & 5 \\
1 & 0 & 4 & 2 & 0
\end{array}
$$

As five lines are required, the minimal solution can be found.

The solution for the present problem is

$$
\begin{aligned}
x_{12} &= 1 \\
x_{23} &= 1 \\
x_{31} &= 1 \\
x_{44} &= 1 \\
x_{55} &= 1.
\end{aligned}
$$

The value of this solution is equal to the total of the numbers subtracted, i.e.,

$$
x_0^* = 1 + 3 + 3 + 3 + 2 + 1 + 0 + 1 + 0 + 0 + 1 = 15.
$$

This value can be checked by inspecting the original (c_{ij}) matrix.

2.8 Exercises

(I) Computational

1. Solve the following problems graphically.
    (a) A baker bakes two types of cakes each day, one chocolate and one banana. He makes a profit of $0.75 for the chocolate cake and $0.60 for the banana cake The chocolate cake requires 4 units of flour and 2 units of butter and the banana

cake requires 6 units of flour and 1 unit of butter. However, only 96 units of flour and 24 units of butter are available each day. How many of each type of cake should he bake each day so as to maximize profit?

(b) A bakery produces two types of bread. Each type requires two grades of flour. The first type requires 5 kg of grade 1 flour and 4 kg of grade 2 flour per batch. The second type requires 4 kg of grade 1 and 6 kg of grade 2 per batch. The bakery makes a profit of $10 and $20 per batch on the first and second types, respectively. How many batches of each type should be made per day if 200 kg of grade 1 and 240 kg of grade 2 flour can be supplied per day?

(c) In the production of wool yarn by carding it is found that the waste produced is dependent on the quantity by weight of a lubricant/water emulsion added before processing. Because of pumping restrictions the concentration of the emulsion should not exceed 1 part lubricant to 2 parts water. The application of the emulsion should be at a rate so that no more than 5% dry wool weight of emulsified wool is emulsion. Assume that the densities of water and lubricant are the same. Quality control measures stipulate that the lubricant should not be more than 4% (dry wool weight) of the emulsified wool. It is found that the waste produced decreases by 8 kg per kg lubricant added and decreases by 5 kg per kg water added. Find the amounts of lubricant and water to apply to 100 kg of dry wool so as to minimize the waste produced.

(d) A company makes two types of brandy: The Seducer (S), and Drunkard's Delight (D). Each barrel of S requires 5 hours in the fermenter and 2 hours in the distiller, while each barrel of D requires 3 hours in the fermenter and 4 hours in the distiller. Because of various restrictions the fermenter and the distiller can be operated for no more than 15 and 8 hours per day, respectively. The company makes a profit of $210 for a barrel of S and $140 for a barrel of D. How many barrels of each type should be produced to maximize daily profit?

(e) A farmer produces potatoes at a profit of $200 per unit and pumpkins at a profit of $140 per unit. It takes him 5 days to crop a unit of potatoes and 7 days to crop a unit of pumpkins. Earlier in the year it takes him 5 days to prepare the land and plant seeds for a unit of potatoes and 3 days for a unit of pumpkins. He has 90 cropping days and 50 preparation days available. What amount of each vegetable should he plan on in order to maximize profit?

2. Solve the following problems by the simplex method.

(a) A plant manufactures three types of vehicle: automobiles, trucks, and vans, on which the company makes a profit of $4,000, $6,000, and $3,000, respectively, per vehicle. The plant has three main departments: parts, assembly, and finishing. The labour in these departments is restricted, with parts, assembly, and finishing operating 120, 100, and 80 hours, respectively, each two-week period. It takes 50, 40, and 30 hours, respectively, to manufacture the parts for an automobile, truck, and van. Assembly takes 40, 30, and 20 hours, respectively, for an automobile, truck, and van. Finishing takes 20, 40 and 10 hours, respectively, for an automobile, truck, and van. How many of each type of vehicle should the company manufacture in order to maximize profit for a two-week period?

(b) A manufacturer produces three soft drink cocktails: Fruito, Fifty/fifty, and Sweeto. The amounts of sugar and extract in one barrel of each are shown in Table 2.87. The manufacturer can obtain 6 kg, 4 kg, and 3 kg per day of sugar,

Table 2.87. Data for Exercise 2(b).

Cocktail	Profit/barrel	Sugar	Orange extract	Lemon extract
Fruito	$30	1 kg	2 kg	1 kg
50/50	$20	2 kg	1 kg	1 kg
Sweeto	$30	3 kg	1 kg	1 kg

orange extract, and lemon extract, respectively. The profit is proportional if fractional quantities of a barrel are produced. How much of each cocktail should be produced in order to maximize daily profit?

(c) The problem is to maximize the satisfaction gained on a 140 km journey when only 3 constant speeds are permitted: 0, 50, and 80 km/hr. The satisfaction gained from stopping and resting (0 km/hr), travelling slow (50 km/hr) and travelling fast (80 km/hr) is rated at 5, 9, and 2 units/hr, respectively. Restrictions imply that the journey must be completed in no longer than 4 hours, the total time spent stationary or travelling at high speed must not exceed 1 hour, and the average speed of the journey must not be less than 40 km/hr.

(d) A small company makes 3 types of biscuits: A, B, and C. 10 kg of biscuit A requires 5 kg of sugar, 3 kg of butter, and 2 kg of flour. 10 kg of biscuit B requires 4 kg of sugar, 3 kg of butter, and 3 kg of flour. 10 kg of biscuit C requires 3 kg of sugar, 4 kg of butter, and 3 kg of flour. The company has available per day 40 kg of sugar, 33 kg of butter, and 24 kg of flour. The company can sell all it produces, and makes a profit of $60, $50, and $30 from 10 kg of biscuits A, B, and C, respectively. How much of each biscuit should the company make to maximize daily profit?

(e) Recall the farmer in Exercise 1(e). He discovers he can now make a profit of $160/kg from beets. These take 4 days for planting and 4 days for cropping per kilogram. He also considers the time it takes to sell his produce in the market. It takes 2 days to sell one kilogram of any vegetable. He has 30 days to sell his vegetables. What weights of the three crops should he now plan for in order to maximize profit?

(f) A man has approximately 100 m^2 of garden space. He decides to grow corn (C), tomatoes (T), and lettuce (L) in the 20 week growing season. He estimates that on average for every expected kg of yield from the crops it takes 0.5, 1.0, and 0.5 minutes each week to cultivate the corn, tomatoes, and lettuces respectively. He does not want to spend more than 3 hours per weekend cultivating. He will spend up to $2.00/week for seeds. The seed costs (on a weekly basis) per kg yield for C, T, and L are 0.5, 1.5, and 1.0 cents, respectively. Each crop, C, T, and L requires $\frac{1}{4}$, $\frac{1}{6}$, and $\frac{1}{8}$ m^2, respectively, in space per kg yield. He can sell the vegetables for 0.40, 1.00, and 0.50 dollars per kg for C, T, and L respectively. What amounts should he plan for in order to maximize revenue?

(g) An ice cream factory makes 3 different types of ice cream: plain (P), hokey pokey (H), and chocolate (C). Profits for one unit of each type are $5, $2, and $1 for P, H, and C, respectively. Time constraints for producing a unit of each are shown in Table 2.88. Available hours per day are 8, 10, and 4 for machining, men, and, packing, respectively. What amounts of the different ice creams should be manufactured to maximize daily profit?

Table 2.88. Data for Exercise 2(g).

Product	Machine hours	Man hours	Packing hours
P	4	3	1
H	2	2	1
C	1	2	1

3. Solve the following problems graphically and by the simplex method and compare your solutions.

  (a) A housewife makes sauce (S) and chutney (C) which she sells to the local store each week. She obtains a profit of 40 and 50 cents for a pound of C and S, respectively. C requires 3 lb tomatoes and 4 cups of vinegar; requires 5 lb tomatoes and 2 cups of vinegar. She can buy 24 lb tomatoes and 3 bottles of vinegar at discount price each week. The 3 bottles provide 16 cups of vinegar. In order to make it worthwhile, the store insists on buying at least 3 lbs of goods each week. What combination should be made in order to maximize profit?

  (b) A man makes glue in his backyard shed. Glue A requires 2 g of oyster shell and 4 g of a special rock to produce a 2 kg package. Glue B requires 3 g of shell and 2 g of rock for each 2 kg package. He must produce at least 8 kg of glue per day to stay in business. His son scours the sea shore for the shell and rock and can gather 12 kg of each per day. If the profit is $3 and $4 on a 2 kg package of type A and B, respectively, what is the maximum profit he can hope to make?

  (c) An orchard which grows apples (A) and pears (B) wishes to know how many pickers to employ to maximize the quantity of fruit picked in a given period. The average quantity of fruit a picker can gather is 14 kg of A or 9 kg of B. The orchard can afford to employ no more than 18 people. There cannot be more than 9 picking apples or the supply will be exhausted too soon, flooding the market and reducing returns. But there must be more than half as many picking apples as there are picking pears or costs are increased because of fallen fruit being wasted.

  (d) Recall Exercise 1(e). Suppose in time of war the government insists that the farmer produces at least 5 kilograms of vegetables. What should he do to maximize profits now?

  (e) Consider the gardener who decides to eat some of his corn and tomatoes. A 100 g serving of corn will add 80 calories; a 100 g serving of tomatoes will add 20 calories. He does not want to take in more than 200 calories from this part of his diet. He needs at least 50 mg of vitamin C and at least 1.8 mg of iron from these vegetables to make up his daily intake. A 100 g serving of corn yields 10 mg of vitamin C and 0.6 mg of iron, while a 100 g serving of tomatoes yields 18 mg of vitamin C and 0.8 mg of iron. With corn and tomatoes costing 4 and 10 cents per 100 g, how should he achieve his dietary needs while minimizing costs?

4. The following problems have multiple optimal solutions. Solve each graphically and by the simplex method. Define the set of all optimal solutions.

  (a) A builder finds he is commonly asked to build two types of buildings, A and B. The profits per building are $4,000 and $5,000 for A and B respectively. There are certain restrictions on available materials. A requires 4,000 board feet of timber, 4 units of steel, 3 units of roofing iron, and 2 units of concrete. *B* requires 5,000 board feet of timber, 3 units of steel, 2 units of roofing iron, and 1 ton of concrete. However only 32,000 board feet of timber, 24 units of steel, 20 units of roofing iron, and 16 units of concrete are available per year. What combination of *A* and *B* should he build per year to maximize profit?

(b) Recall the glue manufacturer of Exercise 3(b). In order to remain competitive he finds he must add resin and filler to his glues. 3 g of resin must be included in each packet of each glue and 4 g of filler must be included in glue *A* and 2 g of filler in glue *B*. His son can manufacture only 15 g of resin and 9 g of filler each day. He can now make a profit of $8 and $4 for a package of glues *A* and *B*, respectively. What is the maximum profit he can hope to make?

(c) A recording company is going to produce an hour-long recording of speeches and music. The problem is to fully utilize the 60 available minutes. There can be no more than 3 speeches and 5 musical items. The time allotted for speeches must be no less than one-eighth of the time allotted to music. The gaps between items or speeches must be filled with commentary, which must be no more than 12 minutes in total. The speeches are 5 minutes long, the items 8 minutes. How many of each should be included so as to minimize the commentary time on the recording?

(d) A bakery makes 2 types of cakes, *A* and *B*. 10 lb of cake *A* requires 2 lb of flour, 3 lb of sugar, 3 eggs, and 4 lb of butter. 10 lb of cake *B* requires 4 lb of flour, 3 lb of sugar, 6 eggs, and 1 lb of butter. The bakery can afford to purchase 24 lb of flour, 27 lb of sugar, 24 eggs, and 20 lb of butter per day. The bakery makes a profit of $3 for 10 lb cake *A* and $6 for 10 lb of *B*. How much of each cake should be made daily in order to maximize profit?

(e) Melt-In-Your-Mouth Biscuit Co. finds that its two best sellers are Coco Delights (*C*) and Cheese Barrel Crackers (*B*). *C* and *B* produce a profit of $10 and $15 per carton sold to the supermarkets. Some ingredients are common to each biscuit. Each week no more than 500 kg of flour, 360 kg of sugar, 250 kg of butter and 180 kg of milk can be used effectively. Every 100 kg of *C* requires 20 kg of flour, 16 kg of sugar, 18 kg of butter, and 15 kg of milk. Every 100 kg of *B* requires 30 kg of flour, 20 kg of sugar, 12 kg of butter, and 10 kg of milk. Find the weekly combination of production which maximizes profit.

(f) A man finds he is eating a lot of corn and no cheese and decides to do something about it. Being very careful of his dietary considerations he realises that he needs 600 I.U. of vitamin A, 1 mg of iron, 0.12 mg of calcium, and no more than 400 calories per day. Now 100 g of cheese gives 400 I.U. Vitamin A, 0.3 mg iron, 0.2 mg calcium, and 120 calories. Also, 100 g of corn gives 160 I.U. vitamin A, 0.6 mg iron, no calcium, and 80 calories. Moreover, cheese costs 10 cents and corn 4 cents per 100 g. What should his daily intake of these two items be if he is to satisfy the requirements above at minimal cost?

5. The following problems have degenerate optimal solutions. Solve them by the simplex method and interpret the final tableau.

(a) Consider a farmer who wishes to plant 4 types of grain: oats, barley, wheat, and corn. The profits he can make from an acre of corn, barley, wheat and oats are $300, $200, $400, and $100, respectively. However, there are a number of restrictions regarding fertilizing, spraying, and cultivation. These are as follows: the corn, barley, wheat, and oats require 8, 2, 5, and 4 cwt of fertilizer per acre, respectively, but only 16 cwt is available for the season. Similarly corn, barley, wheat, and oats require 6, 4, 3, and 2 gallons of insecticide per acre, respectively, but only 10 gallons are available. Also it takes 3, 3, 2, and 1 day to cultivate 1 acre of corn, barley, wheat, and oats, respectively, but the farmer can spare a total of only $6\frac{5}{7}$ days. What crop combination maximizes profit?

(b) A man operates a small warehouse to store goods for other companies on a temporary basis. His warehouse is limited to 150 m^2 in usable space. He can afford to employ up to 10 men. Each load of product A, B, C, and D requires 16, 15, 20, and 30 m^2 of space, respectively. Each load of A, B, C, and D keeps 1, 9, 1, and 2 men fully occupied, respectively. His storage charges are $200, $300, $400, and $700 per load for A, B, C, and D respectively. What combination of goods should he attempt to store in order to maximize revenue?

(c) In a carpet wool spinning plant four blends of wool can be produced and are worth $80, $60, $50, and $20 per kg for blends 1, 2, 3, and 4, respectively. Assuming that all the yarn produced can be sold, find the amount of each blend necessary to maximize profit. Because of certain restrictions with shiftwork and staff regulations, the carding, spinning, twisting, and hanking machinery can only be operated for a maximum of 18, 15, 10, and 12 hours a day, respectively. The hours each machine takes to process 10^3 kg of each blend are shown in Table 2.89. A further restriction limits the quantity of blends 3 and 4 to 5 × 10^3 kg per day.

Table 2.89. Data for Exercise 5(c).

Blend	Carding	Spinning	Twisting	Hanking
1	4	4	4	4
2	4	3	4	2
3	3	3	2	4
4	2	2	0	1

(d) A company produces 4 types of fertilizer: A, B, C, and D. 10 lb of A requires 3 lb of potash (P), 4 lb of phosphate (H), and 3 lb of nitrogen (N). 10 lb of B requires 3 lb of P, 3 lb of H, and 4 lb of N. 10 lb of C requires 5 lb of P, 2 lb of H, and 3 lb of N. 10 lb of D requires 4 lb of P, 4 lb of H, and 2 lb of N. The company can produce 40 lb, 40 lb, and 60 lb of P, H, and N, respectively, per day. The company makes a profit of $20, $40, $50, and $30 per 10 lb of A, B, C, and D respectively. Determine the amount of each type that should be produced each day so as to maximize profit.

(e) A brick manufacturer produces red (R), white (W), brown (B), and grey (G) bricks at profits of $100, $200, $300, and $300 per ton, respectively. These are all produced using the same equipment, which can operate continuously. It takes 2, 3, 5, and 4 equipment hours to produce a ton of R, W, B, and G, respectively. He has a maximum electric power allocation of 252 units because of shortages. It takes 3, 4, 5, and 6 units to produce a ton of R, W, B, and G respectively. Find his maximum weekly profit.

(f) Recall Exercise 2(f). The man now decides to plant pumpkins as well. To produce one kg of yield cultivation will take 0.5 minutes, 0.5 cents will be spent on seed each per week, and 0.5 m^2 of garden space is required. Pumpkins can be sold for 40 cents per kg. Also the cultivation time for corn can now be reduced to $\frac{4}{9}$ of a minute per week. Solve 2(f) over.

6. The following L.P. problems exhibit temporary degeneracy during the simplex iterations. Solve each by this method and comment on this phenomenon.
   (a) A dairy factory is about to start production. The manager wishes to know what lines of production—butter, cheese, milk powder, or yoghurt—would be most profitable. The various restrictions, requirements and unit profits are shown in Table 2.90. Solve this problem by the simplex method.

   Table 2.90. Data for Exercise 6(a).

	Profit	Milk	Labour	Electricity
Butter	3	3	1	2
Cheese	4	2	3	3
Milk powder	2	4	4	5
Yoghurt	1	2	2	1
Amount available	—	8	9	9

   (b) Recall the warehouse problem of 5(b). The manager finds that he receives too many orders for storage of A. So he streamlines the process for A by reducing its storage requirement per load to 120 m^2 and increases the storage fee to $360 per load. What is his optimal strategy now?
   (c) Recall the wool spinning plant in Exercise 5(c). A competing plant realises it has to match the efficiency of the first plant if it is to survive. It produces the same 4 blends, but has a profit of $120, 60, 60, and 30 per kg for each type. The plant can obtain 23 hours and 14 hours carding and hanking time per day, respectively. The other times are identical. However this factory has older twisting machines and it takes 4 hours to produce 10^3 kg of blend 3. Also no more than 2.5 × 10^3 kg of blends 3 and 4 are to be produced per day. What is the best policy for the plant?
   (d) Recall the fertilizer problem of Exercise 5(d). A rival company has the data shown in Table 2.91. What is the best way for this company to operate?

   Table 2.91. Data for Exercise 6(d).

Fertilizer	P	H	N	Profit
A (10 lb)	4	1	5	$3
B (10 lb)	3	2	5	$4
C (10 lb)	5	2	3	$4
D (10 lb)	4	4	2	$5
Availability	40	40	30	—

   (e) An ice cream manufacturer makes 2 types of ice cream—creamy and ordinary. Creamy sells at a profit of $5 per unit, ordinary at $4 per unit. Each requires 4 tanks of milk per unit. Creamy requires 5 tanks of cream and 5 bags of sugar per unit. Ordinary requires 2 tanks of cream and 3 bags of sugar. Also 10 tanks of cream and 10 bags of sugar are available each day. How does the manufacturer maximize profit?

(f) Recall Exercise 2(f). The man decides to plant cucumbers as well. Corn can be cut down to $\frac{1}{3}$ minute cultivation time per week per kg yield. One kg yield of cucumbers require $\frac{7}{12}$ minutes of cultivation per week, 0.5 cents per week on seeds and $\frac{13}{12}$ m^2 of garden space. They sell for 40 cents per kg. The garden has been reduced to $88\frac{8}{9}$ m^2. Also the gardener decides he cannot spend more than 150 minutes in the garden each weekend. Solve the problem over with the new data.

7. The following L.P. problems have no feasible solutions. Prove that this is so by use of the simplex method. Also attempt to solve the problems graphically where possible.

   (a) A small clothing factory makes shirts and skirts for a boutique in town. A profit of $4 and $3 is made from a shirt and skirt respectively. A shirt requires 3 yards of material and a skirt 4, with only 12 yards available daily. It takes 5 hours of total time to make a shirt and 2 hours to make a skirt, with 8 hours available daily. At least 5 garments must be made per day. Attempt to maximize profit.

   (b) A man has a part time job making chairs (C), deck chairs (D), and stools (S). Each chair takes 5 hours to complete and weighs 2 kg. Each deck chair takes 3 hours and weighs 1.5 kg. Each stool takes 2 hours and weighs 1 kg. He has only 10 hours to spend each weekend on this work. Now his employers suddenly state that in order to make it worth their while he must produce at least 20 kg of furniture per week. Can he continue?

   (c) Assuming an unlimited supply of paint and turpentine, attempt to maximize the coverage of a mixture of the two when the addition of an equal quantity of turpentine to the paint increases the coverage by 50%, the coverage of paint alone being 8 m^2/litre. Paint costs $3.00 a litre and turpentine $0.50. The total cost of the mixture must be no more than $21.00. To aid spraying, the volume of turpentine plus $\frac{5}{3}$ times the volume of paint must be greater than 50 litres.

   (d) A nursery covers 5,000 m^2. It grows trees at a profit of 35 cents each and shrubs at a profit of 20 cents each. At least 2,000 plants must be grown. A tree needs 4 m^2 to grow, a shrub 1 m^2. Each tree requires 2 g of fertilizer, each shrub 3 g, while 4 kg is available. Attempt to maximize profit.

   (e) Recall Exercise 2(f). Suppose that seeds costs are to be neglected. However, it is vital that the energy value gained from the crop should be greater than 300 calories. Now corn, lettuce, and tomatoes will yield 0.8, 0.1, and 0.2 calories per gram, respectively. Attempt to solve 2(f) over with the new data.

8. Create and solve the dual for each of the following problems. Find the optimal solution to the original solution by interpreting the optimal dual tableau.

   (a) A local vintner makes two types of wine, medium white (M) and dry white (D), to sell to the local shop. He makes $5 profit per gallon from M and $4 a gallon from D. Now M requires 3 boxes of grapes, 4 lb of sugar, and 2 pints of extract per gallon. Also, D requires 4 boxes of grapes, 2 lb of sugar, and 1 pint of extract per gallon. He has 14 boxes of grapes, 8 lb of sugar, and 6 pints of extract left before selling his business. How should he use these resources to maximize profit?

   (b) A turning workshop manufactures two alloys, A and B, at a profit of $5 and $2 a kg, respectively. Alloy A requires 2, 5, 5, and 2 g of nickel, chrome, germanium, and magnesium, respectively. Alloy B requires 3, 2, 3, and 1 g of the metals in the same order. Supplies of the metals are reduced to 7, 11, 10, and 6 kg of the metals in the same order. The furnace cannot be operated for more than 6 hours per day. Alloys A and B require 1 and 2 hours of furnace time, respectively, to produce 1 kg of alloy. How can profits be maximized?

(c) Recall Exercise 7(c). Suppose now the total cost of paint and turpentine cannot exceed $100. The paint now used is of lower quality and costs only $2/litre. However, because of price rises and the decision to use a better grade, the price of turpentine has risen to $2/litre. A new sprayer has been purchased, and now the volume of turpentine plus twice the volume of paint need exceed only 20 litres. However, the ratio of turpentine to paint must be between 1:4 and 3:1. Solve Exercise 7(c) over with the new data.

(d) Recall Exercise 7(d). Having discovered that this problem was infeasible, the nursery removed the restriction that 2,000 plants must be grown. During the summer each plant requires one litre of water, but because of restrictions brought on by the annual drought only 6,000 gallons can be used per day. Also 4 g of beetle powder must be used on each shrub each day and 1 g on each tree. There are 4 kg of powder available per day. Solve Exercise 7(d) over with the new data.

(e) A person has the option of eating chocolate, oranges, or ice cream as a means of obtaining at least 10% of the minimum recommended daily vitamin intake. At least 0.1 g of calcium, 1 mg of iron, 8 mg of vitamin C, 0.2 mg of riboflavin, and 2 mg of niacin are required daily. The three foods would provide these, as shown in Table 2.92. The problem is to keep calorie intake down to a minimum where 100 gm of chocolate, oranges, or ice cream provide 400, 40, and 160 calories respectively. What combination of the foods should be eaten to achieve these objectives?

Table 2.92. Data for Exercise 8(e).

	100 g of:		
	Chocolate	Oranges	Ice cream
Calcium	0.5 gm	0.03 gm	0.1 gm
Iron	1.0 mg	0.4 mg	0.1 mg
Vitamin C	—	40 mg	1 mg
Riboflavin	0.2 mg	.02 mg	0.1 mg
Niacin	1.0 mg	0.2 mg	0.1 mg

9. In each of the following problems an objective function coefficient has been changed. Examine the effect that this has on the optimal solution and its value by solving the original problem and then performing sensitivity analysis.

(a) Consider the primal L.P. problem of 8(a):

$$\text{Maximize:} \quad 5x_1 + 4x_2$$
$$\text{subject to:} \quad 3x_1 + 4x_2 \leq 14$$
$$4x_1 + 2x_2 \leq 8$$
$$2x_1 + x_2 \leq 6$$
$$x_1, x_2 \geq 0.$$

Table 2.93. Data for Exercise 13(a).

		Taverns			
		1	2	3	4
	1	8	14	12	17
	2	11	9	15	13
Breweries	3	12	19	10	6
	4	12	5	13	8

and 4 are 20, 10, 10, and 5 barrels per day, respectively. The demands of taverns 1, 2, 3, and 4 are 5, 20, 10, and 10 barrels per day respectively. Find the minimum cost schedule.

(b) A bread manufacturer has 4 factories. He supplies 4 towns. The unit transportation costs are shown in Table 2.94. The demands of towns 1, 2, 3, and 4 are 6,000, 12,000, 5,000, and 8,000 loaves per day, respectively. The daily production capacities of the factories 1, 2, 3, and 4 are 7,000, 8,000, 11,000, and 5,000 loaves, respectively. Find the minimum cost schedule.

Table 2.94. Data for
Exercise 13(b).

		Town			
		1	2	3	4
	1	7	5	4	3
	2	5	4	4	3
Factory	3	6	5	6	7
	4	3	4	7	9

(c) An effluent treatment plant has 4 independent oxidation systems with capacities of 15, 18, 20, and 30 (in millions of litres) per day. These systems can be interconnected in any combination to any of 5 effluent mains by intermediate pumping stations. The outputs of the mains are 12, 17, 15, 19, and 14 ($\times 10^6$) litres per day. The cost involved in pumping 10^6 litres from any of the mains to any of the systems is shown in Table 2.95. Find the least cost flow.

Table 2.95. Data for
Exercise 13(c).

		System			
		1	2	3	4
	1	4	6	7	5
	2	3	2	2	1
Main	3	1	7	4	3
	4	7	3	0	4
	5	2	3	8	2

(d) Four large farms produce all the potatoes to satisfy the demands of markets in four towns. The monthly production of the farms and the demand of the towns are shown in Table 2.96, and the transportation costs per ton are shown in Table 2.97. Find the minimum cost schedule.

Table 2.96. Production and Demand in Exercise 13(d).

Farm	Production	Town	Demand
1	30	1	20
2	40	2	35
3	25	3	50
4	45	4	35

Table 2.97. Transportation Cost in Exercise 13(d).

		Town			
		1	2	3	4
Town	1	7	7	10	8
	2	6	6	9	6
	3	8	7	9	5
	4	11	10	12	8

(e) The roads board is about to complete four urgent tasks on state highways in the Wellington province. Costs must be minimized to the satisfaction of the audit team from the treasury. A costly part of the operation involves the transportation of suitable base course and sealing metal from screening plants at Masterton, Otaki, Bulls, Raetihi, and the Desert Road to the tasks at Levin, Palmerston North, Taihape, and Wanganui. In the time available the plants can supply in (1,000-ton units) Masterton, 10; Otaki, 18; Bulls, 12; Raetihi, 14; Desert Road, 24. The demand is: Levin, 20; Palmerston North, 10; Taihape, 30; and Wanganui, 15. Unit costs of loading, transportation and unloading in terms of man hours are shown in Table 2.98. Find the minimum cost schedule.

Table 2.98. Transportation Costs in Exercise 13(e).

	Masterton	Otaki	Bulls	Raetihi	Desert Road
Levin	6	2	2	8	7
Palmerston North	4	4	1	7	6
Taihape	8	7	3	3	2
Wanganui	7	6	2	5	7

14. Each of the following problems is an assignment problem. Solve each one by the Hungarian method.

(a) Consider a collection of six students and six assignments. Each student must be assigned a different assignment. The time (in hours) it is likely to take each student to complete each assignment is given in Table 2.99. Find the minimum time assignment.

Table 2.99. Assignments in Exercise 14(a).

		Tasks					
		1	2	3	4	5	6
	1	7	5	3	9	2	4
	2	8	6	1	4	5	2
Students	3	2	3	5	6	8	9
	4	6	8	1	3	7	2
	5	4	5	6	9	4	7
	6	9	2	3	5	1	8

(b) A factory manager has a table (Table 2.100) which shows how much profit is accomplished in an hour when each of six men operate each of six machines. Note that: man 6 cannot work on machine 1 because this task requires good eyesight; man 2 cannot work on machine 3 because he is allergic to dust; and man 5 cannot work on machine 6 because this job requires two hands. You are required to find the maximum profit assignment.

Table 2.100. Assignments in Exercise 14(b).

		Men					
		1	2	3	4	5	6
	1	7	7	8	6	7	—
	2	8	5	8	6	5	5
Machines	3	6	—	7	5	6	5
	4	5	4	5	5	4	4
	5	6	6	7	7	6	6
	6	7	8	7	6	—	6

(c) In carpet manufacture the carpets are inspected for faults and repaired by hand sewing, called picking. In a certain factory there are 6 picking boards and the management wishes to assign 6 rated workers to these boards so that the total time to repair any quantity of carpet is minimized. The rates of the workers on the different picking boards are shown in Table 2.101; they vary because the boards handle different sizes and types of carpets depending upon their location.

(d) An air force has six pilots which it wishes to assign to six different types of aircraft. Each pilot has been rated on each one and given a numerical rating in terms of

Table 2.101. Data for Exercise 14(c).

		Workers					
		1	2	3	4	5	6
Boards	1	8	6	1	4	9	4
	2	3	4	10	2	3	2
	3	4	5	6	7	4	6
	4	1	8	5	1	5	6
	5	7	2	3	7	10	3
	6	2	5	5	3	7	5

Table 2.102. Ratings in Exercise 14(d).

		Aircraft					
		1	2	3	4	5	6
Pilots	1	7	4	8	2	3	5
	2	8	3	3	6	2	4
	3	2	5	3	7	4	9
	4	5	2	6	6	7	2
	5	6	4	2	8	3	1
	6	3	5	6	4	5	7

errors in operation (Table 2.102). Make an assignment of pilots to aircraft so as to minimize the culmulative rating of those assigned.

(e) Table 2.103 gives the standardized times of seven workers on seven machines. Find a minimum time assignment.

Table 2.103. Data for Problem 14(e).

		Machines						
		1	2	3	4	5	6	7
Workers	1	6	4	4	5	6	7	4
	2	7	5	1	1	3	9	2
	3	3	3	7	1	9	6	6
	4	4	6	5	8	1	5	8
	5	6	1	4	4	2	1	4
	6	6	6	9	8	8	2	9
	7	5	7	3	9	1	8	2

(f) A novelty atheletics meeting is to be held for teams of eight. There are eight events, and one man from each team is to enter each event. A certain team has a member who predicts where each man would be placed if he entered each event (Table 2.104). Given that the team accepts his predictions, and that each finisher's score is inversely proportional to the place he gets, find the optimal allocation.

Table 2.104. Predictions of Exercise 14(f).

Athletes	100 m	100 m hurdles	400 m	1500 m	Long jump	High jump	Javelin	Shot put
							Events	
Allan	4	3	3	4	3	2	4	6
Big Billy	3	2	2	1	3	1	2	4
Chris	5	3	5	7	2	4	2	2
Dangerous Dan	3	2	2	5	4	3	3	2
Ewen R.	2	3	1	2	1	2	3	4
Freddy	1	1	2	3	2	4	4	6
George	2	4	3	3	2	5	1	1
Harry	4	6	4	6	4	6	3	5

(II) Theoretical

15. Formulate a number of real-world problems as linear programming problems.

16. Show that the set of feasible solutions to an L.P. in standard form is convex.

17. Prove that, if more than one basic feasible solution is optimal for a linear programming problem in standard form, then any convex combination of those basic feasible solutions is also optimal.

18. Attempt to solve the problem of Section 2.5.7 by the two-phase method. Compare the efficiency of that approach with using the big M method.

19. Attempt to solve the problems of Section 2.5.8 by the two-phase method. Draw conclusions about the use of that method on problems with unbounded optima in general. Prove your conclusions.

20. Solve the transportation problem of Section 2.7 as a linear programming problem by the simplex method. Compare the process, step by step, with that obtained by the stepping stone method.

21. Formulate a 3×3 assignment problem as a linear programming problem. Solve it by the simplex method. Formulate the same problem as a transportation problem and solve it by the stepping stone method. Compare these processes with solving the problem by the Hungarian method.

22. If a linear programming problem has multiple optima, then its objective function hyperplane is parallel to that of a binding constraint. State conditions which must hold when the converse is not true.

23. Prove that if a linear programming problem has an unbounded optimum its dual cannot have any feasible solutions.

24. Prove that a variable is unrestricted in sign in a L.P. if and only if the corresponding constraint in the dual is an equality.

Chapter 3

Advanced Linear Programming Topics

3.1 Efficient Computational Techniques for Large L.P. Problems

We shall now discuss how large linear programming problems may be solved on a digital computer with the aid of properly organized calculations. In spite of the recent tremendous advancement in the computational power and memory size of modern computers, computational difficulties still arise in solving large L.P. problems. New techniques have been developed to overcome some of these. The techniques that we shall discuss are: the revised simplex method, the dual simplex method, the primal–dual algorithm, and Wolfe–Dantzig decomposition.

3.2 The Revised Simplex Method

We turn now to improving the efficiency of the simplex method presented in the previous chapter. Although fairly small problems can be solved by hand using the method, realistic industrial problems are too large for even the most patient arithmetician. As they are to be solved using scarce, expensive computer time, it is desirable to make the simplex method as efficient as possible.

Anyone who has used the simplex method on a nontrivial problem will have noticed that most tableau entries have their values calculated and re-calculated. Often, many such values are never actually used to make decisions about entering and leaving basic variables and may just as well never have been computed. At any iteration, what entries are necessary in order to know how to proceed? The x_0-row coefficients of nonbasic variables are needed to decide whether or not to continue, and if so what variable enters the basis. The other coefficients of entering variable and the r.h.s. entries are needed to take ratios to decide which variable should leave the basis. It is desirable to have these values available without having to calculate all the others, which are of no immediate interest.

The revised simplex method achieves this. It is in essence no different from the simplex method; it is simply a more efficient way of going about things when using a computer. As fewer numbers are calculated at each iteration, less storage is required by the computer, which may be an important factor in dealing with relatively large problems.

3.2.1 A Numerical Example

Let us return once more to Problem 2.1 and find how we can solve it in an efficient manner. The first two tableaux generated in solving the problem by the regular simplex method are given in Tables 3.1 and 3.2.

Table 3.1

Constraints	x_1	x_2	x_3	x_4	x_5	r.h.s.	Ratio
(2.11)	3	4	1	0	0	12	$\frac{12}{3}$
(2.12)	3	3	0	1	0	10	$\frac{10}{3}$
(2.13)	④	2	0	0	1	8	$\frac{8}{4}$
x_0	-4	-3	0	0	0	0	

Table 3.2

Constraints	x_1	x_2	x_3	x_4	x_5	r.h.s.	Ratio
(2.11)	0	$\frac{5}{2}$	1	0	$-\frac{3}{4}$	6	$\frac{12}{5}$
(2.12)	0	$\frac{3}{2}$	0	1	$-\frac{3}{4}$	4	$\frac{8}{3}$
(2.13)	1	$\frac{1}{2}$	0	0	$\frac{1}{4}$	2	$\frac{4}{1}$
x_0	0	-1	0	0	1	8	

Given Table 3.1, what information is needed to generate the next iteration, which produces Table 3.2? In order to decide whether any further iterations are necessary, the x_0 row is required: $(-4, -3, 0, 0, 0, 0)$. The column $(3, 3, 4, \ )^T$ of the incoming variable x_1 and the r.h.s. column $(12, 10, 8, \ )^T$ are required to decide upon the outgoing basic variable. None of the other information is relevant at this moment.

Recall that the simplex method starts with an initial basis and a corresponding basic feasible solution and generates a sequence of improving basic feasible solutions by replacing one basic variable at a time. The kernel of the revised simplex method is that the basic feasible solution corresponding to any basis can be calculated from the original tableau by a correct sequence of row operations. In order to motivate this, consider Problem 2.1 in matrix form:

$$\text{Maximize:} \qquad x_0 = (4, 3, 0, 0, 0)(x_1, x_2, x_3, x_4, x_5)^T$$

$$\text{subject to:} \qquad \begin{pmatrix} 3 & 4 & 1 & 0 & 0 \\ 3 & 3 & 0 & 1 & 0 \\ 4 & 2 & 0 & 0 & 1 \end{pmatrix} \begin{pmatrix} x_1 \\ x_2 \\ x_3 \\ x_4 \\ x_5 \end{pmatrix} = \begin{pmatrix} 12 \\ 10 \\ 8 \end{pmatrix} \qquad (3.1)$$

$$x_i \geq 0, \qquad i = 1, 2, \ldots, 5. \qquad (3.2)$$

In this problem, three equations in five unknowns form the constraints. Hence any basic feasible solution is found by setting two variables equal to zero and solving the remaining three equations in three unknowns. This creates a *basis matrix*, a submatrix of the original constraint matrix, found by deleting the columns corresponding to nonbasic variables.

The initial basis is $\beta_1 = \{x_3, x_4, x_5\}$. In Table 3.1 it can be seen that x_1 should replace x_5 in the basis, creating a basis $\beta_2 = \{x_3, x_4, x_1\}$. The basis matrix for β_2 is found by deleting columns 2 and 5 from the constraint matrix in (3.1) and rearranging the order of the remaining columns if necessary, i.e.

$$B = \begin{pmatrix} 1 & 0 & 3 \\ 0 & 1 & 3 \\ 0 & 0 & 4 \end{pmatrix}.$$

And, as

$$x_2 = x_5 = 0,$$

(3.1) can be abbreviated as

$$\begin{pmatrix} 1 & 0 & 3 \\ 0 & 1 & 3 \\ 0 & 0 & 4 \end{pmatrix} \begin{pmatrix} x_3 \\ x_4 \\ x_1 \end{pmatrix} = \begin{pmatrix} 12 \\ 10 \\ 8 \end{pmatrix},$$

so that the basic feasible solution (b.f.s.) corresponding to β_2 is

$$\begin{pmatrix} x_3 \\ x_4 \\ x_1 \end{pmatrix} = \begin{pmatrix} 1 & 0 & 3 \\ 0 & 1 & 3 \\ 0 & 0 & 4 \end{pmatrix}^{-1} \begin{pmatrix} 12 \\ 10 \\ 8 \end{pmatrix} = \begin{pmatrix} 1 & 0 & -\frac{3}{4} \\ 0 & 1 & -\frac{3}{4} \\ 0 & 0 & \frac{1}{4} \end{pmatrix} \begin{pmatrix} 12 \\ 10 \\ 8 \end{pmatrix} = \begin{pmatrix} 6 \\ 4 \\ 2 \end{pmatrix}.$$

Once B^{-1} has been found, any column in the tableau representing the b.f.s. based on β_i can be calculated. This is achieved by multiplying the original column by B^{-1}. For instance, if it is desired to find the x_2 column

$\bar{C}_2$ in the tableau corresponding to β_2,

$$\bar{C}_2 = B^{-1}C_2$$

$$= \begin{pmatrix} 1 & 0 & -\frac{3}{4} \\ 0 & 1 & -\frac{3}{4} \\ 0 & 0 & \frac{1}{4} \end{pmatrix} \begin{pmatrix} 4 \\ 3 \\ 2 \end{pmatrix} = \begin{pmatrix} \frac{5}{2} \\ \frac{3}{2} \\ \frac{1}{2} \end{pmatrix},$$

where

$$C_i = x_i \text{ column in original tableau,}$$

$$\bar{C}_i = \text{updated } x_i \text{ column.}$$

It is necessary to calculate the x_0 row in any new tableau to find whether the new tableau is optimal and, if not, which variable should enter the basis. The x_0-row coefficients of the basic variables will be zero. How are the x_0-row coefficients of the nonbasic variables calculated in the regular simplex method? For instance, let us discover the steps taken to calculate $\bar{c}_2$, the x_0-row coefficient of x_2 in Table 3.2. Suppose that rows (2.11), (2.12), and (2.13) have been updated, and now the x_0 row is to be revised. One can form a vector of the coefficients of the basic variables in the original tableau:

$$c_B = (0, 0, -4).$$

The scalar

$$c_B\bar{C}_2 = (0, 0, -4) \begin{pmatrix} \frac{5}{2} \\ \frac{3}{2} \\ \frac{1}{2} \end{pmatrix}$$

is the quantity that has been subtracted from the original x_0-row coefficient of x_2 when Table 3.2 has been arrived at. Hence

$$\bar{c}_2 = c_2 - c_B\bar{C}_2 = -3 - (0, 0, -4) \begin{pmatrix} \frac{5}{2} \\ \frac{3}{2} \\ \frac{1}{2} \end{pmatrix} = -1.$$

But we have seen how to deduce $\bar{C}_2$ from C_2, i.e.,

$$\bar{C}_2 = B^{-1}C_2$$

$$\bar{c}_2 = c_2 - c_B B^{-1}C_2.$$

For brevity let

$$\pi = c_B B^{-1}.$$

Then

$$\bar{c}_2 = c_2 - \pi C_2.$$

The entries in π are called *simplex multipliers*.

We are now in a position to calculate all the nonbasic variable coefficients, given that the new basis is β_2:

$$\pi = c_B B^{-1} = (0, 0, -4) \begin{pmatrix} 1 & 0 & -\frac{3}{4} \\ 0 & 1 & -\frac{3}{4} \\ 0 & 0 & \frac{1}{4} \end{pmatrix} = (0, 0, -1).$$

Hence

$$\bar{c}_2 = -1, \quad \text{as before}$$

$$\bar{c}_5 = c_5 - \pi C_5$$

$$= 0 - (0, 0, -1)\begin{pmatrix} 0 \\ 0 \\ 1 \end{pmatrix}$$

$$= 1.$$

On examining $\bar{c}_2$ and $\bar{c}_5$ we see that $\bar{c}_2$ alone is negative and therefore x_2 enters the basis. We must now decide which variable leaves the basis. The information required for this is the x_2 column entries and the r.h.s. entries in the new tableau, Table 3.2. These can be obtained, as shown earlier, by multiplying the original x_2 column and r.h.s. by B^{-1}. On taking ratios it is seen that x_3 should leave the basis. The new basis becomes

$$\beta_3 = \{x_2, x_4, x_1\}.$$

Thus the new B is

$$B = \begin{pmatrix} 4 & 0 & 3 \\ 3 & 1 & 3 \\ 2 & 0 & 4 \end{pmatrix}.$$

We could calculate the new B^{-1} by directly inverting B. However, because of the nature of the simplex iteration it is computationally more efficient to calculate each new B^{-1} from the previous one. In order to understand how this can be achieved it is necessary to realise that each entry b_{ij}^{-1}, $i \neq j$, in B^{-1} is simply the multiple of the original constraint (j) which has been finally added to the original constraint (i) to obtain the ith row in present tableau. For instance, if we wished to create row (2.12) in the next tableau from Table 3.2 using the regular simplex method we would subtract $\frac{3}{2}/\frac{5}{2}$ times row (2.11) from row (2.12). Thus the middle row in the new B^{-1} (corresponding to (2.12)) can be obtained from the previous B^{-1} in the same way, i.e.,

$$(b_{21}^{-1}, b_{22}^{-1}, b_{23}^{-1}) = (0, 1, -\tfrac{3}{4}) - (\tfrac{3}{2}/\tfrac{5}{2})(1, 0, -\tfrac{3}{4})$$

$$= (-\tfrac{3}{5}, 1, -\tfrac{3}{10}).$$

The bottom row can be found in a similar manner:

$$(b_{31}^{-1}, b_{32}^{-1}, b_{33}^{-1}) = (0, 0, \tfrac{1}{4}) - (\tfrac{1}{2}/\tfrac{5}{2})(1, 0, -\tfrac{3}{4})$$

$$= (-\tfrac{1}{5}, 0, \tfrac{2}{5}).$$

Of course, the top row can be found by dividing the top row of the previous B^{-1} by $\frac{5}{2}$. Therefore,

$$(b_{11}^{-1}, b_{12}^{-1}, b_{13}^{-1}) = \tfrac{2}{5}(1, 0, -\tfrac{3}{4})$$

$$= (\tfrac{2}{5}, 0, -\tfrac{3}{10}).$$

Hence the new B^{-1} is

$$B^{-1} = \begin{pmatrix} \frac{2}{5} & 0 & -\frac{3}{10} \\ -\frac{3}{5} & 1 & -\frac{3}{10} \\ -\frac{1}{5} & 0 & \frac{2}{5} \end{pmatrix}.$$

Now the very first B is I, so

$$B^{-1} = I, \quad \text{initially.}$$

The next B^{-1} can be obtained from this one, using the method illustrated above. And indeed each B^{-1} can be found from the one before.

Going back to the example, we can now calculate the new simplex multipliers associated with the new basis β_3:

$$\pi = c_B B^{-1} = (-3, 0, -4) \begin{pmatrix} \frac{2}{5} & 0 & -\frac{3}{10} \\ -\frac{3}{5} & 1 & -\frac{3}{10} \\ -\frac{1}{5} & 0 & \frac{2}{5} \end{pmatrix} = (-\tfrac{2}{5}, 0, -\tfrac{7}{10}).$$

We now calculate the nonbasic $\bar{c}_j$ to discover whether or not β_3 is an optimal basis:

$$\bar{c}_3 = c_3 - \pi C_3$$

$$= 0 - (-\tfrac{2}{5}, 0, -\tfrac{7}{10}) \begin{pmatrix} 1 \\ 0 \\ 0 \end{pmatrix}$$

$$= \tfrac{2}{5}$$

$$\bar{c}_5 = c_5 - \pi C_5$$

$$= 0 - (-\tfrac{2}{5}, 0, -\tfrac{7}{10}) \begin{pmatrix} 0 \\ 0 \\ 1 \end{pmatrix}$$

$$= \tfrac{7}{10}.$$

As both entries are nonnegative, the optimal solution has been found. This solution is

$$\begin{pmatrix} x_2 \\ x_4 \\ x_1 \end{pmatrix} = \begin{pmatrix} \frac{2}{5} & 0 & -\frac{3}{10} \\ -\frac{3}{5} & 1 & -\frac{3}{10} \\ -\frac{1}{5} & 0 & \frac{2}{5} \end{pmatrix} \begin{pmatrix} 12 \\ 10 \\ 8 \end{pmatrix} = \begin{pmatrix} \frac{12}{5} \\ \frac{2}{5} \\ \frac{4}{5} \end{pmatrix}$$

$$x_i = 0, \qquad i = 3, 5.$$

The actual solution value can be found by substituting these values in the original objective function.

$$x_0^* = (3, 0, 4) \begin{pmatrix} \frac{12}{5} \\ \frac{2}{5} \\ \frac{4}{5} \end{pmatrix}$$

$$= \tfrac{52}{5}.$$

3.2.2 Summary of the Revised Simplex Method

Consider the following problem:

$$\text{Maximize:} \quad x_0 = C^T X \tag{3.3}$$

$$\text{subject to:} \quad AX = B$$
$$X \geq 0, \tag{3.4}$$

with m constraints and n variables.

Let

$$A_{m \times n} = \begin{pmatrix} a_{11} & a_{12} & \cdots & a_{1n} \\ a_{21} & a_{22} & \cdots & a_{2n} \\ \vdots & & & \vdots \\ a_{m1} & a_{m2} & \cdots & a_{mn} \end{pmatrix}$$

$$B_{m \times 1} = \begin{pmatrix} b_1 \\ b_2 \\ \vdots \\ b_m \end{pmatrix}, \quad X_{n \times 1} = \begin{pmatrix} x_1 \\ x_2 \\ \vdots \\ x_n \end{pmatrix}, \quad C_{n \times 1} = (c_1, c_2, \ldots, c_n)^T.$$

Let the jth column of A be denoted by C_j, i.e.,

$$C_j = \begin{pmatrix} a_{1j} \\ a_{2j} \\ \vdots \\ a_{mj} \end{pmatrix}.$$

Suppose at some point in the implementation of the revised simplex method a basis β_i has been identified corresponding to a basic feasible solution to the problem. Without loss of generality, let this basis be given by the first m variables, i.e.

$$\beta_i = \{x_1, x_2, \ldots, x_m\}.$$

A *basis matrix* B_i is defined for each basis β_i. B_i is the matrix formed by ordering the columns of A corresponding to the variables in β_i in the order in which they would form the columns of an identity matrix in the regular simplex tableau.

Suppose that for β_i this order is $1, 2, \ldots, m$; then

$$B_i = \begin{pmatrix} a_{11} & a_{12} & \cdots & a_{1m} \\ a_{21} & a_{22} & \cdots & a_{2m} \\ \vdots & \vdots & & \vdots \\ a_{m1} & a_{m2} & \cdots & a_{mn} \end{pmatrix} = (C_1 \quad C_2 \cdots C_m).$$

For the first basic feasible solution, with a basis of slack and artificial variables,

$$B_1 = I. \tag{3.5}$$

In order to discover whether or not the b.f.s. corresponding to β_i is optimal, it is necessary to calculate B_i^{-1}. With the first basis matrix, because of (3.5),

$$B_1^{-1} = I.$$

However, in general B_i^{-1} can be calculated from B_{i-1}^{-1} as follows.

Suppose that the last variable to enter β_i is x_p, and that x_p is the basic variable for the qth constraint. That is, in terms of the regular simplex method, the element in the qth row and pth column was the pivot element. Now let $(a'_{kp}), k = 1, 2, \ldots, m$ be the column entries according to x_p in the tableau corresponding to β_{i-1} and

$$B_{i-1}^{-1} = \begin{pmatrix} \bar{a}_{11} & \bar{a}_{12} & \cdots & \bar{a}_{1m} \\ \bar{a}_{21} & \bar{a}_{22} & \cdots & \bar{a}_{2m} \\ \vdots & \vdots & & \vdots \\ \bar{a}_{m1} & \bar{a}_{m2} & \cdots & \bar{a}_{mm} \end{pmatrix}.$$

Then the entry in the kth row and jth column of B_i^{-1} is

$$\bar{a}_{kj} - \frac{a'_{kp}}{a'_{qp}}(\bar{a}_{qj}), \quad \text{for } k \neq q,$$

or

$$\frac{\bar{a}_{qp}}{a'_{qp}}, \quad \text{for } k = q.$$

Once B_i^{-1} is calculated as described above, the x_0 row corresponding to β_i is found.

Let c_B be the row vector of the negative of the basic variable coefficients. Define the *simplex multipliers* π_i as

$$\pi_i = c_B B_i^{-1}.$$

Once the row vector π_i has been found, the nonbasic variable coefficients of the x_0 row are calculated.

Let $\bar{c}_j$ be the x_0 row coefficient of each nonbasic variable x_j. Then

$$\bar{c}_j = c_j - \pi_i C_j, \quad \text{for all nonbasic variables } x_j.$$

If

$$\bar{c}_j \geq 0, \quad \text{for all nonbasic variables } x_j, \tag{3.6}$$

the basis β_i corresponds to an optimal solution. This solution x_B can be found as follows:

$$x_B = B_i^{-1} b, \tag{3.7}$$

and the optimal solution value is

$$x_0^* = -c_B x_B.$$

If (3.6) is not satisfied, the column corresponding to the entry which is largest in magnitude is identified. Let this be column p. x_p will enter the next basis, β_{i+1}.

In order to determine which basic variable x_p replaces it is necessary to calculate the r.h.s. and pth column in the tableau according to β_i. The r.h.s. column is given in (3.7); the pth column, C'_p is

$$C'_p = B_i^{-1} C_p.$$

Ratios of corresponding elements of x_B and C'_p are formed to decide which variable leaves β_i, say the basic variable for the qth constraint. Now β_{i+1} can be identified, and the previous steps can be repeated with β_{i+1} replacing β_i. The process stops, as with the regular simplex method, when (3.6) is satisfied.

3.2.3 The Calculations in Compact Form

Problem 2.1 will now be reworked using the revised simplex method with the calculations laid out in the normal compact form. The reader should compare this with the tableaux necessary for the regular simplex method, shown in Tables 2.2–2.8.

From (3.1):

$$A = \begin{pmatrix} 3 & 4 & 1 & 0 & 0 \\ 3 & 3 & 0 & 1 & 0 \\ 4 & 2 & 0 & 0 & 1 \end{pmatrix}, \qquad b = \begin{pmatrix} 12 \\ 10 \\ 8 \end{pmatrix},$$

$$c = (4, 3, 0, 0, 0).$$

The iterations are shown in Tables 3.3–3.5. Table 3.5 reveals the same optimal solution as that found by the regular simplex method in Table 2.8, namely,

$$x_1^* = \tfrac{4}{5}$$
$$x_2^* = \tfrac{12}{5}$$
$$x_4^* = \tfrac{2}{5}$$
$$x_3^* = x_5^* = 0$$
$$x_0^* = \tfrac{52}{5}.$$

Table 3.3

β_1	C_B		B_1^{-1}		π_1	b	C_j	Ratio	Entering variable
x_3	0	1	0	0	0	12	3	$\tfrac{12}{3}$	
x_4	0	0	1	0	0	10	3	$\tfrac{10}{3}$	
x_5	0	0	0	1	0	8	4	$\tfrac{8}{4}$	x_1
$(-\bar{c}_j)$	-4	-3	0	0	0	0			

Table 3.4

β_2	C_B	B_2^{-1}			π_2	b	C_j	Ratio	Entering variable
x_3	0	1	0	$-\frac{3}{4}$	0	6	$\frac{5}{2}$	$\frac{12}{5}$	x_2
x_4	0	0	1	$-\frac{3}{4}$	0	4	$\frac{3}{2}$	$\frac{8}{3}$	
x_1	-4	0	0	$\frac{1}{4}$	-1	2	$\frac{1}{2}$	$\frac{4}{1}$	
$(-\bar{c}_j)$	0	-1	0	0	0	1	8		

Table 3.5

β_3	C_B	B_3^{-1}			π_3	b
x_2	-3	$\frac{2}{5}$	0	$-\frac{3}{10}$	$-\frac{2}{5}$	$\frac{12}{5}$
x_4	0	$-\frac{3}{5}$	1	$-\frac{3}{10}$	0	$\frac{2}{5}$
x_1	-4	$-\frac{1}{5}$	0	$\frac{2}{5}$	$-\frac{7}{10}$	$\frac{4}{5}$
$(-\bar{c}_j)$	0	0	$\frac{2}{5}$	0	$\frac{7}{10}$	$\frac{52}{5}$

To sum up, the advantages of the revised over the regular simplex method are:

1. Fewer calculations are required.
2. Less storage is required when implementing the revised simplex method on a computer.
3. There is less accumulation of round-off error, as tableau entries are not repeatedly recalculated. An updated column of entries is not calculated until its variable is about to become basic.

3.3 The Dual Simplex Method

3.3.1 Background

Consider the application of the simplex method to an L.P. problem. When an optimal solution has been found (assuming its existence), the optimal solution to the dual problem can be found by inspecting the optimal primal tableau. However, each tableau generated by the simplex method in solving the primal can be inspected to yield a solution to the dual. What is the nature of this sequence of solutions to the dual? It can be shown that all except the last are infeasible and have solution values which are better than the optimum. As an example of this, consider Problem 2.1 and its solution by the simplex method:

$$\text{Maximize:} \quad 4x_1 + 3x_2 \qquad\qquad\qquad = x_0 \qquad (2.10)$$

$$\text{subject to:} \quad 3x_1 + 4x_2 + x_3 \qquad\qquad = 12 \qquad (2.11)$$

$$3x_1 + 3x_2 \qquad\quad + x_4 \qquad = 10 \qquad (2.12)$$

$$4x_1 + 2x_2 \qquad\qquad\quad + x_5 = 8 \qquad (2.13)$$

$$x_1, x_2, x_3, x_4, x_5 \geq 0.$$

This problem has the following dual:

$$\text{Minimize:} \quad 12y_1 + 10y_2 + 8y_3 \qquad\qquad = y_0$$

$$\text{subject to:} \quad 3y_1 + 3y_2 + 4y_3 - y_4 \qquad = 4$$

$$4y_1 + 3y_2 + 2y_3 \qquad - y_6 = 3$$

$$y_1, y_2, y_3, y_4, y_6 \geq 0.$$

The initial tableau for the primal is shown in Table 2.6, repeated here for convenience.

Table 2.6

Constraints	x_1	x_2	x_3	x_4	x_5	r.h.s.
(2.11)	3	4	1	0	0	12
(2.12)	3	3	0	1	0	10
(2.13)	④	2	0	0	1	8
(2.10)	−4	−3	0	0	0	0

Using the summary given in Section 2.6.1.3 on interpreting the primal tableau to find a solution to the dual, it can be seen that Table 2.6 corresponds to the following dual solution:

$$y_1 = 0$$
$$y_2 = 0$$
$$y_3 = 0$$
$$y_4 = -4$$
$$y_6 = -3$$

and

$$y_0 = 0.$$

When we solved the dual in Section 2.6.1.2 we found that the optimal solution had value

$$y_0^* = \tfrac{52}{5}.$$

Hence this present solution is better in value ($y_0 = 0$), as we are minimizing.

Table 2.7

Constraints	x_1	x_2	x_3	x_4	x_5	r.h.s.
(2.11)	0	$\frac{5}{2}$	1	0	$-\frac{3}{4}$	6
(2.12)	0	$\frac{3}{2}$	0	1	$-\frac{3}{4}$	4
(2.13)	1	$\frac{1}{2}$	0	0	$\frac{1}{4}$	2
(2.10)	0	-1	0	0	1	8

The next simplex iteration for the primal produces Table 2.7, repeated here for convenience.

This corresponds to the following dual solution:

$$y_1 = 0$$
$$y_2 = 0$$
$$y_3 = 1$$
$$y_4 = 0$$
$$y_6 = -1$$

and

$$y_0 = 8.$$

This solution is still infeasible, and is worse in value than the last produced, but still better than the optimal solution which is generated in the next iteration (see Section 2.6.1.3).

It is true in general that the sequence of solutions (all but the last) for a dual problem generated by interpreting successive primal tableaux have the following properties:

1. They are infeasible.
2. Each has a solution value worse than the last.

The very last such solution is optimal, as has been seen in the previous chapter. So the possibility presents itself of a new approach to solving L.P. problems. Rather than start with a feasible solution and produce a sequence of feasible and improving (better solution values) solutions, why not start with an infeasible solution and produce a sequence of infeasible solutions with worsening solution values ultimately terminating with the optimal solution? The dual simplex method does just that.

When is such a strategy likely to produce a more efficient procedure? When a problem contains many "$\geq$" constraints, many artificial variables have to be introduced in the regular simplex method. Considerable effort may be expended in reaching a solution in which all of these have zero value. In such circumstances it is usually better to start with an initial solution of slack variables. Such a solution will be infeasible, as each slack in a "$\geq$" constraint will have negative value. However, usually fewer iterations are needed to attain optimality than in the two-phase method or the big M method.

A second situation in which one is trying to transform an infeasible solution into a feasible solution with a worse value is in postoptimal analysis. The optimal solution to an L.P. problem may no longer be feasible once changes are made to the parameters of the problem. The dual simplex method can be applied to transform this basic, infeasible solution into the optimal one. The mechanics of the method will be explained by means of an example in the next section.

3.3.2 The Dual Simplex Method Applied to a Numerical Example

One of the problems of the previous section will be solved by the dual simplex method:

$$\text{Minimize:} \quad 12y_1 + 10y_2 + 8y_3 \qquad\qquad = y_0 \qquad\qquad (3.8)$$

$$\text{subject to:} \quad 3y_1 + 3y_2 + 4y_3 - y_4 \quad\quad = 4 \qquad\qquad (3.9)$$

$$4y_1 + 3y_2 + 2y_3 \qquad - y_6 = 3 \qquad\qquad (3.10)$$

$$y_1, y_2, y_3, y_4, y_6 \geq 0. \qquad\qquad (3.11)$$

As usual we shall adopt the criterion of maximization:

$$\text{Maximize:} \quad -12y_1 - 10y_2 - 8y_3 = y_0'. \qquad\qquad (3.12)$$

Consider problem (3.9)–(3.12) as the primal L.P. The initial step is to find a basic solution in the first tableau in which:

1. The criterion for optimality is satisfied, (all nonbasic variables have non-negative y_0'-row coefficients); and
2. All basic variables have zero y_0'-row coefficients.

Table 3.6 displays the problem. It can be seen that criteria 1 and 2 would be satisfied if the nonzero entries in the y_4 and y_6 columns were of opposite sign. Then $\{y_4, y_6\}$ would be a suitable basic set. This is achieved by multiplying the two constraints by negative one, as shown in Table 3.7, i.e.,

$$y_4 = -4$$
$$y_6 = -3.$$

Table 3.6

Constraints	y_1	y_2	y_3	y_4	y_6	r.h.s.
(3.9)	3	3	4	−1	0	4
(3.10)	4	3	2	0	−1	3
y_0'	12	10	8	0	0	0

Table 3.7

Constraints	y_1	y_2	y_3	y_4	y_6	r.h.s.
(3.9)	-3	-3	-4	1	0	-4
(3.10)	-4	-3	-2	0	1	-3
y_0'	12	10	8	0	0	0

This solution corresponds to a basis of all slack variables, which is usually the case. If it was feasible it would be optimal, as the criterion for optimality is satisfied. Alas, this is not the case, as both basic variables are negative. However, the solution value is

$$y_0' = 0,$$

which is better than the known optimum of $\frac{52}{5}$. This solution corresponds to the solution in Table 2.8, as discussed in the previous section. (The reader is urged to compare the discussion concerning the numerical example in Section 3.3.1 with the similar steps of the present section.) As the present solution is infeasible, a change of basis is made in order to reduce this infeasibility, so it must be decided which variable leaves the basis and which enters.

First the question of the leaving variable is settled. Recall that when the regular simplex method is applied to the dual the variable with the most negative x_0-row coefficient is selected to enter the basis. This is shown in Section 3.3.1, where x_1 is selected as the incoming basic variable in Table 2.6. This most negative x_0-row coefficient corresponds to a value of one of basic variables in the problem. For instance, the most negative x_0-row coefficient of variable x_1, equalling -4, corresponds to the present value of y_4. (Compare Tables 2.6 and 3.7). It is this basic variable which is to leave the basis. This is intuitively quite reasonable, as it is natural to remove the most negative variable when trying to attain feasibility by eventually making all variables nonnegative.

Next the question of which variable enters the basis is settled. Once again, let us consider the mechanics of the regular simplex method in solving the dual. Having selected x_1 to enter the basis, one then takes the ratios

$$\left(\frac{b_1}{a_{11}}, \frac{b_2}{a_{12}}, \frac{b_3}{a_{13}}\right) = \left(\frac{12}{3}, \frac{10}{3}, \frac{8}{4}\right)$$

and selects the minimum. The ratios correspond to a set of ratios in Table 3.7:

$$\left(\frac{12}{-3}, \frac{10}{-3}, \frac{8}{-4}\right).$$

Note that when taking ratios with the regular simplex method, ratios with negative denominators are ignored. Now, as all equations in our primal in

Table 3.7 have been multiplied by -1, only ratios with negative denominators are taken into account and the variable corresponding to the largest ratio enters the basis. Thus y_3 enters the basis.

Note that the above criteria for deciding which variables enter and leave the basis represent a departure from the regular simplex method. However, the mechanics of the regular and dual simplex methods are otherwise the same.

Once it is determined that y_3 enters the basis and y_4 leaves, this transformation is carried out by the usual Gauss–Jordan elimination, which produces Table 3.8. This solution corresponds to the second one found in the previous section, i.e.,

$$y_3 = 1$$
$$y_6 = -1$$
$$y_0' = -8, \quad \text{i.e.,} \quad y_0 = 8.$$

Note that the value needs to be multiplied by -1, as we took the negative of the objective function in (3.12) in order to maximize.

Table 3.8

Constraints	y_1	y_2	y_3	y_4	y_6	r.h.s.
(3.9)	$\frac{3}{4}$	$\frac{3}{4}$	1	$-\frac{1}{4}$	0	1
(3.10)	$-\frac{5}{2}$	$-\frac{3}{2}$	0	$-\frac{1}{2}$	1	-1
y_0'	6	4	0	2	0	-8

The process is repeated once more. The leaving basic variable is y_6, as it is the only one with a negative value. Taking the ratios, we obtain

$$6/(-\tfrac{5}{2}), \qquad 4/(-\tfrac{3}{2}), \qquad 2/(-\tfrac{1}{2}).$$

Thus y_1 enters the basis. This produces Table 3.9, which represents the optimal solution to the problem, as it is feasible and satisfies the criterion for optimality. The solution is identical to that found previously.

Table 3.9

Constraints	y_1	y_2	y_3	y_4	y_6	r.h.s.
(3.9)	0	$\frac{3}{10}$	1	$-\frac{2}{5}$	$-\frac{3}{10}$	$\frac{7}{10}$
(3.10)	1	$\frac{3}{5}$	0	$\frac{1}{5}$	$-\frac{2}{5}$	$\frac{2}{5}$
y_0'	0	$\frac{2}{5}$	0	$\frac{4}{5}$	$\frac{12}{5}$	$-\frac{52}{5}$

3.3.3 Summary of the Dual Simplex Method

When solving an L.P. problem with the dual simplex method the following steps are carried out.

Step 1. Equality constraints are split into pairs of inequality constraints with opposite sense. For example, the equality constraint

$$a_{i1}x_1 + a_{i2}x_2 + \cdots + a_{im}x_n = b_i$$

is replaced by

$$a_{i1}x_1 + a_{i2}x_2 + \cdots + a_{im}x_n \leq b_i$$

and

$$a_{i1}x_1 + a_{i2}x_2 + \cdots + a_{im}x_n \geq b_i,$$

which on the introduction of slack variables become

$$a_{i1}x_1 + a_{i2}x_2 + \cdots + a_{im}x_n + x_{n+j} = b_i$$

and

$$a_{i1}x_1 + a_{i2}x_2 + \cdots + a_{im}x_n - x_{n+j+1} = b_i.$$

Step 2. A basic solution (normally comprising exactly the set of slack variables) is found which satisfies the criterion for optimality. This criterion is that all x_0-row coefficients for nonbasic variables are nonnegative. Of course all basic variable x_0-row coefficients must be zero, as usual.

Step 3

3.1. Determination of the feasibility of the present solution. Each solution generated satisfies the condition for optimality. Thus if it is feasible it will be optimal. A solution will be feasible if all its variable values are nonnegative. If this is so, the process is terminated and the present solution is optimal. Otherwise, proceed.

3.2. Determination of variable to leave the basis. Among all variables with negative values, the one with the value which is largest in magnitude is selected to leave the basis.

3.3. Determination of variable to enter the basis. Identify the equation which contains a unit coefficient for the leaving variable discovered in step 3.2. Identify all variables which have negative coefficients in this equation, say (j). For each such variable, form a ratio of its current x_0-row coefficient divided by its coefficient in equation (j). The variable with the ratio which is largest enters the basis.

3.4. Make the change of basis according to the variables found in steps 3.2 and 3.3 by Gauss–Jordan elimination and create a new tableau. Go to step 3.1.

It is important that the reader realises that the dual simplex method performs corresponding iterations on the L.P. problem as the regular simplex method would perform on the dual problem.

3.4 The Primal–Dual Algorithm

The dual simplex method was presented in Section 3.3 as a way of overcoming the inefficiency brought about by introducing a relatively large number of artificial variables when solving an L.P. problem. There is another approach possible. Rather than begin with an infeasible solution with value better than the optimum, as in the dual simplex method, why not begin with an infeasible, worse than optimal solution? At least such an initial solution should be easy to find. This is the essence of the *primal–dual algorithm*.

The algorithm begins by constructing the dual L.P. problem. A feasible solution is then found for the dual. On the basis of this solution, the original primal L.P. is modified and this modified problem is used to create a new feasible solution to the dual with an improved value. The process continues in this manner, examining solutions to the dual and a modified primal alternately until the optimal dual solution is produced. (Convergence must take place.) Loosely speaking the successive solutions to the dual correspond to primal solutions which are successively less infeasible for the primal, until the optimal dual solution corresponds to a primal solution which is not only feasible but optimal.

The algorithm is fully described in Hadley (1962) and Dantzig (1963).

3.5 Dantzig–Wolfe Decomposition

3.5.1 Background

Nearly all real-world L.P. problems have far more variables and constraints than the small problems concocted for illustrative purposes so far in this book. In fact some industrial problems are so large that it is not very practical to consider solving them by the methods presented up to this point. One line of approach is to ask what special structure an L.P. must possess in order for it to be possible to break it up into a number of smaller, hopefully easier subproblems. The idea is to somehow combine the solutions of the subproblems in order to find the solution for the original problem.

It has been found that many realistic L.P. problems possess a matrix A of l.h.s. coefficients which has the property called *block angular* structure. What this structure is will be described a little later. The point is that block angular L.P. problems can be decomposed into smaller subproblems. By solving these in a special way it is possible to identify an optimal solution to the original problem. This is achieved by *Dantzig–Wolfe decomposition*, which is due to Dantzig and Wolfe (1960). Their method is now introduced by means of a numerical example. This section requires a more thorough understanding of matrix theory. The unprepared reader is referred to the Appendix.

3.5.2 Numerical Example

We shall now consider an expanded version of the coal mining problem which was introduced at the beginning of the previous chapter. Suppose now that the coal mine mentioned earlier (mine No. 1) is taken over by another company which already owns a mine (No. 2). Thus the company now has two mines. For simplicity, constraints generated by screening the coal are neglected. However the company is anxious to maintain good labour relations with its new miners, and so maintains the same restriction of 12 and 8 hours of cutting and washing, respectively, per day, with unit consumption being 2 and 1 hour for lignite and 1 hour for anthracite for both cutting and washing. A restriction of 24 and 14 hours per day of cutting and washing are in force at mine 2. Electricity and gas are purchased by the company and supplied to its mines. Because the mines employ different processes, the same type of coal in different mines requires different amounts of electricity and gas to produce the same quantity. Thus, to produce one ton of lignite and anthracite requires 2 and 3 units of electricity, respectively, in mine 1 and 4 and 1 units, respectively, in mine 2. The corresponding figures for gas consumption are $\frac{46}{19}$, $\frac{46}{19}$, 6, and 8. Due to energy shortages the company is allocated a maximum of 20 and 30 units daily of electricity and gas, respectively. The unit profit for mine 1 lignite and anthracite was \$4 and \$3, respectively (in hundreds of dollars), and this can be maintained. Because of increased shipping costs, as mine 2 is in a remote area, unit profit for mine 2 lignite and anthracite is \$$\frac{57}{23}$ and \$$\frac{76}{23}$ respectively. Let

x_1 = daily production of mine 1 lignite in tons
x_2 = daily production of mine 1 anthracite in tons
x_3 = daily production of mine 2 lignite in tons
x_4 = daily production of mine 2 anthracite in tons.

Then the problem can be expressed as follows:

$$\text{Maximize:} \quad 4x_1 + 3x_2 + \tfrac{57}{23}x_3 + \tfrac{76}{23}x_4 = x_0 \tag{3.13}$$

$$2x_1 + 3x_2 + 4x_3 + x_4 \le 20 \tag{3.14}$$

$$\text{subject to:} \quad \tfrac{46}{19}x_1 + \tfrac{46}{19}x_2 + 6x_3 + 8x_4 \le 30 \tag{3.15}$$

$$3x_1 + 4x_2 \le 12 \tag{3.16}$$

$$4x_1 + 2x_2 \le 8 \tag{3.17}$$

$$2x_3 + x_4 \le 24 \tag{3.18}$$

$$x_3 + x_4 \le 14 \tag{3.19}$$

$$x_1, x_2, x_3, x_4 \ge 0. \tag{3.20}$$

Matrix A for this problem is

$$\begin{pmatrix} 2 & 3 & 4 & 1 \\ \frac{46}{19} & \frac{46}{19} & 6 & 8 \\ \hline 3 & 4 & 0 & 0 \\ 4 & 2 & 0 & 0 \\ \hline 0 & 0 & 2 & 1 \\ 0 & 0 & 1 & 1 \end{pmatrix}.$$

Now, letting

$$A_1 = \begin{pmatrix} 2 & 3 \\ \frac{46}{19} & \frac{46}{19} \end{pmatrix}, \qquad A_2 = \begin{pmatrix} 4 & 1 \\ 6 & 8 \end{pmatrix}, \qquad A_3 = \begin{pmatrix} 3 & 4 \\ 4 & 2 \end{pmatrix}, \qquad A_4 = \begin{pmatrix} 2 & 1 \\ 1 & 1 \end{pmatrix},$$

A can be expressed as

$$A = \begin{pmatrix} A_1 & A_2 \\ A_3 & 0 \\ 0 & A_4 \end{pmatrix},$$

where 0 is a 2×2 matrix of zeros. In general, a matrix which can be expressed as

$$\begin{pmatrix} A_1 & A_2 & \cdots & A_N \\ A_{N+1} & 0 & \cdots & 0 \\ 0 & A_{N+2} & \cdots & 0 \\ \vdots & \vdots & & \vdots \\ 0 & 0 & \cdots & A_{2N} \end{pmatrix}$$

is termed *block angular*. Let x_i be the vector of variables corresponding to the columns of submatrix A_i. The constraints:

$$\sum_{i=1}^{N} A_i x_i \leq b_0$$

are called *global constraints*. The constraints:

$$A_{N+i} x_i \leq b_i$$

are called *local constraints*.

Block angular matrices appear in L.P. problems when, as in the present example, the total operation can be divided into groups of activities, each with its own exclusive resources, and there are further resources which must be shared by all activities. Thus, apart from the gas and electricity constraints the problem can be considered as two separate L.P. problems, one for each mine. These subproblems are:

Mine 1		Mine 2	

Maximize: $\quad 4x_1 + 3x_2$ $\qquad$ Maximize: $\quad \frac{57}{23}x_3 + \frac{76}{23}x_4$

subject to: $\quad 3x_1 + 4x_2 \leq 12$ $\qquad$ subject to: $\quad 2x_3 + \quad x_4 \leq 24$

$\qquad\qquad 4x_1 + 2x_2 \leq 8$ $\qquad\qquad\qquad\qquad x_3 + \quad x_4 \leq 14$

$\qquad\qquad\quad x_1, x_2 \geq 0.$ $\qquad\qquad\qquad\qquad\quad x_3, x_4 \geq 0.$

In the general case there will be N subproblems of the form:

$$\text{Maximize:} \quad \mathbf{c}_j^T \mathbf{x}_j$$
$$\text{subject to:} \quad A_{N+j}\mathbf{x}_j \leq \mathbf{b}_j, \quad \text{for } j = 1, 2, \ldots, N,$$
$$\mathbf{x}_j \geq 0$$

where

$\mathbf{c} = (\mathbf{c}_1 \quad \mathbf{c}_2 \quad \cdots \quad \mathbf{c}_N)^T = $ the vector of objective function coefficients.

$\mathbf{x} = (\mathbf{x}_1 \quad \mathbf{x}_2 \quad \cdots \quad \mathbf{x}_N)^T = $ the vector of decision variables.

$\mathbf{b} = (\mathbf{b}_1^T \quad \mathbf{b}_2^T \quad \cdots \quad \mathbf{b}_N^T)^T = $ the vector of r.h.s. constants.

Now suppose the constraints (3.14) and (3.15) are temporarily ignored. Then if the optimal solutions to the two subproblems are feasible for these constraints, their combination represents an optimal solution to the original problem. Hence it appears worthwhile to attempt to solve the original problem by analyzing the subproblems. Of course, there must be some modification to the strategy of simply solving the subproblems, as their combined solutions will seldom satisfy the global constraints (3.14) and (3.15). A technique called the *method of decomposition* developed by Dantzig and Wolfe will now be explained by using it to solve the example problem. The simplified version of the method assumes that each subproblem has a set of feasible solutions which is bounded, i.e., no variable can take on an infinite feasible value. We make that assumption here.

The assumption of boundedness implies that the set of feasible solutions for each subproblem has a finite number of extreme points. Furthermore, any point in such a set can be expressed as a convex combination of these extreme points. More precisely, if subproblem j, $j = 1, 2, \ldots, N$, has m_j extreme points, denoted by $\mathbf{x}_j^k$, $k = 1, 2, \ldots, m_j$, then any feasible solution $\mathbf{x}_j$ to the subproblem j can be expressed as:

$$\mathbf{x}_j = \sum_{k=1}^{m_j} \alpha_j^k \mathbf{x}_j^k$$

where

$$\sum_{k=1}^{m_j} \alpha_j^k = 1$$
$$\alpha_j^k \geq 0, \quad k = 1, 2, \ldots, n_j.$$

Also, no point outside the set can be expressed in this way.

For example, in subproblem 1, Figure 2.1 reveals that there are four extreme points:

$$\mathbf{x}_1^1 = (0,0), \qquad \mathbf{x}_1^2 = (2,0), \qquad \mathbf{x}_1^3 = (0,3), \qquad \mathbf{x}_1^4 = (\tfrac{4}{5}, \tfrac{12}{5}).$$

So $m_1 = 4$. Thus any point, $\mathbf{x}_1$ in the feasible region satisfies

$$\mathbf{x}_1 = \alpha_1^1 \begin{pmatrix} 0 \\ 0 \end{pmatrix} + \alpha_1^2 \begin{pmatrix} 2 \\ 0 \end{pmatrix} + \alpha_1^3 \begin{pmatrix} 0 \\ 3 \end{pmatrix} + \alpha_1^4 \begin{pmatrix} \tfrac{4}{5} \\ \tfrac{12}{5} \end{pmatrix},$$

where

$$\alpha_1^1 + \alpha_1^2 + \alpha_1^3 + \alpha_1^4 = 1$$
$$\alpha_1^k \geq 0, \qquad k = 1, 2, 3, 4.$$

A similar expression for subproblem 2 can be found which also involves four extreme points:

$$\mathbf{x}_1 = \begin{pmatrix} x_1 \\ x_2 \end{pmatrix} = \left(\alpha_1^1 \mathbf{x}_1^1 + \alpha_1^2 \mathbf{x}_1^2 + \alpha_1^3 \mathbf{x}_1^3 + \alpha_1^4 \mathbf{x}_1^4 \right)$$
$$= \sum_{k=1}^{4} \alpha_1^k \mathbf{x}_1^k \tag{3.21}$$

and

$$\mathbf{x}_2 = \begin{pmatrix} x_3 \\ x_4 \end{pmatrix} = \left(\alpha_2^1 \mathbf{x}_2^1 + \alpha_2^2 \mathbf{x}_2^2 + \alpha_2^3 \mathbf{x}_2^3 + \alpha_2^4 \mathbf{x}_2^4 \right)$$
$$= \sum_{k=1}^{4} \alpha_2^k \mathbf{x}_2^k. \tag{3.22}$$

Now problem (3.13)–(3.20) can be written in matrix form as follows:

Maximize: $(4, 3)\begin{pmatrix} x_1 \\ x_2 \end{pmatrix} + (\tfrac{57}{23}, \tfrac{76}{23})\begin{pmatrix} x_3 \\ x_4 \end{pmatrix}$

subject to: $\begin{pmatrix} 2 & 3 \\ \tfrac{46}{19} & \tfrac{46}{19} \end{pmatrix}\begin{pmatrix} x_1 \\ x_2 \end{pmatrix} + \begin{pmatrix} 4 & 1 \\ 6 & 8 \end{pmatrix}\begin{pmatrix} x_3 \\ x_4 \end{pmatrix} \leq \begin{pmatrix} 20 \\ 30 \end{pmatrix}$ \qquad (3.23)

$\begin{pmatrix} 3 & 4 \\ 4 & 2 \end{pmatrix}\begin{pmatrix} x_1 \\ x_2 \end{pmatrix} \qquad\qquad \leq \begin{pmatrix} 12 \\ 8 \end{pmatrix}$ \qquad (3.24)

$\begin{pmatrix} 2 & 1 \\ 1 & 1 \end{pmatrix}\begin{pmatrix} x_3 \\ x_4 \end{pmatrix} \leq \begin{pmatrix} 24 \\ 14 \end{pmatrix}$ \qquad (3.25)

$$x_i \geq 0, \qquad i = 1, 2, 3, 4. \tag{3.26}$$

Now if (3.21) and (3.22) are used to eliminate x_1, x_2, x_3, and x_4 from this formulation, (3.24), (3.25), and (3.26) are no longer needed, as they are implicitly satisfied in (3.21) and (3.22). Making this substitution, the problem becomes:

Maximize: $(4, 3) \displaystyle\sum_{k=1}^{4} \alpha_1^k \mathbf{x}_1^k + (\tfrac{57}{23}, \tfrac{76}{23}) \sum_{k=1}^{4} \alpha_2^k \mathbf{x}_2^k$ \qquad (3.27)

subject to:
$$\begin{pmatrix} 2 & 3 \\ \frac{46}{19} & \frac{46}{19} \end{pmatrix} \sum_{k=1}^{4} \alpha_1^k \mathbf{x}_1^k + \begin{pmatrix} 4 & 1 \\ 6 & 8 \end{pmatrix} \sum_{k=1}^{4} \alpha_2^k \mathbf{x}_2^k \le \begin{bmatrix} 20 \\ 30 \end{bmatrix} \quad (3.28)$$

$$\alpha_j^1 + \alpha_j^2 + \alpha_j^3 + \alpha_j^4 = 1, \quad j = 1, 2 \quad (3.29)$$

$$\alpha_j^k \ge 0, \quad j = 1, 2$$
$$k = 1, 2, 3, 4.$$

The $\mathbf{x}_1^k$ and $\mathbf{x}_2^k$ are constant. The decision variables are the α_i^k. However, the formulation has fewer constraints and should be easier to solve.

It would appear that it is necessary to find all the extreme points of the subproblems before the solving process can be started. This is a substantial task and would negate the gains made by the reduction in the number of constraints. Fortunately, it is unnecessary to find all the extreme points first; they can be found one at a time as needed. The basis for achieving this is the revised simplex method.

We begin by introducing slack variables into (3.28):

$$\begin{pmatrix} 2 & 3 \\ \frac{46}{19} & \frac{46}{19} \end{pmatrix} \sum_{k=1}^{4} \alpha_1^k \mathbf{x}_1^k + \begin{pmatrix} 4 & 1 \\ 6 & 8 \end{pmatrix} \sum_{k=1}^{4} \alpha_2^k \mathbf{x}_2^k + \begin{pmatrix} x_5 \\ x_6 \end{pmatrix} = \begin{pmatrix} 20 \\ 30 \end{pmatrix}.$$

Define the actual variable values at the extreme points as follows:

$$\mathbf{x}_1^1 = \begin{pmatrix} x_1^1 \\ x_2^1 \end{pmatrix}, \quad \mathbf{x}_1^2 = \begin{pmatrix} x_1^2 \\ x_2^2 \end{pmatrix}, \quad \mathbf{x}_1^3 = \begin{pmatrix} x_1^3 \\ x_2^3 \end{pmatrix}, \quad \mathbf{x}_1^4 = \begin{pmatrix} x_1^4 \\ x_2^4 \end{pmatrix},$$

$$\mathbf{x}_2^1 = \begin{pmatrix} x_3^1 \\ x_4^1 \end{pmatrix}, \quad \mathbf{x}_2^2 = \begin{pmatrix} x_3^2 \\ x_4^2 \end{pmatrix}, \quad \mathbf{x}_2^3 = \begin{pmatrix} x_3^3 \\ x_4^3 \end{pmatrix}, \quad \mathbf{x}_2^4 = \begin{pmatrix} x_3^4 \\ x_4^4 \end{pmatrix}.$$

On substituting these into problem (3.27)–(3.29), we obtain the following problem:

Maximize:
$$(4, 3) \left\{ \alpha_1^1 \begin{pmatrix} x_1^1 \\ x_2^1 \end{pmatrix} + \alpha_1^2 \begin{pmatrix} x_1^2 \\ x_2^2 \end{pmatrix} + \alpha_1^3 \begin{pmatrix} x_1^3 \\ x_2^3 \end{pmatrix} + \alpha_1^4 \begin{pmatrix} x_1^4 \\ x_2^4 \end{pmatrix} \right\}$$

$$+ \left(\tfrac{57}{23}, \tfrac{76}{23} \right) \left\{ \alpha_2^1 \begin{pmatrix} x_3^1 \\ x_4^1 \end{pmatrix} + \alpha_2^2 \begin{pmatrix} x_3^2 \\ x_4^2 \end{pmatrix} + \alpha_2^3 \begin{pmatrix} x_3^3 \\ x_4^3 \end{pmatrix} + \alpha_2^4 \begin{pmatrix} x_3^4 \\ x_4^4 \end{pmatrix} \right\}$$

subject to:
$$\begin{pmatrix} 2 & 3 \\ \frac{46}{19} & \frac{46}{19} \end{pmatrix} \left\{ \alpha_1^1 \begin{pmatrix} x_1^1 \\ x_2^1 \end{pmatrix} + \alpha_1^2 \begin{pmatrix} x_1^2 \\ x_2^2 \end{pmatrix} + \alpha_1^3 \begin{pmatrix} x_1^3 \\ x_2^3 \end{pmatrix} + \alpha_1^4 \begin{pmatrix} x_1^4 \\ x_2^4 \end{pmatrix} \right\}$$

$$+ \begin{pmatrix} 4 & 1 \\ 6 & 8 \end{pmatrix} \left\{ \alpha_2^1 \begin{pmatrix} x_3^1 \\ x_4^1 \end{pmatrix} + \alpha_2^2 \begin{pmatrix} x_3^2 \\ x_4^2 \end{pmatrix} + \alpha_2^3 \begin{pmatrix} x_3^3 \\ x_4^3 \end{pmatrix} + \alpha_2^4 \begin{pmatrix} x_3^4 \\ x_4^4 \end{pmatrix} \right\}$$

$$+ \begin{pmatrix} x_5 \\ x_6 \end{pmatrix} = \begin{pmatrix} 20 \\ 30 \end{pmatrix}$$

$$\alpha_1^1 + \alpha_1^2 + \alpha_1^3 + \alpha_1^4 = 1$$
$$\alpha_2^1 + \alpha_2^2 + \alpha_2^3 + \alpha_2^4 = 1$$
$$\alpha_j^k \ge 0, \quad \text{for all } j, k.$$

On rearrangement, this becomes

Maximize:
$$(4x_1^1 + 3x_2^1)\alpha_1^1 + (4x_1^2 + 3x_2^2)\alpha_1^2 + (4x_1^3 + 3x_2^3)\alpha_1^3$$
$$+ (4x_1^4 + 3x_2^4)\alpha_2^4 + (\tfrac{57}{23}x_3^1 + \tfrac{76}{23}x_4^1)\alpha_2^1 + (\tfrac{57}{23}x_3^2 + \tfrac{76}{23}x_4^2)\alpha_2^2 \quad (3.30)$$
$$+ (\tfrac{57}{23}x_3^3 + \tfrac{76}{23}x_4^3)\alpha_2^3 + (\tfrac{57}{23}x_3^4 + \tfrac{76}{23}x_4^4)\alpha_2^4$$

subject to:
$$(2x_1^1 + 3x_2^1)\alpha_1^1 + (2x_1^2 + 3x_2^2)\alpha_1^2 + (2x_1^3 + 3x_2^3)\alpha_1^3$$
$$+ (2x_1^4 + 3x_2^4)\alpha_1^4 + (4x_3^1 + x_4^1)\alpha_2^1 + (4x_3^2 + x_4^2)\alpha_2^2 \quad (3.31)$$
$$+ (4x_3^3 + x_4^3)\alpha_2^3 + (4x_3^4 + x_4^4)\alpha_2^4 + x_5 = 20$$
$$(\tfrac{46}{19}x_1^1 + \tfrac{46}{19}x_2^1)\alpha_1^1 + (\tfrac{46}{19}x_1^2 + \tfrac{46}{19}x_2^2)\alpha_1^2 + (\tfrac{46}{19}x_1^3 + \tfrac{46}{19}x_2^3)\alpha_1^3$$
$$+ (\tfrac{46}{19}x_1^4 + \tfrac{46}{19}x_2^4)\alpha_1^4 + (6x_3^1 + 8x_4^1)\alpha_2^1 + (6x_3^2 + 8x_4^2)\alpha_2^2 \quad (3.32)$$
$$+ (6x_3^3 + 8x_4^3)\alpha_2^3 + (6x_3^4 + 8x_4^4)\alpha_2^4 + x_6 = 30$$
$$\alpha_1^1 + \alpha_1^2 + \alpha_1^3 + \alpha_1^4 = 1$$
$$\alpha_2^1 + \alpha_2^2 + \alpha_2^3 + \alpha_2^4 = 1$$
$$\alpha_j^k \geq 0, \quad \text{for all } j, k.$$

Now on examining the subproblems it can be seen that $(0,0)^T$ is an extreme point for both problems. Let

$$\begin{pmatrix} x_1^1 \\ x_2^1 \end{pmatrix} = \begin{pmatrix} 0 \\ 0 \end{pmatrix}$$

and

$$\begin{pmatrix} x_3^1 \\ x_4^1 \end{pmatrix} = \begin{pmatrix} 0 \\ 0 \end{pmatrix}.$$

These points are associated with α_1^1 and α_2^1, respectively. These values of x_1^1, x_2^1, x_3^1, and x_4^1, cause α_1^1 and α_2^1 to drop out of (3.31) and (3.32). Thus a suitable initial basis is

$$\beta_1 = \{x_5, x_6, \alpha_1^1, \alpha_2^1\}.$$

Hence, in terms of the revised simplex method,

$$B = \begin{pmatrix} 1 & 0 & 0 & 0 \\ 0 & 1 & 0 & 0 \\ 0 & 0 & 1 & 0 \\ 0 & 0 & 0 & 1 \end{pmatrix} = B^{-1},$$

$$x_B = [20, 30, 1, 1]^T$$
$$c_B = [0, 0, 0, 0]^T.$$

The subscript of the B, denoting the iteration number, has been dropped, as we shall use it for another purpose soon.

We must now decide whether or not this solution is optimal. This is done in the revised simplex method by examining the sign of the minimum element in the x_0 row. Let $\bar{c}_{jk}$ be the x_0-row coefficient of α_j^k. Now the evaluation of the $\bar{c}_{jk}$ depends upon the extreme points x_j^k. However, rather than determining all extreme points one can simply find the extreme point for each subproblem j which yields the smallest $\bar{c}_{jk}$. Now remember that the $\mathbf{x}_j^k$ are the extreme points for subproblem j, which has the following feasible region:

$$A_{N+j}\mathbf{x}_j \leq \mathbf{b}_j$$
$$\mathbf{x}_j \geq 0.$$

Thus the problem of finding the extreme point for each subproblem j yielding the smallest $\bar{c}_{jk}$ reduces to solving a number of L.P.'s of the following form:

$$\text{Minimize:} \quad \bar{c}_{jk} \tag{3.33}$$
$$\text{subject to:} \quad A_{N+j}\mathbf{x}_j \leq \mathbf{b}_j$$
$$\mathbf{x}_j \geq 0.$$

At any iteration of the revised simplex method, an expression for the $\bar{c}_{jk}$ can be found as follows. Recall that the x_0-row coefficient $\bar{c}_j$ of each nonbasic variable x_j in an ordinary problem is found by

$$\bar{c}_j = c_j - \pi_i C_j,$$

where

$c_j = $ the original coefficient of x_j

$C_j = $ the jth column of A

$\pi_i = \mathbf{c}_B^T B^{-1}$, the current simplex multipliers.

Now for our original problem let p be the number of global constraints. In the present example, let

$(B^{-1})^p = $ the first p columns of B^{-1}

$B_j^{-1} = $ the jth column of B^{-1}, considered as a vector.

Then

$$\bar{c}_{jk} = \mathbf{c}_B^T(B^{-1})^p A_j \cdot \mathbf{x}_j^k + c_B^T B_{p+j}^{-1} - \mathbf{c}_{jk}^T \mathbf{x}_j^k$$
$$= (\mathbf{c}_B^T(B^{-1})^p A_j - \mathbf{c}_{jk}^T)\mathbf{x}_j^k + \mathbf{c}_B^T B_{p+j}^{-1}.$$

This expression can be substituted into (3.33) to produce the following formulation:

$$\text{Minimize:} \quad (\mathbf{c}_B^T(B^{-1})^p A_j - \mathbf{c}_{jk}^T)\mathbf{x}_j + \mathbf{c}_B^T B_{p+j}^{-1}$$
$$\text{subject to:} \quad A_{N+j}\mathbf{x}_j \leq \mathbf{b}_j$$
$$\mathbf{x}_j \geq 0.$$

The optimal (minimal) solution value of this L.P. corresponds to the minimum x_0-row coefficient in the original problem. If it is negative, the optimal solution to (3.33) corresponds to the extreme point we are trying to find.

Thus an L.P. of the form of (3.33) must be solved for each subproblem. If all solution values are nonnegative, no further iterations are required. Returning to the example problem, the L.P. of the form (3.33) for the first subproblem is

$$\text{Minimize} \cdot \quad \left((0,0,0,0) \begin{pmatrix} 1 & 0 \\ 0 & 1 \\ 0 & 0 \\ 0 & 0 \end{pmatrix} \begin{pmatrix} 2 & 3 \\ \frac{46}{19} & \frac{46}{19} \end{pmatrix} - (4,3) \right) \begin{pmatrix} x_1 \\ x_2 \end{pmatrix} + (0,0,0,0) \begin{pmatrix} 0 \\ 0 \\ 1 \\ 0 \end{pmatrix} = x_0^1$$

subject to:
$$\begin{pmatrix} 3 & 4 \\ 4 & 2 \end{pmatrix} \begin{pmatrix} x_1 \\ x_2 \end{pmatrix} \le \begin{pmatrix} 12 \\ 8 \end{pmatrix}$$

$$\begin{pmatrix} x_1 \\ x_2 \end{pmatrix} \ge \begin{pmatrix} 0 \\ 0 \end{pmatrix}.$$

This problem can be solved graphically by examining Figure 2.1. More general problems can be solved by the simplex method. The optimal solution is

$$x_1^* = \tfrac{4}{5}$$
$$x_2^* = \tfrac{12}{5}$$

and
$$(x_0^1)^* = -\tfrac{52}{5}.$$

This corresponds to an extreme point of subproblem 1, say x_1^2. Therefore
$$\mathbf{x}_1^2 = [\tfrac{4}{5}, \tfrac{12}{5}]^T$$

and
$$\bar{c}_{12} = -\tfrac{52}{5} < 0.$$

Hence optimality has not been reached.

The L.P. associated with the second subproblem is

$$\text{Minimize:} \quad \left((0,0,0,0) \begin{pmatrix} 1 & 0 \\ 0 & 1 \\ 0 & 0 \\ 0 & 0 \end{pmatrix} \begin{pmatrix} 4 & 1 \\ 6 & 8 \end{pmatrix} - (\tfrac{57}{23}, \tfrac{76}{23}) \right) \begin{pmatrix} x_3 \\ x_4 \end{pmatrix} + (0,0,0,0) \begin{pmatrix} 0 \\ 0 \\ 0 \\ 1 \end{pmatrix} = x_0^2$$

subject to:
$$\begin{pmatrix} 2 & 1 \\ 1 & 1 \end{pmatrix} \begin{pmatrix} x_3 \\ x_4 \end{pmatrix} \le \begin{pmatrix} 24 \\ 14 \end{pmatrix}$$

$$\begin{pmatrix} x_3 \\ x_4 \end{pmatrix} \ge \begin{pmatrix} 0 \\ 0 \end{pmatrix}.$$

This problem can also be solved graphically to yield the following optimal solution:

$$x_3^* = 0$$
$$x_4^* = 14$$
$$x_0^{2*} = -\tfrac{1064}{23}.$$

Let this solution correspond to the solution $\mathbf{x}_2^2$ of subproblem 2.

$$\mathbf{x}_2^2 = [x_3^*, x_4^*] = [10,4]^T \quad \text{and} \quad \bar{c}_{22} = -\tfrac{1064}{23} < 0.$$

As this value $\bar{c}_{22}$ represents the minimum x_0-row coefficient, and it is negative, its variable α_2^2 enters the basis. Following the steps of the revised simplex method we next calculate the α_2^2 column in the tableau. On looking at (3.31) and (3.32) it can be seen that this column is.

$$[4x_3^2 + x_4^2, 6x_3^2 + 8x_4^2, 0, 1]^T = [14, 112, 0, 1]^T$$

The r.h.s. column is $(20, 30, 1,)$. On taking the ratios:

$$\tfrac{10}{7}, \tfrac{15}{56}, -, \tfrac{1}{1}$$

it can be seen that the minimum corresponds to variable x_6, which now leaves the basis. Therefore,

$$\beta_2 = \{x_5, \alpha_2^2, \alpha_1^1, \alpha_2^1\}$$

and

$$c_B = (0, 1064, 0, 0)^T.$$

B^{-1} is now updated and becomes

$$B^{-1} = \begin{pmatrix} 1 & -\tfrac{1}{8} & 0 & 0 \\ 0 & \tfrac{1}{112} & 0 & 0 \\ 0 & 0 & 1 & 0 \\ 0 & -\tfrac{1}{112} & 0 & 1 \end{pmatrix}.$$

We must now test whether β_2 corresponds to an optimal solution. The L.P. for subproblem 1 is:

Minimize:
$$\left(0, \tfrac{1064}{23}, 0, 0\right) \begin{pmatrix} 1 & -\tfrac{1}{8} \\ 0 & \tfrac{1}{112} \\ 0 & 0 \\ 0 & -\tfrac{1}{112} \end{pmatrix} \begin{pmatrix} 2 & 3 \\ \tfrac{46}{19} & \tfrac{46}{19} \end{pmatrix}$$

$$- (4, 3) \bigg) \begin{pmatrix} x_1 \\ x_2 \end{pmatrix} + (0, \tfrac{1064}{23}, 0) \begin{pmatrix} 0 \\ 0 \\ 1 \\ 0 \end{pmatrix} = x_0^1$$

subject to:
$$\begin{pmatrix} 3 & 4 \\ 4 & 2 \end{pmatrix} \begin{pmatrix} x_1 \\ x_2 \end{pmatrix} \leq \begin{pmatrix} 12 \\ 8 \end{pmatrix}$$

$$\begin{pmatrix} x_1 \\ x_2 \end{pmatrix} \geq \begin{pmatrix} 0 \\ 0 \end{pmatrix}.$$

Now

$$x_0^1 = \left((0, \tfrac{38}{92}) \begin{pmatrix} 2 & 3 \\ \tfrac{46}{19} & \tfrac{46}{19} \end{pmatrix} - (4, 3) \right) \begin{pmatrix} x_1 \\ x_2 \end{pmatrix}$$

$$= -3x_1 - 2x_2.$$

The optimal solution to this problem is

$$x_1^* = \tfrac{4}{5}$$
$$x_2^* = \tfrac{12}{5}$$

and

$$x_0^{1*} = -\tfrac{36}{5}.$$

As x_0^{1*} is negative, the optimal solution has not been reached. The L.P. problem for subproblem 2 is

Minimize: $\left((0,\tfrac{38}{92})\begin{pmatrix}4 & 1\\ 6 & 8\end{pmatrix} - (\tfrac{57}{23},\tfrac{76}{23})\right)\begin{pmatrix}x_3\\ x_4\end{pmatrix} + (0,\tfrac{1064}{23},0,0)\begin{pmatrix}0\\ 0\\ 0\\ 0\\ 1\end{pmatrix} = x_0^2 = \begin{pmatrix}0\\ 0\end{pmatrix}$

subject to: $\begin{pmatrix}2 & 1\\ 1 & 1\end{pmatrix}\begin{pmatrix}x_3\\ x_4\end{pmatrix} \le \begin{pmatrix}24\\ 14\end{pmatrix}$

$$\begin{pmatrix}x_3\\ x_4\end{pmatrix} \ge \begin{pmatrix}0\\ 0\end{pmatrix}.$$

Hence the minimum solution is provided by subproblem 1, and corresponds to the extreme point

$$\mathbf{x}_1^2 = [x_1^2, x_2^2] = [0,0]^T$$

and

$$\bar{c}_{12} = 0.$$

The corresponding variable α_1^2 enters the basis. We next calculate the α_1^2 column in the tableau, which is

$$B^{-1}\begin{pmatrix}2x_1^2 + 3x_2^2\\ \tfrac{46}{19}x_1^2 + \tfrac{46}{19}x_2^2\\ 1\\ 0\end{pmatrix} = \begin{pmatrix}1 & -\tfrac{1}{8} & 0 & 0\\ 0 & \tfrac{1}{112} & 0 & 0\\ 0 & 0 & 1 & 0\\ 0 & -\tfrac{1}{112} & 0 & 1\end{pmatrix}\begin{pmatrix}\tfrac{44}{5}\\ \tfrac{736}{95}\\ 1\\ 0\end{pmatrix} = \begin{pmatrix}\tfrac{744}{95}\\ \tfrac{46}{665}\\ 1\\ -\tfrac{46}{165}\end{pmatrix}.$$

The r.h.s. column is

$$B^{-1}\mathbf{b}_0 = \begin{pmatrix}1 & \tfrac{1}{8} & 0 & 0\\ 0 & \tfrac{1}{112} & 0 & 0\\ 0 & 0 & 1 & 0\\ 0 & -\tfrac{1}{112} & 0 & 1\end{pmatrix}\begin{pmatrix}20\\ 30\\ 1\\ 1\end{pmatrix} = \begin{pmatrix}\tfrac{65}{4}\\ \tfrac{15}{56}\\ 1\\ \tfrac{41}{56}\end{pmatrix}.$$

On taking the ratios:

$$\left(\tfrac{6175}{2976}, \tfrac{9975}{2576}, \tfrac{1}{1}, \text{—}\right)$$

it can be seen that the minimum corresponds to variable α_1^1, which now leaves the basis. Therefore

$$\beta_3 = \{x_5, \alpha_2^2, \alpha_1^2, \alpha_2^1\}$$
$$C_B = (0, \tfrac{1064}{23}, \tfrac{52}{5}, 0)^T.$$

B^{-1} is now updated and becomes:

$$B^{-1} = \begin{pmatrix} 1 & -\frac{1}{8} & -\frac{744}{95} & 0 \\ 0 & \frac{1}{112} & -\frac{46}{665} & 0 \\ 0 & 0 & 1 & 0 \\ 0 & -\frac{1}{112} & \frac{46}{665} & 1 \end{pmatrix}.$$

We now test whether β_3 corresponds to an optimal solution. The L.P. for subproblem 1 is

Minimize:
$$(0, \tfrac{1064}{23}, \tfrac{52}{5}, 0) \begin{pmatrix} 1 & -\frac{1}{8} \\ 0 & -\frac{1}{112} \\ 0 & 0 \\ 0 & -\frac{1}{112} \end{pmatrix} \begin{pmatrix} 2 & 3 \\ \frac{46}{19} & \frac{46}{19} \end{pmatrix}$$

$$- (4, 3)\begin{pmatrix} x_1 \\ x_2 \end{pmatrix} + (0, \tfrac{1064}{23}, \tfrac{52}{5}, 0) \begin{pmatrix} \frac{744}{95} \\ \frac{46}{665} \\ 1 \\ -\frac{46}{665} \end{pmatrix}$$

subject to:
$$\begin{pmatrix} 3 & 4 \\ 4 & 2 \end{pmatrix}\begin{pmatrix} x_1 \\ x_2 \end{pmatrix} \leq \begin{pmatrix} 12 \\ 8 \end{pmatrix}$$

$$\begin{pmatrix} x_1 \\ x_2 \end{pmatrix} \geq 0.$$

Now

$$x_0^1 = \left((0, \tfrac{38}{92})\begin{pmatrix} 2 & 3 \\ \frac{46}{19} & \frac{46}{19} \end{pmatrix} - (4, 3) \right)\begin{pmatrix} x_1 \\ x_2 \end{pmatrix} + \tfrac{36}{5}$$

$$= -3x_1 - 2x_2 + \tfrac{36}{5},$$

and the optimal solution is

$$x_1^* = \tfrac{4}{5}$$
$$x_2^* = \tfrac{12}{5}$$
$$x_0^{1*} = 0.$$

The L.P. problem for subproblem 2 is

Minimize:
$$\left((0, \tfrac{38}{92})\begin{pmatrix} 4 & 1 \\ 6 & 8 \end{pmatrix} - (\tfrac{57}{23}, \tfrac{76}{23}) \right)\begin{pmatrix} x_3 \\ x_4 \end{pmatrix} + (0, \tfrac{1064}{23}, \tfrac{52}{5}, 0)\begin{pmatrix} 0 \\ 0 \\ 0 \\ 1 \end{pmatrix} = x_0^2 = \begin{pmatrix} 0 \\ 0 \end{pmatrix}$$

subject to:
$$\begin{pmatrix} 2 & 1 \\ 1 & 1 \end{pmatrix}\begin{pmatrix} x_3 \\ x_4 \end{pmatrix} \leq \begin{pmatrix} 24 \\ 14 \end{pmatrix},$$

$$\begin{pmatrix} x_3 \\ x_4 \end{pmatrix} \geq \begin{pmatrix} 0 \\ 0 \end{pmatrix}.$$

Hence the minimum solution is provided by subproblem 1. We have arrived at an optimum as none of the x_0-row coefficients are negative.

The optimal solution is

$$\begin{pmatrix} x_5 \\ \alpha_2^2 \\ \alpha_1^2 \\ \alpha_2^1 \end{pmatrix} = B^{-1}\mathbf{b}_0 = \begin{pmatrix} 1 & -\frac{1}{8} & -\frac{744}{95} & 0 \\ 0 & \frac{1}{112} & -\frac{46}{665} & 0 \\ 0 & 0 & 1 & 0 \\ 0 & -\frac{1}{112} & \frac{46}{665} & 1 \end{pmatrix} \begin{pmatrix} 20 \\ 30 \\ 1 \\ 1 \end{pmatrix} = \begin{pmatrix} \frac{3199}{380} \\ \frac{151}{766} \\ 1 \\ \frac{609}{760} \end{pmatrix}$$

Therefore

$$\mathbf{x}_1^* = \begin{pmatrix} x_1^* \\ x_2^* \end{pmatrix} = \sum_{k=1}^{4} \alpha_1^k \mathbf{x}_1^k = 0 + 1 \times \begin{pmatrix} \frac{4}{5} \\ \frac{12}{5} \end{pmatrix} + 0 + 0 = \begin{pmatrix} \frac{4}{5} \\ \frac{12}{5} \end{pmatrix}$$

$$\mathbf{x}_2^* = \begin{pmatrix} x_3^* \\ x_4^* \end{pmatrix} = \sum_{k=1}^{4} \alpha_2^k \mathbf{x}_2^k = \frac{609}{760}\begin{pmatrix} 0 \\ 0 \end{pmatrix} + \frac{151}{760}\begin{pmatrix} 0 \\ 14 \end{pmatrix} + 0 + 0 = \begin{pmatrix} 0 \\ \frac{1057}{380} \end{pmatrix}$$

$$x_5^* = \frac{3199}{380}$$
$$\alpha_2^{2*} = \frac{151}{760}$$
$$\alpha_1^{2*} = 1$$
$$\alpha_2^{1*} = \frac{609}{760}$$
$$x_6^* = 0$$

$$x_0^* = [0, \tfrac{1064}{23}, \tfrac{52}{5}, 0]\begin{bmatrix} \frac{3199}{380} \\ \frac{151}{760} \\ 1 \\ \frac{609}{760} \end{bmatrix} = \frac{2253}{115} \simeq 19.59.$$

The objective function hyperplane is parallel to that representing the constraint on the gas:

$$\tfrac{46}{19}x_1 + \tfrac{46}{19}x_2 + 6x_3 + 8x_4 + x_6 = 30.$$

As this contraint is binding at the optimum ($x_6^* = 0$), multiple optima exist. The complete set of solutions is given by:

$$x_0^* = \frac{2253}{115}$$
$$x_1^* = \frac{4}{5}$$
$$x_2^* = \frac{12}{5}$$
$$3x_3^* + 4x_4^* = \frac{1057}{95},$$

where

$$0 \le x_3^* \le \tfrac{14}{5}$$

and

$$\tfrac{259}{380} \le x_4^* \le \tfrac{1057}{380}.$$

3.5.3 Summary of the Decomposition Algorithm

Given an L.P. in the following form:

Maximize: $\quad \mathbf{c}^T \mathbf{x}$

subject to:
$$\begin{bmatrix} A_1 & A_2 & \cdots & A_N \\ A_{N+1} & 0 & \cdots & 0 \\ 0 & A_{N+2} & \cdots & 0 \\ \vdots & \vdots & & \vdots \\ 0 & 0 & \cdots & A_{2N} \end{bmatrix} \mathbf{x} \le \mathbf{b}$$

$$\mathbf{x} \ge 0.$$

Let
$$\mathbf{c} = (\mathbf{c}_1^T, \mathbf{c}_2^T, \ldots, \mathbf{c}_N^T)^T$$
$$\mathbf{x} = (\mathbf{x}_1, \mathbf{x}_2, \cdots, \mathbf{x}_N)^T$$
and
$$\mathbf{b} = (\mathbf{b}_0^T, \mathbf{b}_1^T, \ldots, \mathbf{b}_N^T)^T.$$

Define a set of feasible points $\mathbf{x}_j$ satisfying
$$A_{N+j} \mathbf{x}_j \le \mathbf{b}_j$$
$$\mathbf{x}_j \ge 0.$$

Let the jth such set have n_j extreme points $\mathbf{x}_j^1, \mathbf{x}_j^2, \ldots, \mathbf{x}_j^{n_j}$. Then any point $\mathbf{x}_j$ in the jth set can be expressed as

$$\mathbf{x}_j = \sum_{k=1}^{n_j} \alpha_j^k \mathbf{x}_j^k$$

$$\sum_{k=1}^{n_j} \alpha_j^k = 1$$

$$\alpha_j^k \ge 0, \qquad k = 1, 2, \ldots, n_j.$$

The given problem can now be reformulated with the introduction of a vector $\mathbf{x}_s$ of slack variables:

Maximize: $\quad \displaystyle\sum_{j=1}^{N} \sum_{k=1}^{n_j} (\mathbf{c}_j \mathbf{x}_j^k) \alpha_j^k$

subject to: $\quad \displaystyle\sum_{j=1}^{N} \sum_{k=1}^{n_j} (A_j \mathbf{x}_j^k) \alpha_j^k + \mathbf{x}_s = \mathbf{b}_0$

$$\sum_{k=1}^{n_j} \alpha_j^k = 1, \qquad j = 1, 2, \ldots, N,$$

$$\alpha_j^k \ge 0, \qquad j = 1, 2, \ldots, N,$$
$$k = 1, 2, \ldots, n_j.$$

This formation is then solved using the revised simplex method.

In order to calculate the minimum x_0-row coefficient, N linear programming problems of the form

Minimize: $\quad (c_{\mathbf{B}}^T (B^{-1})^p A_j - \mathbf{c}_j) \mathbf{x}_j + \mathbf{c}_B^T B_{p+j}^{-1}$

subject to: $\quad A_{N+j} \mathbf{x}_j \leq b_j$

$$\mathbf{x}_j \geq 0$$

are solved, where the terms above have been defined in the previous section. The minimum solution value obtained in solving the above problems is equal to the minimum x_0-row coefficient. If it is nonnegative, the optimal solution has been found. Otherwise, a substitution in the set of basic variables is made in the usual way.

3.6 Parametric Programming

3.6.1 Background

In Section 2.6 the sensitivity of the optimal solution of an L.P. problem to changes in its coefficients was discussed. It was assumed that these changes were made one at a time. We now look at the possibility of analyzing the effects of simultaneous changes. Only changes in objective function coefficients and r.h.s. constants will be dealt with here. The approach is to develop techniques whereby the investigation can take place in an efficient manner, as opposed to solving the whole problem from scratch with the new values inserted. The techniques are collectively called *parametric linear programming*, although the term *linear* will be dropped, as it is understood that we are dealing solely with L.P. problems. Such methods are useful in situations in which, because of the effects of some predictable process, many of the L.P. parameters vary at constant rates. For example, profits or costs could vary as a result of inflation, or daily consumption of resources may have to be steadily reduced as the supply of raw materials dwindles. It will be assumed that the coefficients vary *linearly* with time.

3.6.2 Numerical Example

3.6.2.1 *Changes in the Objective Function Coefficients*

Consider again problem 2.1.

Maximize: $\quad 4x_1 + 3x_2 \qquad\qquad\qquad\qquad = x_0$

subject to: $\quad 3x_1 + 4x_2 + x_3 \qquad\qquad = 12$

$\qquad\qquad\quad 3x_1 + 3x_2 \qquad\; + x_4 \qquad = 10$

$\qquad\qquad\quad 4x_1 + 2x_2 \qquad\qquad\; + x_5 = 8$

$\qquad\qquad\qquad x_i \geq 0, \qquad i = 1, 2, \ldots, 5.$

Suppose that the x_0-row coefficients, 4 and 3, are changing at the rates of 2 and 3 units per unit of time, respectively. Then if θ is defined as the amount of elapsed time, after θ units of time have elapsed x_0 becomes

$$x_0 = (4 + 2\theta)x_1 + (3 + 3\theta)x_2. \tag{3.34}$$

Given a particular value of θ, it is possible to determine the optimal solution and its value. This has already been done for the case $\theta = 0$. We now address ourselves to the task of using this solution to find the solution for any other positive θ in a way which requires less work than solving the new problem from the beginning using the simplex method.

The optimal solution to the problem when $\theta = 0$ is given Table 2.8, repeated here for convenience. When θ is given any nonnegative real value, the only change in Problem 2.1 occurs in the objective function. Hence the solution in Table 2.8 will be feasible for the problem corresponding to any θ. As θ is increased from zero to a relatively small positive value it is likely that the present solution will remain optimal. However, it may be that as θ is progressively increased in value there will occur a critical point at which the present solution is no longer optimal. A new optimal solution can be established and θ increased further. Later a new critical point may be established. We shall now establish the ranges for θ for which the various possible bases are optimal.

Table 2.8

Constraints	x_1	x_2	x_3	x_4	x_5	r.h.s.
(2.11)	0	1	$\frac{2}{5}$	0	$-\frac{3}{10}$	$\frac{12}{5}$
(2.12)	0	0	$-\frac{3}{5}$	1	$-\frac{3}{10}$	$\frac{2}{5}$
(2.13)	1	0	$-\frac{1}{5}$	0	$\frac{2}{5}$	$\frac{4}{5}$
x_0	0	0	$\frac{2}{5}$	0	$\frac{7}{10}$	$\frac{52}{5}$

Suppose Problem 2.1 has its objective function replaced by (3.34). If the manipulations carried out in Tables (2.1)–(2.8) are applied to this new problem, the final tableau will be as shown in Table 3.10. Transforming this to

Table 3.10

Constraints	x_1	x_2	x_3	x_4	x_5	r.h.s.
(2.11)	0	1	$\frac{2}{5}$	0	$-\frac{3}{10}$	$\frac{12}{5}$
(2.12)	0	0	$-\frac{3}{5}$	1	$-\frac{3}{10}$	$\frac{2}{5}$
(2.13)	1	0	$-\frac{1}{5}$	0	$\frac{2}{5}$	$\frac{4}{5}$
x_0	-2θ	-3θ	$\frac{2}{5}$	0	$\frac{7}{10}$	$\frac{52}{5}$

canonical form, we get Table 3.11. Thus it can be seen that for the present basis to remain optimal all the x_0-row coefficients must be nonnegative, i.e.,

$$\frac{2 + 4\theta}{5} \geq 0$$

and

$$\frac{7 - \theta}{10} \geq 0.$$

As it has been assumed that θ is nonnegative,

$$\theta \leq 7.$$

So setting $\theta = 7$ we have reached the first critical point.

Table 3.11

Constraints	x_1	x_2	x_3	x_4	x_5	r.h.s.
(2.11)	0	1	$\frac{2}{5}$	0	$-\frac{3}{10}$	$\frac{12}{5}$
(2.12)	0	0	$-\frac{3}{5}$	1	$-\frac{3}{10}$	$\frac{2}{5}$
(2.13)	1	0	$-\frac{1}{5}$	0	$\frac{2}{5}$	$\frac{4}{5}$
x_0	0	0	$\frac{2 + 4\theta}{5}$	0	$\frac{7 - \theta}{10}$	$\frac{52 + 44\theta}{5}$

Substituting this value into Table 3.11, we obtain Table 3.12. This, of course, corresponds to a situation with multiple optimal solutions. The objective function for this value of θ is:

$$x_0 = (4 + 2(7))x_1 + (3 + 3(7))x_2 = 18x_1 + 24x_2.$$

It can be seen from Figure 3.1 that the increase in θ from 0 to 7 has changed the slope of the objective function to a point where it is now parallel to (2.11).

Table 3.12

Constraints	x_1	x_2	x_3	x_4	x_5	r.h.s.	Ratio
(2.11)	0	1	$\frac{2}{5}$	0	$-\frac{3}{10}$	$\frac{12}{5}$	—
(2.12)	0	0	$-\frac{3}{5}$	1	$-\frac{3}{10}$	$\frac{2}{5}$	—
(2.13)	1	0	$-\frac{1}{5}$	0	$\frac{2}{5}$	$\frac{4}{5}$	$\frac{2}{1}$
x_0	0	0	6	0	0	72	

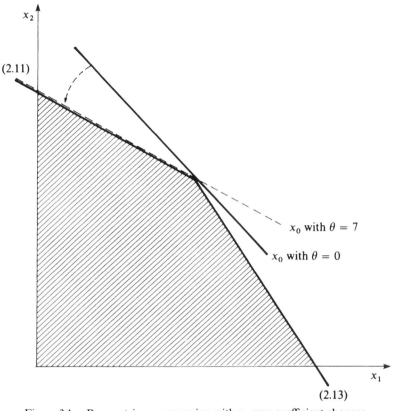

Figure 3.1. Parametric programming with x_0-row coefficient changes.

Table 3.13

Constraints	x_1	x_2	x_3	x_4	x_5	r.h.s.
(2.11)	$\frac{3}{4}$	1	$\frac{1}{4}$	0	0	3
(2.12)	$\frac{3}{4}$	0	$-\frac{3}{4}$	1	0	1
(2.13)	$\frac{5}{2}$	0	$-\frac{1}{2}$	0	1	2
x_0	0	0	$\dfrac{2+4\theta}{5}$	0	$\dfrac{7-\theta}{10}$	$\dfrac{52+44\theta}{5}$

Bringing x_5 into the basis at the expense of x_1 in Table 3.11 produces Table 3.13. Now as it has been assumed that θ is nonnegative, all that is required is that

$$\theta \geq 7.$$

For any values of θ no less than 7 the present basis remains optimal. Thus there is only one critical point. These results are summarized in Table 3.14.

Table 3.14. Results of a
Change in the Objective
Function coefficients.

	$0 \leq \theta \leq 7$	$7 \leq \theta$
	$4 \leq c_1 \leq 18$	$18 \leq c_1$
	$3 \leq c_2 \leq 24$	$24 \leq c_2$
x_0^*	$\dfrac{52 + 44\theta}{5}$	$\dfrac{47 + 45\theta}{5}$
x_1^*	$\frac{4}{5}$	0
x_2^*	$\frac{12}{5}$	3
x_3^*	0	0
x_4^*	$\frac{2}{5}$	1
x_5^*	0	2

3.6.2.2 Changes in the r.h.s. Constants

Suppose now that the r.h.s. constants in Problem 2.1 are increasing from 12, 10, and 8 at the rates of 2, 2, and 3 units per unit of time, respectively. Then after θ units of time have elapsed, these r.h.s. constants are, respectively,

$$12 + 2\theta, \qquad 10 + 2\theta, \quad \text{and} \quad 8 + 3\theta.$$

Once again we wish to identify the optimal solution and its value for increasing θ. The optimal solution when $\theta = 0$ is given in Table 2.8. Now suppose the new r.h.s. constants for a given positive value of the θ are introduced to the problem and this new problem has the same simplex iterations applied to it as produced Table 2.8. Then if the methods of Section 2.6.2.2 are used repeatedly for each r.h.s. constant, the new tableau will be as shown in Table 3.15. For this basis to remain feasible, all r.h.s. values must be nonnegative. Therefore

$$\tfrac{12}{5} + (\tfrac{4}{5} - \tfrac{9}{10})\theta \geq 0$$
$$\tfrac{2}{5} + (-\tfrac{6}{5} + 2 - \tfrac{9}{10})\theta \geq 0$$

and

$$\tfrac{4}{5} + (-\tfrac{2}{5} + \tfrac{6}{5})\theta \geq 0,$$

i.e.,

$$\theta \leq 4.$$

So the first critical point occurs at $\theta = 4$. Substituting this value into Table 3.15, we obtain Table 3.16. If θ is increased beyond 4 the solution in Table 3.16 will become infeasible, as x_4 will become negative. Thus x_4 should leave the basis. This is done using the dual simplex method, which is explained in Section 3.3.

Table 3.15

Constraints	x_1	x_2	x_3	x_4	x_5	r.h.s.
(2.11)	0	1	$\frac{2}{5}$	0	$-\frac{3}{10}$	$\frac{12}{5} + (\frac{2}{5}(2) + 0(2) - \frac{3}{10}(3))\theta$
(2.12)	0	0	$-\frac{3}{5}$	1	$-\frac{3}{10}$	$\frac{2}{5} + (-\frac{3}{5}(2) + 1(2) - \frac{3}{10}(3))\theta$
(2.13)	1	0	$-\frac{1}{5}$	0	$\frac{2}{5}$	$\frac{4}{5} + (-\frac{1}{5}(2) + 0(2) + \frac{2}{5}(3))\theta$
x_0	0	0	$\frac{2}{5}$	0	$\frac{7}{10}$	$\frac{52}{5} + (\frac{2}{5}(2) + 0(2) + \frac{7}{10}(3))\theta$

Table 3.16

Constraints	x_1	x_2	x_3	x_4	x_5	r.h.s.
(2.11)	0	1	$\frac{2}{5}$	0	$-\frac{3}{10}$	2
(2.12)	0	0	$-\frac{3}{5}$	1	$-\frac{3}{10}$	0
(2.13)	1	0	$-\frac{1}{5}$	0	$\frac{2}{5}$	4
x_0	0	0	$\frac{2}{5}$	0	$\frac{7}{10}$	22

Thus x_3 enters the basis in Table 3.15 at the expense of x_4, producing Table 3.17. This basis will remain feasible if all the r.h.s. values are nonnegative:

$$\frac{16 - \theta}{6} \geq 0$$

$$\frac{\theta - 4}{6} \geq 0$$

$$\frac{4 + 5\theta}{6} \geq 0,$$

i.e.,

$$16 \geq \theta \geq 4.$$

Table 3.17

Constraints	x_1	x_2	x_3	x_4	x_5	r.h.s.
(2.11)	0	1	0	$\frac{2}{3}$	$-\frac{1}{2}$	$\dfrac{16 - \theta}{6}$
(2.12)	0	0	1	$-\frac{5}{3}$	$\frac{1}{2}$	$\dfrac{\theta - 4}{6}$
(2.13)	1	0	0	$-\frac{1}{3}$	$\frac{1}{2}$	$\dfrac{4 + 5\theta}{6}$
x_0	0	0	0	$\frac{2}{3}$	$\frac{1}{2}$	$\dfrac{64 + 17\theta}{6}$

Table 3.18

Constraints	x_1	x_2	x_3	x_4	x_5	r.h.s.
(2.11)	0	-2	0	$-\frac{4}{3}$	1	$\dfrac{\theta - 16}{3}$
(2.12)	0	1	1	-1	0	2
(2.13)	1	1	0	$\frac{1}{3}$	0	$\dfrac{10 + 2\theta}{3}$
x_0	0	1	0	$\frac{4}{3}$	0	$\dfrac{40 + 8\theta}{3}$

So the second critical point occurs at $\theta = 16$. If θ is increased beyond 16 the solution in Table 3.17 will become infeasible, as x_2 will become negative. Thus x_2 should leave the basis.

Now x_5 enters the basis at the expense of x_2, producing Table 3.18. This basis will remain feasible if all r.h.s. values are nonnegative:

$$\frac{\theta - 16}{3} \geq 0$$

$$\frac{10 + 2\theta}{3} \geq 0,$$

i.e.,

$$\theta \geq 16.$$

For any value of θ no less than 16 the present basis remains feasible. Thus there are two critical points. These results are summarized in Table 3.19.

Changing r.h.s. constants in an L.P. problem is equivalent to changing objective function coefficients in the dual. Hence let us resolve the problem just analyzed by examining its dual. The dual of the problem is

$$\begin{aligned}
\text{Minimize:} \quad & 12y_1 + 10y_2 + 8y_3 & = y_0 \\
\text{subject to:} \quad & 3y_1 + 3y_2 + 4y_3 - y_4 & = 4 \\
& 4y_1 + 3y_2 + 2y_3 - y_6 & = 3 \\
& y_1, y_2, y_3, y_4, y_6 \geq 0.
\end{aligned}$$

Changing the r.h.s. constants of the primal by 2, 2, and 3 units per unit of time corresponds to changing the dual y_0 row coefficients by the same amounts. The optimal solution to the above problem is given in Table 3.20.

Following the ideas of Section 3.6.2.1, when θ is introduced the objective becomes

Minimize: $\quad (12 + 2\theta)y_1 + (10 + 2\theta)y_2 + (8 + 3\theta)y_3,$

i.e.,

Maximize: $\quad y_0' = -(12 + 2\theta)y_1 - (10 + 2\theta)y_2 - (8 + 3\theta)y_3. \quad (3.35)$

Table 3.19. Results of Changes in the r.h.s. Constants.

	$0 \le \theta \le 4$	$4 \le \theta \le 16$	$16 \le \theta$
	$12 \le b_1 \le 20$	$20 \le b_1 \le 44$	$44 \le b_1$
	$10 \le b_2 \le 18$	$18 \le b_2 \le 42$	$42 \le b_2$
	$8 \le b_3 \le 20$	$20 \le b_3 \le 56$	$56 \le b_3$
x_0^*	$\dfrac{104 + 29\theta}{10}$	$\dfrac{64 + 17\theta}{6}$	$\dfrac{40 + 8\theta}{3}$
x_1^*	$\dfrac{8 - \theta}{10}$	$\dfrac{4 + 5\theta}{6}$	$\dfrac{10 + 2\theta}{3}$
x_2^*	$\dfrac{24 - \theta}{10}$	$\dfrac{16 - \theta}{6}$	0
x_3^*	0	$\dfrac{\theta - 4}{6}$	2
x_4^*	$\dfrac{2 + 4\theta}{5}$	0	0
x_5^*	0	0	$\dfrac{\theta - 16}{3}$

Table 3.20

Constraints	y_1	y_2	y_3	y_4	y_6	r.h.s.
(3.09)	0	$\frac{3}{10}$	1	$-\frac{2}{5}$	$\frac{3}{10}$	$\frac{7}{10}$
(3.10)	1	$\frac{3}{5}$	0	$\frac{1}{5}$	$-\frac{2}{5}$	$\frac{2}{5}$
y_0'	0	$\frac{2}{5}$	0	$\frac{4}{5}$	$\frac{12}{5}$	$-\frac{52}{5}$

When the manipulations applied to the dual to produce Table 3.20 are applied to the problem with (3.35) as an objective function, Table 3.21 is produced. Transforming this into canonical form, we obtain Table 3.22. Hence for the present basis to remain optimal all the y_0'-row coefficients

Table 3.21

Constraints	y_1	y_2	y_3	y_4	y_6	r.h.s.
(3.09)	0	$\frac{3}{10}$	1	$-\frac{2}{5}$	$\frac{3}{10}$	$\frac{7}{10}$
(3.10)	1	$\frac{3}{5}$	0	$\frac{1}{5}$	$-\frac{2}{5}$	$\frac{2}{5}$
y_0'	2θ	$\frac{2}{5} + 2\theta$	3θ	$\frac{4}{5}$	$\frac{12}{5}$	$-\frac{52}{5}$

Table 3.22

Constraints	y_1	y_2	y_3	y_4	y_6	r.h.s.
(3.09)	0	$\frac{3}{10}$	1	$-\frac{2}{5}$	$\frac{3}{10}$	$\frac{7}{10}$
(3.10)	1	$\left(\frac{3}{5}\right)$	0	$\frac{1}{5}$	$-\frac{2}{5}$	$\frac{2}{5}$
y_0'	0	$\frac{2}{5}-\frac{\theta}{10}$	0	$\frac{4}{5}+\frac{4\theta}{5}$	$\frac{12}{5}-\frac{\theta}{10}$	$\frac{52}{5}-\frac{29}{10}\theta$

must be nonnegative:

$$\frac{2}{5}-\frac{\theta}{10}\geq 0$$

$$\frac{4}{5}+\frac{4\theta}{5}\geq 0$$

and

$$\frac{12}{5}-\frac{\theta}{10}\geq 0,$$

i.e.,

$$\theta \leq 4.$$

Thus $\theta = 4$ is the first critical point. When $\theta > 4$ the objective function of y_2 is negative, so y_2 enters the basis, as in Table 3.23. For this basis to remain optimal all the x_0-row coefficients must be nonnegative:

$$\frac{\theta - 4}{6}\geq 0$$

$$\frac{4 + 5\theta}{6}\geq 0$$

$$\frac{16 - \theta}{6}\geq 0,$$

i.e.,

$$16 \geq \theta \geq 4.$$

Table 3.23

Constraints	y_1	y_2	y_3	y_4	y_6	r.h.s.
(3.09)	$-\frac{1}{2}$	0	1	$-\frac{1}{2}$	$\frac{1}{2}$	$\frac{1}{2}$
(3.10)	$\frac{5}{3}$	1	0	$\frac{1}{3}$	$-\frac{2}{3}$	$\frac{2}{3}$
y_0'	$\frac{\theta - 4}{6}$	0	0	$\frac{4 + 5\theta}{6}$	$\frac{16 - \theta}{6}$	$\frac{64 + 17\theta}{6}$

Table 3.24

Constraints	y_1	y_2	y_3	y_4	y_6	r.h.s.
(3.09)	-1	0	2	-1	1	1
(3.10)	1	1	$\frac{4}{3}$	$-\frac{1}{3}$	0	$\frac{4}{3}$
y_0'	2	0	$\dfrac{\theta - 16}{3}$	$\dfrac{10 + 2\theta}{3}$	4	$\dfrac{40 + 8\theta}{3}$

Thus $\theta = 16$ is the second critical point. When $\theta > 16$ the objective function coefficient of y_6 is negative, so y_6 enters the basis, as in Table 3.24. For this basis to remain optimal all the y_0'-row coefficients must be nonnegative, which is certainly true for $\theta \geq 16$. Thus there are only two critical points, at 4 and 16. These results confirm what was discovered by analyzing the primal earlier in this section.

3.6.3 Summary of Parametric Programming

3.6.3.1 *Changes in the Objective Function Coefficients*

Given an L.P. in the following form:

$$\text{Maximize:} \quad x_0 = \sum_{i=1}^{n} c_i x_i \tag{3.36}$$

$$\text{subject to:} \quad \sum_{i=1}^{n} A_{ij} x_i \leq b_j, \quad j = 1, 2, \ldots, m \tag{3.37}$$

$$x_i \geq 0, \quad i = 1, 2, \ldots, n.$$

Suppose that the x_0-row coefficients c_i, $i = 1, 2, \ldots, n$, are changing at the rate of δ_i units per unit of time. Then after θ units of time, x_0 becomes

$$x_0 = \sum_{i=1}^{n} (c_i + \delta_i \theta) x_i.$$

It is obvious that x_0^* is a function of θ. It may be desirable to find $x_0^*(\theta)$ and to find ranges for θ for which the various possible bases are optimal. As θ represents elapsed time it is assumed that $\theta \geq 0$.

First the problem is solved for $\theta = 0$. Then θ at the positive level is introduced. As the only change in the problem comes about in the objective function, the present solution (found when $\theta = 0$) will still be feasible for the problem when $\theta > 0$. Thus if the same manipulations used in solving the problem when $\theta = 0$ are applied to the problem when $\theta > 0$, only changes in the x_0-row will occur. The new x_0 row can be obtained by subtracting

$\delta_i\theta$ from the x_0-row coefficient of x_i, and then transforming the tableau into canonical form, as explained in Section 2.5.2.

If all the new x_0-row coefficients are nonnegative, the present solution is still optimal. Hence the maximum value for θ, say θ_1, for which nonnegativity of all the coefficients occurs can be found; θ_1 is called the first *critical value*. This value is substituted into the x_0-row, producing at least one nonbasic coefficient with value zero. A variable corresponding to this zero is brought into the basis in the usual manner, and θ is introduced once more. The process is repeated to produce further critical values until it is obvious that the increases in the value of θ will not create a situation in which the current basis is suboptimal. Successive tableaux can be examined to find the ranges for θ and their corresponding solutions and values.

3.6.3.2 *Changes in the r.h.s. Constants*

Given an L.P. in the form of (3.36) and (3.37), suppose that the r.h.s. constants $b_j, j = 1, 2, \ldots, m$, are changing at the rate of δ_j units per unit of time. Then after θ units of time the constraints (3.37) become

$$\sum_{i=1}^{n} a_{ij}x_i \le b_j + \delta_j\theta, \qquad j = 1, 2, \ldots, m.$$

First the problem is solved for $\theta = 0$. Then θ at the positive level is introduced. If the present solution, found when $\theta = 0$, is still feasible it will still be optimal. The final tableau, produced by applying the manipulations that created the original optimum to the new problem with $\theta > 0$, is now be deduced.

This final tableau can be obtained from the original optimal tableau by repeatedly using the considerations of Section 2.6.2.2 for each r.h.s. constant. For this new tableau to represent an optimal solution, all the entries in the r.h.s. column must be nonnegative. As they are functions of θ, an upper bound on θ can be obtained. That is, a value θ_1 can be found such that if

$$\theta > \theta_1$$

at least one r.h.s. entry will be negative. θ_1 is the first critical value. The solution and its value, as functions of θ, can be found from the tableau for

$$0 \le \theta \le \theta_1.$$

θ_1 is substituted into the tableau, creating at least one zero entry in the r.h.s. column. The dual simplex method of Section 3.3 is now applied to effect a change of basis, with a basic variable with present value of zero departing. When a nondegenerate basis has been found, the above procedure is repeated and a second critical point is identified.

The process is repeated until a basis is found with the property that further increases in the value of θ will not lead to the basis being suboptimal.

This analysis could also be carried out by taking the dual of the problem, which is

$$\text{Minimize:} \quad \sum_{j=1}^{m} (b_j + \theta\delta_j)y_j$$

$$\text{subject to:} \quad \sum_{j=1}^{m} a_{ji}y_j \geq c_i, \qquad i = 1, 2, \ldots, n$$

$$y_j \geq 0, \qquad j = 1, 2, \ldots, m.$$

Then the procedure of Section 3.6.3.1 can be used.

3.7 Exercises

1. Solve the following problems using the revised simplex method.

(a) Maximize: $3x_1 + 2x_2 + x_3 + 2x_4$

     subject to: $3x_1 + x_2 + x_3 + 2x_4 \leq 9$

            $x_1 + 2x_2 + x_3 + 4x_4 \leq 12$

            $2x_1 + x_2 + 3x_3 + x_4 \leq 8$

            $3x_1 + 3x_2 + 2x_3 + x_4 \leq 10$

            $x_i \geq 0, \qquad i = 1, 2, 3, 4.$

(b) Maximize: $2x_1 - 3x_2 + 2x_3 + 4x_4$

     subject to: $2x_1 + 5x_2 + 3x_3 + 3x_4 \leq 20$

            $2x_1 + 4x_2 + x_3 + 6x_4 \leq 20$

            $2x_1 + 2x_2 + 2x_3 + 3x_4 \leq 12$

            $x_1 + 2x_2 + 2x_3 + 4x_4 \leq 16$

            $x_i \geq 0, \qquad i = 1, 2, 3, 4.$

(c) Maximize: $x_1 + 2x_2 + 3x_3 - x_4$

     subject to: $x_1 + x_2 - x_3 + x_4 \leq 3$

            $2x_1 + 3x_3 \leq 6$

            $3x_1 + x_2 + 2x_3 - 2x_4 \leq 10$

            $2x_2 + 3x_3 + 2x_4 \leq 8$

            $x_i \geq 0, \qquad i = 1, 2, 3, 4.$

(d) Maximize: $4x_1 + 2x_2 + 3x_3 - x_4$

     subject to: $x_1 + 2x_2 + x_4 \leq 8$

            $3x_1 + 2x_3 + x_4 \leq 12$

            $2x_1 + x_2 + 3x_3 \leq 20$

            $2x_2 + 2x_3 + x_4 \leq 10$

            $x_i \geq 0, \qquad i = 1, 2, 3, 4.$

2. Solve the problems of Exercise 1 by the regular simplex method. Compare the amount of computational effort required with that required by the revised simplex method.

3. Solve the duals of the following problems by the dual simplex method.

(a) Maximize: $\quad 3x_1 + 2x_2 + x_3 + 2x_4$

   subject to: $\quad x_1 + 2x_2 + x_3 + 3x_4 \le 6$

   $\qquad\qquad 3x_1 + 4x_2 + 2x_3 + x_4 \le 8$

   $\qquad\qquad 2x_1 + 3x_2 + 3x_3 + x_4 \le 9$

   $\qquad\qquad 2x_1 + x_2 + 2x_3 + 2x_4 \le 12$

   $\qquad\qquad\quad x_i \ge 0, \quad i = 1, 2, 3, 4.$

(b) Maximize: $\quad 2x_1 + 4x_2 + x_3 + 3x_4$

   subject to: $\quad 2x_1 - x_2 + x_3 + 2x_4 \le 6$

   $\qquad\qquad\qquad 2x_2 \quad\quad - x_4 \le 1$

   $\qquad\qquad x_1 + x_2 \quad\quad + 2x_4 \le 4$

   $\qquad\qquad 3x_1 + 2x_2 + 2x_3 + x_4 \le 9$

   $\qquad\qquad\quad x_i \ge 0, \quad i = 1, 2, 3, 4.$

(c) Maximize: $\quad 2x_1 + x_2 + x_3 + x_4$

   subject to: $\quad x_1 - 2x_2 + x_3 + x_4 \le 11$

   $\qquad\qquad -4x_1 - x_2 + 2x_3 \qquad \le 4$

   $\qquad\qquad 2x_1 \quad\quad - 2x_3 + x_4 \le 1$

   $\qquad\qquad\qquad\quad - x_3 + x_4 \ge 2$

   $\qquad\qquad\quad x_i \ge 0, \quad i = 1, 2, 3, 4.$

(d) Minimize: $\quad x_1 + 4x_2$

   subject to: $\quad x_1 + 2x_2 - x_3 + x_4 \ge 3$

   $\qquad\qquad -2x_1 - x_2 + 4x_3 + x_4 \ge 2$

   $\qquad\qquad x_1 + 2x_2 + x_3 \qquad \le 11$

   $\qquad\qquad\qquad 2x_2 + 2x_3 + x_4 \ge 8$

   $\qquad\qquad\quad x_i \ge 0, \quad i = 1, 2, 3, 4.$

4. Solve the problems of Exercise 3 by using the regular simplex method on the duals. Compare the computation step by step for each problem.

5. Solve the following parametric programming problems where the x_0-row coefficient x_i is changing at the rate of S_i units per unit of time, where $S = (S_1, S_2, S_3)$.

(a) Maximize: $\quad 3x_1 + x_2 + 2x_3$

   subject to: $\quad 2x_1 + x_2 + 4x_3 \le 10$

   $\qquad\qquad x_1 + 2x_2 + x_3 \le 4$

   $\qquad\qquad 3x_1 - 2x_2 + x_3 \le 6$

   $\qquad\qquad\quad x_i \ge 0, \quad i = 1, 2, 3$

   $\qquad\qquad\qquad S = (1, 2, 3).$

(b) Minimize: $x_1 + 4x_2 + x_3$

subject to:
$$x_1 + 2x_2 - x_3 \geq 3$$
$$-2x_1 - x_2 + 4x_3 \geq 1$$
$$x_1 + 2x_2 + x_3 \leq 11$$
$$x_i \geq 0, \qquad i = 1, 2, 3$$
$$S = (1, 6, 1).$$

(c) Maximize: $x_1 + 2x_2 + 2x_3$

subject to:
$$2x_1 + 2x_2 - x_3 \leq 8$$
$$2x_1 - x_2 + x_3 \leq 2$$
$$x_i \geq 0, \qquad i = 1, 2, 3$$
$$S = (2, -2, 1).$$

(d) Maximize: $3x_1 + 2x_2 + x_3$

subject to:
$$3x_1 + x_2 + x_3 \leq 9$$
$$x_1 + 2x_2 + x_3 \leq 12$$
$$2x_1 + x_2 + 3x_3 \leq 8$$
$$x_i \geq 0, \qquad i = 1, 2, 3$$
$$S = (2, 3, 4).$$

6. Solve each of the problems of Exercise 5 by taking the dual and using postoptimal analysis on the r.h.s. parameters.

7. Solve each of the parametric programming problems in Exercise 5 when the r.h.s. parameters b_i increase with time at the rate of λ_i units per unit of time, $\lambda = (\lambda_1, \lambda_2, \lambda_3)$.
   (a) $\lambda = (2, 1, 3)$
   (b) $\lambda = (1, -2, 3)$
   (c) $\lambda = (-5, 1, 1)$
   (d) $\lambda = (2, -3, 3).$

Chapter 4

Integer Programming

4.1 A Simple Integer Programming Problem

The Speed of Light Freight Company has just secured a contract from a corporation which wants its big crates of machine parts periodically shipped from its factory to its new mineral exploration site. There are two types of crate; A and B, weighing 3 and 4 units, with volume 4 and 2 units, respectively. The company has one aircraft with a capacity of 12 and 9 units of weight and volume, respectively. The company gains revenue of 4 and 3 units (in hundreds of dollars), respectively, for each crate of A and B flown to the site. As the revenue for road transport is much lower, the company would like to make maximum revenue from its one aircraft, the remaining goods being trucked. We can formulate this problem mathematically as follows. Let

x_1 = the number of crates of type A flown
x_2 = the number of crates of type B flown.

As 4 units are gained for one A crate flown, the revenue for x_1 crates is $4x_1$. Similarly, $3x_2$ is gained for x_2 B crates. Thus the total return for a policy of flying x_1 A crates and x_2 B crates is $4x_1 + 3x_2$, which we denote by x_0. Now as one A crate weighs 3 units, x_1 A crates will weigh $3x_1$. Similarly x_2 B crates weigh $4x_2$. Thus the total weight flown by the policy is $3x_1 + 4x_2$, which must be less than or equal to 12 units. By similar reasoning we can formulate a constraint for volume:

$$4x_1 + 2x_2 \leq 9.$$

We are now in a position to define the problem mathematically.

$$\text{Maximize:} \qquad 4x_1 + 3x_2 = x_0 \qquad\qquad (4.1)$$

$$\text{subject to:} \quad 3x_1 + 4x_2 \leq 12 \tag{4.2}$$

$$4_1 + 2x_2 \leq 9 \tag{4.3}$$

$$x_1, x_2 \geq 0 \tag{4.4}$$

$$x_1, x_2 \text{ integers.} \tag{4.5}$$

This problem is an example of an *integer programming* problem. It would be a linear programming problem if it were not for (4.5). Before going on to develop methods which will solve this problem, let us define the general area of combinatorial optimization of which integer programming is a part.

4.2 Combinatorial Optimization

A combinatorial optimization problem is defined as that of assigning discrete numerical values (from a finite set of values) to a finite set of variables X so as to maximize some function $f(X)$ while satisfying a given set of constraints on the values the variables can assume. Some problems of this type have already been considered: the transportation problem of Section 2.7.1 and the assignment problem of Section 2.7.2. Stated formally the combinatorial optimization problem is

$$\text{Maximize:} \quad f(X)$$
$$\text{subject to:} \quad g_j(X) = 0, \quad j = 1, 2, \ldots, m,$$
$$h_i(X) \leq 0, \quad i = 1, 2, \ldots, k,$$
$$X \text{ a vector of integer values.}$$

Note that there are no restrictions on the functions f, g_j, $j = 1, 2, \ldots, m$, and h_i, $i = 1, 2, \ldots, k$. These functions may be nonlinear, discontinuous, or implicit. This general problem is difficult to solve, and so we confine our attention to a drastic simplification, which is a linear programming problem in which at least one specified variable must have an integer value in any feasible solution.

Let n be the number of decision variables. Without loss of generality, suppose that the first $q (1 \leq q \leq m)$ variables are constrained to be integer. Consider the following problem:

$$\text{Maximize:} \quad C^T X \tag{4.6}$$

$$\text{subject to:} \quad AX = B, \tag{4.7}$$

$$X \geq 0 \tag{4.8}$$

$$x_1, x_2, \ldots, x_q \text{ integer,} \tag{4.9}$$

where $X = (x_1, x_2, \ldots, x_q, \ldots, x_n)^T$ and C is $n \times 1$, B is $m \times 1$, and A is $m \times n$.

If

$$q = n,$$

the problem is termed an *integer linear programming problem*. Our air freight problem comes into this category.

If

$$1 \leq q < n,$$

the problem is termed a *mixed integer linear programming problem*.

If (4.9) is replaced by

$$x_i = 0 \text{ or } 1, \qquad i = 1, 2, \ldots, n,$$

then the problem is termed a *zero–one linear programming problem*.

Of course, if

$$q = 0,$$

(4.9) ceases to be relevant and the problem becomes an ordinary linear programming problem.

The transportation problem is an integer linear programming problem and the assignment problem is a zero–one linear programming problem. Further examples of integer linear programming problems will be given later. Because only *linear* problems will be considered, we will simply refer to an *integer program* (I.P.).

The formulation (4.6)–(4.9) is identical to an L.P. except for the presence of (4.9). Because the simplex method is a very efficient way of solving an L.P., it seems natural to ask whether this method might not be used on the I.P., solving it by ignoring (4.9). If the solution obtained satisfies (4.9) it is optimal. However, suppose the solution contained, for at least one i, $1 \leq i \leq q$,

$$x_i^* = \bar{b}_i,$$

where $\bar{b}_i$ is noninteger. In this case the L.P. solution is infeasible as an I.P. solution. The value for each such x_i could be rounded either up or down as

$$x_i = [\bar{b}_i] \quad \text{or} \quad x_i = [\bar{b}_i] + 1$$

to achieve feasibility, where $[\bar{b}_i]$ denotes the integer part of $\bar{b}_i$. Sometimes this approach yields a satisfactory solution. There are, however, problems.

Consider, for example, Figure 4.1, where the constraints for the following small I.P. problem have been drawn:

$$
\begin{aligned}
\text{Maximize:} \quad & x_0 = x_1 + x_2 \\
\text{subject to:} \quad & 2x_1 + 12x_2 \leq 39 \\
& \qquad\quad 4x_2 \geq 9 \\
& x_1, x_2 \geq 0 \\
& x_1, x_2 \text{ integer.}
\end{aligned}
$$

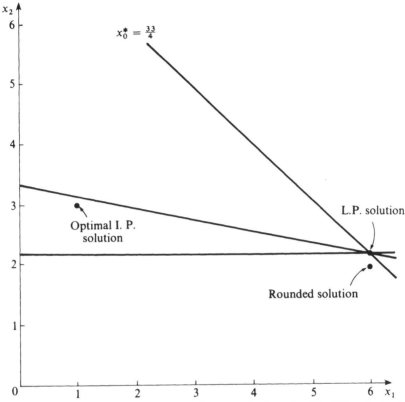

Figure 4.1. An example showing the failure to obtain a feasible I.P. solution by rounding an L.P. solution.

It can be seen in Figure 4.1 that the L.P. solution is

$$x_1^* = 6$$
$$x_2^* = \tfrac{27}{12},$$

which is infeasible for the above I.P. The rounding of x_2, either up or down, does not produce a feasible solution. In fact the optimal I.P. solution, as shown, is not at all close, relatively speaking, to the L.P. solution.

This example points up the pitfalls of rounding L.P. solutions to obtain I.P. solutions. No combination of rounding either up or down of the non-integer variables may be feasible, let alone optimal. Even when rounding does produce feasibility, the solution may be far from optimal.

It is obvious that more sophisticated methods need to be developed if we are to guarantee an optimal solution to an I.P. problem. Some such methods are described in the next two sections.

4.3 Enumerative Techniques

Theoretically any I.P. problem can be solved by simply listing all possible feasible solutions, finding the value of each one and choosing the best. Such a technique is called *exhaustive enumeration*. However even for zero–one problems, in which there are just two possibilities for each variable, as the number of variables increases the number of possibilities quickly becomes very large: 2^n for n variables. The situation is far worse for general problems. Hence it is impractical to solve anything other than trivial problems in this way.

What can be done, however, is to examine the set of all possible solutions in such a way that whole sets of solutions can be discarded without specific evaluation of all the solutions in each of these sets. Thus the enumeration is carried out implicitly, and this approach is termed *implicit enumeration*. Dynamic programming, which will be covered in Chapter 6, is an example of implicit enumeration. An implicit enumeration technique designed especially for integer programming problems, called *branch and bound enumeration*, will be described next.

4.3.1 Branch and Bound Enumeration

Branch and bound enumeration is a sequential technique for solving combinatorial optimization problems. Its use on such problems produces a *decision tree*. The first iteration produces the point at which the tree is rooted. Any subsequent iteration produces a number of new points which are connected to the existing tree by lines which all emanate from one existing point. A set of decisions concerning the values that the variables can assume is associated with each point along with a bound. The bound represents a value which is at least as good as that which could be attained by any feasible solution obeying the set of decisions of that point. The process begins by creating the root of the decision tree, which represents all feasible solutions to the problem. A bounding routine calculates a bound for this point, i.e., a bound on the optimal value. If the solution associated with this bound is feasible it is optimal and the procedure is terminated. If not, a partitioning routine partitions the set of feasible solutions into a number of subsets, each represented by a distinct point in the decision tree, all connected by lines to the parent point. The bounding routine then calculates a bound for each of these points. An elimination routine discards a point from the tree if it can be shown that no solution in its set can be optimal. This would occur, for example, if its bound is worse than the value of a known feasible solution. The process continues generating new points at each iteration. Termination occurs when finally the optimal solution or evidence that no such solution exists has been obtained.

4.3.1.1 *Solving the Numerical Example by Dakin's Method*

Land and Doig (1960) presented a branch and bound algorithm for solving I.P. or mixed I.P. problems. It was found to be very difficult to program a computer to implement it efficiently. However, Dakin (1965) introduced a modification of their algorithm which overcome this restriction. The latter algorithm will be explained here.

The branch and bound decision tree built up in applying Dakin's method to problem (4.1)–(4.5) is shown in Figure 4.2. The algorithm begins by solving (4.1)–(4.4) as an L.P. This has the following optimal solution:

$$x_1^* = \tfrac{12}{10}$$
$$x_2^* = \tfrac{21}{10}$$
$$x_0^* = \tfrac{111}{10}.$$

If this first optimal solution had satisfied (4.5), it would have been optimal for the I.P. and the method would have been terminated. However, as this is not the case, we proceed. The bound of $\tfrac{111}{10}$ is associated with the highest point of the decision tree, labelled a.f.s. (which means that it represents the set of *all* *f*easible solutions). Any feasible solution for the I.P. cannot have a value greater than this bound. As this solution is infeasible with regard to (4.5), one of the variables with a noninteger value is arbitrarily chosen, say x_2. The integer part of its value is identified. That is, we find the greatest integer less than or equal to the current value $(\tfrac{21}{10})$ of x_2. As

$$\tfrac{21}{10} = 2 + \tfrac{1}{10},$$

this integer part is 2. Now, as x_2 must be integral in any feasible solution, either

$$x_2 \leq 2 \tag{4.10}$$

or

$$x_2 \geq 3. \tag{4.11}$$

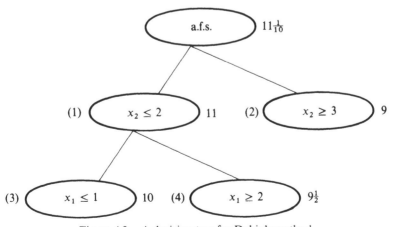

Figure 4.2.   A decision tree for Dakin's method.

We now create two new L.P. problems, I and II:

I: (4.1), (4.2), (4.3), (4.4), and (4.10)
II: (4.1), (4.2), (4.3), (4.4), and (4.11),

that is:

PROBLEM I

$$
\begin{aligned}
\text{Maximize:} \quad & 4x_1 + 3x_2 \\
\text{subject to:} \quad & 3x_1 + 4x_2 \leq 12 \\
& 4x_1 + 2x_2 \leq 9 \\
& x_2 \leq 2 \\
& x_1, x_2 \geq 0.
\end{aligned}
$$

PROBLEM II

$$
\begin{aligned}
\text{Maximize:} \quad & 4x_1 + 3x_2 \\
\text{subject to:} \quad & 3x_1 + 4x_2 \leq 12 \\
& 4x_1 + 2x_2 \leq 9 \\
& x_2 \geq 3 \\
& x_1, x_2 \geq 0.
\end{aligned}
$$

We have in effect partitioned the set of feasible solutions to the original I.P. into two disjoint subsets: one comprising all the solutions where $x_2 \leq 2$, and the other all solutions where $x_2 \geq 3$. Consequently, two new points representing these two sets of solutions are added to the decision tree in Figure 4.2. Problems I and II are now solved. Problem I has optimal solution

$$
\begin{aligned}
x_1^* &= \tfrac{5}{4} \\
x_2^* &= 2 \\
x_0^* &= 11.
\end{aligned}
$$

Problem II has an optimal solution

$$
\begin{aligned}
x_1^* &= 0 \\
x_2^* &= 3 \\
x_0^* &= 9.
\end{aligned}
$$

The solution to II satisfies (4.5) and is thus stored as the best solution found so far for the I.P., with value 9. However the bound 11 is associated with point 1 in the tree. Thus the possibility remains that there may be a better I.P. solution lurking in its set. We choose x_1 as the noninteger valued variable with value $\tfrac{5}{4}$. As the integer part of this value is 1, we create two constraints,

$$
x_1 \leq 1 \quad \text{and} \quad x_2 \geq 2.
$$

We create two new L.P.'s:

PROBLEM III

$$\begin{aligned}
\text{Maximize:} \quad & 4x_1 + 3x_2 \\
\text{subject to:} \quad & 3x_1 + 4x_2 \le 12 \\
& 4x_1 + 2x_2 \le 9 \\
& x_2 \le 2 \\
& x_1 \le 1 \\
& x_1, x_2 \ge 0.
\end{aligned}$$

PROBLEM IV

$$\begin{aligned}
\text{Maximize:} \quad & 4x_1 + 3x_2 \\
\text{subject to:} \quad & 3x_1 + 4x_2 \le 12 \\
& 4x_1 + 2x_2 \le 9 \\
& x_2 \le 2 \\
& x_1 \ge 2 \\
& x_1, x_2 \ge 0.
\end{aligned}$$

These problems are now solved. Problem III has an optimal solution:

$$\begin{aligned}
x_1^* &= 1 \\
x_2^* &= 2 \\
x_0^* &= 10.
\end{aligned}$$

As this satisfies (4.5) and its value exceeds that of the best solution found so far, it is stored as our best solution. Problem IV has value $9\frac{1}{2}$, which is less than the value of present incumbent. We have found a solution whose value exceeds the bound for any other set of feasible solutions. This solution must be optimal.

We have discovered that the company should fly one A crate and two B crates on each trip for a maximum return of 10 units.

4.3.1.2 *Dakin's Method in General*

Dakin's method begins to solve a problem of the form of (4.6)–(4.9) by first ignoring (4.9) and solving the problem as an L.P. using the simplex method. The value of the solution thus found is the bound assigned to the first point of the decision tree, representing all feasible solutions to the original I.P. problem. This makes sense, as (4.6)–(4.9) can be thought of as the equivalent L.P. with the added constraint of (4.9). Hence it cannot have an optimal solution better than the equivalent L.P. If the optimal L.P. solution has integer values for the first q variables, it is optimal for the I.P. and the method terminates. However, suppose at least one variable, x_i $(1 \le i \le q)$ has a non-integer value

$$x_i^* = \bar{b}_i, \quad \bar{b}_i \text{ noninteger.}$$

Now as x_i is constrained to be an integer, values in the range

$$[\overline{b}_i] < x_i < [\overline{b}_i] + 1,$$

are infeasible. Hence x_i must obey exactly one of the following constraints:

$$x_i \leq [\overline{b}_i]$$

or

$$x_i \geq [\overline{b}_i] + 1.$$

Two new L.P. problems are now created:

I: (4.6), (4.7), (4.8), and $x_i \leq [b_i]$
II: (4.6), (4.7), (4.8), and $x_i \geq [b_i] + 1$.

Note that problems I and II differ from the original problem only in the fact that one more constraint has been added. It is thus possible to deduce the optimal solutions to these amended problems with relatively little extra computational effort using the ideas of Section 2.6.2.5 and the dual simplex method of Section 3.3. Constraints of the type $x_i \leq [b_i]$ and $x_1 \geq [b_i] + 1$ are called *Dakin cuts*. Notice that it is no longer possible for x_i to take on the offending value $\overline{b}_i$ in either problem I or II. Two new points are created in the decision tree, both joined by lines to the original point. The first represents all feasible solutions to problem I, the second to problem II. The optimal solution to the original I.P. (if such a solution exists) must lie in one of these sets. In fact the set, S of feasible solutions to (4.6)–(4.9) has been partitioned into these two sets S_I and S_{II} in the sense that

$$S_I \cup S_{II} = S$$

and

$$S_I \cap S_{II} = \varnothing, \quad \text{the empty set.}$$

Both L.P. problems I and II are solved. Their optimal solution values are bounds assigned to the corresponding points in the decision tree. The better of the two bounds is identified. As the objective is one of maximization the larger bound will be selected. Ties can be settled arbitrarily. If this better bound corresponds to a feasible solution to (4.6)–(4.9) this solution is declared optimal and the procedure is terminated. If it corresponds to an infeasible solution another of the variables constrained to be integral with a noninteger value is identified. Two more cuts are defined based on this variable. The partitioning (branching) routine is repeated, creating two more decision tree points.

The algorithm is continued until either (a) a feasible solution with value no less than that for any other bound is found (this solution is then pronounced optimal), or (b) it is found that no feasible solution exists (all points have been eliminated from the tree). When a solution is found to be feasible, its point is never selected for branching, and the point is said to be *fathomed*. The point is eliminated unless it is the best feasible solution so far found, in which case it is recorded as the incumbent. Of course any point with a bound worse than that of the incumbent is eliminated.

4.3.1.3 *The Zero–One Method of Balas*

We now examine the zero–one programming problem and a method for its solution. Although the model assumes that each variable is binary it is still useful, as many I.P. problems are formulated this way. Often the variables represent decisions as to whether to adopt a particular policy or not, i.e.

$$x_i = \begin{cases} 1, & \text{if policy } i \text{ is adopted} \\ 0, & \text{otherwise.} \end{cases}$$

Further, any I.P. can be converted into a zero–one problem by redefining each nonbinary variable x_j as follows.

Let u_j be the largest possible integer value that x_j could possibly assume in any feasible solution. This bound u_j is usually deduced by examining the constraints.

Let N_j be the smallest integer such that

$$2^{N_j + 1} > u_j.$$

Then x_j can be expressed in terms of the binary variables $y_1^i, y_2^j, y_{N_j+1}^j$ as

$$x_j = \sum_{i=1}^{N_j+1} (2^{i-1}) y_i^j.$$

Examples of this conversion will be given in the next section.

Balas (1965) has developed a method for solving zero–one problems which involves branch and bound enumeration. His approach differs from that of Dakin's in that it does not require the simplex method as a subroutine. Balas describes the method as "additive," as it requires only the addition and not the multiplication of numbers. The method is applicable only to problems with nonnegative objective function coefficients. Any zero–one programme can be converted into this form by replacing any variable x_i with negative c_i by $\bar{x}_i = (1 - x_i)$.

The method partitions the variables into three sets:

W: the set of variables which have been assigned a value 1
V: the set of variables which have been assigned a value 0
F: the set of unassigned (free) variables.

Initially all variables are assigned to F. For maximization problems all variables in F are next temporarily assigned a value of 1. If this solution is feasible it is clearly optimal, as $c_i \geq 0$, $i = 1, 2, \ldots n$. If this solution is infeasible an upper bound on the value of the optimal solution can be obtained. This bound is equal to the sum of the c_i, neglecting the minimum c_i, for all $x_i \in F$. After this first iteration, a bound can be found for any partition of the variables among W, V, and F as follows.

The bound is equal to the sum of the c_i for all $x_i \in W$ plus the sum of c_i, neglecting the minimum c_i, for all $x_i \in F$, i.e.,

$$\sum_{x_i \in W} c_i + \sum_{x_i \in F} c_i - \min_{x_i \in F} \{c_i\}.$$

When a solution is found to be infeasible its corresponding point in the decision tree sprouts two new points, effecting the branching step. Suppose W, V, and F denote the partition at the parent point and the variable corresponding to the minimum c_i among variables in F is x_i. Then the partition at one of the new points is:

F becomes $F \backslash \{x_i\}$, W becomes $W \cup \{x_i\}$, V remains the same

and at the other new point:

F becomes $F \backslash \{x_i\}$, W remains the same, V becomes $V \cup \{x_i\}$.

Bounds are then calculated for these two new nodes as just described. When the partition of a particular point cannot possibly lead to an optimal solution, the point is eliminated from the tree. When the partition of a particular point corresponds to a feasible solution the point is fathomed and no further branching takes place from it. When a point corresponds to a feasible solution and has a value no less than that for any other node, this solution is declared optimal.

4.3.1.4 Numerical Example

The method will be illustrated using the problem (4.1)–(4.5). First upper bounds on x_1 and x_2 must be found, as they are not binary:

$$u_1 = \min_{i=1,2} \{[b_i/a_{1i}]\} = \min \{[\tfrac{12}{3}], [\tfrac{9}{4}]\} = 2$$
$$u_2 = \min_{i=1,2} \{[b_i/a_{2i}]\} = \min \{[\tfrac{12}{4}], [\tfrac{9}{2}]\} = 3.$$

Therefore

$$N_1 = 1, \qquad N_2 = 1.$$

Let

$$x_1 = 2^0 y_1^1 + 2^1 y_2^1$$
$$x_2 = 2^0 y_1^2 + 2^1 y_2^2.$$

Then the problem becomes

Maximize: $x_0 = 4(y_1^1 + 2y_2^1) + 3(y_1^2 + 2y_2^2) = 4y_1^1 + 8y_2^1 + 3y_1^2 + 6y_2^2$

subject to: $3(y_1^1 + 2y_2^1) + 4(y_1^2 + 2y_2^2) \le 12$

$4(y_1^1 + 2y_2^1) + 2(y_1^2 + 2y_2^2) \le 9$

$y_1^1, y_2^1, y_1^2, y_2^2 = 0$ or 1.

The decision tree built up by the method is shown in Figure 4.3. The method begins by partitioning the variables into

$$W = \varnothing, \qquad V = \varnothing, \qquad F = \{y_1^1, y_2^1, y_1^2, y_2^2\}.$$

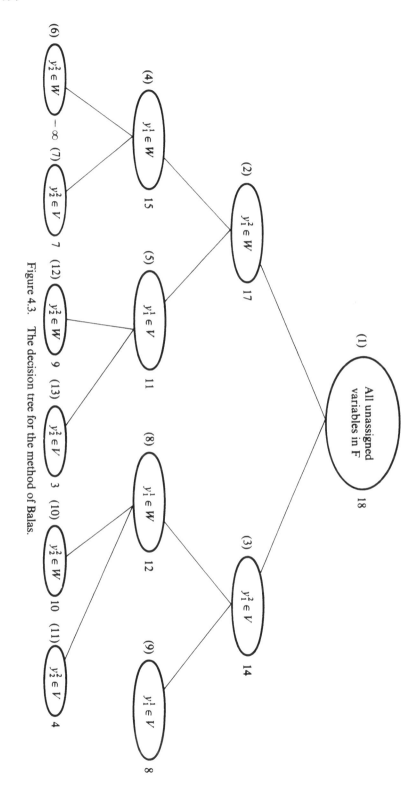

Figure 4.3.   The decision tree for the method of Balas.

Next all the variables in F are temporarily assigned a value 1. This is obviously infeasible. The bound on the first point is calculated as $4 + 8 + 6 = 18$, the sum of all c_i of the variables in F except the minimum, which is 3, corresponding to y_1^2. Branching now takes place and y_1^2 is transferred to W in point (2) and V in point (3). Bounds are calculated for these nodes as explained earlier. For instance, the bound for point (2) is arrived at by setting all the variables in F, y_1^1, y_2^1, y_2^2, equal to 1. This is infeasible, as $y_1^2 = 1$ because it is in W. Hence the variable with the minimum coefficient in F, y_1^1, is discounted when calculating the bound, which is $8 + 6$ plus the 3 from y_1^2 For point (3) all the free variables are set equal to 1, but this is infeasible. The variable discarded is y_1^1 (y_1^2 is unavailable, as it is in V) and the bound is $8 + 6$.

As we are maximizing we branch from point (2) because it has the higher bound. This produces points (4) and (5). Branching from point (4) produces points (6) and (7). However, point (6) cannot represent any feasible solutions, as y_1^2, y_1^1, and y_2^2 cannot all be equal to 1, so it is eliminated. (Hence the "$-\infty$" symbol.)

At this stage point (3) has the largest bound. This eventually produces point (10) with a bound of 10 which represents a feasible solution. All points with inferior bounds can be eliminated. This leaves point (5), which spawns points (12) and (13) both with bounds less than 10.

Therefore point (10) is declared optimal, with the solution

$$y_1^1, y_2^2 = 1$$
$$y_1^2, y_2^1 = 0$$
$$x_0^* = 10.$$

This solution corresponds to that found by Dakin's method:

$$x_1^* = 2^0(1) + 2^1(0) = 1$$
$$x_2^* = 2^0(0) + 2^1(1) = 2.$$

4.4 Cutting Plane Methods

Gomory (1958) developed *cutting plane* algorithms for solving all-integer and mixed-integer programming problems. He proved that these methods will produce an optimal solution in a finite number of iterations when applied to problems with rational data. The methods revolve around the idea of introducing new constraints (or cuts) to the problem. These cuts slice away noninteger optimal solutions to the associated L.P. problem, but leave all feasible integer solutions untouched. This is similar to what is done in Dakin's method, but there are fundamental differences between the two approaches. In cutting plane methods successive constraints are added to just one problem, whereas in branch and bound methods many different (linear programming) problems may be created. Thus in cutting plane methods the original feasible region for the associated L.P. problem is gradually reduced as extra constraints are added. In contrast, in branch and bound methods the original feasible region is often broken up into disconnected subregions. In Dakin's method cuts are parallel to the axes; this seldom happens in cutting plane methods. Finally, cutting plane methods always preserve all feasible integer solutions, while some feasible integer solutions are usually eliminated from some of the problems created by the branch and bound method.

The all-integer and mixed-integer methods will be explained in the next two sections.

4.4.1 Gomory's All-Integer I.P. Method

The way in which the Gomory all-integer cutting plane algorithm solves (4.6)–(4.9) will now be explained. It will be assumed that all variables are constrained to be integers in (4.6)–(4.9):

$$q = n.$$

The outline of the algorithm is as follows. Problem (4.6)–(4.8) is solved by the simplex method. If the optimal solution is all-integer the problem is solved and the algorithm is terminated. If at least one variable is noninteger a new constraint is added to the problem. This constraint is derived by choosing a noninteger valued variable and examining the tableau row in which it appears. The problem is then resolved with this new constraint.

It has been assumed that all variables, including slack variables, are to be integer in any feasible solution. This assumption can be made workable by clearing fractions from the constraint coefficients *before* introducing the slack variables. That is, if one is confronted with a constraint like

$$\tfrac{5}{11}x_1 + \tfrac{6}{9}x_2 \le 1,$$

one can multiply the constraint by the lowest common denominator of the coefficients (99) to obtain

$$45x_1 + 66x_2 \le 99.$$

Once this has been done for all necessary constraints the initial L.P. problem is then solved by the simplex method.

The way in which a new constraint is constructed from a noninteger tableau will now be explained. Suppose the associated L.P. problem has been solved and at least one variable, say, x_i has a noninteger value. The row in the optimal tableau in which x_i has a unit entry is found, say the jth row. Let it correspond to the equation

$$x_i + \bar{a}_{j1}y_1 + \bar{a}_{j2}y_2 + \cdots + \bar{a}_{jp}y_p = \bar{b}_j, \tag{4.12}$$

where y_k, $k = 1, 2, \ldots, p$, are the nonbasic variables, $\bar{a}_{jk}$, $k = 1, 2, \ldots, p$ is the coefficient of y_k in this jth row; and $\bar{b}_j$ is the value of x_i. Now (4.12) is solved for x_i:

$$x_i = \bar{b}_j - \bar{a}_{j1} y_1 - \bar{a}_{j2} y_2 - \cdots - \bar{a}_{jp} y_p. \tag{4.13}$$

For any $\alpha \in \mathbf{R}$, let $[\alpha]$ denote the largest integer no greater than α. Then

$$\alpha = [\alpha] + \alpha', \tag{4.14}$$

where α' is the fractional part of α; for example,

$$\alpha = 3\tfrac{1}{7} \Rightarrow [\alpha] = 3 \quad \text{and} \quad \alpha' = \tfrac{1}{7}$$
$$\alpha = -\tfrac{4}{5} \Rightarrow [\alpha] = -1 \quad \text{and} \quad \alpha' = \tfrac{1}{5}$$
$$\alpha = 2 \Rightarrow [\alpha] = 2 \quad \text{and} \quad \alpha' = 0.$$

Each rational number in the r.h.s. of (4.13) can be expressed in the following format:

$$x_i = [\bar{b}_j] + \bar{b}'_j - ([\bar{a}_{j1}] + \bar{a}'_{j1}) y_1 - \cdots - ([\bar{a}_{jp}] + \bar{a}'_{jp}) y_p.$$

On collecting integer terms, this becomes

$$x_i = \{[\bar{b}_j] - [\bar{a}_{j1}] y_1 - \cdots - [\bar{a}_{jp}] y_p\} + \{\bar{b}'_j - \bar{a}'_{j1} y_1 - \cdots - \bar{a}'_{jp} y_p\}.$$

Now the first part:

$$\{[\bar{b}_j] - [\bar{a}_{j1}] y_1 - \cdots - [\bar{a}_{jp}] y_p\}$$

will be an integer if all the variables $y_1, y_2, \ldots, y_p$ are integers, which is true by assumption. Hence for x_i to be an integer, the second part:

$$\{\bar{b}'_j - \bar{a}'_{j1} y_1 - \cdots - \bar{a}'_{jp} y_p\} \tag{4.15}$$

must be an integer. But

$$0 < \bar{b}'_j < 1,$$

as $\bar{b}$ was assumed to be noninteger. Also

$$0 \le \bar{a}'_{ji} < 1, \qquad i = 1, 2, \ldots, p$$

because of the definition (4.14). Hence, as the $y_1, y_2, \ldots, y_p$ are constrained to be nonnegative integers, (4.15) cannot be a positive integer. Hence (4.15) must be a nonpositive integer. So the constraint:

$$\bar{b}'_j - \bar{a}'_{j1} y_1 - \cdots - \bar{a}'_{jp} y_p \le 0 \tag{4.16}$$

must hold in any feasible integer solution.

Let the slack variable x_r be introduced into (4.16):

$$\bar{b}'_j - \bar{a}'_{j1} y_1 - \cdots - \bar{a}'_{jp} y_p + x_r = 0. \tag{4.17}$$

As (4.15) must be an integer, then x_r must of necessity be an integer also. This constraint (4.17) is now added to the final simplex tableau and an optimal solution to the amended L.P. solution is found using the dual simplex method of Section 3.3.

The constraint (4.16) represents a *Gomory cut*. The process is repeated until the dual simplex method either produces an all-integer solution (which will be an optimal solution for the original I.P. problem) or evidence that no feasible solution exists (in which there are no feasible all-integer solutions). The algorithm will now be illustrated by solving a numerical example.

4.4.1.1 Numerical Example

The method will be illustrated on the problem (4.1)–(4.5). The problem is first solved by the simplex method, ignoring (4.5). This produces Table 4.1, where x_3 and x_4 are the slack variables introduced in (4.2) and (4.3).

Table 4.1

Constraints	x_1	x_2	x_3	x_4	r.h.s.
(4.19)	0	1	$\frac{2}{5}$	$-\frac{3}{10}$	$\frac{21}{10}$
(4.20)	1	0	$-\frac{1}{5}$	$\frac{2}{5}$	$\frac{6}{5}$
x_0	0	0	$\frac{2}{5}$	$\frac{7}{10}$	$\frac{111}{10}$

This solution is noninteger, so we must introduce a cut. Consider the second row in Table 4.1, corresponding to the noninteger-valued variable x_1. This row corresponds to the equation:

$$x_1 - \tfrac{1}{5}x_3 + \tfrac{2}{5}x_4 = \tfrac{6}{5}.$$

Therefore

$$x_1 = \tfrac{6}{5} - (-\tfrac{1}{5})x_3 - \tfrac{2}{5}x_4$$
$$= (1 + \tfrac{1}{5}) - (-1 + \tfrac{4}{5})x_3 - (0 + \tfrac{2}{5})x_4.$$

The fractional part of this expression is

$$\tfrac{1}{5} - \tfrac{4}{5}x_3 - \tfrac{2}{5}x_4,$$

which cannot be a positive integer; hence

$$\tfrac{1}{5} - \tfrac{4}{5}x_3 - \tfrac{2}{5}x_4 \leq 0. \tag{4.18}$$

As the problem has only two structural variables, it is instructive to follow the progress of the method graphically. Figure 4.4 shows the graphical solution to the original L.P. problem. Using the equations

$$3x_1 + 4x_2 + x_3 = 12 \tag{4.19}$$

$$4x_1 + 2x_2 + x_4 = 9, \tag{4.20}$$

one can substitute for x_3 and x_4 in (4.18), producing

$$4x_1 + 4x_2 \leq 13, \tag{4.21}$$

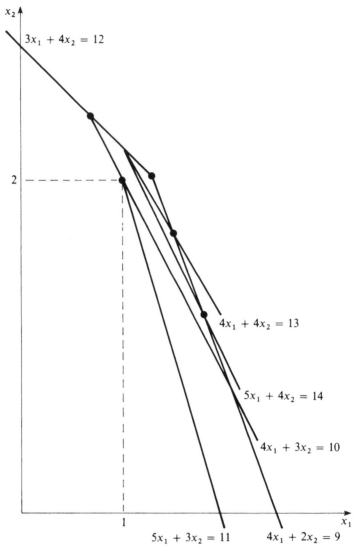

Figure 4.4. Graphical solution to the example problem using Gomory's method.

which is shown in Figure 4.4. Note that this solution cuts away part of the feasible region, including the optimal solution to the present, but leaves all feasible integer solutions still in the region. This will always happen.

On adding a slack variable x_5 to (4.18) and taking the constant to the r.h.s., we have

$$-\tfrac{4}{5}x_3 - \tfrac{2}{5}x_4 + x_5 = -\tfrac{1}{5}. \tag{4.22}$$

This constraint is added to Table 4.1, and the dual simplex method is used to produce a new optimum, given in Table 4.2.

Table 4.2

Constraints	x_1	x_2	x_3	x_4	x_5	r.h.s.
(4.19)	0	1	0	$-\frac{1}{2}$	$\frac{1}{2}$	2
(4.20)	1	0	0	$\frac{1}{2}$	$-\frac{1}{4}$	$\frac{5}{4}$
(4.22)	0	0	1	$\frac{1}{2}$	$-\frac{5}{4}$	$\frac{1}{4}$
x_0	0	0	0	$\frac{1}{2}$	$\frac{1}{2}$	11

As can be seen in Table 4.2, x_3 has a noninteger value. Hence the equation

$$x_3 + \tfrac{1}{2}x_4 - \tfrac{5}{4}x_5 = \tfrac{1}{4}$$

is expressed in terms of x_3 with integer and noninteger parts:

$$x_3 = (0 + \tfrac{1}{4}) - (0 + \tfrac{1}{2})x_4 - (-2 + \tfrac{3}{4})x_5.$$

So the new cut is:

$$\tfrac{1}{4} - \tfrac{1}{2}x_4 - \tfrac{3}{4}x_5 \leq 0. \tag{4.23}$$

As an aside, we can use (4.19), (4.20), and (4.22) to show that (4.23) is equivalent to

$$5x_1 + 4x_2 \leq 14.$$

This constraint is plotted in Figure 4.4. On adding the slack variable x_6 to (4.23) we have

$$-\tfrac{1}{2}x_4 - \tfrac{3}{4}x_5 + x_6 = -\tfrac{1}{4}. \tag{4.24}$$

(4.24) is added to Table 4.2 and the dual simplex method is used to produce the optimal tableau shown in Table 4.3.

Table 4.3

Constraints	x_1	x_2	x_3	x_4	x_5	x_6	r.h.s.
(4.19)	0	1	0	$-\frac{5}{6}$	0	$\frac{2}{3}$	$\frac{11}{6}$
(4.20)	1	0	0	$\frac{2}{3}$	0	$-\frac{1}{3}$	$\frac{4}{3}$
(4.22)	0	0	1	$\frac{4}{3}$	0	$-\frac{5}{3}$	$\frac{2}{3}$
(4.24)	0	0	0	$\frac{2}{3}$	1	$-\frac{4}{3}$	$\frac{1}{3}$
x_0	0	0	0	$\frac{1}{6}$	0	$\frac{2}{3}$	$\frac{65}{6}$

Now all the basic variables have noninteger values. The reader who thinks we are chasing our tails is asked not to despair. The "optimal" solution value is steadily being reduced at each iteration: from an initial $\frac{111}{10}$ to 11 to $\frac{65}{6}$. The Gomory cuts are slicing away nonoptimal parts of the original feasible region, as can be seen in Figure 4.4. Applying the technique to row (4.19) in

Table 4.3 produces

$$x_2 = \tfrac{11}{6} - (-\tfrac{5}{6})x_4 - \tfrac{2}{3}x_6$$
$$= (1 + \tfrac{5}{6}) - (-1 + \tfrac{1}{6})x_4 - (-1 + \tfrac{1}{3})x_6.$$

Therefore

$$\tfrac{5}{6} - \tfrac{1}{6}x_4 - \tfrac{2}{3}x_6 \le 0, \tag{4.25}$$

which corresponds to

$$4x_1 + 3x_2 \le 10,$$

shown in Figure 4.4.

Adding the slack variable x_7 to (4.25), which is then added to Table 4.3, allows the dual simplex method to produce the optimal tableau shown in Table 4.4. Applying the technique to (4.20) in Table 4.4 produces

$$x_1 = \tfrac{7}{4} - \tfrac{3}{4}x_4 - (-\tfrac{1}{2})x_7$$
$$= (1 + \tfrac{3}{4}) - (0 + \tfrac{3}{4})x_4 - (-1 + \tfrac{1}{2})x_7.$$

Therefore

$$\tfrac{3}{4} - \tfrac{3}{4}x_4 - \tfrac{1}{2}x_7 \le 0, \tag{4.26}$$

which corresponds to

$$5x_1 + 3x_2 \le 11$$

shown in Figure 4.4.

Adding the slack variable x_8 to (4.26), which is then added to Table 4.4, allows the dual simplex method to produce Table 4.5, which displays the

Table 4.4

Constraints	x_1	x_2	x_3	x_4	x_5	x_6	x_7	r.h.s.
(4.19)	0	1	0	-1	0	0	1	1
(4.20)	1	0	0	$\tfrac{3}{4}$	0	0	$-\tfrac{1}{2}$	$\tfrac{7}{4}$
(4.22)	0	0	1	$\tfrac{7}{4}$	0	0	$-\tfrac{5}{2}$	$\tfrac{11}{4}$
(4.24)	0	0	0	1	1	0	-2	2
(4.25)	0	0	0	$\tfrac{1}{4}$	0	1	$-\tfrac{3}{2}$	$\tfrac{5}{4}$
x_0	0	0	0	0	0	0	1	$\tfrac{60}{6}$

Table 4.5

Constraints	x_1	x_2	x_3	x_4	x_5	x_6	x_7	x_8	r.h.s.
(4.19)	0	1	0	0	0	0	$\tfrac{5}{3}$	$-\tfrac{4}{3}$	2
(4.20)	1	0	0	0	0	0	-1	1	1
(4.22)	0	0	1	0	0	0	$-\tfrac{11}{3}$	$\tfrac{7}{3}$	1
(4.24)	0	0	0	0	1	0	$-\tfrac{8}{3}$	$\tfrac{4}{3}$	1
(4.25)	0	0	0	0	0	1	$-\tfrac{5}{3}$	$\tfrac{1}{3}$	1
(4.26)	0	0	0	1	0	0	$\tfrac{2}{3}$	$-\tfrac{4}{3}$	1
x_0	0	0	0	0	0	0	1	0	$\tfrac{60}{6}$

optimal solution to the original problem, (4.1)–(4.5):

$$x_1^* = 1$$
$$x_2^* = 2$$
$$x_0^* = 10.$$

4.4.2 Gomory's Mixed-Integer I.P. Algorithm

Consider now a mixed-integer programming problem, i.e., some but not all of the variables are constrained to be integer. In terms of (4.6)–(4.9),

$$0 < q < n.$$

Gomory's mixed-integer I.P. algorithm follows the same initial pattern as the all-integer algorithm. Suppose the initial simplex solution contains a noninteger-valued variable x_j which is one of those which is constrained to be integer. Then its tableau equation (4.12) can be rewritten as

$$[\bar{b}_j] + \bar{b}_j' - x_j = \sum_{k=1}^{p} \bar{a}_{jk} y_k. \tag{4.27}$$

At this point the analysis takes a different path from that of Section 4.4.1, because not all of the variables y_k, $k = 1, \ldots, p$ may be constrained to be integer. Let

$$S_+ = \{k : \bar{a}_{jk} \geq 0\}$$
$$S_- = \{k : \bar{a}_{jk} < 0\}.$$

Then (4.27) can be written as

$$[\bar{b}_j] + \bar{b}_j' - x_j = \sum_{k \in S_+} \bar{a}_{jk} y_k + \sum_{k \in S_-} \bar{a}_{jk} y_k. \tag{4.28}$$

Case I. Assume

$$[\bar{b}_j] + \bar{b}_j' - x_j < 0.$$

As $[\bar{b}_j]$ is an integer, x_j is constrained to be an integer in any feasible solution, and $\bar{b}_j'$ is a nonnegative fraction. Hence

$$[\bar{b}_j] - x_j$$

must be a negative integer, say $-u$. Therefore

$$[\bar{b}_j] + \bar{b}_j' - x_j = \bar{b}_j' - u,$$

where $u \in \{1, 2, 3, \ldots\}$. Substituting this into (4.28) produces

$$\bar{b}_j' - u = \sum_{k \in S_+} \bar{a}_{jk} y_k + \sum_{k \in S_-} \bar{a}_{jk} y_k.$$

Now, since

$$u \geq 1,$$

we have

$$\bar{b}'_j - 1 \geq \sum_{k \in S_+} \bar{a}_{jk} y_k + \sum_{k \in S_-} \bar{a}_{jk} y_k,$$

And, from the definition of S_+ and the fact that $y_k \geq 0$ for all k,

$$\bar{b}'_j - 1 \geq \sum_{k \in S_-} \bar{a}_{jk} y_k.$$

Now, as

$$\bar{b}'_j - 1 < 0,$$

we have

$$1 \leq (\bar{b}'_j - 1)^{-1} \sum_{k \in S_-} \bar{a}_{jk} y_k.$$

Multiplying both sides by $\bar{b}'_j$, we obtain

$$\bar{b}'_j \leq \bar{b}'_j(\bar{b}'_j - 1)^{-1} \sum_{k \in S_-} \bar{a}_{jk} y_k. \tag{4.29}$$

Case II. Assume

$$[\bar{b}_j] + \bar{b}'_j - x_j \geq 0.$$

As x_j is constrained to be an integer in any feasible solution, we have

$$[\bar{b}_j] + \bar{b}'_j - x_j = \bar{b}'_j + v$$

for some v, where $v \in \{0, 1, 2, 3, \ldots\}$. Substituting this into (4.28), we get

$$\bar{b}'_j + v = \sum_{k \in S_+} \bar{a}_{jk} y_k + \sum_{k \in S_-} \bar{a}_{jk} y_k.$$

Now, since

$$v \geq 0,$$

we have

$$\bar{b}'_j \leq \sum_{k \in S_+} \bar{a}_{jk} y_k + \sum_{k \in S_-} \bar{a}_{jk} y_k,$$

and, from the definition of S_- and the fact that $y_k \geq 0$ for all k

$$\bar{b}'_j \leq \sum_{k \in S_+} \bar{a}_{jk} y_k. \tag{4.30}$$

Combining (4.29) and (4.30), we obtain

$$\bar{b}'_j \leq \bar{b}'_j(\bar{b}'_j - 1)^{-1} \sum_{k \in S_-} \bar{a}_{jk} y_k + \sum_{k \in S_+} \bar{a}_{jk} y_k. \tag{4.31}$$

This inequality must be satisfied if x_j is to be an integer. The constraint (4.31) is the Gomory cut, which is introduced into the final tableau.

A slack variable x_r is now added to (4.31):

$$\bar{b}'_j = \bar{b}'_j(\bar{b}'_j - 1)^{-1} \sum_{k \in S_-} \bar{a}_{jk} y_k + \sum_{k \in S_+} \bar{a}_{jk} y_k - x_r. \tag{4.32}$$

Now, as

$$y_k = 0, \qquad k = 1, 2, \ldots, p$$

we have

$$x_r = -\bar{b}'_j,$$

which is infeasible. The dual simplex method is used to remedy this situation. The above process is repeated until either:

1. A tableau is produced in which x_i, $i = 1, 2, \ldots, q$ are integer, in which case the corresponding solution is optimal; or
2. The use of the dual simplex method leads to the conclusion that no feasible solution exists, in which case one can conclude that the original mixed-integer problem has no feasible solution.

4.4.2.1 Numerical Example

The method will be illustrated on the problem (4.1)–(4.4), with the following additional constraint:

$$x_1 \text{ must be an integer,}$$

i.e.,

$$q = 1.$$

On examining Table 4.1 it can be seen that x_1 is noninteger and can be expressed as

$$1 + \tfrac{1}{5} - x_1 = -\tfrac{1}{5}x_3 + \tfrac{2}{5}x_4.$$

Therefore, in terms of (4.27),

$$[\bar{b}_j] = 1$$
$$\bar{b}'_j = \tfrac{1}{5}$$
$$j = 2$$
$$i = 1$$
$$p = 2$$
$$\bar{a}_{j1} = -\tfrac{1}{5}$$
$$\bar{a}_{j2} = \tfrac{2}{3}$$
$$y_1 = x_3$$
$$y_2 = x_4.$$

Also,

$$S_+ = \{4\}$$
$$S_- = \{3\}.$$

Letting

$$x_r = x_5,$$

in terms of (4.32) the cut becomes

$$\tfrac{1}{5} = \tfrac{1}{5}(\tfrac{1}{5} - 1)^{-1}(-\tfrac{1}{5})x_3 + \tfrac{2}{5}x_4 - x_5. \tag{4.33}$$

Adding the negative of this constraint to Table 4.1 yields Table 4.6. The application of the dual simplex method to Table 4.6 yields Table 4.7, which displays the optimal solution to the problem, as x_1 is now integer-valued.

Table 4.6

Constraints	x_1	x_2	x_3	x_4	x_5	r.h.s.
(4.19)	0	1	$\frac{2}{5}$	$-\frac{3}{10}$	0	$\frac{21}{10}$
(4.20)	1	0	$-\frac{1}{5}$	$\frac{2}{5}$	0	$\frac{6}{5}$
(4.33)	0	0	$-\frac{1}{20}$	$-\frac{2}{5}$	1	$-\frac{1}{5}$
x_0	0	0	$\frac{2}{5}$	$\frac{7}{10}$	0	$\frac{111}{10}$

Table 4.7

Constraints	x_1	x_2	x_3	x_4	x_5	r.h.s.
(4.19)	0	1	$\frac{47}{80}$	0	$-\frac{3}{4}$	$\frac{9}{4}$
(4.20)	1	0	$-\frac{9}{20}$	0	1	1
(4.33)	0	0	$\frac{1}{8}$	1	$-\frac{5}{2}$	$\frac{1}{2}$
x_0	0	0	$\frac{67}{80}$	0	$\frac{7}{4}$	$\frac{43}{4}$

This solution is

$$x_1^* = 1$$
$$x_2^* = \frac{9}{4}$$
$$x_0^* = \frac{43}{4}.$$

4.5 Applications of Integer Programming

In the sections that follow we shall outline some real-world problems that can be formulated in terms of integer programming. There is quite an art in this. On the surface it does not seem possible to describe many of the problems as integer programs. However with imaginative definition of variables and construction of constraints it can be done. Once it has been recognized that a problem is amenable to I.P. formulation there is a great deal to the task of making the formulation efficient. That is, it is one matter to be able to formulate a problem, it is another matter to endow the formulation with a structure or size that can be solved efficiently.

4.5.1 The Travelling Salesman Problem

The travelling salesman problem is one of the classical problems of combinatorial optimization. It is concerned with a salesman who must visit a number of cities once each and return to the city from whence he started. The problem is to assign an itinerary to the salesman which minimizes the total distance travelled in order to accomplish this circuit. It is assumed that the distance travelled in proceeding directly from one city to any other is known for all pairs of cities. Note that it is not assumed that the distance from city i to city j is necessarily the same as the distance from city j to city i. These two distances may differ for example when the "cities" are intersections in a one-way street network. When all such i, j-pairs of distances are equal the problem is called the *symmetric* travelling salesman problem (T.S.P.), otherwise it is called the *asymmetric* travelling salesman problem.

The T.S.P. can be formulated as a zero–one I.P. problem. Let

n = the number of cities,

c_{ij} = the cost of travelling from city i to city j.

Note that if one does not wish the salesman to travel directly from a certain town to another one can assign a prohibitively large value (denoted by "∞") to the appropriate c_{ij} value. This will ensure that such a path is never selected in any optimal solution. For instance, one sets

$$c_{ii} = \infty, \qquad 1 = 1, 2, \ldots, n.$$

Let

$$x_{ij} = \begin{cases} 1, & \text{if the salesman is to proceed directly from city } i \text{ to city } j \\ 0, & \text{otherwise.} \end{cases}$$

Because each city i must be left exactly once,

$$\sum_{j=1}^{n} x_{ij} = 1, \qquad i = 1, 2, \ldots, n. \tag{4.34}$$

Also, because each city j must be visited exactly once,

$$\sum_{i=1}^{n} x_{ij} = 1, \qquad j = 1, 2, \ldots, n. \tag{4.35}$$

For any given circuit defined by x_{ij} the objective is to

$$\text{Minimize:} \quad \sum_{i=1}^{n} \sum_{j=1}^{n} c_{ij} x_{ij}. \tag{4.36}$$

The reader will recognize that minimizing (4.36) subject to (4.34) and (4.35) is the assignment problem of Section 2.72. Unfortunately, extra constraints are needed in order to formulate the T.S.P. This is because (4.34), (4.35), (4.36) do not exclude the possibility of subtours being formed.

For instance, in a six-city problem one might make the assignments $x_{12} = x_{23} = x_{31} = x_{45} = x_{56} = x_{64} = 1$, all other $x_{ij} = 0$. That is, the "circuit" is $1 \to 2 \to 3 \to 1$ and then $4 \to 5 \to 6 \to 4$. This is a feasible solution for (4.34) and (4.35), as each city is left once and arrived at once. However, it represents two disjoint subtours. (A subtour is a circuit which does not involve all cities). Hence such a solution is not feasible for the T.S.P. Hence we need an extra family of constraints which prevent subtours from being formed. In order to develop this, we notice that there is a partition T, T' of the set of cities N:

$$T = \{1, 2, 3\}, \text{ and } T' = N - T = \{4, 5, 6\}$$

such that $x_{ij} = 0$ for all $i \in T$ and all $j \in T'$. This occurs if and only if subtours exist. Thus the following constraint will prevent subtours:

$$\sum_{i \in T} \sum_{j \in T'} x_{ij} \geq 1, \quad \text{for all proper partitions } T, T' \text{ of } N. \qquad (4.37)$$

(A proper partition T, T' of N is a partition such that $T \neq \emptyset$ or N.) Thus the T.S.P. can be expressed as the following zero–one I.P.: minimize (4.36) subject to (4.34), (4.35), and (4.37).

Of course, (4.37) involves a relatively large number of constraints for non-trivial n. Hence it is not practical to use the above formulation on anything other than very small problems. However before the reader despairs, one can consider solving the problem ignoring (4.37). If the resulting solution is a feasible circuit it is optimal; if not, its value represents a valid lower bound on the value of the optimal T.S.P. solution. This suggests that one could use a branch and bound approach calculating bounds in this way. This has indeed been done initially by Little et al (1963) and Eastman (1958). There have been a number of improvements to this approach, including those by Bellmore and Malone (1971), which have been adopted by Garfinkel and Nemhauser (1972).

4.5.2 The Vehicle Scheduling Problem

The travelling salesman problem of the previous section can be extended in a number of ways. Suppose that there are now a number of salesmen, all operating from one base, which is one of the cities. All of the other cities must be visited by one salesman who delivers a quantity of goods. Each city has a known demand for the goods and each salesman has a capacity for carrying goods. The problem is to assign each salesman a circuit of cities, starting and ending at the base where total demand on a circuit must not exceed the salesman's capacity. All cities must have their demand met and the total cost of travel is to be minimized.

This problem can be made more realistic by thinking of the "salesmen" as representing vehicles (say delivery vans) and the "cities" as demand points within one city. This problem has a number of important applications, such as school bus scheduling (Foulds et al. 1977a), milk tanker scheduling (Foulds et al. 1977b), municipal waste collection (Beltrami and Bodin 1974), fuel oil delivery (Garvin et al. 1957) and newspaper distribution (Golden et al. 1975). Surveys of literature on the problem have been carried out by Turner, Ghare, and Foulds (1974) and Watson–Gandy and Foulds (1981).

The problem will now be formulated in terms of integer programming. The first formulation is due to Balinski and Quandt (1964). First all feasible circuits which begin and end at the base are identified. This may be an

extremely difficult task for problems with 20 or more demand points. However, the formulation is still useful as a conceptual tool. Let

m = the total number of feasible circuits

$\delta_{ij} = \begin{cases} 1, & \text{if the } j\text{th demand point is on the } i\text{th feasible circuit} \\ 0, & \text{otherwise} \end{cases}$

c_i = the total cost of travelling the ith feasible circuit

$x_i = \begin{cases} 1, & \text{if the feasible circuit } i \text{ is chosen} \\ 0, & \text{otherwise.} \end{cases}$

Then the problem is to

$$\text{Minimize:} \quad \sum_{i=1}^{m} c_i x_i$$

$$\text{subject to:} \quad \sum_{i=1}^{m} \delta_{ij} x_i = 1, \qquad j = 1, 2, \ldots, n$$

$$x_i = 0 \text{ or } 1, \qquad i = 1, 2, \ldots, m.$$

The following formulation, due to Garvin et al. (1957), is more explicit
and is far more amenable to integer programming techniques. Let

p_k = the demand at point k

C = the capacity of each vehicle (assumed to be identical for
all vehicles)

d_{ij} = the cost of travelling from point i to point j

y_{ijk} = the quantity shipped from point i to point j which is destined
for point k

$x_{ij} = \begin{cases} 1, & \text{if a vehicle travels directly from point } i \text{ to point } j \\ 0, & \text{otherwise.} \end{cases}$

The base shall be denoted by the subscript 0.

Consider two distinct demand points, j and k. Then y_{ijk} denotes the
quantity arriving at point j from point i which is destined for point k. Thus

$$\sum_i y_{ijk}$$

denotes the total quantity arriving at point j destined for point k. Also,
y_{jrk} denotes the quantity leaving point j for point r which is destined for
point k. Thus

$$\sum_r y_{jrk}$$

denotes the total quantity leaving point j for point r which is destined for
point k. Now because all goods arriving at point j, destined for point k,
should leave point j, we have:

$$\sum_i y_{ijk} = \sum_r y_{jrk}, \quad \text{for all points } j, k, j \neq k. \tag{4.38}$$

Also, y_{ikk} denotes the quantity arriving at point k from point i which is destined for point k. Thus

$$\sum_i y_{ikk}$$

denotes the total quantity arriving at point k which is destined for point k. Now because this total quantity must equal the demand of point k, we have:

$$\sum_i y_{ikk} = p_k, \quad \text{for all points } k. \tag{4.39}$$

Also, y_{0jk} denotes the quantity leaving the base for point j which is destined for point k. Thus

$$\sum_j y_{0jk}$$

denotes the total quantity leaving the base destined for point k and

$$\sum_j \sum_k y_{0jk}$$

denotes the total quantity leaving the base. Also

$$\sum_k q_k$$

denotes the total demand. Now, as the total quantity leaving the base must equal the total demand, we have:

$$\sum_j \sum_k y_{0jk} = \sum_k q_k. \tag{4.40}$$

It is usually assumed in formulating vehicle scheduling models that only one vehicle will visit each point. The problem of having points with demand greater than vehicle capacity can be overcome by distributing the demand of such a point between a number of artificial points all at the same location, one vehicle visiting each. The assumption implies that only one vehicle will leave each point. Thus we have:

$$\sum_i x_{ij} = \sum_r x_{jr} = 1, \quad \text{for all points } j. \tag{4.41}$$

Also

$$\sum_k y_{ijk}$$

denotes the total quantity carried by the vehicle (if any) which leaves point i for point j. This quantity cannot exceed vehicle capacity, and if no vehicle travels on this segment, the quantity is zero. Thus we have:

$$\sum_k y_{ijk} \leq x_{ij}C, \quad \text{for all points } i, j, i \neq j. \tag{4.42}$$

Of course it is implicit that

$$y_{ijk} \geq 0$$

and

$$x_{ij} = 0 \text{ or } 1, \quad \text{for all points } i, j, k. \tag{4.43}$$

Then the objective is to

$$\text{Minimize:} \quad \sum_i \sum_j d_{ij} x_{ij}, \tag{4.44}$$

which is the total cost of all travel.

Thus it can be seen that the problem of minimizing (4.44) subject to (4.38)–(4.43) is a mixed 0–1 programming problem. It would be difficult to solve such problems when there are more than about 10 points, as the number of constraints would be prohibitive. What is usually done is to solve realistically sized vehicle scheduling problems by heuristic techniques, such as those of Clarke and Wright (1964), or Foster and Ryan (1976). A heuristic technique is a solution procedure represented by a series of rules which, although not guaranteed to find the optimum, usually produce relatively good solutions. Techniques guaranteed to produce the optimal solution, such as branch and bound enumeration, can at present be used only on small problems because of the amounts of computer time and storage they require. Hence most people studying the vehicle scheduling problem prefer to concentrate on heuristic techniques. The heuristic of Foster and Ryan mentioned above does actually use an integer programming formulation.

4.5.3 Political Redistricting

Consider the problem of finding a just method of assigning the census tracts of a region to a number of electorates (voting districts) for the purposes of voting. The assignment must satisfy a number of criteria, including approximate population equality between electorates and connectedness and compactness of electorates. Each tract is indivisible in the sense that all of it must be included in exactly one electorate. The number of electorates created must be equal to the given number of members of parliament (congressmen) for the region. Each electorate should be connected in the sense that it is possible to travel between any two points of the electorate without leaving the electorate. Each electorate should be relatively compact in the sense that its physical shape should be somewhat circular or square rather than long and thin.

Some of the above criteria will now be expressed in mathematical form. Let

$m = $ the number of tracts in the region,

$n = $ the number of electorates to be created,

$x_{ij} = \begin{cases} 1, & \text{if tract } i \text{ is assigned to electorate } j, \\ 0, & \text{otherwise,} \end{cases}$

$p_i = $ the population of tract i.

Let

$$\bar{p} = \frac{1}{n} \sum_{i=1}^{m} p_i$$

be the mean electorate population. In any true democratic system each electorate should have a population $\bar{p}$ to ensure voting equality. However, this is usually impossible because of the indivisibility of each tract.

The population of electorate j is

$$\sum_{i=1}^{m} p_i x_{ij}.$$

Let its deviation from $\bar{p}$ be defined as

$$d_j = \left| \sum_{i=1}^{m} p_i x_{ij} - \bar{p} \right|.$$

Then one might attempt to make the maximum deviation over all electorates as small as possible:

$$\text{Minimize:} \qquad \underset{j=1,2,\ldots,n}{\text{Max}} \; d_j. \qquad (4.45)$$

Each tract i must belong to precisely one electorate:

$$\sum_{j=1}^{n} x_{ij} = 1, \qquad i = 1, 2, \ldots, m. \qquad (4.46)$$

Also, there must be exactly n electorates created. That is, each electorate must have at least one tract assigned to it:

$$\sum_{i=1}^{m} x_{ij} \geq 1, \qquad j = 1, 2, \ldots, n. \qquad (4.47)$$

As

$$x_{ij} = 0 \text{ or } 1, \qquad \begin{aligned} &i = 1, 2, \ldots, m \\ &j = 1, 2, \ldots, n, \end{aligned} \qquad (4.48)$$

(4.45)–(4.48) would be a zero–one I.P. except for the form of (4.45). However, all is not lost, as one can convert the problem into a standard zero–one I.P. as follows. Let

$$v = \underset{j=1,2,\ldots,n}{\text{Max}} \; d_j,$$

i.e.,

$$d_j \leq v, \qquad j = 1, 2, \ldots, n.$$

Hence

$$\left| \sum_{i=1}^{m} p_i x_{ij} - \bar{p} \right| \leq v, \qquad j = 1, 2, \ldots, n,$$

and therefore

$$\left.\begin{array}{c} \sum_{i=1}^{m} p_i x_{ij} - \bar{p} \leq v \\[2ex] \sum_{i=1}^{m} p_i x_{ij} - \bar{p} \geq -v \end{array}\right\} \quad j = 1, 2, \ldots, n. \qquad (4.49)$$

Now the problem becomes

$$\begin{array}{ll} \text{Minimize:} & v \\ \text{subject to:} & (4.46)\text{--}(4.49), \end{array}$$

which is a straightforward I.P.

One can introduce the concept of tract area and develop further constraints concerning the connectedness and compactness of the electorates. This has been done by Smith, Foulds, and Read (1976) and others, including Garfinkel and Nemhauser (1970); Hess et al. (1965); and Wagner (1968), who used integer programming to solve his model.

4.5.4 The Fixed Charge Problem

Consider the problem of a factory which must produce at least M units of a certain commodity and there are n machines available. Let

p_i = the unit cost of producing one article on machine i, $i = 1, 2, \ldots, n$

F_i = the positive fixed cost of setting up machine i for production $i = 1, 2, \ldots, n$

x_i = the number of units produced on machine i, $i = 1, 2, \ldots, n$.

Then the production cost for producing x_i units on machine i is

$$C_i(x_i) = \begin{cases} p_i x_i + F_i, & \text{if } x_i > 0 \\ 0, & \text{otherwise,} \end{cases}$$

where we have assumed that production costs for each article are additive. The problem is to minimize the total production cost:

$$\text{Minimize:} \quad \sum_{i=1}^{n} C_i(x_i). \qquad (4.50)$$

At least M units must be produced, hence

$$\sum_{i-1}^{n} x_i \geq M. \qquad (4.51)$$

Also,

$$x_i \text{ is a nonnegative integer, } i = 1, 2, \ldots, n. \qquad (4.52)$$

Now (4.50), (4.51), (4.52) would be an I.P. apart from the nonlinearity of (4.50). However, this nonlinearity can be overcome by defining

$$y_i = \begin{cases} 1, & \text{if machine } i \text{ is set up} \\ 0, & \text{otherwise.} \end{cases}$$

Also, let

$u_i =$ the maximum possible number of units that machine i could possibly produce.

Then

$$C_i(x_i) = p_i x_i + F_i y_i, \qquad i = 1, 2, \ldots, n.$$

So (4.50) becomes

$$\text{Minimize:} \qquad \sum_{i=1}^{n} p_i x_i + \sum_{i=1}^{n} F_i y_i. \tag{4.53}$$

Some extra constraints need to be added:

$$x_i \leq u_i y_i, \qquad i = 1, 2, \ldots, n. \tag{4.54}$$

(4.54) ensures that

$$x_i > 0 \Rightarrow y_i = 1$$

and

$$x_i = 0 \Rightarrow y_i = 0,$$

the latter implication arising from the facts that (4.53) has the objective of minimization and all $F_i > 0$. So, with the proviso

$$y_i = 0 \text{ or } 1, \qquad i = 1, 2, \ldots, n, \tag{4.55}$$

the problem (4.51)–(4.55) is a mixed integer programming problem.

4.5.5 Capital Budgeting

Consider a company which has the opportunity to initiate a number of projects. Let

$n =$ the number of projects available

$m =$ the number of time periods, during which funds will have to be injected into the projects

$p_i =$ the ultimate profit of project i

$f_{ij} =$ the level of funds that needs to be allocated to project i in time period j

$c_j =$ the total capital available for distribution in time period j

$$x_i = \begin{cases} 1, & \text{if project } i \text{ is selected,} \\ 0, & \text{otherwise.} \end{cases}$$

Then the objective is to maximize ultimate profit, i.e.,

$$\text{Maximize:} \quad \sum_{i=1}^{n} p_i x_i \tag{4.56}$$

subject to the fact that the total capital available in each period j:

$$\sum_{i=1}^{n} f_{ij} x_i$$

cannot exceed the amount available, i.e.,

$$\sum_{i=1}^{n} f_{ij} x_i \le c_j, \qquad j = 1, 2, \ldots, m. \tag{4.57}$$

Also

$$x_i = 0 \text{ or } 1, \qquad i = 1, 2, \ldots, n. \tag{4.58}$$

Problem (4.56)–(4.58) is a standard zero–one I.P. If

$$m = 1,$$

the variables can be redefined as follows:

$$f_{i1} = f_i, \qquad i = 1, 2, \ldots, n$$
$$c_1 = c.$$

Consider now the problem of deciding which items to take on a hiking trip. Let

$n = $ the number of different types of possessions to be taken

$p_i = $ the value assigned to an item of type i

$f_i = $ the weight of an item of type i

$c = $ the total weight that can be carried

$x_i = $ the number of items of type i to be taken.

Then let us assume the objective of maximizing the total value of all possessions taken, i.e.,

$$\text{Maximize:} \quad \sum_{i=1}^{n} p_i x_i. \tag{4.59}$$

The total weight of all items:

$$\sum_{i=1}^{n} f_i x_i$$

cannot exceed the total allowable weight, i.e.,

$$\sum_{i=1}^{n} f_i x_i \le c. \tag{4.60}$$

Of course, only integer quantities of each item can be taken along:

$$x_i = \text{a nonnegative integer}, \qquad i = 1, 2, \ldots, n. \qquad (4.61)$$

Problem (4.59), (4.60), (4.61) is called the *knapsack* problem for obvious reasons, and will be further examined in Chapter 6.

4.6 Exercises

(I) Computational

1. Solve the following integer programming problems by Dakin's method.

(a) Maximize: $3x_1 + 5x_2 + 4x_3$
     subject to: $2x_1 + 6x_2 + 3x_3 \le 8$
     $5x_1 + 4x_2 + 4x_3 \le 7$
     $6x_1 + x_2 + x_3 \le 12$
     x_1, x_2, x_3 nonnegative integers.

(b) Maximize: $4x_1 + 3x_2 + 3x_3$
     subject to: $4x_1 + 2x_2 + x_3 \le 10$
     $3x_1 + 4x_2 + 2x_3 \le 14$
     $2x_1 + x_2 + 3x_3 \le 7$
     x_1, x_2, x_3 nonnegative integers.

(c) Maximize: $2x_1 + 4x_2 + 5x_3$
     subject to: $x_1 + x_2 + 2x_3 \le 9$
     $2x_1 + x_2 + 3x_3 \le 13$
     $3x_1 + 2x_2 + x_3 \le 11$
     x_1, x_2, x_3 nonnegative integers.

(d) Maximize: $5x_1 + 4x_2 + 3x_3$
     subject to: $3x_1 + 4x_2 + x_3 \le 12$
     $4x_1 + 2x_2 + x_3 \le 9$
     $2x_1 + 3x_2 + 2x_3 \le 15$
     x_1, x_2, x_3 nonnegative integers.

(e) Maximize: $4x_1 + 6x_2 + x_3$
     subject to: $2x_1 + x_2 + 2x_3 \le 16$
     $x_1 + 2x_2 + x_3 \le 10$
     $3x_1 + x_2 + x_3 \le 13$
     x_1, x_2, x_3 nonnegative integers.

(f) A food factory produces three types of fruit salad: A, B, C. Each type requires a different amount of three varieties of fruits: peaches, pears, and apples as summarized in Table 4.8. No more than 5, 4, and 6 pounds of pears, peaches, and apples can be used in producing a can. How many of each type of can should be produced in order to maximize profits?

Table 4.8. Data for Exercise 1(f).

Type	Weight in pounds:			
	Pears	Peaches	Apples	Profit per can
A	2	3	4	6
B	2	2	4	5
C	3	3	2	4

(g) Maximize: $x_1 + 2x_2 + x_3$

subject to: $2x_1 + x_2 + 3x_3 \leq 12$

$x_1 + 4x_2 + 2x_3 \leq 10$

$x_1 + 3x_2 + x_3 \leq 14$

x_1, x_2, x_3 nonnegative integers.

(h) Maximize: $x_1 + 3x_2 + 2x_3$

subject to: $x_1 + 2x_2 + 2x_3 \leq 9$

$2x_1 + x_2 + x_3 \leq 18$

$2x_1 + 2x_2 + x_3 \leq 20$

x_1, x_2, x_3 nonnegative integers.

(i) Maximize: $x_1 + 3x_2 + 2x_3$

subject to: $2x_1 + 4x_2 + x_3 \leq 7$

$3x_1 + 2x_2 + 2x_3 \leq 5$

$x_1 + x_2 + 3x_3 \leq 6$

x_1, x_2, x_3 nonnegative integers.

(j) Maximize: $x_1 + 2x_2 + 3x_3$

subject to: $3x_1 + 2x_2 + x_3 \leq 5$

$4x_1 + 3x_3 \leq 7$

$2x_1 + 4x_2 + x_3 \leq 4$

x_1, x_2, x_3 nonnegative integers.

(k) Maximize: $3x_1 + 4x_2 + x_3$

subject to: $x_1 + x_2 + x_3 \leq 8$

$x_1 + 3x_2 + 4x_3 \leq 15$

$x_2 + 2x_3 \leq 12$

x_1, x_2, x_3 nonnegative integers.

(l) Maximize: $3x_1 + 4x_2 + x_3$

    subject to: $x_1 + 2x_2 - 2x_3 \leq 9$

                    $2x_1 - x_2 + 4x_3 \leq 15$

                    $3x_1 + 3x_2 - x_3 \leq 0$

        x_1, x_2, x_3 nonnegative integers.

(m) Maximize: $2x_1 + 2x_2 + 4x_3$

    subject to: $x_1 + x_2 + x_3 \leq 9$

                    $3x_1 + 4x_2 + 2x_3 \leq 10$

                  $-2x_1 + 4x_2 + 4x_3 \leq 8$

        x_1, x_2, x_3 nonnegative integers.

(n) Maximize: $3x_1 + 5x_2 + 2x_3$

    subject to: $2x_1 + x_2 + 5x_3 \leq 12$

                    $x_1 + 3x_2 + x_3 \leq 8$

                  $5x_1 + 2x_2 + 3x_3 \leq 9$

        x_1, x_2, x_3 nonnegative integers.

(o) Maximize: $5x_1 + 7x_2 + 4x_3$

    subject to: $x_1 + x_2 - x_3 \leq 0$

                    $2x_1 + x_2 + 4x_3 \leq 32$

                  $6x_1 + 9x_2 \leq 50$

        x_1, x_2, x_3 nonnegative integers.

(p) Maximize: $2x_1 + 4x_2 + 5x_3$

    subject to: $x_1 + x_2 + 2x_3 \leq 9$

                    $2x_1 + x_2 + 3x_3 \leq 13$

                  $3x_1 + 2x_2 + x_3 \leq 11$

        x_1, x_2, x_3 nonnegative integers.

(q) A surfboard manufacturer wants to know how many of each type of surfboard he should make per week in order to maximize profits. He makes three types of board: the knee board (K), the beacher (B), and the cruiser (C), which are 4, 6, and 8 feet long, respectively, but he can blow only 50 feet of foam per week. The profits are \$40, \$60, and \$30 for K, B, and C, and they require 10, 15, and 25 feet of fibreglass cloth respectively. He has 140 feet of cloth available per week, and 70 pounds of resin per week. K, B, and C need 6, 10, and 14 pounds of resin each, respectively.

(r) Maximize: $x_1 + 2x_2 + 3x_3$

    subject to: $x_2 + 2x_3 \leq 6$

                    $x_1 + x_2 + x_3 \leq 5$

                  $3x_1 + 2x_2 \leq 4$

        x_1, x_2, x_3 nonnegative integers.

(s) Maximize: $16x_1 + 10x_2 + 12x_3$

   subject to: $2x_1 + 3x_2 + 4x_3 \le 10$

   $4x_1 + 3x_2 + 2x_3 \le 12$

   $x_1 + 2x_2 + 3x_3 \le 6$

   x_1, x_2, x_3 nonnegative integers.

(t) A hobbyist making cane baskets (B), trays (T), and plant holders (P), makes a profit of \$10 on each item, and incorporates three colours: white (W), red (R), and yellow (Y). He has a maximum of 6, 9, and 10 yards of W, R, and Y cane per week respectively. B, T, and P require 2, 1, 1; 1, 3, 1; and 1, 2, 2 of W, R, and Y cane, respectively. How many items of each type of product should he make per week in order to maximize profit?

(u) Maximize: $2x_1 + 3x_2 + x_3$

   subject to: $x_1 + 2x_2 + x_3 \le 17$

   $3x_1 + x_2 \qquad \le 15$

   $x_2 + 4x_3 \le 12$

   x_1, x_2, x_3 nonnegative integers.

(v) Maximize: $x_1 + x_2 + 2x_3$

   subject to: $\frac{1}{3}x_1 + \frac{1}{5}x_2 + \frac{1}{4}x_3 \le \frac{18}{5}$

   $\frac{1}{2}x_1 + \frac{1}{4}x_2 + \frac{1}{3}x_3 \le \frac{15}{7}$

   $\frac{1}{8}x_1 + \frac{2}{3}x_2 + \frac{1}{7}x_3 \le \frac{19}{6}$

   x_1, x_2, x_3 nonnegative integers.

(w) Maximize: $2x_1 + 4x_2 + 6x_3$

   subject to: $2x_1 + x_2 + x_3 \le 3$

   $x_1 \qquad - 2x_3 \le 6$

   $4x_2 + 6x_3 \le 10$

   x_1, x_2, x_3 nonnegative integers.

(x) Maximize: $6x_1 + 5x_2 + 4x_3$

   subject to: $5x_1 + 4x_2 + 2x_3 \le 40$

   $3x_1 + 3x_2 + 4x_3 \le 30$

   $2x_1 + 3x_2 + 3x_3 \le 20$

   x_1, x_2, x_3 nonnegative integers.

(y) A jeweller makes three types of silver rings. Ring A takes 3 hours, 20 g of silver and 1 hour of polishing. These quantities are 3, 10, and 3; 1, 20, and 1 for rings B and C, respectively. The polishing machine is available to him for 2 hours per day and he can work for another 11 hours per day and can afford to buy 60 g of silver per day. Profits are \$30, \$20, and \$10 for A, B, and C rings, respectively. How many of each type of ring should he make per day in order to maximize profit?

(z) Maximize: $10x_1 + 12x_2 + 16x_3$

   subject to: $2x_1 + 3x_2 + 4x_3 \leq 20$

                   $3x_1 + 3x_2 + 4x_3 \leq 30$

                   $4x_1 + 3x_2 + 2x_3 \leq 25$

           x_1, x_2, x_3 nonnegative integers.

2. Assume for each problem in Exercise 1 that

$$x_1, x_2, x_3 = 0 \text{ or } 1.$$

   Find the new optimal solution for each problem using the method of Balas.

3. By converting to zero–one variables solve each problem in Exercise 1 by the method of Balas.

4. Solve each problem in Exercise 1 by the Gomory cutting plane method.

5. Solve each problem in Exercise 1, assuming that only x_2 must be integral, by the Gomory mixed integer method.

(II) Theoretical

6. Formulate the N-city travelling salesman problem as an I.P. in a way that requires fewer constraints than the formulation given in Section 4.4.1.

7. Construct a branch and bound algorithm for the travelling salesman problem along the lines of the approach suggested at the end of Section 4.4.1.

8. List at least three realistic applications for the vehicle scheduling problem not listed in Section 4.4.2.

9. Construct a branch and bound algorithm for the vehicle scheduling problem.

10. Construct a branch and bound algorithm for the fixed charge problem.

11. Construct a branch and bound algorithm for the knapsack problem.

12. Construct a branch and bound algorithm for the assignment problem of Chapter 2.

13. Solve each of the problems of Exercise 2 by exhaustive enumeration. Compare the amount of computation involved with that required by the method of Balas.

Chapter 5

Network Analysis

5.1 The Importance of Network Models

Many important decision-making problems can be described in terms of networks. Some obvious examples are concerned with traffic and the shipment of goods. However, there are many other examples with less obvious links with network modelling such as production planning, capital budgeting, machine replacement, and project scheduling.

One of the basic network optimization problems is concerned with finding the shortest path between two given points in a network, the *shortest path problem*. A second problem arises in connection with finding a subset of links of the network which has the property that there is a path between every pair of points in the network and the total length of the links in the subset is minimal. This problem is called the *minimal spanning tree problem*. A third problem is connected with maximizing the flow of some commodity through the links of a network from a given origin to a given destination where each link has a capacity of flow. This is the *maximal flow problem*. A fourth problem is related to minimizing the cost of transporting a given quantity of a commodity from a given origin to a given destination: the *minimum cost flow problem*. A fifth problem, *critical path scheduling*, is concerned with scheduling the activities of a project.

Because these and other basic network problems can be modelled as L.P. problems requiring integer solutions, network analysis has strong links with integer programming. In the next section the basic mathematical notions necessary to study networks are introduced. The underlying mathematical subject is called graph theory.

5.2 An Introduction to Graph Theory

What most people normally think of as a network (as in road, communication, or telephone networks) is a special example of a mathematical entity called a *graph*. In order to analyze network problems efficiently it is necessary to master some graph theoretic concepts, which are presented in this section. The discussion here is chiefly for reference that is, the reader should proceed directly to section 5.3, returning to this section for clarification of terminology when needed. The interested reader who wishes a more detailed exposition of graph theory is directed to any of a number of excellent texts on graph theory, including Busacker and Saaty (1965), Deo (1974), and Harary (1969).

We begin by defining the term *graph*, and we use the terminology of the Harary.

A *graph* $G = (P, L)$ is an ordered pair where P is a nonempty set of *points* (sometimes called vertices, nodes, or junctions) and L is a set of unordered pairs of distinct points of P, called *lines* (sometimes also called *links*, *edges*, or *branches*). Although a graph is an abstract mathematical concept, it is usual to represent a graph by a picture. For instance, the graph $G = (P, L)$ where

$$P = \{p_1, p_2, p_3, p_4\}$$
$$L = \{\{p_1, p_2,\}, \{p_2, p_3\}, \{p_3, p_4,\}, \{p_1, p_4,\}, \{p_3, p_1,\}\}$$

is represented in Figure 5.1. It is important to realise that pictures like Figure 5.1 are only diagrams of graphs, not the graphs themselves, which are defined abstractly by the specification of P and L. A similar relationship holds between Venn diagrams and formally defined sets.

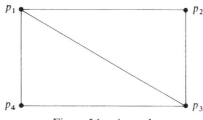

Figure 5.1.   A graph.

Further terminology is now introduced.

A *walk* is an alternating sequence of points and lines of the form:

$$p_0, \{p_0, p_1\}, p_1, \{p_1, p_2\}, p_2, \ldots, \{p_{n-1}, p_n\}, p_n.$$

For example, the sequence

$$p_1, \{p_1, p_4\}, p_4, \{p_4, p_3\}, p_3, \{p_3, p_1\}, p_1, \{p_1, p_4\}, p_4$$

is a walk for the graph in Figure 5.1. A walk is termed *closed* if

$$p_0 = p_n$$

and *open* if

$$p_0 \neq p_n.$$

The sample walk above is open, but if the last two elements—$\{p_1, p_4\}$, p_4—are removed it becomes closed.

A *trail* is a walk in which all the lines are distinct. Hence the walk

$$p_1, \{p_1, p_4\}, p_4, \{p_4, p_3\}, p_3, \{p_3, p_1\}, p_1, \{p_1, p_2\} \, p_2$$

in Figure 5.1 is a trail. A *path* is a trail in which all the points are distinct. Hence the trail

$$p_1, \{p_1, p_4\}, p_4, \{p_4, p_3\}, p_3, \{p_3, p_2\}, p_2$$

in Figure 5.1 is a path. Of course if all the points are distinct in any trail, all its lines are also distinct. A *cycle* is a closed walk of at least three points with all its points distinct except that the first and the last are the same. Hence the walk

$$p_1, \{p_1, p_2\}, p_2, \{p_2, p_3\}, p_3, \{p_3, p_1\}, p_1$$

in Figure 5.1 is a cycle. A graph is said to be *connected* if there exists a path between every pair of points. The graph is Figure 5.1 is certainly connected. However, if lines $\{p_1, p_2\}$ and $\{p_2, p_3\}$ are removed the graph is no longer connected, as there are no paths from p_2 to any of the other points.

A *tree* is a connected graph without any cycles. If the lines $\{p_1, p_2\}$ and $\{p_3, p_4\}$ are removed from the graph in Figure 5.1 it becomes a tree. The concept of a tree is one of the most important in graph theory. We can make some interesting observations about trees. If a graph $G = (P, L)$ is a tree then

1. Every two distinct points of G are joined by exactly one path.
2. The number of lines in L is one less than the number of points in P.
3. If a line not present in L is added to G, then exactly one cycle is created.

The reader should construct a number of trees according to the definition and verify that these properties are true for those trees.

A graph G' is said to be a *subgraph* of a graph G if G' has all its points and lines in G and G' is a graph. Hence the graph (P', L') defined by

$$P' = \{\{p_1, p_2\}, \{p_2, p_3\}, \{p_1, p_3\}\}$$

is a subgraph of the graph in Figure 5.1. A subgraph (P', L') is said to *span* a graph (P, L) if

$$P' = P,$$

i.e., all the points of the graph are part of the spanning subgraph. A graph that is a tree and a spanning subgraph of some graph (P, L) is said to be a *spanning* tree (of (P, L)).

In some applications it is desirable to orient each line of a graph with a direction. Graphs with directed lines are called *digraphs* (short for directed graphs). Pictures of digraphs are drawn in the same manner as those of graphs, except that each line has an arrow attached to it to signify its direction. For example, Figure 5.2 depicts a digraph obtained from the graph in Figure 5.1 by orienting its lines.

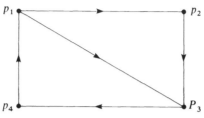

Figure 5.2.   A digraph.

More formally, a *digraph* $D = (P, A)$ is an ordered pair where P is a nonempty set of points and A is a set of ordered pairs of distinct points of P, called *arcs* (sometimes called *directed lines*). The digraph in Figure 5.2 can be expressed formally as follows:

$$P = \{p_1, p_2, p_3, p_4\}$$
$$A = \{(p_1, p_2), (p_2, p_3), (p_1, p_3), (p_3, p_4), (p_4, p_1)\}.$$

Many of the concepts of graphs can be defined in an analogous fashion for digraphs. A *directed walk* is an alternating sequence of points and arcs of the form:

$$p_0, (p_0, p_1), p_1, (p_1, p_2), p_2, \ldots, (p_{n-1}, p_n), p_n.$$

For example, the sequence:

$$p_1, (p_1, p_2), p_2, (p_2, p_3), p_3, (p_3, p_4), p_4, (p_4, p_1), p_1, (p_1, p_3), p_3$$

in Figure 5.2 is a directed walk. A *directed path* is a directed walk in which all the points are distinct. As an example the directed walk

$$p_1, (p_1, p_2), p_2, (p_2, p_3), p_3, (p_3, p_4), p_4$$

is a directed path A *cycle* is a directed walk of at least two points with all its points distinct except that the first and the last are the same. Hence the directed walk:

$$p_1, (p_1, p_2), p_2, (p_2, p_3), p_3, (p_3, p_4), p_4, (p_4, p_1), p_1$$

is a cycle.

If p_i and p_j are points in a digraph D and there is a directed walk from p_i to p_j in D, then p_j is said to be *reachable* from p_i. A *network* is a digraph with at least one point α (called the *source*) such that every point is reachable from it, and another point w (called the *sink*) which is reachable from every other point. It is usual to associate flows of some commodity with the arcs of a network.

5.3 The Shortest Path Problem

Consider a digraph D in which each arc has a given traversal cost, i.e., for each arc (i, j) in D let c_{ij} be the cost of travelling from point i to point j along arc (i, j). This traversal cost may be in terms of distance, time, money, or some other optimality criterion. Graphs or digraphs of this nature are called *weighted*. Figure 5.3 shows a weighted digraph.

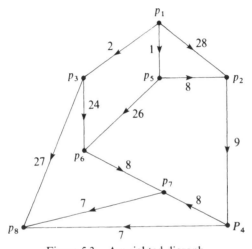

Figure 5.3. A weighted digraph.

The cost of a path in a digraph is defined to be the sum of the costs of the arcs of the path. In the *shortest path problem* one must find the path of least cost which joins one given point to another given point. The cost c_{ij} of a point pair i, j for which there is no arc (i, j) in A is set equal to a prohibitively large number.

5.3.1 Dijkstra's Method

Suppose one is given a weighted digraph and a source–sink pair of points such that it is desired to find the shortest path from the source to the sink. At each iteration Dijkstra's method identifies a new point which is the closest to the source among all those points which are currently not yet identified. The length of the path from the source to this point is calculated and associated with the point. The method builds up a series of shortest paths from the source to successive points until the sink is included in this set, at which stage the problem is solved. All that remains is to find the actual arcs making up this shortest path by a backtracking process. The procedure can be continued until all points have been identified if it is desired to find shortest paths from the origin to all other points.

The method will now be illustrated by finding the shortest path from point p_1 to point p_8 in the digraph in Figure 5.3. We begin by partitioning the set of points into two sets: A containing the origin and B containing all other points. So

$$B = \{p_2, p_3, p_4, p_5, p_6, p_7, p_8\}.$$

In the course of the method some of the points are going to be labeled, a label of $d(i)$ for point p_i representing the shortest distance from the source to point p_i. First the origin is assigned a label $d(1) = 0$. Next the point in B which is closest to the origin is found. We require

$$\text{Min } \{d(i) + c_{ij}\}.$$
$$i \in A$$
$$j \in B.$$

The point, j in B satisfying the minimization is the point required. It has its label $d(j)$ set equal to the minimum quantity found. In the present example this minimum is

(I) $$d(1) + c_{15} = 0 + 1 = 1.$$

Then the point, j in B found is removed from B and placed in A. So in the present example:

$$A = \{p_1, p_5\},$$
$$B = \{p_2, p_3, p_4, p_6, p_7, p_8\}$$

and

$$d(5) = 1.$$

This series of steps is now repeated until the sink (in the present example, P_8) is transferred from B to A. These steps are now carried out

(II) $$\text{Min } \{d(i) + c_{ij}\} = d(1) + c_{13} = 0 + 2 = 2$$
$$i \in A$$
$$j \in B$$
$$A = \{p_1, p_3, p_5\},$$
$$B = \{p_2, p_4, p_6, p_7, p_8\}.$$
$$d(3) = 2;$$

(III) $\text{Min} \{d(i) + c_{ij}\} = d(5) + c_{52} = 1 + 8 = 9$

$i \in A$

$j \in B$

$A = \{p_1, p_2, p_3, p_5\},$

$B = \{p_4, p_6, p_7, p_8\}$

$d(2) = 9;$

(IV) $\text{Min} \{d(i) + c_{ij}\} = d(2) + c_{24} = 9 + 9 = 18$

$i \in A$

$j \in B$

$A = \{p_1, p_2, p_3, p_4, p_5\},$

$B = \{p_6, p_7, p_8\}$

$d(4) = 18;$

(V) $\text{Min} \{d(i) + c_{ij}\} = d(4) + c_{48} = 18 + 7 = 25$

$i \in A$

$j \in B$

$A = \{p_1, p_2, p_3, p_4, p_5, p_8\}$

$B = \{p_6, p_7\}$

$d(8) = 25.$

Now as p_8, the sink, is included in A the repetition of the above series of steps in terminated (If the shortest paths from the origin to all points were required the steps would be repeated until all points were in A.) To find the sequence(s) arcs making up the shortest path(s) from source to sink we must work backwards (backtrack) through the digraph as follows. One forms a list of values of the form

$$d(j) - c_{ij} - d(i), \tag{5.1}$$

where p_j is the sink and p_i are labelled points connected directly to p_j.
Now

$$d(8) - c_{38} - d(3) = 25 - 27 - 2 \neq 0,$$

hence arc (p_3, p_8) is not on the shortest path. However

$$d(8) - c_{48} - d(4) = 25 - 7 - 18 = 0,$$

so arc (p_4, p_8) is.

Next replace p_8 in (5.1) by p_4, the point just found to be on the shortest path. A new list of values of the form of (5.1) is found.

$$d(4) - c_{24} - d(2) = 18 - 9 - 9 = 0$$

so arc (p_2, p_4) is on the shortest path. Also

$$d(2) - c_{52} - d(15) = 9 - 8 - 1 = 0$$

and
$$d(5) - c_{15} - d(1) = 1 - 1 - 0 = 0,$$

so arcs (p_5, p_2) and (p_1, p_5) are also on the shortest path. Unravelling this information we conclude that the shortest path is

$(p_1, p_5), (p_5, p_2), (p_2, p_4), (p_4, p_8)$, with a length of $d(8)$ which is 25.

There are a number of related shortest path problems including that of finding a shortest path between each pair of points. The above procedure (Dijkstra (1959)) could, in theory, be used to solve this problem, however more efficient procedures have been developed; see for example Floyd (1962) and Murchland (1967).

5.4 The Minimal Spanning Tree Problem

The following problem is somewhat similar to the shortest path problem however it is concerned with graphs rather than digraphs. Given a weighted graph one desires to find among all its subgraphs a spanning tree (see section 5.1) of minimum total weight. In other words we wish to find a subset of lines forming a tree which spans the graph and which has a sum of the weights of the individual lines which is a minimum among all such spanning trees. This problem is called the *minimal spanning tree problem*.

There are many applications of the problem. Examples are transportation planning problems where the points represent cities or distribution centres and the lines represent air lanes, railway lines, or roads. In these cases one is trying to design a system in which it is possible for travel between all pairs of centres at minimum total outlay. A less direct application arises in finding a lower bound for the length of a travelling salesman's circuit (see section 4.41). One can represent symmetric T.S.P.'s by weighted graphs. In solving such problems by branch and bound enumeration (see section 4.2) one needs the minimum distance the salesman would be required to travel given that certain lines must be used and others must not. The weight of a minimum spanning tree incorporating such decisions provides this information. Other applications occur in project planning and communications network design.

5.4.1 Kruskal's Algorithm

Given a weighted, connected graph, suppose it is desired to find a spanning tree of minimum total weight. Kruskal (1956) showed that the following algorithm always produces such a tree. One begins by ordering all the lines in the graph in order of nondecreasing weight, i.e., least weight first. Each

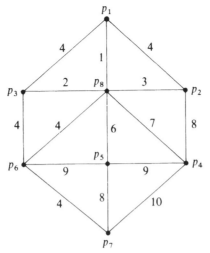

Figure 5.4. A weighted graph.

line is then examined in this order in turn. When a line is examined it is accepted as part of the spanning tree unless it would form a cycle with those lines already accepted, in which case it is rejected and the next line is examined. The examination process is terminated when all the accepted lines form a spanning tree. This tree constitutes a minimal spanning tree.

As an example, consider the weighted graph in Figure 5.4. The minimal spanning tree is found for this graph using Kruskal's algorithm as follows. The order of lines is:

$$\{p_1,p_8\}, \{p_3,p_8\}, \{p_2,p_8\}, \{p_1,p_3\}, \{p_1,p_2\}, \{p_6,p_8\}, \{p_6,p_7\}, \{p_3,p_6\},$$
$$\{p_5,p_8\}, \{p_4,p_8\}, \{p_2,p_4\}, \{p_4,p_5\}, \{p_5,p_6\}, \{p_4,p_7\}, \{p_5,p_7\}.$$

The lines with weight 4—$\{p_1,p_3\}$, $\{p_1,p_2\}$, $\{p_3,p_6\}$, $\{p_6,p_8\}$, and $\{p_6,p_7\}$— have been assembled in arbitrary order. We now start building the tree by examining each line in this order. Lines $\{p_1,p_8\}$, $\{p_3,p_8\}$, and $\{p_2,p_8\}$ are all accepted. Next, $\{p_1,p_3\}$ is rejected, as it would create a cycle with $\{p_1,p_8\}$ and $\{p_3,p_8\}$. Then $\{p_1,p_2\}$ is rejected as it would create a cycle with $\{p_1,p_8\}$ and $\{p_2,p_8\}$. Moving on, $\{p_6,p_8\}$ is accepted and $\{p_3,p_6\}$ is rejected. Then $\{p_5,p_8\}$ and $\{p_4,p_8\}$ are accepted. At this point a spanning tree has been created and the examination process stops. The minimal spanning tree, with weight

$$1 + 2 + 3 + 4 + 4 + 6 + 7 = 27$$

is shown in Figure 5.5.

It is true that a graph with n points will result in a spanning tree with $n - 1$ lines. Thus, once $n - 1$ lines, creating no circuits, been accepted, the minimal spanning tree algorithm has constructed a spanning tree.

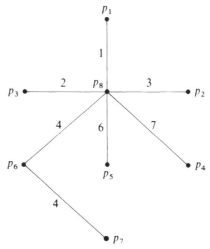

Figure 5.5. A minimal spanning tree found by Kruskal's algorithm.

5.4.2 Prim's Algorithm

We now consider a second algorithm, due to Prim (1957), which also guarantees to find a minimal spanning tree in any connected, weighted graph. Despite refinements to increase efficiency in Kruskal's algorithm, Prim's approach is superior for all but very sparse (few lines) graphs.

Prim's algorithm does not require that the lines of the graph are ordered in advance. It builds up a single connected component (which is actually a tree) until this component spans the original graph. This component then represents a minimal spanning tree. One begins by selecting the line of least weight say $\{p_i, p_j\}$. This line and its two incident points forms the initial component. One then finds the line of minimum weight among all those that connect a point in the component to a point that is not. This line and its noncomponent point then become part of the component.

Both Kruskal's and Prim's algorithms involve making the best (in this case least weight) decision at each stage with little regard to previous decisions. Combinational optimization procedures with this philosophy are termed *greedy*. Greedy procedures seldom guarantee optimal solutions as they do in the two algorithms for the minimal spanning tree problem. However, a greedy procedure is often used to find relatively good (near optimal) solutions with little computational effort for many combinatorial optimization problems.

In order to illustrate Prim's algorithm, let us apply it to the graph in Figure 5.4. The least-weight line is $\{p_1, p_8\}$ with weight 1. So the initial component is $[p_1, p_8; \{p_1, p_8\}]$. We now look for points which are directly connected to the component. There are two: p_2 and p_3. The lines $\{p_8, p_3\}$, $\{p_8, p_2\}$, $\{p_1, p_2\}$, and $\{p_1, p_3\}$ connect them to the component, $\{p_8, p_3\}$ being the smallest. Hence this and p_3 are added to the component, which becomes $[p_1, p_8, p_3; \{p_1, p_8\}, \{p_8, p_3\}]$. Now p_2 and p_6 are directly connected to the component. The least-weight line is $\{p_8, p_2\}$ which is added, along with p_2, to the component. Now p_4, p_5, and p_6 are directly connected to the component. However, there is a tie among the weights of the connecting lines: $\{p_3, p_6\}$ and $\{p_8, p_6\}$ are both of weight 4. Let us arbitrarily choose $\{p_3, p_6\}$, which is added to the component, along with p_6. Next $\{p_6, p_7\}$ and p_7 are added to the component, then $\{p_8, p_5\}$ and p_5, and finally $\{p_8, p_4\}$ and p_4. The component is now:

$$[p_1, p_8, p_3, p_2, p_6, p_7, p_5, p_4; \{p_1, p_8\}, \{p_8, p_3\}, \{p_8, p_2\}, \{p_3, p_6\},$$
$$\{p_6, p_7\}, \{p_8, p_5\}, \{p_8, p_4\}].$$

The component now contains all the points of the graph and hence represents a minimal spanning tree. The minimal spanning tree is given by the lines present in the component. This tree is shown in Figure 5.6.

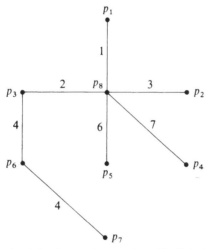

Figure 5.6. A minimal spanning tree found by Prim's algorithm.

Although this tree has a different set of lines from that found by Kruskal's algorithm, it has the same weight (27). Differences between the two trees are due only to the way in which ties between line weights were settled. Indeed, both algorithms are capable of producing both trees.

5.5 Flow Networks

Recall that in Section 5.2 a network was defined as a digraph with a source and a sink. Many flow network problems are concerned with optimizing some parameter of a network system where there is a flow of material or

goods from its source to its sink. A network of pipes carrying crude petro-
leum from an oil field to a port is an example. It is assumed that there is no
loss of the commodity being transported at the intermediate points. This
assumption is called *conservation of flow*. In effect it means that, for points
other than the source and sink, the total flow travelling into each point is
equal to the total flow travelling out of it. Associated with each arc is a
capacity, which represents the maximum amount of flow that the arc can
accommodate. Many flow networks are such that each of their arcs has a *unit
transportation cost* representing the cost of shipping one unit of the commodity along the arc.

Until now we have implicitly assumed that a network has exactly one
source and exactly one sink, and the algorithms to be presented in the next
two sections are designed for networks of this nature only. Any network
with multiple sources and sinks can easily be converted into one with a
single source and a single sink using the following artificial device. If more
than one source is present, a *supersource* s_0 is created and represented by a
new point. This new point is connected to each source s_i by an arc (s_0, s_i).

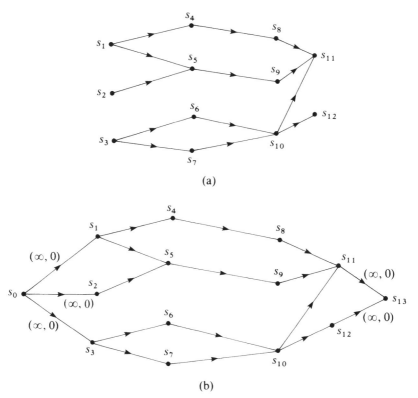

Figure 5.7. The conversion from multiple sources and sinks (a) to a unique source
and sink (b).

For networks with multiple sinks a supersink s_{n+1} is created. Each sink s_j is connected to s_{n+1} by an arc (s_j, s_{n+1}). Arcs of the form (s_0, s_i) and (s_j, s_{n+1}) in multisource–multisink networks are assigned zero unit transportation costs and infinite capacity. An example of the conversion from multisource to multisink network to a single source–single sink network is given in Figure 5.7. The capacity and unit transportation costs are given as an ordered pair for each new arc. In Figure 5.7(a) the sources are s_1, s_2, and s_3, and the sinks are s_{11} and s_{12}. In Figure 5.7(b) the supersource is s_0 and the supersink is s_{13}. Once the conversion has been made and a solution to the problem has been found, arcs from s_0 and to s_{n+1} are ignored.

Two network flow problems and solution procedures for them are presented in the next few sections. In the *maximal* flow problem one must maximize the total rate of flow from source to sink neglecting unit transportation costs. In the *minimal cost flow problem* one must minimize the cost of shipping a given quantity of a commodity from source to sink.

5.5.1 The Maximal Flow Problem

Consider a network with arc capacities but no unit transportation costs. The maximal flow problem is concerned with finding an assignment of flow to each arc so that the total flow from source to sink is maximized. The problem can be formulated in mathematical terms. Let

$n = $ the number of points in the network

$c_{ij} = $ the capacity of arc (p_i, p_j)

$f_{ij} = $ the flow assigned to arc (p_i, p_j)

$p_1 = $ the source

$p_n = $ the sink.

Given a set of flow assignments f_{ij}, the flow out of point i is

$$\sum_{\substack{\text{all arcs} \\ (p_i, p_j)}} f_{ij}.$$

The flow into a point i is

$$\sum_{\substack{\text{all arcs} \\ (p_j, p_i)}} f_{ji}.$$

Hence the assumption of the conservation of flow implies:

$$\sum_{\substack{\text{all arcs} \\ (p_i, p_j)}} f_{ij} - \sum_{\substack{\text{all arcs} \\ (p_j, p_i)}} f_{ji} = 0, \qquad i \neq 1, i \neq n. \tag{5.2}$$

Note that the restriction on i in (5.2) is important. Conservation of flow does not hold for the source or sink. Let F denote the total amount of flow travelling through the network. This amount of flow F must leave the source and arrive at the sink. Thus

$$\sum_{\substack{\text{all arcs} \\ (p_1, p_j)}} f_{1j} = \sum_{\substack{\text{all arcs} \\ (p_i, p_n)}} f_{in} = F.$$

In the maximal flow problem one must

Maximize: F

subject to: $\sum_{\substack{\text{all arcs} \\ (p_i, p_j)}} f_{ij} - \sum_{\substack{\text{all arcs} \\ (p_j, p_i)}} f_{ji} = \begin{cases} F, & \text{if } i = 1 \\ 0, & \text{if } i \neq 1, i \neq n \\ -F, & \text{if } i = n \end{cases}$

$$0 \leq f_{ij} \leq c_{ij}, \quad \text{for all arcs } (p_i, p_j).$$

We turn now to developing methods for solving the maximal flow problem. Consider the network in Figure 5.8, where arc capacities are shown. If the arcs (p_4, p_6), (p_3, p_6), (p_5, p_6) were removed from the network it would be disconnected, in the sense that there would no longer be any paths from source, p_1 to sink, p_6. A set of arcs with at least one element in every source-sink path is called a *cut*. Thus the removal of the arcs in any cut disconnects every source–sink path. Hence the set $\mathbf{C} = \{(p_4, p_6), (p_3, p_6), (p_5, p_6)\}$ is a cut. The *capacity of a cut* is defined to be the sum of the capacities of the in-dividual arcs in the cut. Thus the capacity of $\mathbf{C}$ is

$$c_{46} + c_{36} + c_{56} = 1 + 2 + 2 = 5.$$

The cut with the smallest capacity is called the *minimum cut*. The reader should verify that $\mathbf{C}$ is the minimum cut for the network in Figure 5.8.

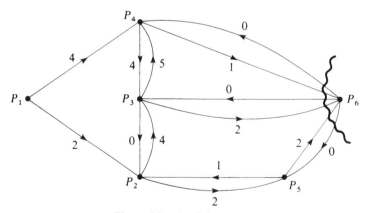

Figure 5.8.   A minimum cut.

The following important theorem is very useful in the designing of an algorithm to solve the maximal flow problem:

Theorem 5.1 (The maximum-flow, minimum-cut theorem). *In any network the value of the maximum flow from source to sink equals the capacity of the minimum cut.*

This theorem was proved by Ford and Fulkerson (1962) in an excellent text which made a substantial contribution to the theory of network flows. The book deals with the maximal flow problem and many of the other topics of this chapter.

When confronted with a maximal flow problem one can begin by identifying the minimum cut. The network can then be gradually loaded with flow that satisfies the assumption of conservation of flow. When the flow from source to sink has total volume equal to the capacity of the minimum cut, we know because of Theorem 5.1 that no further addition of flow is possible. Thus the loading can be stopped and the present assignment is optimal. The strategy by which the network is loaded is called the *labelling method*, and is due to Ford and Fulkerson (1962). This method is explained in the next section.

5.5.2 The Labelling Method

Consider the network in Figure 5.8. Flows in opposite directions in a single arc have their magnitudes subtracted to produce a single flow in the direction of the larger flow. For instance, if arc $(3, 4)$ had a flow of 5 from 3 to 4 and a flow of 4 from 4 to 3, the net result would be a flow of 1 from 3 to 4. In order to be able to change flows already assigned we allow the possibility of a *notional* flow in an arc in a direction in which it cannot receive further flow. This is brought about by the concept of excess capacity. The *excess capacity* e_{ij} of an arc (i, j) for a given assigned flow f_{ij} is initially defined as

$$e_{ij} = c_{ij} - f_{ij}, \tag{5.3}$$

i.e., the amount of extra flow that an arc could accommodate, over and above what it is now assigned. Suppose an arc (i, j), has a present flow of f_{ij} and a capacity of c_{ij}. If a further flow f'_{ij} is assigned to it, its excess capacity is reduced by f'_{ij}, but the excess capacity of arc (j, i) is increased by f'_{ij}. This allows us the possibility of later changing our minds and reducing the flow in (i, j) by f'_{ij} to get back to the original flow of f_{ij}. For example, suppose initially:

$$f_{36} = 0$$
$$f_{63} = 0.$$

Then

$$e_{36} = 2 - 0 = 2$$
$$e_{63} = 0 - 0 = 0.$$

Now suppose a flow of 1 unit is assigned to $(3, 6)$; then

$$f_{36} = 1$$
$$e_{36} = 1,$$

but e_{63} is increased to 1. Although in reality it is impossible for arc $(6, 3)$ to accommodate any flow, this positive excess flow is a useful tool. It allows us to *notionally* assign a flow of 1 to arc $(6, 3)$ (since it has excess capacity of 1). This unit of flow cancels with the unit flowing along (3, 6), leaving no flow at all. Also, e_{63} is reduced to 0, e_{36} is increased to 2, and we are back where we started.

Armed with the above ideas we shall now explain the labelling method by using it to solve the problem defined by the network in Figure 5.8. We begin by labelling the source with the symbol $b_1 = \infty$, to indicate that it is theoretically the source of an infinite amount of flow as far as the method is concerned. All arcs are initially assigned an excess capacity as defined by (5.3), with $f_{ij} = 0$. Any unlabelled points directly connected to a labelled point by arcs with positive excess capacity are identified. Thus, if unlabelled point j is such that

$$e_{ij} > 0$$

for some arc (i, j) and some labelled point i, point j is then labelled with the ordered pair (a_j, b_j), where

$$a_j = i, \qquad \text{the starting point for } (i, j)$$
$$b_j = \min \{e_{ij}, b_i\}, \quad \text{the maximum possible flow.}$$

This represents the fact that it is possible to find a path from the source to point j which can carry an extra b_j units of flow. Thus points p_2 and p_4 are unlabelled and connected to the labelled point p_1. Hence they are labelled $(1, 2)$ and $(1, 4)$, respectively. The labelled point with the smallest index which is connected to an unlabelled point is identified. This is point p_2, connected to point p_3. Thus point p_3 is labelled $(2, 2)$. Next the sink is labelled $(3, 2)$ by the same reasoning.

Once the sink has been labelled, *breakthrough* has been achieved. We have now discovered a path from source to sink which is capable of carrying b_n additional units of flow; b_n is the second label associated with the sink, point p_n. In the present case the path we have found is capable of carrying $b_6 = 2$ extra units of flow. This path can be traced back to the source by examining the a_i values of point labels. For instance, $a_6 = 3$, hence the path proceeds $p_3 \to p_6$; $a_3 = 2$, hence the path is $p_2 \to p_3 \to p_6$; and $a_2 = 1$, hence the complete path is $p_1 \to p_2 \to p_3 \to p_6$. The flows in the arcs of this path are increased by $b_n (= 2)$; i.e., $f_{12} = f_{23} = f_{36} = 2$.

The excess capacity in these arcs is reduced by the amount of flow just assigned, i.e.,

$$e_{12} = 2 - 2 = 0$$
$$e_{23} = 4 - 2 = 2$$
$$e_{36} = 2 - 2 = 0.$$

The excess capacity of arcs in the opposite direction to those on the path have their excess capacities increased by the amount of flow just assigned, i.e.,

$$e_{21} = 0 + 2 = 2$$
$$e_{32} = 0 + 2 = 2$$
$$e_{36} = 0 + 2 = 2.$$

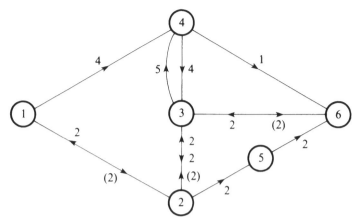

Figure 5.9.   An initial flow assignment.

All labels except that of the source are then removed. This completes one iteration of the method. Figure 5.9 indicates the present flow assignment. Actual flows assigned are shown in parentheses, excess capacities without parentheses.

The process is then repeated. This time only point p_4 can be initially labelled from the source, as the arc connecting point p_2 has zero excess capacity. It is once again labelled $(1, 4)$. Next points p_3 and p_6 can be labelled from point p_4. The maximum amount of extra flow that can travel via the path $p_1 \rightarrow p_4 \rightarrow p_6$ is the minimum of two quantities: the amount that can arrive at point p_4 (4 units) and the excess capacity of arc $(4, 6)$, namely 1. Hence the points p_3 and p_6 are labelled $(1, 4)$ and $(4, 1)$, respectively. Break-through has once again been achieved. We have identified a path: $p_1 \rightarrow p_4 \rightarrow p_6$ to which we can assign a flow of 1. We now perform the necessary book-keeping tasks to keep track of present flow assignments:

$$f_{14} = f_{46} = 1$$
$$e_{14} = 4 - 1 = 3$$
$$e_{46} = 1 - 1 = 0$$
$$e_{41} = 0 + 1 = 1$$
$$e_{64} = 0 + 1 = 1.$$

Figure 5.10 indicates the present flow assignments. It now looks as if we have reached a stalemate and cannot assign any further flows by this method. Arc $(1, 2)$ has zero excess capacity. Arc $(1, 4)$ has positive excess capacity (3), but arc $(4, 6)$ has zero excess capacity. Hence we would have to send any flow arriving at p_4 from p_1 on to p_3. But arcs $(3, 6)$ has zero excess capacity so this flow would have to be sent along arcs $(3, 2), (2, 5),$ and $(5, 6)$. All the arcs on this path have excess capacity. Hence this path represents a possibility for increasing flow. We have already assigned a flow of 2 along arc $(2, 3)$.

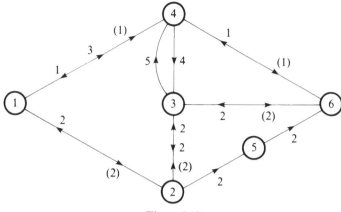

Figure 5.10

Hence if we send any flow along arc $(3, 2)$, this flow along arc $(2, 3)$ would be correspondingly reduced. This possibility allows us to change our minds and remove the allocation of 2 units along arc $(2, 3)$.

In practice, the next iteration of the labelling method achieves what we have just discussed: p_4 is labelled $(1, 3)$, p_3 is labelled $(4, 3)$, p_2 is labelled $(3, 2)$, p_5 is labelled $(2, 2)$, and p_6 is labelled $(5, 2)$. Breakthrough has been achieved. We have discovered the path $p_1 \rightarrow p_4 \rightarrow p_3 \rightarrow p_2 \rightarrow p_5 \rightarrow p_6$ along which it is possible to send an extra 2 units. When we perform the necessary book-keeping, what happens to arc $(2, 3)$? The flow of 2 presently in arc $(2, 3)$ is cancelled with the flow of 2 presently in arc $(3, 2)$, leaving zero flow in both $(2, 3)$ and $(3, 2)$. The excess capacity of $(2, 3)$ is increased:

$$e_{23} = 2 + 2 = 4$$

and

$$e_{32} = 2 - 2 = 0.$$

Hence we are back to the original situation of zero flow between points p_2 and p_3. The rest of the bookkeeping is recorded:

$$
\begin{aligned}
f_{14} &= 1 + 2 = 3, & e_{14} &= 3 - 2 = 1, & e_{41} &= 1 + 2 = 3 \\
f_{43} &= 0 + 2 = 2, & e_{43} &= 4 - 2 = 2, & e_{34} &= 5 + 2 = 7 \\
f_{32} &= 2 - 2 = 0 \\
f_{23} &= 2 - 2 = 0 \\
f_{25} &= 0 + 2 = 2, & e_{25} &= 2 - 2 = 0, & e_{52} &= 0 + 2 = 2 \\
f_{56} &= 0 + 2 = 2, & e_{56} &= 2 - 2 = 0, & e_{65} &= 0 + 2 = 2.
\end{aligned}
$$

The present flow assignment is shown in Figure 5.11.

When the next iteration is performed it is found that the sink cannot be labelled, as there are no arcs incident with the sink with positive excess capacity. When this occurs the present flow assignment is optimal. As the

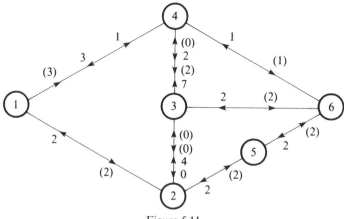

Figure 5.11

value of the present assignment is equal to the capacity of the minimum cut (5), we could have stopped before this last iteration, knowing the optimum was at hand by the maximum-flow, minimum-cut theorem.

The labelling method in algorithmic form is given below.

Labelling Method

1. Label point p_1, the source, $b_1 = \infty$. Set

$$f_{ij} = 0$$
$$e_{ij} = c_{ij}, \quad \text{for all arcs } (i, j).$$

2. If there is no unlabelled point p_j connected to a labelled point p_i by an arc with positive excess capacity, terminate—the present assignment of flow is optimal. Otherwise go to step 3.

3. Choose the smallest index i of those found in step 2. Set

$$a_j = i$$
$$b_j = \min \{e_{ij}, b_i\}.$$

4. If the sink point p_n is unlabelled, go to step 2. Otherwise go to step 5.

5. Identify a path of labelled points from source to sink. For each arc (i, j) on this path; let

$$e_{ij} \quad \text{become} \quad e_{ij} - b_n$$
$$e_{ji} \quad \text{become} \quad e_{ji} + b_n.$$

If $f_{ji} = 0$, let

$$f_{ij} \quad \text{become} \quad f_{ij} + b_n.$$

If $f_{ji} > 0$, let

$$\left.\begin{array}{l} f_{ij} \quad \text{become} \quad b_n - f_{ji} \\ f_{ji} \quad \text{become} \quad 0 \end{array}\right\} \quad \text{if } b_n \geqslant f_{ji}$$

and

$$\left.\begin{array}{ll} f_{ij} & \text{become} \quad 0 \\ f_{ji} & \text{become} \quad f_{ji} - b_m \end{array}\right\} \quad \text{if } b_n < f_{ji}.$$

6. Erase all point labels except that of the source. Go to step 2.

5.5.3 The Minimal Cost Flow Problem

Suppose now that a network has not only a capacity but also a unit cost associated with each arc. The minimal cost flow problem involves finding the flow assignment for transporting a given quantity F from source to sink at minimal cost. Using the terminology of Section 5.5.1, the problem can be formulated mathematically as follows:

Minimize: $\displaystyle\sum_{\substack{\text{all arcs} \\ (i,j)}} d_{ij} f_{ij}$

subject to:

$$\sum_{\substack{\text{all arcs} \\ (p_i, p_j)}} f_{ij} - \sum_{\substack{\text{all arcs} \\ (p_j, p_i)}} f_{ji} = \begin{cases} F, & \text{if } i = 1 \\ 0, & \text{if } i \neq 1, i \neq n \\ -F, & \text{if } i = n \end{cases}$$

$$0 \le f_{ij} \le c_{ij}, \quad \text{for all arcs } (p_i, p_j),$$

where p_1 corresponds to the source, p_n corresponds to the sink, and d_{ij} is the unit traversal cost of arc (p_i, p_j).

A multiple-source, multiple-sink minimal cost flow problem with no intermediate nodes is the transportation problem studied in Chapter 2. Also, if

$$F = 1$$

and

$$c_{ij} = 1, \quad \text{for all arcs } (p_i, p_j),$$

then the minimal cost flow problem reduces to the shortest path problem of Section 5.3.

5.5.4 An Algorithm for the Minimal Cost Flow Problem

The following algorithm, due to Busacker and Gowan (1961) will be explained by using it to solve a minimal cost flow problem concerned with the network shown in Figure 5.12. Each arc has an ordered pair associated with it. The first entry in the ordered pair specifies the capacity of the arc, the second the unit cost. Suppose it is desired to assign a total flow of 5 from source to sink with minimal cost.

Basically, the algorithm identifies at each iteration a least cost path which can accommodate further flow. The maximum possible flow is added to the path. This is repeated until the total flow from source to sink is built up to F.

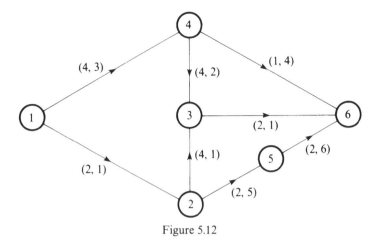

Figure 5.12

In calculating the cost of each path which can be assigned further flow one adds the cost of arcs oriented in the direction of the path and subtracts costs oriented in the opposite direction.

The method begins with zero flow in each arc. The least cost path from source to sink for the network in Figure 5.12 is $p_1 \to p_2 \to p_3 \to p_6$. The maximum flow which can be assigned to this path is the smallest arc capacity, namely 2, due to arcs $(1, 2)$ and $(3, 6)$. This flow is duly assigned. The capacity of the arcs involved is correspondingly reduced. Arcs with zero capacity are given a unit cost of ∞. Arcs in the opposite direction to those on the path are assigned a capacity equal to that just assigned, and a unit cost equal to the negative of that originally belonging to that of the arc concerned. This is shown in Figure 5.13. The flows assigned are written without parentheses.

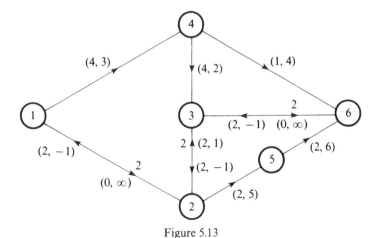

Figure 5.13

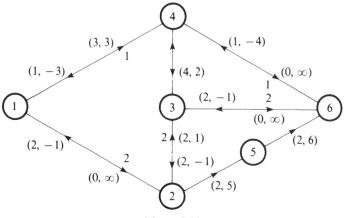

Figure 5.14

The process is now repeated. This time the shortest path is $p_1 \rightarrow p_4 \rightarrow p_6$. A flow of 1 can be assigned, as this is the minimum arc capacity belonging to $(4, 6)$. The arc labels are adjusted, and the result is shown in Figure 5.14.

The next shortest path is $p_1 \rightarrow p_4 \rightarrow p_3 \rightarrow p_2 \rightarrow p_5 \rightarrow p_6$, with a length of

$$d_{14} + d_{43} + d_{25} + d_{56} = 3 + 2 - 1 + 5 + 6 = 15.$$

The maximum that can be assigned to this is 2 units. When this assignment is made, the 2 units assigned to $(3, 2)$ cancel with the 2 units assigned to arc $(2, 3)$ to produce a label for $(2, 3)$ of $(4, 1)$. Arc $(2, 3)$ is now in the same state as it was originally. This has been brought about by the fact that we changed our minds about the assignment of the 2 units originally made to $(2, 3)$ and withdrew that allocation. The current assignments and arc labels are shown in Figure 5.15.

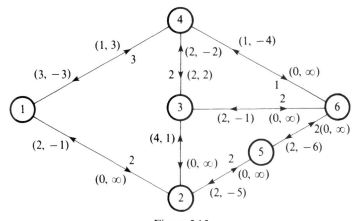

Figure 5.15

The flow now assigned is

1 unit on path	$p_1 \rightarrow p_4 \rightarrow p_6$
2 units on path	$p_1 \rightarrow p_4 \rightarrow p_3 \rightarrow p_6$
2 units on path	$p_1 \rightarrow p_2 \rightarrow p_5 \rightarrow p_6$.

which sums to the total of 5 units to be assigned. The cost is:

$$
\begin{aligned}
1 \times (\text{cost of path } p_1 \rightarrow p_4 \rightarrow p_6) &= 1 \times (3 + 4) &= 7 \\
+2 \times (\text{cost of path } p_1 \rightarrow p_4 \rightarrow p_3 \rightarrow p_6) &= 2 \times (3 + 2 + 1) &= 12 \\
+2 \times (\text{cost of path } p_1 \rightarrow p_2 \rightarrow p_5 \rightarrow p_6) &= 2 \times (1 + 5 + 6) &= \underline{24} \\
& & 43
\end{aligned}
$$

Note that if the flow to be assigned had been more than 5, the problem would have had no feasible solution, as there are no $p_1 \rightarrow p_6$ paths of finite cost and positive excess capacity left.

5.5.5 The Out-of-Kilter Method*

The procedure of Section 5.5.4 requires the search for a source-to-sink path prior to labelling and flow assignment at each iteration. Thus the procedure is not very suitable for large networks. A more general algorithm has been developed Ford and Fulkerson (1962) and is efficient when applied to large networks. This algorithm, called the *out-of-kilter method*, is presented in this section.

As with the algorithm of the previous section, flow is progressively added to the network in the out-of-kilter method. Flow may be added to existing flow in an arc in the direction of the arc (termed *forward flow*) and this increases the flow in the arc. Flow may be added to existing flow in an arc in the opposite direction to the arc (termed *backward flow*), and this decreases the flow in the arc. For instance, the arc (i, j) in Figure 5.16(a) has 6 units of flow. An addition of 3 units of forward flow increases the flow in the $i \rightarrow j$ direction to 9, as shown in Figure 5.16(b). An addition of 4 units of backward flow decreases the flow in the $i \rightarrow j$ direction to 2, as shown in Figure 5.16(c).

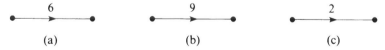

6	9	2
(a)	(b)	(c)

Figure 5.16. Forward and backward flow.

It is assumed here that no arc can accomodate a positive flow in the direction opposite to its orientation. The out-of-kilter algorithm will handle the case where each arc (p_i, p_j) has a positive lower bound b_{ij} on the amount of flow it carries.

---

* This section is based on pp. 132–145 of Plane and McMillan (1971).

The reader may have wondered about the unusual name of the method. It comes about as follows. During the course of the method each arc is assigned a definite *state*. The state of a particular arc may change from time to time. There are two possible states: *in kilter*, signifying that a change of flow in the arc will not bring about an improvement; and *out of kilter*, signifying that a change of flow in the arc will bring about an improvement. When all the arcs are in kilter, no further improvement is possible, and the optimal solution is at hand.

In order to decide how to assign states to the arcs, *modified costs* are assigned to them. Recall that each arc has a unit transportation cost, representing the cost of shipping one unit of flow in the direction of the arc. These unit costs are modified by adding and subtracting *tolls* from them. Let t_i be the toll for one unit arriving at point i. When a number of units arrives at point i, let us suppose that t_i must be paid for each unit. When the units are shipped along arc (i,j), t_i is charged for each unit. Because conservation of flow is assumed, this shipping causes no profit nor no loss. However, the total unit cost of shipping along arc (i,j), including tolls, is denoted by

$$\bar{d}_{ij} = d_{ij} + t_i - t_j. \tag{5.4}$$

Thus $\bar{d}_{ij}$ is the modified cost for each arc. The values assigned to the t_i change from time to time during the course of the method, according to strict rules. These rules will be explained later.

Given a set of modified costs and flows for the arcs, one is in a position to discover how the flows might be rearranged in order to save costs. For instance, if a modified cost is negative, the flow in the appropriate arc should be increased until it reaches capacity or until the modified cost becomes zero. An arc with a negative modified cost will be out of kilter unless its flow is at capacity. Also, if a modified cost is positive, the flow in the appropriate arc should be reduced until it becomes equal to the lower bound b_{ij}. Once all arcs are in kilter, no more savings can be made and an optimal solution has been found.

One can also ascertain whether it is possible to add forward flow, backward flow, both or neither to a particular arc. We associate with the arcs the following symbols,

I $\sim$ in-kilter,
O $\sim$ out-of-kilter,
F $\sim$ forward flow possible,
B $\sim$ backward flow possible,

depending upon its status. An arc will be endowed with either an I or O depending upon whether it is in or out of kilter, and with either F, B, FB, or no further symbol depending upon whether forward flow, backward flow, both or neither is capable of being assigned. The particular mix of symbols assigned to an arc depends upon its current level of flow relative to its

Table 5.1. Assignment rules for out of kilter method

Modified cost	Flow Level				
	$f_{ij} < b_{ij}$	$f_{ij} = b_{ij}$	$b_{ij} < f_{ij} < c_{ij}$	$f_{ij} = c_{ij}$	$f_{ij} > c_{ij}$
$\bar{d}_{ij} > 0$	OF	I	OB	OB	OB
$\bar{d}_{ij} = 0$	OF	IF	IFB	IB	OB
$\bar{d}_{ij} < 0$	OF	OF	OF	I	OB

capacity and its modified cost. The rules of assignment are summarized in Table 5.1. An explanation of how some of the symbols in the table are arrived at has been given. The reader should satisfy himself that, in view of the previous discussion, the other entries in the table make sense.

Before stating the complete method in algorithmic form we shall outline the out-of-kilter method in general terms. As with the previous methods in Section 5.5, we usually begin with all flow assignments set at zero. (However, if a feasible set of flow assignments is known this could be used instead.) All point tolls t_i are initially set at zero, and then the modified costs $\bar{d}_{ij}$ can be calculated using (5.4). One can then assign a state to each arc according to Table 5.1. Next, a path of labelled points is built up. Once breakthrough is achieved, additional flow is added to the labelled path. Point tolls are adjusted, all labels are removed, modified costs are recalculated, and the states of certain arcs may be altered. The process then begins all over again, building up a new labelled path. When all arcs are in kilter the method is terminated.

Let us take each of the processes of the method in turn, beginning with the labelling of the points.

The Labelling Process

To begin the labelling process, when all points are unlabelled, one arbitrarily chooses an arc (p_i, p_j) that is out of kilter: point p_j is labelled if forward flow is possible in the arc; point p_i is labelled if backward flow is possible in the arc. Having done this one searches for other points to be labelled. A point p_i can be labelled if it is either:

1. Directly connected by an arc (p_i, p_j) to a labelled point p_j and backward flow is possible in (p_i, p_j); or
2. Directly connected by an arc (p_j, p_i) to a labelled point p_j and forward flow is possible in (p_j, p_i).

A label for point p_i connected to labelled point p_j is of the form:

$$[j^\alpha, A_i],$$

where

$$\alpha = \begin{cases} +, \text{ if extra forward flow is possible in arc } (p_i, p_j), \\ -, \text{ if extra backward flow is possible in arc } (p_i, p_j) \end{cases}$$

and A_i is the maximum amount of flow (either forward or backward) that can be added to the arc joining p_i and p_j which arrives at p_i from p_j. Thus A_i will be the smaller of the following two amounts:

L1. A_j, the value which is part of the label of point p_j.
L2. (a) $b_{ij} - f_{ij}$, if $\bar{d}_{ij} > 0$ and $f_{ij} < b_{ij}$;
     (b) $f_{ij} - c_{ij}$, if $\bar{d}_{ij} < 0$ and $f_{ij} > c_{ij}$;
     (c) $c_{ij} - f_{ij}$, if forward flow is possible and neither (a) nor (b);
     (d) $f_{ij} - b_{ij}$, if backward flow is possible and neither (a) nor (b).

When the very first point at each iteration is labelled, no other points will have been labelled. In this case A_i is assigned a value according to the second alternative.

It may be that no further labelling of points can be carried out but some arcs are still out of kilter. When this occurs all the tolls of unlabelled points must be adjusted. This means that some modified costs must be recomputed. This leads to a change of state of at least one arc, making either forward or backward flow possible. Thus further labelling will be possible. This toll adjustment is carried out as follows.

Toll Adjustment

T1. Identify all arcs which connect a labelled point and an unlabelled point.
T2. Among all such arcs found in T1, identify those arcs (p_i, p_j) such that:
     (a) $\bar{d}_{ij} > 0$, p_i is labelled, and $f_{ij} \leq c_{ij}$, or
     (b) $\bar{d}_{ij} < 0$, p_i is not labelled, and $f_{ij} \geq b_{ij}$.
     If no arc meets conditions (a) or (b), the problem has no feasible solution.
T3. Among all arcs identified in T2, find the one with minimum $|\bar{d}_{ij}|$ (absolute value of $\bar{d}_{ij}$).
T4. Increase tolls of all unlabelled points by the amount found in T3.

The out-of-kilter method operates by considering *circulation flows*, rather than source-to-sink flows. A circulation flow is one in which flow travels round a cycle in the network, returning to the point from which it started out. In order to be able to use the method on a minimal cost flow problem, we have to make a minor addition to the network concerned. A sink-to-source arc (p_t, p_s) is added, where p_t is the sink and p_s is the source. The arc is assigned unit cost $d_{ts} = 0$ (so as to not affect the cost of the final solution) and bounds $b_{ts} = c_{ts} = F$, the amount of source-to-sink flow required. Because of conservation of flow, any feasible solution must allow F units to flow from p_s to p_t (and back via (p_t, p_s).)

The out-of-kilter method is now stated in algorithmic form.

1. Add arc (p_t, p_s) to network with $d_{ts} = 0$, $b_{ts} = c_{ts} = F$, where s is the source and t is the sink. Set

$$t_i = 0, \quad \text{for all points } p_i \text{ in the network}$$
$$f_{ij} = 0, \quad \text{for all arcs } (p_i, p_j) \text{ in the network.}$$

2. Calculate $\bar{d}_{ij} = d_{ij} + t_i - t_j$, for all arcs. Assign a state to each arc.
3. If all arcs are in kilter, go to step 13; otherwise, continue.
4. Choose arbitrarily an arc (p_i, p_j) which is out of kilter, and label p_i and p_j according to the point labelling procedure.
5. If there is a path of labelled points including the arc (p_i, p_j) found in step 4, go to step 11; otherwise, continue.
6. If another point can be labelled, label it according to the point labelling procedure and go to step 5; otherwise, continue.
7. Change the tolls according to the toll adjustment procedure. If no tolls can be adjusted, no feasible solution exists; terminate.
8. Calculate new modified costs according to (5.4), for all arcs with only one unlabelled point.
9. Assign new states for arcs where necessary.
10. If all arcs are in kilter, go to step 13; otherwise, go to step 6.
11. Adjust the flow in each arc on the path found in step 5 by the minimum A_i among its point labels.
12. Remove all labels and arc states and go to step 2.
13. The present flow assignment is optimal. Terminate the algorithm.

5.5.6 Numerical Example Illustrating the Out-of-Kilter Method

We shall now solve again the minimal cost flow problem of Figure 5.12 using the out-of-kilter method. The out-of-kilter method has a rather elaborate mechanism, and its use on such a relatively small problem is rather like using a sledgehammer to crack a peanut. The method is designed for large problems; we use it on a small one only so that the explanation will be brief.

Following the algorithm, we begin in step 1 by setting all tolls and flows equal to zero. Thus each modified cost calculated in step 2 will equal the corresponding unit cost. These modified costs and the arc states are shown in Figure 5.17(a), as well as the source-to-sink arc (p_6, p_1) with capacity and lower bound 5. (Variables with current value zero are not shown in the figures accompanying this discussion.)

All arcs are found to be in state I except (p_6, p_1), which is in state OF. This arc is chosen as in step 4, and point p_1 is labelled $[6^+, 5]$, indicating that a forward flow of 5 is possible from p_6 to p_1 according to part L2(c) of the point labelling process. No further labelling can take place, as neither of the arcs out of p_1—(p_1, p_4) and (p_1, p_2)—have forward flow possible. But we have not been able to find a path of labelled points including (p_6, p_1), the original out-of-kilter arc. Hence, according to step 6, we go to step 7

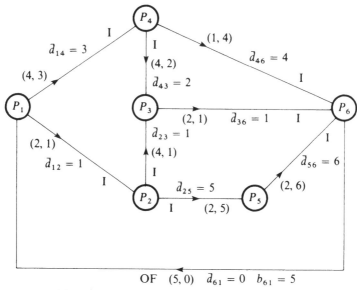

OF   (5, 0) $\bar{d}_{61} = 0$ $b_{61} = 5$

Figure 5.17(a).   Applying the out-of-kilter method.

and adjust the tolls. Arcs connecting a labelled point to an unlabelled point are (p_1, p_4) and (p_1, p_2). Hence, according to T2(a), both arcs can be identified and the tolls of unlabelled points should be increased by $|\bar{d}_{12}| = 1$. New modified costs and states are computed as in steps 8 and 9, and this is shown in Figure 5.17(b).

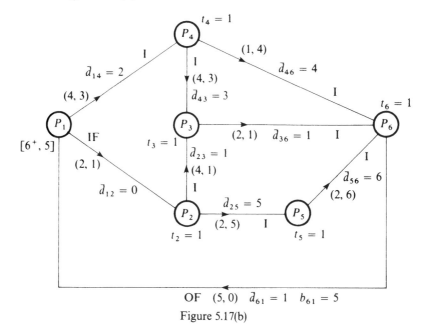

OF   (5, 0) $\bar{d}_{61} = 1$ $b_{61} = 5$

Figure 5.17(b)

Going back to step 6, we can now label point p_2, as forward flow is possible in arc (p_1, p_2). The label is $[1^+, 2]$, where $A_2 = 2 = \min\{A_1, c_{12}\} = \min\{5, 2\}$. Once again no further labelling can take place, so the tolls are adjusted. Arcs connecting a labelled point to an unlabelled one are (p_1, p_4), (p_2, p_3), (p_6, p_1), and (p_2, p_5). The tolls of unlabelled points are increased by $|\bar{d}_{23}| = 1$. New modified costs and states are computed as in steps 8 and 9, and this is shown in Figure 5.17(c).

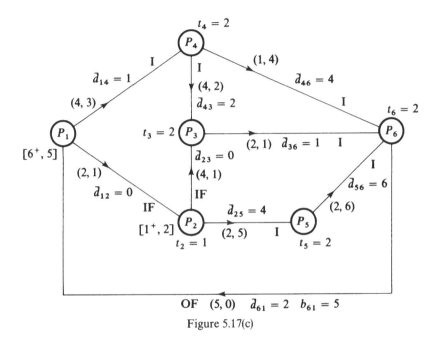

Figure 5.17(c)

Going back to step 6, we can now label point p_3, as forward flow is possible in arc (p_2, p_3). Once again no further labelling can take place, so the tolls are adjusted. Arcs connecting a labelled point to an unlabelled one are (p_1, p_4), (p_2, p_5), and (p_3, p_6). The tolls of unlabelled points are increased by $|\bar{d}_{36}| = 1$. New modified costs and states are computed as in steps 8 and 9 and this is shown in Figure 5.17(d).

Going back to step 6, we can now label points p_4 and p_6, as forward flow is possible in arcs (p_1, p_4) and (p_3, p_6). We have now created a cycle of labelled points $\langle p_1, p_2, p_3, p_6, p_1 \rangle$, as required in step 5. Going to step 11, the flow in the arcs of this path is adjusted by the minimum A_i among the labels of the points on the path, namely $A_6 = 2$. All arc states and point labels are removed, as in step 12. States are calculated as in step 2. These are shown in Figure 5.17(e).

The only arc out of kilter is arc (p_6, p_1). This arc is chosen as in step 4, and point p_1 is labelled. Next point p_4 is labelled, as it is connected by arc (p_1, p_4) in which forward flow is possible. As no further labelling is possible,

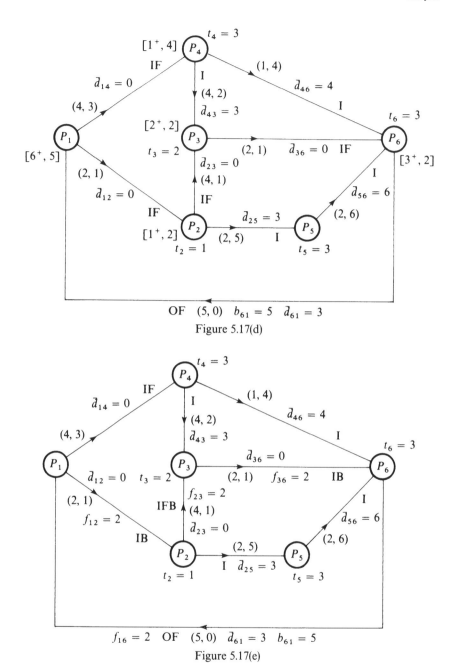

Figure 5.17(d)

Figure 5.17(e)

the tolls of unlabelled points are increased by $|\bar{d}_{43}| = 3$. New modified costs and states are computed and are shown in Figure 5.17(f).

It is now possible to label point p_3 and then p_2. Once again the tolls of unlabelled points are adjusted. This time they are increased by $|\bar{d}_{46}| = 1$. New modified costs and states are computed and are shown in Figure 5.17(g).

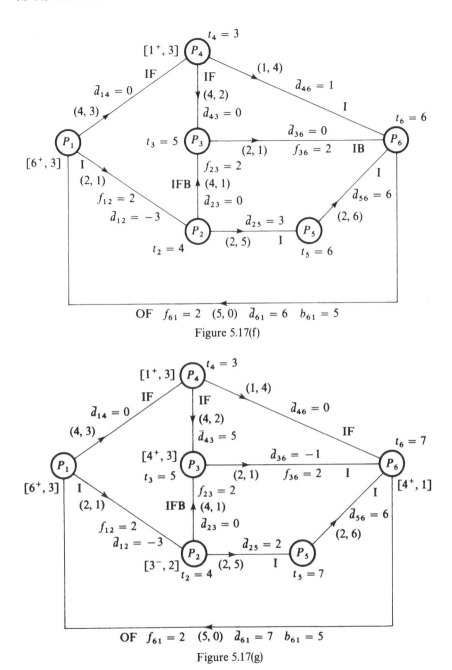

OF $f_{61} = 2$ (5, 0) $\bar{d}_{61} = 6$ $b_{61} = 5$

Figure 5.17(f)

OF $f_{61} = 2$ (5, 0) $\bar{d}_{61} = 7$ $b_{61} = 5$

Figure 5.17(g)

It is now possible to label point p_6. We have now created a path of labelled points $\langle p_1, p_4, p_6, p_1 \rangle$, as required in step 5. Going to step 11, the flow in the arcs of this path is increased by the minimum $A_i = A_6 = 1$. All arc states and point labels are removed as in step 12. New modified costs and states are calculated as in step 2, and are shown in Figure 5.17(h).

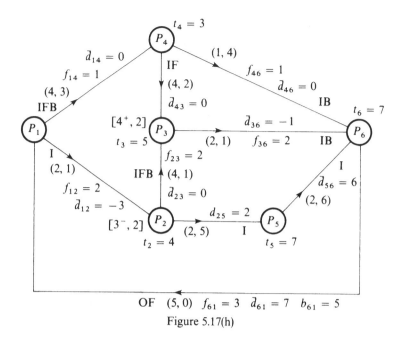

Figure 5.17(h)

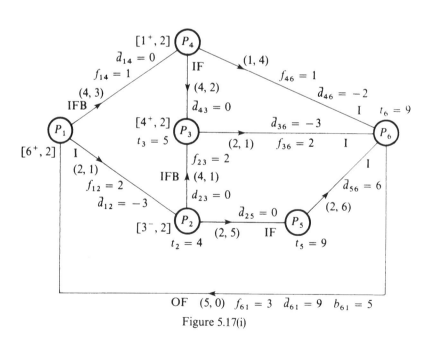

Figure 5.17(i)

The only out-of-kilter arc is (p_6, p_1). This arc is chosen as in step 4, and point p_1 is labelled. Next point p_4 is labelled, as it is connected to p_1 by arc (p_1, p_4) in which forward flow is possible. Then point p_3 is labelled, as it is connected to p_4 by arc (p_4, p_3). Then point p_2 can be labelled, as it is connected to the labelled point p_3 by arc (p_2, p_3) in which backward flow is possible. We cannot label any more points, so the tolls of unlabelled points are changed. They are increased by $|\bar{d}_{25}| = 2$. New modified costs and states are computed and are shown in Figure 5.17(i).

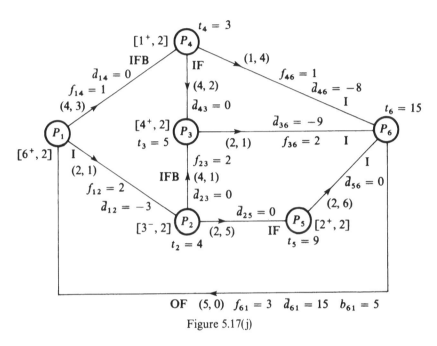

OF (5, 0) $f_{61} = 3$ $\bar{d}_{61} = 15$ $b_{61} = 5$

Figure 5.17(j)

It is now possible to label point p_5, but no further labels can be attached. Once again the tolls of unlabelled points are adjusted, and are increased by $|\bar{d}_{56}| = 6$. New modified costs and states are computed and are shown in Figure 5.17(j).

It is now possible to label p_6. We have created a path points $\langle p_1, p_4, p_3,$ $p_2, p_5, p_6, p_1 \rangle$ as required in step 5. Going to step 11, the flow in the arcs of this path are adjusted by the minimum $A_i = A_6 = 2$. All arc states and point labels are removed as in step 12. New modified costs and states are calculated as in step 2. These are shown in Figure 5.17(k).

The final solution, as shown in Figure 5.17(k) is identical to that found in section 5.5.4, as the arc (p_6, p_1) can now be ignored. It will be noticed that once an arc was in kilter, it never became out of kilter. This is no coincidence and will always happen. In fact, the method adopts the strategy of changing the status of out of kilter arcs to in kilter, while keeping the status of all in kilter arcs unchanged.

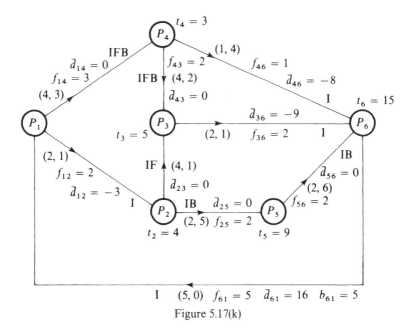

Figure 5.17(k)

5.6 Critical Path Scheduling

The reader has no doubt come across industrial or other real-life projects which on analysis can be seen to be made up of a number of individual *activities*, many of which may possibly be carried out simultaneously, assuming sufficient resources. Usually there are certain pairs of activities $\{a_i, a_j\}$ with the property that a_j cannot be started before a_i is completed. For example, in the project of building a house it may not be wise to lay the carpet before the interior walls are painted.

It is often desirable to represent the interrelationships between the activities by a *network*, that is, a digraph with a source and a sink. In order to see how this can be done we need to develop the notion of *precedence*. We say that activity a_i *precedes* a_j if a_i must be completed before a_j can begin. Of course, certain activities may be preceded by more than one other. We concern ourselves only with direct precedence. If a_i precedes a_j and a_j precedes a_k then strictly speaking a_i also precedes a_k; however, in the construction of a network to represent the project we shall not take notice of this last fact: the a_i–a_j and a_j–a_k precedences imply the a_i—a_k precedence.

We associate with each activity a_i a *duration time* t_i, which is the estimated time to complete a_i. The network for a given project is constructed as follows. Each activity is represented by a point in the network. There is also a unique source α, which represents the start of the project and a unique sink ω, which represents the completion of the project. It is assumed that the activity α

precedes any activities with no other precedents. Also the activity ω is preceded by any activities which precede no other activities. The duration time of the activities represented by α and ω are defined to be zero. Whenever activity a_i precedes activity a_j, join point a_i to point a_j by arc (a_i, a_j). Associate with each point the duration time of its activity. It should be noted that the network will not possess any cycles, for if it did, no activity on a cycle could ever be started.

In any project there will be a number of activities with the following property: If the start of the activity is delayed any later than it strictly has to be, or if the duration time of the activity is prolonged, then the completion time of the whole project will be extended. Such activities are termed *critical*. Because of the nature of the precedence relationships and the way we have constructed the network, there will be at least one source-to-sink path of critical activities—the longest path from source to sink (in terms of the sum of the duration times of its points) in the network. The aim of our analysis is to identify all such critical paths. Then a schedule can be devised giving the recommended starting and finishing times for each activity. Then if an activity looks like it is falling behind, extra resources may possibly be channelled into it from other activities with a comfortable margin.

There is another approach to modelling projects of this sort by digraphs. This uses arcs to represent activities and the points represent *events* that certain activities has been completed. Coverage of this approach is beyond the scope of this book and the interested reader is referred to Taha (1976). That author also covers the case where the duration time estimates are probabilistic in nature; in this case a technique called PERT (Program Evaluation and Review Technique) is explained. We confine ourselves in this chapter to constant, given duration times and present what is called the Critical Path Method (C.P.M.) which will find all critical paths.

We shall explain C.P.M. by using it on a numerical example which has been streamlined in a rather simple-minded way for expository purposes. Let us construct the network for the project of building a house with the activities shown in Table 5.2. Activities 1 and 2 have no precedents, so we create arcs $(\alpha, 1)$ and $(\alpha, 2)$ as in Figure 5.18. Then we see that activities 3 and 4 are preceded by these two, and that 3 precedes 4. Thus arcs $(2, 3), (1, 3), (1, 4)$, and $(3, 4)$ are created. Proceeding in this way, as 5, 7, 8, and 11 depend upon 3 and 4, arcs $(3, 11), (3, 7), (3, 8)$, and $(4, 5)$ are drawn. No arcs are drawn between any pair of 5, 7, 8, and 11, as they are not related. The next iteration creates arcs $(11, 12), (7, 14), (8, 9), (8, 10)$, and $(5, 6)$. Then arcs $(12, 13)$ and $(6, 14)$ are drawn. Then arcs $(13, 14)$ and $(13, 15)$ come into being, where the points of arc $(13, 14)$ were already present. Finally points 14 and 15 are connected to ω, as they do not precede any activities.

We shall now find all critical paths in the network, whose length represents the minimum possible completion time of the project. Secondly we shall discover for each activity the *earliest start time* it could possibly be begun and the *latest finish time* it could possibly be finished if the whole project is to be

Table 5.2. House Building Projects

Activity	Precedence	Time (days)
1. Excavate to prepare for foundations	—	10
2. Establish driveway	—	2
3. Deliver building materials	1, 2	3
4. Establish foundations	1, 3	15
5. Build walls and interior	4	40
6. Build roof	5	10
7. Build separate garage	3	10
8. Hook up power supply	3	1
9. Reticulate house with water, gas and electricity	8	5
10. Wallpaper interior	8	2
11. Fence property	3	2
12. Landscape section	11	5
13. Build swimming pool	12	4
14. Spray paint inside and out	6, 7, 9, 10, 13	8
15. Plant garden	13	2

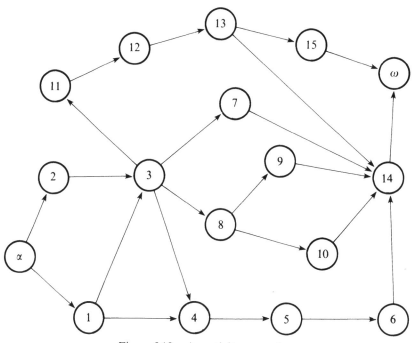

Figure 5.18.   An activity network.

completed at the earliest possible instant. Naturally, a critical activity will have its earliest start time plus its duration time equal to its latest finish time, as there is no leeway. Let

es_i = earliest start time for activity a_i

lf_i = latest finish time for activity a_i

t_i = duration time of activity a_i.

Then

$$es_i + t_i = lf_i, \quad \text{if } a_i \text{ is critical.} \tag{5.5}$$

However, if a_i is not critical,

$$es_i + t_i < lf_i. \tag{5.6}$$

The actual leeway is called the *total float* tf_i for a_i:

$$tf_i = lf_i - t_i - es_i. \tag{5.7}$$

Thus a critical activity has zero total float. Given the possibility of re-allocating manpower to speed up ailing activities, it is desirable to define two further variables for each activity a_i:

ls_i = latest start time of a_i if project is to be completed on time

ef_i = earliest possible finish time of a_i given its precedence.

For each activity a_i,

$$ef_i = es_i + t_i \tag{5.8}$$

$$lf_i = ls_i + t_i. \tag{5.9}$$

Therefore,

$$tf_i = ls_i - es_i = lf_i - ef_i. \tag{5.10}$$

We associate es_i, lf_i, ls_i, ef_i, and t_i with each point a_i in the network, as shown in Figure 5.19. We fill in the four numbers in the interior of each circle by a two-pass process.

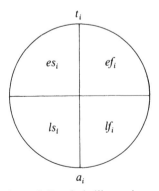

Figure 5.19.   Labelling point a_i.

PASS I

(a) Define $es_\alpha = 0$, the earliest start time of the source.

(b) ef_i is defined by (5.8).

(c)
$$es_j = \max_{(a_i, a_j)} \{ef_i\},$$

where this maximum is taken over all ef_i where arc (a_i, a_j) exists.

Using (a), (b), and (c), es_i and ef_i can be calculated for all points in the network. We now illustrate pass I on our example, calculating the top two numbers in each circle in Figure 5.20.

(a) $es_\alpha = 0.$

(b) $ef_\alpha = 0,$ by (5.8).

(c) $es_1 = 0$

                                   $es_2 = 0.$

(b) $ef_1 = 0 + 2 = 2$

                                   $ef_2 = 0 + 10 = 10.$

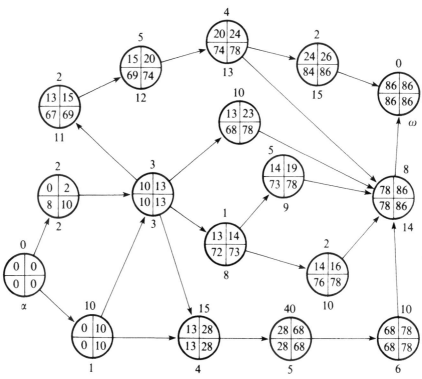

Figure 5.20.   Calculating start and finish times.

(c)
$$es_3 = \max \{ef_1, ef_2\}$$
$$= \max \{2, 10\}$$
$$= 10.$$

(b)
$$ef_3 = 10 + 3 = 13.$$

Proceeding in this way, we eventually calculate

$$ef_\alpha = 86.$$

This establishes that the earliest possible finish time for building the house is 86 days.

We now make a backwards pass through the network, filling in the bottom two numbers in each circle. This is done as follows:

PASS II

(d) Define
$$lf_\omega = ef_\omega.$$

(e) ls_i is defined by (5.9), i.e.,

$$ls_i = lf_i - t_i$$

(f)
$$lf_i = \min_{(a_i, a_j)} \{ls_i\}$$

is taken over all ls_i where arc (a_i, a_j) exists.

Using (d), (e), and (f), ls_i and lf_i can be calculated for all points in the network. For our example, as shown in Figure 5.20,

(d) $$lf_\omega = ef_\omega = 86.$$

(e) $$ls_\omega = lf_\omega - tw = 86, \quad \text{by} \quad (5.9).$$

(f) $$lf_{14} = 86$$
         $$lf_{15} = 86.$$

(e) $$ls_{14} = 86 - 8 = 78$$
         $$ls_{15} = 86 - 2 = 84$$
         $$lf_4 = \min \{ls_{14}, ls_{15}\} = \min \{78, 84\} = 78.$$

(e) $$ls_4 = 78 - 4 = 74.$$

Proceeding in this way we eventually calaculate

$$ls_\alpha = lf_\alpha = 0, \tag{5.11}$$

which must be true for any network. In fact, (5.11) is a good check on the accuracy of one's arithmetric. Having calculated the four numbers in each point it can be seen, for $i = \alpha, 1, 3, 4, 5, 6, 8, \omega$, whether

$$es_i = ls_i \qquad (5.12)$$

and

$$ef_i = lf_i. \qquad (5.13)$$

Points for which (5.12) and (5.13) hold are critical. Thus the critical path is $\langle \alpha, 1, 3, 4, 5, 6, 8, \omega \rangle$.

We can calculate the total float for each activity using (5.10); the results are shown in Table 5.3. For example, the total float of a_2 is 8 days. This means that as long as activity 2 is started within 8 days of the earliest possible time it can be started (day 0), and there are no other critical delays, then the whole project will still be completed on time.

Table 5.3.

Activity	Precedence	t_i	es_i	ls_i	ef_i	lf_i	tf_i	ff_i
α	—	0	0	0	0	0	0	0
1	—	10	0	0	10	10	0	0
2	—	2	0	8	2	10	8	8
3	1, 2	3	10	10	13	13	0	0
4	1, 3	15	13	13	28	28	0	0
5	4	40	28	28	68	68	0	0
6	5	10	68	68	78	78	0	0
7	3	10	13	68	23	78	55	55
8	3	1	13	72	14	73	59	0
9	8	5	14	73	19	78	59	59
10	8	2	14	76	16	78	62	62
11	3	2	13	67	15	69	54	0
12	11	5	15	69	20	74	54	0
13	12	4	20	74	24	78	54	0
14	6, 7, 9, 10, 13	8	78	78	86	86	0	0
15	13	2	24	84	26	86	60	60
ω	14, 15	0	86	86	86	86	0	0

There is another type of float called *free float*. Total float is a global concept, in the sense that it defines the leeway in getting an activity started with regard to the project as a whole. Free float is a local concept, in the sense that it defines the leeway in getting an activity started *with regard only to the activities it precedes*. For example, consider activity 8 with total float 58. Assume we wish to start the activities which a_8 precedes (a_9 and a_{10}) as early as possible. Activities 9 and 10 both have earliest start times of 14. As the earliest finish time of a_8 is 14, it cannot be delayed. In this case the free float of a_8 is zero. However consider activity 2. Its earliest finish time is 2 but the earliest start time of the only project it precedes (a_3) is 10. Thus

a_2 could be delayed $10 - 2 = 8$ days and project 3 would still be started as early as possible. In this case the free float of a_2 is 8. In general we define free float ff_i for activity a_i to be

$$ff_i = \min_{(a_i, a_j)} \{es_j - ef_i\},$$

where the minimum is taken over all $es_j - ef_i$ where arc (a_i, a_j) exists. The free floats for all activities are also listed in Table 5.3.

5.7 Exercises

(I) Computational

1. Solve the following shortest path problems using Dijkstra's method. The entry i, j in each matrix is the cost of traversing arc (i, j); a dash or a blank space indicates the fact that there is no arc present. In all cases the i, j and j, i entries are equal.
   (a) From 1 to 11.

	1	2	3	4	5	6	7	8	9	10	11
1	—	12	12	—	—	—	—	—	—	—	—
2	12	—	—	6	11	—	—	—	—	—	—
3	12	—	—	3	—	—	9	—	—	—	—
4	—	6	3	—	—	5	—	—	—	—	—
5	—	11	—	—	—	9	—	10	—	—	—
6	—	—	—	5	9	—	6	—	12	—	—
7	—	—	9	—	—	6	—	—	—	11	—
8	—	—	—	—	10	—	—	—	8	—	7
9	—	—	—	—	—	12	—	8	—	9	12
10	—	—	—	—	—	—	11	—	9	—	10
11	—	—	—	—	—	—	—	7	12	10	—

(b) From 1 to 10.

	1	2	3	4	5	6	7	8	9	10
1	—	—	25	14	—	—	—	—	—	—
2		—	7	2	8	—	—	—	—	—
3			—	—	—	—	—	—	—	—
4				—	—	12	—	—	—	—
5					—	—	18	13	—	—
6						—	—	20	—	—
7							—	16	7	—
8								—	—	4
9									—	6

(c) From 1 to 11.

	1	2	3	4	5	6	7	8	9	10	11
1	—	6	2	3	—	—	—	—	—	—	—
2		—	6	—	7	—	—	—	—	—	—
3			—	7	—	8	2	—	—	—	—
4				—	—	—	6	—	—	—	—
5					—	5	—	8	—	—	—
6						—	7	1	—	—	—
7							—	—	—	4	—
8								—	5	—	4
9									—	4	1
10										—	5

(d) From 1 to 10.

	1	2	3	4	5	6	7	8	9	10
1	—	21	1	—	—	—	—	—	—	—
2		—	5	8	—	—	—	—	—	—
3			—	—	16	17	24	—	—	—
4				—	13	—	—	—	—	—
5					—	—	—	—	10	—
6						—	—	—	12	—
7							—	—	—	18
8								—	—	20
9									—	19
10										—

(e) From 1 to 12.

	2	3	4	5	6	7	8	9	10	11	12
1	3	2	1	—	—	—	—	—	—	—	—
2		—	—	5	6	—	—	—	—	—	—
3			—	—	—	6	7	—	—	—	—
4				7	7	5	—	—	—	—	—
5					2	—	—	—	4	—	—
6						—	—	1	3	—	—
7							4	3	—	—	2
8								—	—	—	6
9									—	9	4
10										5	—
11											—

2. Find a minimal spanning tree for each of the problems in Exercise 1 using the method of Prim.

3. Find a minimal spanning tree for each of the problems in Exercise 1 using the method of Kruskal.

4. In the following maximum flow problems, the source is point 1 and the sink is the point with the largest number as its label. The i, j entry in each matrix represents the capacity of arc (i, j). Find the minimum source–sink cut.

(a)

	1	2	3	4	5	6	7	8
1	—	7	—	—	12	—	—	—
2	—	—	6	4	—	—	—	—
3	—	—	—	3	—	—	—	3
4	—	—	—	—	—	—	—	8
5	—	—	—	—	—	9	5	—
6	—	2	—	—	—	—	3	4
7	—	—	—	—	—	—	—	5
8	—	—	—	—	—	—	—	—

(b)

	1	2	3	4	5	6	7	8	9	10
1	—	∞	—	∞	—	—	—	—	—	—
2	—	—	—	—	7	—	—	—	—	—
3	—	—	—	—	—	5	—	—	—	—
4	—	—	—	—	—	6	—	—	—	—
5	—	—	—	—	—	7	—	4	—	—
6	—	—	—	5	8	—	—	—	—	2
7	—	—	—	—	—	—	—	4	—	—
8	—	—	—	—	—	—	—	—	—	1
9	—	—	—	—	—	—	—	—	—	4
10	—	—	—	—	—	—	—	—	—	—

(c)

	1	2	3	4	5	6	7	8
1	—	2	3	—	—	—	—	—
2	—	—	—	4	8	—	—	—
3	—	—	—	2	1	—	—	—
4	—	—	—	—	—	2	6	—
5	—	—	—	—	—	5	4	—
6	—	—	—	—	—	—	—	8
7	—	—	—	—	—	—	—	9
8	—	—	—	—	—	—	—	—

(d)

	1	2	3	4	5	6	7	8	9
1	—	4	1	—	—	—	—	—	—
2	—	—	—	3	—	—	—	—	—
3	—	3	—	2	5	—	—	—	—
4	—	—	2	—	—	—	—	2	—
5	—	—	5	—	—	4	—	—	—
6	—	—	—	—	4	—	—	—	1
7	—	—	—	—	—	—	—	1	3
8	—	—	—	2	—	—	1	—	1
9	—	—	—	—	—	1	3	3	—

(e)

	1	2	3	4	5	6	7	8
1	—	3	—	—	—	—	2	—
2	—	—	1	—	—	1	—	—
3	—	—	—	2	2	—	—	—
4	—	—	—	—	—	—	—	1
5	—	—	—	—	—	—	—	3
6	—	—	—	2	—	—	—	—
7	—	—	—	—	5	—	—	—
8	—	—	—	—	—	—	—	—

(f)

	1	2	3	4	5	6	7	8
1	—	1	1	—	—	2	—	—
2	1	—	3	—	—	—	—	—
3	1	3	—	2	2	2	—	—
4	—	—	2	—	1	—	—	—
5	—	—	2	1	—	—	—	4
6	2	—	2	—	—	—	1	—
7	—	—	—	—	—	1	—	2
8	—	—	—	—	4	—	2	—

(g)

	1	2	3	4	5	6	7	8
1	—	1	2	—	—	—	—	—
2	1	—	—	3	2	—	—	—
3	2	—	—	1	2	—	—	—
4	—	3	1	—	—	2	1	—
5	—	2	2	—	—	3	4	—
6	—	—	—	2	3	—	—	1
7	—	—	—	1	4	—	—	1
8	—	—	—	—	—	1	1	—

(h)

	1	2	3	4	5	6	7	8
1	—	3	—	3	—	—	—	—
2	3	—	5	—	—	—	3	—
3	—	5	—	2	2	3	—	—
4	3	—	2	—	4	4	—	—
5	—	—	—	4	—	—	—	2
6	—	—	1	2	—	—	2	1
7	—	3	—	—	—	2	—	4
8	—	—	—	—	2	1	4	—

5. Solve each of the problems in Exercise 4 by the labelling method.

6. Solve each of the following minimal cost flow problems using the out of kilter method. The networks with their arc capacities are given in Exercise 4. Each matrix

below indicates the arc costs. Assume that the amount of flow to be transported is the maximum amount possible, as found in Exercise 4.

(a)

	1	2	3	4	5	6	7	8
1	—	5	—	—	3	—	—	—
2	—	—	3	3	—	—	—	—
3	—	—	—	1	—	—	—	8
4	—	—	—	—	—	—	—	7
5	—	—	—	—	—	2	3	—
6	—	2	—	—	—	—	6	4
7	—	—	—	—	—	—	—	5
8	—	—	—	—	—	—	—	—

(b)

	1	2	3	4	5	6	7	8	9	10
1	—	2	5	7	—	—	—	—	—	—
2	—	—	—	—	—	—	1	—	—	—
3	—	—	—	—	—	2	—	—	—	—
4	—	—	—	—	—	—	2	—	—	—
5	—	—	—	—	—	1	—	7	—	—
6	—	—	—	1	1	—	—	—	—	10
7	—	—	—	—	—	—	—	—	1	—
8	—	—	—	—	—	—	—	—	—	6
9	—	—	—	—	—	—	—	—	—	5
10	—	—	—	—	—	—	—	—	—	—

(c)

	1	2	3	4	5	6	7	8
1	—	1	1	—	—	—	—	—
2	—	—	—	2	3	—	—	—
3	—	—	—	4	1	—	—	—
4	—	—	—	—	—	2	4	—
5	—	—	—	—	—	4	1	—
6	—	—	—	—	—	—	—	1
7	—	—	—	—	—	—	—	1

(d)

	1	2	3	4	5	6	7	8	9
1	—	1	1	—	—	—	—	—	—
2	—	—	—	1	—	—	—	—	—
3	—	—	—	1	2	—	—	—	—
4	—	—	—	—	—	—	—	2	—
5	—	—	—	—	—	2	—	—	—
6	—	—	—	—	—	—	—	—	1
7	—	—	—	—	—	—	—	—	2
8	—	—	—	—	—	—	—	—	1
9	—	—	—	—	—	—	—	—	—

(e)

	1	2	3	4	5	6	7	8
1	—	1	—	—	—	—	2	—
2	—	—	1	—	—	1	—	—
3	—	—	—	2	3	—	—	—
4	—	—	—	—	—	—	—	1
5	—	—	—	—	—	—	—	2
6	—	—	—	2	—	—	—	—
7	—	—	—	—	2	—	—	—
8	—	—	—	—	—	—	—	—

(f)

	1	2	3	4	5	6	7	8
1	—	4	3	—	—	3	—	—
2	4	—	6	—	—	—	—	—
3	3	6	—	1	6	7	—	—
4	—	—	1	—	3	—	—	—
6	3	—	7	—	—	—	2	—
7	—	—	—	—	—	2	—	5
8	—	—	—	—	2	—	—	5

(g)

	1	2	3	4	5	6	7	8
1	—	1	2	—	—	—	—	—
2	1	—	—	2	2	—	—	—
3	2	—	—	1	2	—	—	—
4	—	2	2	—	—	1	2	—
5	—	2	2	—	—	2	2	—
6	—	—	—	1	2	—	—	1
7	—	—	—	2	2	—	—	1
8	—	—	—	—	—	1	1	—

(h)

	1	2	3	4	5	6	7	8
1	—	2	—	4	—	—	—	—
2	2	—	8	—	—	—	9	—
3	—	—	—	6	1	1	—	—
4	4	—	6	—	5	3	—	—
5	—	—	1	5	—	—	—	7
6	—	—	1	3	—	—	1	2
7	—	9	—	—	—	1	—	3
8	—	—	—	—	7	2	3	—

7. For each of the following projects identify critical activities, earliest completion time and activity float.

(a)

Activity	Precedence	Duration time
1	14, 16, 13	3
2	16	7
3	1, 2	9
4	9, 10, 3	4
5	9, 10, 3	6
6	4, 5	1
7	—	1
8	7	8
9	8	7
10	8, 12	4
11	7	5
12	7	2
13	12	3
14	11, 15	16
15	12	20
16	17	11
17	11	19

(b)

Activity	Precedence	Duration time
1	—	5
2	—	10
3	—	8
4	1	6
5	1	12
6	2, 4	7
7	3	4
8	5, 6, 7	6
9	3	10

(c)

Activity	Precedence	Duration time
1	—	6
2	—	4
3	2	5
4	2	6
5	2	4
6	3	3
7	4, 5	10
8	7	12
9	6, 8	4

8. Consider Exercise 7(c). Suppose the duration time of activity 4 is reduced from 6 to 4 units. How does this affect the outcome?

9. Consider the project of painting the exerior of a house with two coats of paint. Assume a team of three men is to carry out the task. Construct a list of about 10 activities with their duration times. Analyze the project using critical path scheduling.

10. Carry out critical path scheduling on each of the following tasks: making jam, bottling fruit, making a cup of coffee, laying a concrete path.

11. Critical path scheduling assumes there is sufficient manpower to do as many activities simultaneously as is necessary. Examine the solutions obtained to exercise 10 to determine the smallest number of people necessary to carry out task in minimum time.

(II) Theoretical

12. Prove observations 1–3 of Section 5.2.

13. A graph is termed *simple* if it has no *loops* (lines of the form $\{p_i, p_i\}$) or *parallel lines* (lines connecting the same pair of points). Show that a simple graph with n points can have no more than $n(n-1)/2$ lines.

14. Prove that a simple graph (see Exercise 13) with n vertices must be connected if it has more than $(n-1)(n-2)/2$ lines.

15. Prove that if G_1 and G_2 are the two subgraphs resulting from any decomposition of a connected graph G, that there must be a least one point which is in both G_1 and G_2.

16. Prove that a line in a graph G belongs to at least one circuit in G if and only if G remains connected after the removal of the line.

17. Prove that all trees are simple (see Exercise 13).

18. Suppose that it is desired to find the shortest tour for a travelling salesman in a connected, weighted graph G. Prove that the weight of a minimal spanning tree of G is a lower bound on the weight of the minimal tour.

19. Two distributors, A and B, have 6 and 4 units, respectively of a commodity on hand. Warehouses C and D require 3 and 4 units, respectively. Unit shipping costs to supply C and D from A are \$1.00 and \$2.00, respectively, and from B are \$4.00 and \$3.00 respectively.
    (a) Devise a network representation of this situation, regarded as a minimal cost flow problem. Add a supersource S_0, a supersink S_i, and an arc (S_i, S_0) to the network, taking care to label all arcs as is necessary for implementation of the out-of-kilter algorithm.
    (b) Implement the out-of-kilter algorithm on the problem until toll adjustment occurs for the first time.
    (c) State why it is no longer possible to proceed with the implementation of the algorithm.
    (d) Devise a new network formulation with a single node representing both S_i and S_0, making it possible to solve the problem using the out-of-kilter algorithm.
    (e) Implement the algorithm on the new network up to and including the first toll adjustment.
    (f) State the difference in conditions between the situations reached for (b) and (e), and explain why it is possible to proceed with the algorithm from the former situation.

Chapter 6

Dynamic Programming

6.1 Introduction

Dynamic programming is a technique for formulating problems in which decisions are to be made in stages—*a multistage decision problem*. This represents a departure from the types of problems we have analyzed so far, where it has been assumed that all decisions are made at one time. It is not difficult to think of real world scenarios which are multistage decision problems. Many construction projects can be divided up into stages corresponding to the completion of events. However, there are also many such problems in which different stages are not identified with different time periods. For instance, many problems involving the investment of funds to maximize return can be formulated with the different investment options being represented by different stages.

Dynamic programming (D.P.) has been used to solve successfully problems from a wide variety of areas including all branches of engineering, operations research, and business. It is an implicit enumeration approach (as was branch and bound enumeration, presented in Chapter 4) and can be very useful in reducing the computational effort required to solve a problem by other means. However, before the reader begins to think that he has found the answer to all his planning problems let us sound a note of caution. There are weaknesses with the D.P. approach, including the large number of intermediate calculations that have to be recorded. This is summed up as "the curse of dimensionality," which will be referred to later in this chapter.

The name of the technique was coined by Richard Bellman (1957), who developed D.P. and also wrote the first book on the subject. Since that time many books on D.P. have appeared, including those by Bellman and Dreyfus (1962), Hadley (1964), Nemhauser (1966) and White (1969). This vast and ever expanding field could not be explained in any depth in a single chapter of a book of the present size. Hence all that is attempted here is to introduce some of the basic D.P. ideas with view to stimulating the reader to attempt some of the more specialized texts mentioned earlier. In particular Hadley's book is recommended for techniques and White's for the mathematical theory of D.P. A knowledge of the calculus is required to comprehend the remainder of this book. The unprepared reader is referred to the appendix.

6.2 A Simple D.P. Problem

Consider a tramper who wishes to walk from a national park hut to the coast. On studying the map of the area he finds that there is quite a network of paths linking the intermediate huts one day's walk apart. He rates each path with a number which represents the enjoyment to be gained by walking along it, based upon scenery, the likely number of users, and travel time. He wishes to select a route with maximum enjoyment. The network is shown in Figure 6.1, where point 1 represents his present hut and points 8, 9, and 10 each represent coastal huts. The rating for each path is shown alongside

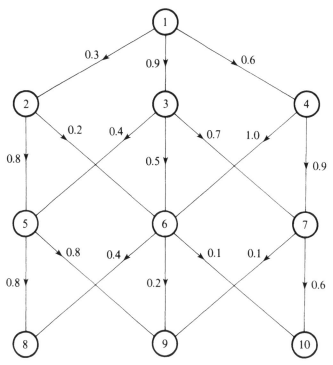

Figure 6.1. The network for the tramper's problem.

its arc. The problem is to find the longest path from point 1 to any of points 8, 9, or 10.

As in most combinatorial optimization problems, it is theoretically possible to evaluate all solutions to this problem and select the best; this is exhaustive enumeration, as discussed in Chapter 4. As more points are introduced into the network, however, the amount of computational effort required quickly becomes enormous. Clearly, a method which reduces the number of calculations required for exhaustive enumeration must be employed for networks with a reasonably large number of points. Dynamic programming offers such a reduction and will now be applied to the present problem.

It can be seen from Figure 6.1 that the tramper must pass through exactly one of the points from each of the following sets:

$$
\begin{array}{ll}
\{1\} & \text{(stage 0)} \\
\{2,3,4\} & \text{(stage 1)} \\
\{5,6,7\} & \text{(stage 2)} \\
\{8,9,10\} & \text{(stage 3).}
\end{array}
$$

When the tramper is currently at a point in one of these sets he is at a particular *stage* of his journey. The stages are numbered so that the number of a stage represents the number of paths walked to get to it from point 1. When the tramper is at a particular stage he will be in a particular *state* (apart from probably being cold, wet, tired, or hungry!), defined to be the particular point of that stage at which he is located. Associated with each state there is a *return*, which represents the *maximum possible enjoyment* the tramper could have experienced so far in arriving at that point. These concepts will now be used to solve the problem.

Initially the tramper leaves point 1 (stage 0), walks to one of points 2, 3, or 4, and finds himself at stage 1. He is now in either state 2, with a return of 0.3; state 3, with a return of 0.9; or state 4, with a return of 0.6. He now leaves stage 1 and proceeds to stage 2, ending up in one of states 5, 6, or 7. If he proceeds to state 5, which route is best in the sense of affording maximum enjoyment? He could have come from state 2, with cumulative enjoyment of 1.1 (0.3 + 0.8); or from state 3, with cumulative enjoyment of 1.3 (0.4 + 0.9). Thus the return at state 5 is the maximum of these two, which is 1.3. By the same reasoning, the return at state 6 is the maximum of (0.2 + 0.3), (0.5 + 0.9), and (0.6 + 1.0), which is 1.6. Similarly, the return at state 7 is the maximum of (0.7 + 0.9) and (0.9 + 0.6), which is 1.6. In sum, the returns at states 5, 6, and 7 are 1.3, 1.6, and 1.6, respectively.

The tramper now leaves stage 2 and arrives in stage 3 in one of states 8, 9, or 10. The return at state 8 can be calculated by adding the returns for states from which state 8 is accessible to the gains incurred in making the transition to state 8. For instance, if the tramper arrived to state 8 from state 5, the maximum enjoyment would be the return at state 5 (1.3) plus 0.8, i.e., 2.1. Note that one does not need to know how the return of 1.3 for state 5 was arrived at. It is sufficient to know that the state 5 return is 1.3. This fact embodies a very important assumption made in the problems to be solved by D.P. in this chapter. This assumption is that the return for a state depends only upon the optimal path to that state from a previous state and the previous state. If the tramper arrived at state 8 from state 6 the maximum enjoyment would be the return at state 6 plus 0.4, i.e., 1.6 + 0.4. Thus the return at state 8 is the maximum of 1.3 + 0.8 and 1.6 + 0.4, i.e., 2.1. Similarly the return at state 9 is the maximum of 1.3 + 0.8, 1.6 + 0.2, and 1.6 + 0.1, i.e., 2.1. The return at state 10 is the maximum of 1.6 + 0.1 and 1.6 + 0.6, i.e., 2.2.

So state 10 has the largest return, and we now know that this return of 2.2 represents the maximum enjoyment that can be attained. The actual path to be traversed in attaining this maximum can be found by unravelling the information contained in the state returns. To begin with it was the return at state 7 (1.6) plus the gain from the state 7 to state 10 transition (0.6) that produced the return of 2.2 at state 10. Hence point 7 and arc (7, 10) is on the longest path. By the same token it was the return at state 3 (0.9) plus the gain from the state 3 to state 7 transition (0.7) that produced the return of 1.6 at state 7. Hence point 3 and arc (3, 7) is on the path. Thus arc (1, 3) must also be included. The optimal path is then

$$\langle 1, 3, 7, 10 \rangle.$$

There are 17 paths from point 1 to points 8, 9, and 10. We could have evaluated them all and chosen the longest. The above approach involves less calculation and benefits become more and more apparent as the network size increases. The solution procedure just unfolded contains the basic approach of dynamic programming. The next section sets the stage in a more general fashion.

6.3 Basic D.P. Structure

The longest path problem of the previous section has the following property: In finding the return at a particular point p_j by arriving from a given point p_i, all one needed to know was the return at p_i and the gain in the p_i to p_j transition. This latter return was *independent of the way in which the system arrived at* p_i. Systems with this property are called *serial systems*. Thus a *serial system* is one in which the return at a stage i of the system depends only upon the returns at the stage $(i - 1)$ immediately preceding it and the gains in transforming the system from stage $(i - 1)$ into stage i.

In *nonserial systems* this property is not present and *feedback loops* or dependence upon earlier stages may occur. D.P. can be extended to analyze these problems; the resulting theory is related to another area called *optimal control*. This topic is outside the scope of the present book. We shall deal with only serial systems.

As can be seen by the example problem, in a serial system one is required to make a number of sequential, interrelated decisions. A complete set of decisions for a serial system problem, representing a solution to the problem, is called a *policy*. A single decision of how to transform the system from one stage to the next is called a *policy choice*. And a set of policy choices which transform the system from some intermediate stage to the final stage is called a *subpolicy*.

We now introduce notation which will allow us to express the D.P. approach to the example problem in general terms. Let

N = the number of the last stage in the problem

s_n = the state of the system at the nth stage

c_{ij} = the benefit gained in transforming the system from state i to state j

$f_n(s)$ = the return when the system is in state s at the nth stage

(i.e., $f_n(s)$ is the optimal benefit gained in transforming the system from the initial stage to state s at the nth stage). Let us now explain this notation in terms of the longest path problem. We wish to find the longest path from stage 0 to stage 3. That is, we require a path of arcs of the form

$$\langle (s_0, s_1), (s_1, s_2), (s_2, s_3) \rangle$$

whose total benefit

$$\sum_{i=1}^{N} c_{s_{i-1} s_i}, \qquad N = 3$$

is a maximum.

We will begin to solve the problem by using the above machinery. Initially the system is in state 1 at stage 0, having accrued no benefit so far. Thus

$$s_0 = 1$$

and

$$f_0(s_0) = 0.$$

The returns at the next stage are calculated as simple additions: the return at stage 1 in state 2 is

$$f_1(2) = f_0(1) + c_{12} = 0 + 0.3 = 0.3.$$

Similarly,

$$f_1(3) = 0.9$$

and

$$f_1(4) = 0.6.$$

Let us now calculate the return at stage 2, state 6. The tramper can arrive at state 6 from one of states 2, 3, or 4. The respective benefits are

$$f_1(2) + c_{26}$$
$$f_1(3) + c_{36}$$
$$f_1(4) + c_{46}.$$

The return at stage 2, state 6 ($f_2(6)$) is the maximum of these. Thus

$$f_2(6) = \max_{s_i = 2,3,4} \{f_1(s_1) + c_{s_i 6}\}$$

$$= \max \{(0.3 + 0.2), (0.9 + 0.5), (0.6 + 1.0)\} = 1.6.$$

The other returns at stage 2, $f_2(5)$ and $f_2(7)$, can be found in the same way.

Having calculated the three stage 2 returns, we can then use them to find the stage 3 returns. In general the return for stage n, state s is:

$$f_n(s) = \max_{s_{n-1}} \{f_{(n-1)}(s_{n-1}) + c_{(n-1)s}\}, \qquad n = 1, 2, \ldots . \tag{6.1}$$

We can now use (6.1) to solve the longest path problem:

$$f_0(1) = 0$$

$$f_1(2) = 0.3, \qquad f_1(3) = 0.9, \qquad f_1(4) = 0.6$$

$$f_2(5) = \max_{s_1 = 2,3} \{f_1(s_1) + c_{s_1 5}\} = \max \{(0.3 + 0.8), (0.9 + 0.4)\} = 1.3$$

$$f_2(6) = 1.6, \quad \text{as found before}$$

$$f_2(7) = \max_{s_1 = 3,4} \{f_1(s_1) + c_{s_1 7}\} = \max \{(0.9 + 0.7), (0.6 + 0.9)\} = 1.6.$$

Using these values in (6.1) recursively, we can calculate the stage 3 returns:

$$f_3(8) = \max_{s_2 = 5,6} \{f_2(s_2) + c_{s_2 8}\} = \max \{(1.3 + 0.8), (1.6 + 0.4)\} = 2.1$$

$$f_3(9) = \max_{s_2 = 5,6,7} \{f_2(s_2) + c_{s_2 9}\} = \max \{(1.3 + 0.8), (1.6 + 0.2), (1.6 + 0.1)\} = 2.1$$

$$f_3(10) = \max_{s_2 = 6,7} \{f_2(s_2) + c_{s_2 10}\} = \max \{(1.6 + 0.1), (1.6 + 0.6)\} = 2.2.$$

Thus the optimal solution has value 2.2, with the actual longest path being $\langle 1, 3, 7, 10 \rangle$, as found before.

Equations of the form of (6.1) are called *recursive equations*. Such equations, in one form or another, are usually used in solving a problem by dynamic programming. The family of equations in (6.1) underline the key fact that an optimal subpolicy at any stage of a multistage decision problem depends upon the state at that stage, and does not depend upon policy choices made at earlier stages. This can be started as follows:

The Dynamic Programming Principle of Optimality. When a system is at a given stage, the decisions of the optimal policy for future stages will constitute an optimal subpolicy regardless of how the system entered that stage.

Any system optimization problem for which the above principle is true can be attacked using D.P. Such systems are the serial systems, as described earlier.

We now examine some of the implications of this principle. In the longest path problem, when the tramper left stage 2 and walked to one of points,

8, 9, or 10 (stage 3), the return at each state of stage 3 was calculated without regard to states prior to stage 2. This allowed us to solve the problem one stage at a time. For example, we could temporarily "forget" about earlier decisions and find the best returns for the stage 3 states by examining only the stage 2 state returns and the stage 2 to stage benefits.

In the example problem the returns at each stage were calculated by finding the maximum among sums of pairs of numbers. In other uses of D.P. different ways of calculating optima must be used. How this is done is not part of the D.P. approach in itself. The user of D.P. is very much alone in finding returns and must use what ingenuity and knowledge of the particular system that he has.

6.4 Multiplicative and More General Recursive Relationships

As was stated in the previous section, the returns at each stage were calculated by finding the maximum among sums of pairs of numbers. Dynamic programming can also be applied to serial problems in which returns are calculated in other ways. Such a problem will now be presented.

Consider once again the network of Figure 6.1, but suppose it represents a different scenario. We now have a spy who wishes to send a confidential document from his present station (point 1) to any of the three receiving stations in his home base (points 8, 9, and 10). The number c_{ij} attached to the arc (p_i, p_j) in the network represents the probability that the document will be safely transmitted from the station represented by p_i to the station represented by p_j without falling into enemy hands. The problem is to find the route from p_1 to one of p_8, p_9, and p_{10} which affords the highest probability of a safe trip.

Using the notation of the previous section, we wish to find a path of arcs of the form:

$$\langle (s_0, s_1), (s_1, s_2), (s_2, s_3) \rangle$$

whose total probability

$$\prod_{i=1}^{N} c_{s_{i-1}s_i}, \qquad N = 3$$

is a maximum.

Notice here that we are multiplying relevant c_{ij} values together, rather than adding them as we did in the longest path problem. This is because of a basic property of probability theory: If events A and B are independent with probabilities $p(A)$ and $p(B)$, then the probability of both A and B occurring is $p(A)p(B)$. Thus (6.1) has a different form for this problem, namely

$$f_n(s) = \max_{s_{n-1}} \{ f_{(n-1)}(s_{n-1}) \times c_{(n-1)} \times c_{(n-1)s} \}, \qquad n = 1, 2, \ldots, \qquad (6.2)$$

where

$$f_0(1) = 1.0. \tag{6.3}$$

Note that (6.3) is true because the system begins in state 1 with probability one.

The only difference between (6.1) and (6.2) occurs in the replacement of the "$+$" in (6.1) by the "$\times$" in (6.2). It is conceivable that other operations may be involved in the interaction between $f_{(n-1)}(s_{n-1})$ and $c_{(n-1)s}$, such as

$$f_{n(s)} = \max_{s_{n-1}} \{f_{(n-1)}(s_{(n-1)}) \pm \sqrt{c_{(n-1)s}}\}, \qquad n = 1, 2, \ldots .$$

Hence the *general recursive equation* (forward form) is:

$$f_{n(s)} = \text{optimum}_{s_{n-1}} \{f_{(n-1)}(s_{n-1}) \oplus c_{(n-1)s}\}, \qquad n = 1, 2, \ldots, \tag{6.4}$$

where "$\oplus$" is an operation on $f_{(n-1)}(s_{n-1})$ and $c_{(n-1)s}$ depending upon the particular system being analyzed, and the objective may be one of maximization or minimization.

The problem of the spy will now be solved using (6.2) and (6.3):

$$f_1(2) = 1.0 \times 0.3 = 0.3$$
$$f_1(3) = 1.0 \times 0.9 = 0.9$$
$$f_1(4) = 1.0 \times 0.6 = 0.6$$

$$f_2(5) = \max_{s_1 = 2,3} \{f_1(s_1) \times c_{s_1 5}\} = \max \{((0.3) \times (0.8)), ((0.9) \times (0.4))\} = 0.36$$

$$f_2(6) = \max_{s_1 = 2,3,4} \{f_1(s_1) \times c_{s_1 6}\}$$

$$= \max \{((0.3) \times (0.2)), ((0.9) \times (0.5)), ((0.6) \times (1.0))\} = 0.6$$

$$f_2(7) = \max_{s_1 = 3,4} \{f_1(s_1) \times c_{s_1 7}\} = \max \{((0.9) \times (0.7)), ((0.6) \times (0.9))\} = 0.63.$$

Using these values in (6.2) we can calculate the stage 3 returns:

$$f_3(8) = \max_{s_2 = 5,6} \{f_2(s_2) \times c_{s_2 8}\} = \max \{((0.36) \times (0.8)), ((0.6) \times (0.4))\} = 0.288$$

$$f_3(9) = \max_{s_2 = 5,6,7} \{f_2(s_2) \times c_{s_2 9}\}$$

$$= \max \{((0.36) \times (0.8)), ((0.6) \times (0.2)), ((0.63) \times (0.1))\} = 0.288$$

$$f_3(10) = \max_{s_2 = 6,7} \{f_2(s_2) \times c_{s_2 10}\} = \max \{((0.6) \times (0.1)), ((0.63) \times (0.6))\} = 0.378.$$

Thus the optimal solution has probability 0.378. (Let's hope it isn't vital that the documents arrive safely, as the chances aren't too high!) The actual route is

$$\langle 1, 3, 7, 10 \rangle.$$

6.5 Continuous State Problems

In the problems analyzed so far the state and decision variables have been allowed to assume only values from a finite, discrete set. In this section this assumption is relaxed, and we allow the variables to assume any feasible real value: the *continuous state serial system problems*. Such a problem is presented below:

$$\text{Maximize:} \qquad x_0 = \sum_{i=1}^{N} \sqrt{x_i} \tag{6.5}$$

$$\text{subject to:} \qquad \sum_{i=1}^{N} x_i = d, \quad \text{a positive real constant} \tag{6.6}$$

$$x_i > 0, \qquad i = 1, 2, \ldots, N. \tag{6.7}$$

That is, it is desired to subdivide a given positive real number d into N positive parts, $x_1, x_2, \ldots, x_N$ (N being given), so that the sum of the square roots of the parts is a maximum. We now approach this problem with dynamic programming.

The problem can be looked upon as a serial system problem in which it is desired to assign a value to each x_i one at a time in the order $x_1, x_2, \ldots, x_N$. Thus the problem has N stages. When the system is at the nth stage, the state of the system s_n is defined to be the amount of the number d which has been assigned to $x_1, x_2, \ldots, x_n$ so far, i.e.,

$$s_n = x_1 + x_2 + \cdots + x_n, \qquad n = 1, 2, \ldots, N. \tag{6.8}$$

Because the only restrictions on the decision variables, x_i are those of (6.6) and (6.7), there is an infinite number of possibilities at each stage.

It so happens, because of the addition of the terms $\sqrt{x_i}$ in (6.4), that the recursive relationship is additive. So (6.5) becomes

$$f_n(s_n) = \max_{\substack{x_n \\ 0 < x_n \le s_n}} \{ f_{(n-1)}(s_{n-1}) + \sqrt{x_n} \}, \qquad n = 2, 3, \ldots, N.$$

But, from (6.8),

$$s_{n-1} + x_n = s_n, \qquad n = 2, 3, \ldots, N.$$

Thus

$$f_n(s_n) = \max_{\substack{x_n \\ 0 < x_n \le s_n}} \{ f_{(n-1)}(s_n - x_n) + \sqrt{x_n} \}, \qquad n = 2, 3, \ldots, N \tag{6.9}$$

and

$$f_1(s_1) = \sqrt{s_1}. \tag{6.10}$$

Now, from (6.8),

$$s_1 = x_1,$$

so that, from (6.10),

$$f_1(s_1) = \sqrt{x_1}. \tag{6.11}$$

Setting $n = 2$ in (6.9), we obtain

$$f_2(s_2) = \max_{\substack{x_2 \\ 0 < x_2 \le s_2}} \{f_1(s_2 - x_2) + \sqrt{x_2}\}.$$

By (6.11),

$$f_2(s_2) = \max_{\substack{x_2 \\ 0 < x_2 \le s_2}} \{\sqrt{s_2 - x_2} + \sqrt{x_2}\}.$$

Thus, in order to find $f_2(s_2)$ we must find the maximum value of $(\sqrt{s_2 - x_2} + \sqrt{x_2})$, where x_2 can range between 0 and s_2. To do this we use basic differential calculus. Let

$$F_2(x_2) = \sqrt{s_2 - x_2} + \sqrt{x_2}.$$

Then

$$\frac{\partial F_2}{\partial x_2} = \tfrac{1}{2}(s_2 - x_2)^{-1/2}(-1) + \tfrac{1}{2}x_2^{-1/2}.$$

For a stationary point,

$$\frac{\partial F_2}{\partial x_2} = 0.$$

Hence

$$\tfrac{1}{2}(s_2 - x_2)^{-1/2}(-1) + \tfrac{1}{2}x_2^{-1/2} = 0.$$

Therefore

$$x_2^* = s_2/2,$$

which is certainly in the range $(0, s_2]$. Also,

$$\frac{\partial^2 F_2(s_2/2)}{\partial x_2} = \left[-\tfrac{1}{4}(s_2 - x_2)^{-3/2} - \tfrac{1}{4}x_2^{-3/2}\right]_{x_2 = s_2/2}$$

$$= -\tfrac{1}{4}(s_2/2)^{-3/2} - \tfrac{1}{4}(s_2/2)^{-1/3} < 0,$$

indicating a maximum. Thus

$$f_2(s_2) = \sqrt{s_2 - (s_2/2)} + \sqrt{s_2/2} = \sqrt{2s_2}. \tag{6.12}$$

Now, setting $n = 3$ in (6.9),

$$f_3(s_3) = \max_{\substack{x_3 \\ 0 < x_3 \le s_3}} \{f_2(s_3 - x_3) + \sqrt{x_3}\}.$$

By (6.12),

$$f_3(s_3) = \max_{\substack{x_3 \\ 0 < x_3 \le s_3}} \{\sqrt{2(s_3 - x_3)} + \sqrt{x_3}\}.$$

We repeat the calculus technique just used in order to find $f_3(s_3)$. Let

$$F_3(x_3) = \sqrt{2(s_3 - x_3)} + \sqrt{x_3}$$

$$\frac{\partial F_3}{\partial x_3} = \tfrac{1}{2}[2(s_3 - x_3)]^{-1/2}(-2) + \tfrac{1}{2}x_3^{-1/2}.$$

For a stationary point,

$$\frac{\partial F_3}{\partial x_3} = 0,$$

hence

$$\tfrac{1}{2}[2(s_3 - x_3)]^{-1/2}(-2) + \tfrac{1}{2}x_3^{-1/2} = 0.$$

Therefore

$$x_3^* = s_3/3, \tag{6.13}$$

which is certainly in the range $(0, s_3]$. Also,

$$\frac{\partial_2 F_3(s_3/3)}{\partial x_3^2} = \left[-\tfrac{1}{4}[2(s_3 - x_3)]^{-3/2}(4) - \tfrac{1}{4}x_3^{-3/2} \right]_{x_3 = s_3/3}$$

$$= -\tfrac{1}{4}[2(s_3 - (s_3/3))]^{-3/2}(4) - \tfrac{1}{4}(s_3/3)^{-3/2} < 0,$$

indicating a maximum. Thus

$$f_3(s_3) = \sqrt{2(s_3 - (s_3/3))} + \sqrt{s_3/3} = \sqrt{3s_3}.$$

Let us now review what has been achieved so far by way of temporarily setting $N = 3$. In this case the problem has been solved, and

$$s_3 = x_1 + x_2 + x_3 = d$$
$$x_3^* = s_3/3 = d/3, \quad \text{by (6.13)}.$$

Thus

$$x_1^* + x_2^* = s_2 = 2d/3,$$

hence

$$x_2^* = s_2/2 = d/3$$

and therefore

$$x_1^* = d/3.$$

Also,

$$f_n(s_n) = \sqrt{ns_n}, \qquad n = 1, 2, 3$$

and the optimal solution has value $f_3(s_3) = \sqrt{3d}$.

The reader has no doubt suspected by now that the above results are true for general N. That this is so will be proved by induction:

$$x_n^* = d/n, \qquad n = 1, 2, 3, \ldots, N \tag{6.14}$$

$$f_n(s_n) = \sqrt{ns_n}, \qquad n = 1, 2, 3, \ldots, N. \tag{6.15}$$

Now (6.14) and (6.15) are certainly true for $n = 1, 2$, and 3. Assume that they are true for $n = k$, i.e.,

$$x_k^* = d/k$$
$$f_k(s_k) = \sqrt{ks_k}.$$

By (6.9),

$$f_{k+1}(s_{k+1}) = \max_{\substack{x_{k+1} \\ 0 < x_{k+1} \le s_{k+1}}} \left\{ f_k(s_{k+1} - x_{k+1}) + \sqrt{x_{k+1}} \right\}$$

$$= \max_{\substack{x_{k+1} \\ 0 < x_{k+1} \le s_{k+1}}} \left\{ \sqrt{k(s_{k+1} - x_{k+1})} + \sqrt{x_{k+1}} \right\}.$$

Let

$$F_{k+1}(x_{k+1}) = \sqrt{k(s_{k+1} - x_{k+1})} + \sqrt{x_{k+1}}.$$

For a stationary point,

$$\frac{\partial F_{k+1}}{\partial x_{k+1}} = 0,$$

hence

$$\tfrac{1}{2}[k(s_{k+1} - x_{k+1})]^{-1/2}(-k) + \tfrac{1}{2}x_{k+1}^{-1/2} = 0.$$

Therefore

$$x_{k+1}^* = s_{k+1}/(k+1),$$

which is certainly in the range $(0, s_{k+1}]$. Also,

$$\frac{\partial^2 F_{k+1}}{\partial x_{k+1}^2} < 0 \quad \text{at} \quad x_{k+1} = \frac{s_{k+1}}{k+1},$$

indicating a maximum:

$$f_{k+1}(s_{k+1}) = \sqrt{k(s_{k+1} - s_{k+1}/(k+1))} + \sqrt{s_{k+1}/(k+1)}$$
$$= \sqrt{(k+1)s_{k+1}}$$

which completes the proof that (6.14) and (6.15) are true. The solution to the problem is

$$x_n^* = d/n, \qquad n = 1, 2, \ldots, N$$

and the optimal solution value is

$$\sqrt{nd}.$$

6.6 The Direction of Computations

The problems solved so far in this chapter have all been approached by finding values for the return functions f_i, in the order $f_1, f_2, \ldots, f_N$. This is called *forward recursion*. In terms of the longest path problem and the spy problem this approach seems logical. However, there exist some serial systems for which the reverse order is more straightforward. That is, it is sometimes desirable to calculate the f_i in the order $f_N, f_{N-1}, \ldots, f_1$. This is called *backward recursion*, and it involves "working back" through the problem in the opposite direction to the actual sequence of events as they will take place when a feasible solution is implemented.

Let us approach the longest path problem using backward recursion. First we redefine $f_n(s)$ as follows:

$f_n(s)$ is the return (optimal benefit *to be gained*) from the remaining stages, $(n+1), (n+2), \ldots, N$, given that the system is in state s at the nth stage.

This represents a departure from the definition of $f_n(s)$ in Section 6.3, where $f_n(s)$ was the return gained *so far* from the previous stages, $1, 2, \ldots, n$ given that the system is in state s at the nth stage.

In terms of the longest path problem, we know that there is no further benefit to be gained once the system is in any of states 8, 9, or 10 at stage 3. Thus

$$f_3(8) = f_3(9) = f_3(10) = 0.$$

Suppose the system is in state 6, stage 2. What further benefit can the tramper look forward to? If he proceeds to state 8, 9, or 10, the extra benefits are 0.4, 0.2, and 0.1, respectively. Thus the extra benefit to be gained in leaving state 6 and arriving at the coast is $(f_3(8) + 0.4)$, $(f_3(9) + 0.2)$, or $(f_3(10) + 0.1)$, depending upon which choice is made. However, each first term in these expressions is zero; thus the maximum addition benefit to be had in leaving state 6 is 0.4:

$$f_2(6) = \max_{s_3 = 8,9,10} \{f_3(s_3) + c_{6s_3}\} = f_3(8) + 0.4 = 0.4.$$

The state returns can be similarly calculated for each other stage 2 state. In general the backward recursive equations for this problem are

$$f_n(s) = \max_{s_{n+1}} \{f_{n+1}(s_{n+1}) + c_{s(s_{n+1})}\}, \qquad n = 0, 1, 2 \qquad (6.16)$$

where

$$f_3(8) = f_3(9) = f_3(10) = 0. \qquad (6.17)$$

The reader may find it instructive to actually use (6.16) and (6.17) to verify that this approach produces the same optimal solution as is obtained by forward recursion.

There is no difference in the computational effort required to solve the problem by forward or backward recursion. This is because the benefit c_{ij} gained from transforming the system from state i to state j is a given constant which is simply added to the return of the present state (state i for forward recursion and state j for backward recursion). However, not all serial systems rejoice in such simplicity. Indeed, for problems with more complicated state transformations there may be a very marked difference in the amount of computational effort required depending upon whether forward or backward recursion is used. In fact some problems can be solved in only one direction. For instance, serial systems in which the state variable at each stage is random can be solved only by backward recursion.

We end this section by stating the general *recursive equation* (backward form):

$$f_n(s) = \operatorname*{optimum}_{s_{n+1}} \{f_{n+1}(s_{n+1}) \oplus c_{s(n+1)}\}.$$

6.7 Tabular Form

In the problems examined so far it has been a relatively easy matter to keep track of all the intermediate information necessary to use the recursive equations, as in each case there have not been many stages. For problems with many more stages and many possibilities at each stage one needs an efficient way of recording state returns, such as in tables. We illustrate this now on the following problem.

The Easy Tread shoe company has received an order for 20 truckloads of its biggest seller, the Won't Get Wet Tennis Shoe. The warehouse wishes to receive the total order all at one time. As the company can make at most 5 truckloads in any one production period, production must be scheduled over a number of periods. In fact, production costs vary over the next 6 periods, the time during which the order must be filled or it will be lost. Table 6.1 gives the total production cost for different numbers of truckloads produced in the different periods, $1, 2, \ldots, 6$. In addition to production costs there also storage (inventory) costs of \$1.00 per truckload stored per period. These inventory costs are incurred for complete periods only. Thus if 3 loads are produced in period 5 they incur only the cost of storage in period 6, i.e. \$(3 \times 1)$. The problem is to schedule production over the 6 periods so as to guarantee that the 20 loads are ready at the end of the sixth period and the total costs (production and inventory) are minimized.

Table 6.1.

Production number	Cost in period					
	1	2	3	4	5	6
0	3	2	4	1	5	2
1	4	6	6	7	11	6
2	9	11	11	12	12	9
3	16	15	12	19	14	10
4	19	18	14	27	19	15
5	20	21	20	32	23	20

This problem will now be formulated as a serial system and solved by D.P. Each production period will constitute a stage. At the nth stage, the state variable s_n is defined as the total number of loads produced so far by the end of that stage. Thus, assuming the system starts out at stage 0,

$$s_0 = 0.$$

Suppose, for instance, that three loads were produced in the first period; then

$$s_1 = 3.$$

As 20 loads must be produced by the end of the sixth period,

$$s_6 = 20.$$

Setting the problem up in general terms, let

N = the number of the last stage (in this case, $N = 6$)

c_{ij} = the production cost if i loads are produced in period j

x_i = the number of loads produced in period i

$f_n(s)$ = the return (minimum cost that can be incurred) when the system is in state s at the nth stage.

Assume that after $n - 1$ periods s_{n-1} loads have been produced, i.e.,

$$x_1 + x_2 + \cdots + x_{n-1} = s_{n-1},$$

and the return for the system to be in state s_{n-1} at the $(n - 1)$th stage is known, i.e., $f_{(n-1)}(s_{n-1})$ has been calculated. Suppose now that it is decided to produce x_n loads in period n. The costs involved are $c_{x_n n}$ for production and $1.0(x_n)(N - n)$ for inventory. (We have assumed forward recursion is to be adopted.) Thus the complete cost involved in this decision is

$$f_{(n-1)}(s_n - x_n) + c_{x_n n} + 1.0(x_n)(N - n),$$

as

$$s_{n-1} + x_n = s_n.$$

Thus

$$f_n(s_n) = \min_{\substack{0 \le x_n \le 5 \\ x_n \le s_n}} \{f_{(n-1)}(s_n - x_n) + c_{x_n n} + 1.0x_n(N - n)\}, \qquad n = 1, 2, 3, \ldots, N \tag{6.18}$$

where

$$f_0(s_0) = 0 \tag{6.19}$$

and

$$s_0 = 0.$$

We now use (6.18) and (6.19) to solve the problem, storing information calculated in tables. We begin by creating a table of inventory costs, where the entry in the ith row, jth column represents the total inventory cost if i loads are produced in the jth period. (See Table 6.2.) Using (6.18), (6.19) and Tables 6.1 and 6.2 we can calculate the stage 1 returns $f_1(s_1)$ for each possible state, $s_1 = 0, 1, 2, \ldots, 5$. For instance, if nothing is produced at stage 1, $s_1 = 0$. The contribution from Table 6.1 is $c_{01} = 3$ and the contribution from Table 6.2 is zero, i.e.,

$$f_1(0) = 3.$$

However, if one load is produced at stage 1, then $s_1 = 1$. The production cost is $c_{11} = 4$ and the inventory cost of Table 6.2 is 5, i.e.,

$$f_1(1) = 9.$$

The complete set of values is given in Table 6.3.

Table 6.2

| | Cost in period: | | | | | |
Number	1	2	3	4	5	6
0	0	0	0	0	0	0
1	5	4	3	2	1	0
2	10	8	6	4	2	0
3	15	12	9	6	3	0
4	20	16	12	8	4	0
5	25	20	15	10	5	0

Table 6.3

s_1	x_1	$f_1(s_1)$
0	0	3
1	1	9
2	2	19
3	3	31
4	4	39
5	5	45

We now calculate the stage 2 returns. Recall that

$$s_2 = x_1 + x_2.$$

As

$$0 \le x_1 \le 5$$
$$0 \le x_2 \le 5,$$

then

$$0 \le s_2 \le 10.$$

Suppose, for instance, that

$$s_2 = 6.$$

The (x_1, x_2) pairs which result in this value of s_2 are: $(1, 5)$, $(2, 4)$, $(3, 3)$, $(4, 2)$, $(5, 1)$. Using (6.18) and Table 6.3, we obtain the following costs:

$$x_1 = 1, x_2 = 5: \quad \text{cost} = f_1(1) + 20 + 21 = 50$$
$$x_1 = 2, x_2 = 4: \quad \text{cost} = f_1(2) + 16 + 18 = 53$$
$$x_1 = 3, x_2 = 3: \quad \text{cost} = f_1(3) + 12 + 15 = 58$$
$$x_1 = 4, x_2 = 2: \quad \text{cost} = f_1(4) + \ 8 + 11 = 58$$
$$x_1 = 5, x_2 = 1: \quad \text{cost} = f_1(5) + \ 4 + \ 6 = 55.$$

Taking the minimum of these costs, we obtain

$$f_2(6) = 50.$$

Table 6.4

| | x_2 | | | | | | | |
s_2	0	1	2	3	4	5	x_2^*	$f_2(s_2)$
0	5	—	—	—	—	—	0	5
1	11	13	—	—	—	—	0	11
2	21	19	22	—	—	—	1	19
3	33	29	28	30	—	—	2	28
4	41	41	38	36	37	—	3	36
5	47	49	50	46	43	44	4	43
6	—	55	58	58	53	50	5	50
7	—	—	64	66	65	60	5	60
8	—	—	—	72	73	72	3, 5	72
9	—	—	—	—	79	80	4	79
10	—	—	—	—	—	86	5	86

We now lay out the calculations for all possible stage 2 states and their returns in Table 6.4. As 20 loads have to be produced by stage 6, at least $20 - 5(6 - n)$ loads have to be produced by stage n, where $n = 3, 4, 5, 6$. Hence at least 5 loads must be produced by stage 3, i.e.,

$$5 \le s_3 \le 15.$$

The stage 3 returns are calculated as shown in Table 6.5 by the previous method, the stage 4 returns are calculated in Table 6.6, remembering that $10 \le s_4 \le 20$; the stage 5 returns are calculated in Table 6.7, remembering that $15 \le s_5 \le 20$; and the stage 6 returns, where $s_6 = 20$, are shown in Table 6.8.

The optimal solution is $x_1^* = 1$, $x_2^* = 3$, $x_3^* = 4$, $x_4^* = 4$, $x_5^* = 3$, $x_6^* = 5$, with minimum cost \$124.

Table 6.5

s_3	0	1	2	3	4	5	x_3^*	$f_3(s_3)$
5	47	45	45	40	37	40	4	37
6	54	52	53	49	45	46	4	45
7	64	59	60	57	54	54	4, 5	54
8	76	69	67	64	62	63	4	62
9	83	81	87	71	69	71	4	69
10	90	88	89	81	76	78	4	76
11	—	—	96	93	86	85	5	85
12	—	—	103	100	98	95	5	95
13	—	—	—	116	105	107	4	105
14	—	—	—	—	112	114	4	112
15	—	—	—	—	—	121	5	121

Table 6.6

			x_4					
s_4	0	1	2	3	4	5	x_4^*	$f_4(s_4)$
10	77	78	78	79	80	79	0	77
11	86	85	85	89	89	87	1, 2	85
12	96	94	92	94	87	96	4	87
13	106	104	101	101	94	104	4	94
14	113	114	111	110	101	111	4	101
15	122	121	121	120	120	118	5	118
16	—	130	128	130	130	127	5	127
17	—	—	137	137	140	137	2, 3, 5	137
18	—	—	—	146	147	147	3	146
19	—	—	—	—	156	154	5	154
20	—	—	—	—	—	163	5	163

Table 6.7.

			x_5					
s_5	0	1	2	3	4	5	x_5^*	$f_5(s_5)$
15	123	113	108	104	108	105	3	104
16	132	130	115	111	110	113	4	110
17	142	139	132	118	117	115	5	115
18	151	149	141	135	124	122	5	122
19	159	158	151	144	141	129	5	129
20	168	175	160	154	150	146	5	146

Table 6.8

			x_6					
s_6	0	1	2	3	4	5	x_6^*	$f_6(s_6)$
20	148	135	131	125	125	124	5	124

6.8 Multi–state Variable Problems and the Limitations of D.P.

The dynamic programming formulation of each serial system problem examined so far has the property of possessing just one state variable at each stage. Many problems cannot be adequately formulated without defining two or more state variables per stage. (There are techniques available to reduce the number of state variables. See, for example, Bellman and Dreyfus (1962). Coverage of these is, however, beyond the scope of this book.)

Although such problems can in theory be solved by dynamic programming, in practice the amount of computational effort is often enormous. For example, consider a 5-stage serial system with a single state variable, capable of assuming 10 different states at each stage. At each stage no more than 10 calculations are necessary to evaluate the return for each state, i.e., there are 10^2 calculations per stage. Thus a maximum of 5×10^2 calculations are needed to solve the problem. Suppose now that a new state variable is added at each stage, which also is capable of assuming 10 states at each stage. A maximum of 10^3 calculations are necessary per stage and thus 5×10^3 calculations may be necessary to solve the problem. Thus the number of calculations has increased by a factor of 10. For problems with more possibilities per stage, the increases are enormous, hence the term "the curse of dimensionality," coined by R. Bellman (1957) for this problem. This is a severe limitation to the ability of dynamic programming to solve realistic serial system problems.

We now present and formulate a two-state variable serial system problem with dynamic programming.

The Quick As A Flash freight company carries cargo in its aircraft between two cities. Each aircraft has 1,000 cubic feet of capacity and can carry 5,500 lb of freight. The company accepts three commodities for carriage, c_1, c_2, and c_3, with unit volumes of 100, 200, and 50 cu. ft, respectively, and unit weights of 1,000, 500, and 1,500 lb, respectively. The profits for transporting one item of each of c_1, c_2, and c_3 are \$90, \$100, and \$150, respectively. The problem is to decide how many of each of c_1, c_2, and c_3 will be flown per trip in order to maximize profit.

The decision variable is x_n, the number of c_i accepted, and the stages correspond to an allocation of each of the three commodities, c_1, c_2, c_3. Since the problem has two constraints (volume and weight) at each stage the system will be in two states, s_n and t_n, corresponding to the total volume and weight, respectively, which has been allocated to the nth stage. Let $f_n(s_n, t_n)$ be the return at the nth stage when the system is in states s_n and t_n. The problem can be stated as:

$$\text{Maximize:} \quad 90x_1 + 100x_2 + 150x_3 = x_0$$
$$\text{subject to:} \quad 100x_1 + 200x_2 + 50x_3 \leq 1000$$
$$1000x_1 + 500x_2 + 1500x_3 \leq 5500$$
$$x_1, x_2, x_3 \text{ nonnegative integers.}$$

This is, of course, an integer programming problem and could be solved by the methods of Chapter 4. This particular scenario, of deciding how many of a number of commodities to select subject to various restrictions, is another example of the knapsack problem, also called the *fly-away kit problem*.

Assume that at the nth stage s_n and t_n units of volume and weight, respectively, have been allocated:

$$\sum_{i=1}^{n} v_i x_i = s_n$$

$$s_{i+1} = s_i + v_{i+1}x_{i+1}, \qquad i = 1, 2$$

and

$$\sum_{i=1}^{n} w_i x_i = t_n$$

$$t_{i+1} = t_i + w_{i+1}x_{i+1}, \qquad i = 1, 2,$$

where

$$(v_1, v_2, v_3) = (100, 200, 50)$$

and

$$(w_1, w_2, w_3) = (1000, 500, 1500).$$

We shall adopt forward recursion. Suppose that the return of the system in states s_n and t_n is known, i.e., $f_n(s_n, t_n)$ has been calculated. Suppose now that it has been decided to allocate x_{n+1} of c_{n+1}. The profit at the $(n+1)$th stage is then

$$f_n(s_n, t_n) + d_{n+1}x_{n+1},$$

where

$$(d_1, d_2, d_3) = (90, 100, 150).$$

Thus the return at the $(n+1)$th stage is

$$f_{n+1}(s_{n+1}, t_{n+1}) = \max_{x_{n+1}} \{ f_n(s_n, t_n) + d_{n+1}x_{n+1} \}$$

$$= \max_{x_{n+1}} \{ f_n((s_{n+1} - v_{n+1}x_{n+1}), (t_{n+1} - w_{n+1}x_{n+1})) + d_{n+1}x_{n+1} \}, \tag{6.20}$$

where x_{n+1} is an integer such that

$$0 \le x_{n+1} \le \left[\frac{1000}{v_{n+1}} \right] \tag{6.21}$$

and

$$0 \le x_{n+1} \le \left[\frac{5500}{w_{n+1}} \right], \tag{6.22}$$

where $[a]$ is the integer part of real number a.

If the reader solves this problem using (6.20) and (6.21) he will become convinced that the addition of the extra constraint (6.22) has created a great deal of extra computation. If further constraints of the form of (6.21) and (6.22) were added to the problem, it would become increasingly unattractive to solve it using dynamic programming.

The optimal solution to problem (6.20), (6.21), (6.22) is

$$x_1^* = 0$$
$$x_2^* = 4$$
$$x_3^* = 2$$

with value

$$x_0^* = 700.$$

6.9 Exercises

1. Find the shortest path from the point with the lowest index number to the point with the highest index number in the networks in Exercise 1, Chapter 5, using dynamic programming.

2. Solve the following problem by dynamic programming using (a) forward recursion and (b) backward recursion, and compare the computational effort involved in the two approaches.

   Given a total resource of 8 units and a benefit of $7x_n - nx_n^2$ at stage n ($n = 1, 2, 3, 4$), where x_n is the allocation made at the nth stage, find the optimal allocation policy to maximize total return if all 8 units must be allocated. Assume that each x_n, $n = 1, 2, \ldots, 4$, is a nonnegative real number.

3. Solve Exercise 2 if x_n must be a nonnegative integer, $n = 1, 2, 3, 4$.

4. A production process produces integer numbers of units of a single commodity over 4 periods I, II, III, and IV, where the maximum number produced in any period is 6. There is a storage cost of $1.00 per unit per complete period. Table 6.9 gives the production cost for different numbers of units in the different periods. Find the minimum total cost of production and storage if 19 units must be produced by the end of period IV. Solve this problem by dynamic programming.

Table 6.9. Data for Exercise 4.

Number produced	Cost in period:			
	I	II	III	IV
0	2	6	5	4
1	4	7	8	5
2	8	9	11	9
3	9	11	15	13
4	11	15	16	15
5	12	19	17	17
6	14	20	20	22

5. Solve the following problem by dynamic programming:

$$\text{Maximize:} \quad x_0 = x_1 x_2 x_3 x_4$$
$$\text{subject to:} \quad x_1 + x_2 + x_3 + x_4 = 9$$
$$x_i \geq 0, \quad i = 1, 2, 3, 4.$$

6. Solve the following problem by dynamic programming:

$$\text{Minimize:} \quad x_1^2 + x_2^2 + x_3^2 + x_4^2 + x_5^2$$
$$\text{subject to:} \quad x_1 x_2 x_3 x_4 x_5 = 11$$
$$x_i \geq 0, \quad i = 1, 2, 3, 4, 5.$$

7. Solve the following linear integer programming problem by dynamic programming:

$$\text{Minimize:} \quad x_1 + x_2$$
$$\text{subject to:} \quad 3x_1 + 4x_2 \geq 12$$
$$x_1, x_2 \text{ nonnegative integers.}$$

8. Solve the following nonlinear integer programming problem by dynamic programming using (a) forward recursion and (b) backward recursion. Compare the amount of computational difficulty involved in the two approaches.

$$\text{Maximize:} \quad 8x_1^2 + 4x_2^2 - 3x_1 - 4x_2$$
$$\text{subject to:} \quad 3x_1 + 4x_2 \leq 24$$
$$4x_1 + 5x_2 \leq 20$$
$$x_1, x_2 \text{ nonnegative integers.}$$

9. (A knapsack problem.) A burglar is confronted with seven objects with respective weights and values $(40, 50, 30, 10, 10, 40, 30)$ and $(40, 60, 10, 10, 3, 20, 60)$. He can carry away 100 units in weight. Solve by dynamic programming the problem of determining which objects he should remove given an objective of maximum value and that only one item of each object is available.

10. Solve Exercise 9 with the extra proviso that each object is now also assigned a volume of $(25, 50, 25, 25, 50, 0, 75)$, respectively, and the total volume that the burglar can remove is 100 units.

11. Solve the problem posed in Section 6.8.

12. (The farmer's problem.) At the beginning of a certain year a farmer has 20 tons of seed potatoes. In five years' time he is going to sell all the potatoes, if any, that he then has. If he keeps a ton of seed potatoes it will produce 3 tons of seed potatoes in a year's time. He estimates that the selling price of a ton of seed potatoes over the next five years is going to be 400, 330, 44, 15, and 5 units, respectively. The problem is to decide how many tons of seed to keep and plant each year and how many to sell. Formulate and solve this problem using dynamic programming, assuming that only integer numbers of tons of potatoes are considered.

Chapter 7

Classical Optimization

7.1 Introduction

Until now we have considered the optimization of a linear function subject to linear constraints. This assumption of linearity is now relaxed and we examine the complex problems of optimizing a function which is not necessarily linear which may possibly be subject to constraints which are also not necessarily linear. This present chapter is concerned with the calculus necessary to identify the optimal points of a continuous function or a functional. This is often called *classical optimization*, even though many of the results are of relatively recent origin. Occasionally these methods can be used to solve real-world problems. However, it is usual that too many variables are present for the methods to be at all efficient from the point of view of numerical computation. In these cases nonlinear programming algorithms must be developed and some of these are presented in the next chapter. However, most of these algorithms rely on the theoretical development of the present chapter.

7.2 Optimization of Functions of One Variable

7.2.1 Definitions

Consider a continuous function,

$$f : I \to \mathbf{R},$$

where $I = (a, b)$ is some open interval on the real line and $\mathbf{R}$ is the set of real numbers. We now present some definitions concerning properties of the values that $f(x)$, $x \in I$ can assume.

Definition 7.1. f has a *global minimum* at $x_1 \in I$ if

$$f(x_1) \le f(x) \quad \text{for all } x \in I.$$

(A global minimum is sometimes called an *absolute minimum*.)

Definition 7.2. f has a *global maximum* at $x_1 \in I$ if

$$f(x_1) \ge f(x) \quad \text{for all } x \in I.$$

(A global maximum is sometimes called an *absolute maximum*.)

Definition 7.3. f has a *global extremum* at $x_1 \in I$ if f has either a global minimum or a global maximum at x_1. (A global extremum is sometimes called an *absolute extremum*.)

Definition 7.4. f has a *local minimum* at $x_1 \in I$ if there exists a $\delta \in \mathbf{R}^+$ such that

$$f(x_1) \le f(x)$$

for all $x \in I$ satisfying

$$|x - x_1| < \delta.$$

(A local minimum is sometimes called a *relative minimum*.)

Definition 7.5. f has a *local maximum* at $x_1 \in I$ if there exists a $\delta \in \mathbf{R}^+$ such that

$$f(x_1) \ge f(x)$$

for all $x \in I$ satisfying

$$|x - x_1| < \delta.$$

(A local maximum is sometimes called a *relative maximum*.)

Definition 7.6. f has a *local extremum* at $x_1 \in I$ if f has either a local minimum or a local maximum at x_1. (A local extremum is sometimes called a *relative extremum*.)

Figure 7.1 serves to illustrate the concepts just defined.

Definition 7.7. f has a *stationary point* at $x_1 \in I$ if f is differentiable at x_1 and

$$f'(x_1) = 0.$$

(A stationary point is called a *critical point* by some authors.)

There may be points which are local or even global extrema of f but which are not stationary points. Such a point is x_6 in Figure 7.1, where f is

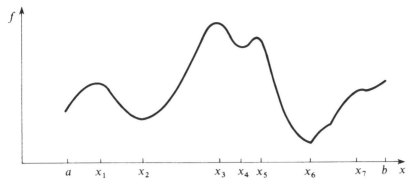

Figure 7.1. Points x_1, x_3, and x_5 are local maxima. Point x_3 is a global maximum. Points x_2, x_4, and x_6 are local minima. Point x_6 is a global minimum.

not differentiable. Further if f is defined on the *closed* interval, $[a, b]$, definitions 7.4 and 7.5 have to be modified for the special cases $x_1 = a$ or $x_1 = b$. In these cases the neighbourhoods $|x_1 - x| < \delta$ are defined as

$$|x_1 - x| < \delta, \qquad x_1 \neq a \text{ or } b$$
$$0 < x - x_1 < \delta, \quad \text{if } x_1 = a,$$
$$0 < x_1 - x < \delta, \quad \text{if } x_1 = b.$$

It is possible that $x_1 = a$ or $x_1 = b$ are global extrema even though f does not have a stationary point at either.

7.2.2 A Necessary Condition for Local Extrema

Given a function $f : I \to \mathbf{R}$ with I open, it is often of interest to find the global extrema of f. Unfortunately it is not easy to find the global extrema directly. Thus we set our sights a little lower and develop ways of finding all local extrema.

Theorem 7.1. *If* $f : I \to \mathbf{R}$ *is differentiable at* x_1, *in an open interval* I, *then* f *has a local extremum at* $x_1 \in I \Rightarrow f'(x_1) = 0$.

PROOF. Suppose $x_1 \in I$ is a local minimum. By assumption $f'(x_1)$ exists, and

$$\lim_{\Delta x \to 0^+} \frac{f(x_1 + \Delta x) - f(x_1)}{\Delta x} = \lim_{\Delta x \to 0^-} \frac{f(x_1 + \Delta x) - f(x_1)}{\Delta x} = f'(x_1). \quad (7.1)$$

But, as x_1 is a local minimum, there exists a $\delta \in \mathbf{R}^+$ such that, for all Δx where $|\Delta x| \leq \delta$,

$$\frac{f(x_1 + \Delta x) - f(x_1)}{|\Delta x|} \geq 0.$$

Hence the first limit in (7.1) is nonnegative. However, if

$$\Delta x < 0,$$

as it is in the middle expression in (7.1), we have

$$\frac{f(x_1 + \Delta x) - f(x_1)}{\Delta x} \le 0.$$

Hence the second limit in (7.1) is nonpositive. As $f'(x_1)$ must equal both of these limits, one nonnegative and one nonpositive, they must both be zero. Hence $f'(x_1) = 0$.

The proof when x_1 is a local maximum is similar.   □

The necessary condition for a local extremum is not sufficient, as evidenced by point x_7 in Figure 7.1, where

$$f'(x_7) = 0$$

but x_7 is not a local extremum. Indeed x_7 is a *point* of *inflection* which is defined as follows.

Definition 7.8. f has a *point of inflection* at x_1 if f has a stationary point at x_1 but f does not have a local extremum at x_1 and f' has a local extremum at x_1.

This leads to a distinction between the stationary points of f:

Definition 7.9. f has a *critical point* at x_1 if f has a stationary point at x_1 and x_1 is a local extremum for f but not for f'.

Thus points of inflection are stationary points which are not critical points.

7.2.3 Sufficient Conditions for Local Extrema

In view of Theorem 7.1, it is desirable to develop sufficient conditions for local extrema to exist. Then the stationary points found can be examined to see if any of them are local extrema. To this end, consider the Taylor series expansion about a point $x_1 \in I$ (see Section 9.2 of the Appendix):

$$f(x_1 + h) = f(x_1) + hf'(x_1)$$

$$+ \frac{h^2}{2} f''(\theta x_1 + (1 - \theta)(x_1 + h)), \quad \text{for some } \theta, 0 < \theta < 1, \quad (7.2)$$

where it is assumed that

$$(x_1 + h) \in I,$$

and f has first and second derivatives f', f'' for all points in I. Suppose that f has a stationary point at x_1. Then, by Theorem 7.1,

$$f'(x_1) = 0.$$

Then (7.2) can be rearranged as:

$$f(x_1 + h) - f(x_1) = \frac{h^2}{2} f''(\theta x_1 + (1 - \theta)(x_1 + h)), \qquad 0 < \theta < 1. \quad (7.3)$$

Assume

$$f''(x_1) > 0.$$

Now if f'' is continuous on I, at all points sufficiently near x_1, f'' will have the same sign as it does at x_1, i.e., positive. Thus, for all h sufficiently small in magnitude,

$$\frac{h^2}{2} f''(\theta x_1 + (1 - \theta)(x_1 + h)) > 0.$$

Using this result in (7.3), we obtain

$$f(x_1 + h) - f(x_1) > 0.$$

We conclude that if

$$f'(x_1) = 0,$$

and f'' is continuous in a neighbourhood of x_1, then

$$f''(x_1) > 0$$

is a *sufficient* condition for x_1 to be a local minimum.

It can be shown analogously that if

$$f'(x_1) = 0,$$

and f'' is continuous in a neighbourhood of x_1, then

$$f''(x_1) < 0$$

is a *sufficient* condition for x_1 to be a local maximum.

The preceding deductions cannot be used to come to any conclusions about the character of x_1 if

$$f'(x_1) = f''(x_1) = 0.$$

Indeed, it may be that

$$f^{(n)}(x_1) = 0, \qquad n = 1, 2, \ldots, k$$

for some integer $k > 2$. The following theorem settles such cases.

Theorem 7.2. *If*

$$f^{(n)}(x_1) = 0, \qquad n = 1, 2, \ldots, k, \qquad\qquad (7.4)$$

and

$$f^{(k+1)}(x_1) \neq 0 \qquad\qquad (7.5)$$

and $f^{(k+1)}$ is continuous in a neighbourhood of x_1, then f has a local extremum at x_1 if and only if $(k + 1)$ is even. If

$$f^{(k+1)}(x_1) > 0,$$

x_1 is a local minimum. If

$$f^{(k+1)}(x_1) < 0,$$

x_1 is a local maximum.

PROOF. *Sufficient condition.* We assume the hypothesis of Theorem 7.2. We wish to show that f has a local extremum at x_1. Taylor's theorem about point x_1 yields

$$f(x_1 + h) = f(x_1) + hf'(x_1) + \frac{h^2}{2} f''(x_1) + \cdots$$

$$+ \frac{h^{k+1}}{(k+1)!} f^{(k+1)}(\theta x_1 + (1 - \theta)(x_1 + h)),$$

for some $\theta, 0 \leq \theta \leq 1$. Using (7.4) and rearranging, this becomes

$$f(x_1 + h) - f(x_1) = \frac{h^{k+1}}{(k+1)!} f^{(k+1)}(\theta x_1 + (1 - \theta)(x_1 + h)), \qquad 0 \leq \theta \leq 1. \quad (7.6)$$

It is assumed that $f^{(k+1)}$ is continuous at x_1. This fact can be used to show that at all points sufficiently near x_1, $f^{(k+1)}$ will have the same sign as $f^{(k+1)}(x_1)$. Hence if h is sufficiently small, $f^{(k+1)}(\theta x_1 + (1 - \theta)(x_1 + h))$ will have the same sign as $f^{(k+1)}(x_1)$. In view of this, on examining (7.6) we can see that for odd $k + 1$ and sufficiently small positive h, $f(x_1 + h) - f(x_1)$ will have the same sign as $f^{(k+1)}(x_1)$. However, for sufficiently small negative h, $f(x_1 + h) - f(x_1)$ has the opposite sign to $f^{(k+1)}(x_1)$. Hence, for odd $k + 1$, x_1 is not a local extremum. However if $k + 1$ is assumed to be even, $f(x_1 + h) - f(x_1)$ has the same sign as $f^{(k+1)}(x_1)$, independently of the sign of h. If

$$f^{(k+1)}(x_1) > 0,$$

then

$$f(x_1 + h) - f(x_1) > 0$$

for all h sufficiently small in magnitude, and thus x_1 is a local minimum. If

$$f^{(k+1)}(x_1) < 0$$

then

$$f(x_1 + h) - f(x_1) < 0$$

for all h sufficiently small in magnitude and thus x_1 is a local maximum.

Necessary condition. We assume (7.4) and (7.5) and that f has a local extremum at x_1. We wish to show that $k + 1$ is even. Let us suppose for definiteness that f has a local minimum at x_1, i.e.,

$$f(x_1 + h) - f(x_1) > 0$$

for all h sufficiently small in magnitude. Using (7.6), we obtain

$$\frac{h^{k+1}}{(k+1)!} f^{(k+1)}(\theta x_1 + (1-\theta)(x_1+h)) > 0, \qquad 0 < \theta < 1, \qquad (7.7)$$

i.e., the expression on the left-hand side of (7.7) is of constant sign, namely positive. However, from the arguments mounted earlier in the proof, $f^{(k+1)}(\theta x_1 + (1-\theta)(x_1+h))$ will have constant sign (it cannot be zero if (7.7) is to hold) for h sufficiently small in magnitude. Now when h is negative the expression in (7.7) can have constant sign only if $k+1$ is even.

A similar argument follows when f has a local maximum at x_1. This completes the proof. ☐

7.2.4 Examples

Consider

$$f(x) = x^3 - 9x^2 + 27x - 27.$$

We use the previous results to find the extrema of this function. The first derivative is

$$f'(x) = 3x^2 - 18x + 27,$$

which has a unique zero at $x_1 = 3$, which, by Theorem 7.1, is the only candidate for an extremum. However,

$$f''(x) = 6x - 18,$$

so that

$$f''(3) = 0.$$

Now

$$f^{(3)}(x) = 6 \neq 0,$$

but, as $k+1 = 3$ is odd, $x_1 = 3$ is not an extremum. Indeed, f has a point of inflection at $x_1 = 3$.

Consider

$$f(x) = x^4 - 8x^3 + 24x^2 - 32x + 16.$$

The first derivative is

$$f'(x) = 4x^3 - 24x^2 + 48x - 32,$$

which has a unique zero at $x_1 = 2$, which is thus the only candidate for an extremum.

However,

$$f''(x) = 12x^2 - 48x + 48$$

and

$$f''(2) = 0.$$

Also,
$$f^{(3)}(x) = 24x - 48$$
and
$$f^{(3)}(2) = 0.$$
But
$$f^{(4)}(x) = 24.$$
Hence
$$f^{(4)}(2) = 24 \neq 0.$$
Now as
$$(k + 1) = 4, \quad \text{which is even, hence}$$
by theorem 7.02
$$x_1 = 2 \text{ is a local extremum.}$$
As
$$f^{(4)}(2) > 0,$$
$$x_1 = 2$$

is a local minimum.

7.2.5 The Solution of Nonlinear Equations

It can be seen in the previous example that in order to locate the extrema of a function f it is necessary to find the roots of

$$f'(x) = 0. \tag{7.8}$$

This is often a difficult task when f is of high order. There are many numerical methods which exist for locating the roots. Some of these are presented in Conte and de Boor (1972). We present one simple method here; the interested reader should seek further advice if he suspects his function is ill-behaved. The method presented here is called *Newton's method* and is motivated as follows.

We assume that f has continuous second derivatives and that some estimate x_1 of a solution to (7.8) is available. If no such estimate is known, x_1 is chosen at random. If x_1 is a reasonably good estimate, the Taylor series expansion of f' about x_1 can be approximated as:

$$f'(x) = f'(x_1) + (x - x_1)f''(x_1).$$

Hence if x is a solution to (7.8),

$$0 = f'(x_1) + (x - x_1)f''(x_1)$$
$$x = x_1 - f'(x_1)/f''(x_1). \tag{7.9}$$

Now unless f is a quadratic, x will not in general be an exact solution to (7.8). However, x can be used as an improved estimate. Indeed, (7.9) can be looked upon as the first equation in a family which generates successive improved estimates of a solution to (7.8). The family has the following general form:

$$x_{n+1} = x_n - \frac{f'(x_n)}{f''(x_n)}, \qquad n = 1, 2, \ldots . \tag{7.10}$$

Once an estimate is finally found which is sufficiently close to a root, a new starting point can be selected in an effort to find a new root. This procedure is repeated until all roots are found. However there is no guarantee that this method will be successful. The reader is referred to Himmelblau (1972) for a more complete treatment of this problem.

7.2.6 Global Extrema

Let us now return to the classical optimization of a function of one variable. As was explained in Section 7.2.2, if I is an open interval Theorem 7.1 is used to identify local extrema. However, if I is closed the possibility exists that a global extremum occurs at one or both of the endpoints of I. Hence when I is closed its endpoints must be considered candidates for global extrema. This possibility will occur when f has no critical points in the interior of I.

For example, let

$$f(x) = x^3 - 9x^2 + 24x + 1, \qquad x \in I = (0, 3).$$

Then

$$f'(x) = 3x^2 - 18x + 24,$$

which has zeros at $x_1 = 2$ and 4. By Theorem 7.1, as I is open, these points are the only candidates for extrema. However, $4 \notin (0, 3)$, hence $x_1 = 4$ can be disregarded. Now,

$$f''(x) = 6x - 18.$$

Therefore

$$f''(2) = -6 < 0.$$

Thus $x_1 = 2$ is a local maximum. No local (or global) minimum exists within I. This is because, as 0 is approached from the right, values of f become successively lower, without ever attaining the limit of $f(0)$.

However, if I is redefined as

$$I = [0, 6],$$

it is now closed and the endpoints $x_1 = 0$ and $x_1 = 6$ must be checked. Indeed,

$$f(0) = 1$$
$$f(6) = 37.$$

But

$$f(2) = 21$$
$$f(4) = 17.$$

Hence

$$f(0) < f(4) < f(2) < f(6).$$

Thus

$x_1 = 0$ is the global minimum

$x_1 = 4$ is a local minimum

$x_1 = 2$ is a local maximum

$x_1 = 6$ is the global maximum.

7.2.7 Concave and Convex Functions

It was pointed out in Section 7.2.2 that the necessary condition of Theorem 7.1 is not always sufficient. However, there are two classes of functions for which the condition is sufficient. These are *concave* and *convex* functions, which are defined next.

Definition 7.10. A function, f defined on a closed interval, I is said to be *concave on I* if for all $\alpha \in \mathbf{R}$, $0 \le \alpha \le 1$, and for all $x_1, x_2 \in I$,

$$f(\alpha x_1 + (1 - \alpha)x_2) \ge \alpha f(x_1) + (1 - \alpha)f(x_2). \tag{7.11}$$

Definition 7.11. A function f defined on a closed interval I is said to be *convex on I* if $-f$ is concave on I.

Some examples of concave and convex functions are given in Figures 7.2(a) and (b), respectively.

We now build up a series of results which amount to a somewhat stronger result than the converse of Theorem 7.1. This is that, for f concave (convex),

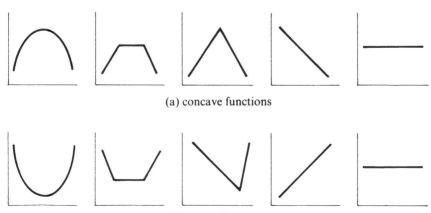

(a) concave functions

(b) convex functions

Figure 7.2.   Concave and convex functions.

$f'(x^*) = 0$ is sufficient for x^* to be a global maximum (minimum). We begin with:

Theorem 7.3. *If f is concave on a closed interval I with a local maximum at $x^* \in I$, then f must have a global maximum at x^*.*

PROOF. As f has a local maximum at x^* there exists $\delta \in \mathbf{R}^+$ such that for all $x \in I$ satisfying

$$|x^* - x| < \delta, \tag{7.12}$$

we have

$$f(x^*) \geq f(x). \tag{7.13}$$

Hence if we can show for all $x \in I$ that (7.13) holds we have shown that x^* is a global maximum. We prove this by contradiction. Let $x_1 \in I$ be such that

$$f(x^*) < f(x_1). \tag{7.14}$$

Now, as f is concave, we can invoke (7.11), with

$$x_2 = x^*.$$

That is, for all $\alpha \in \mathbf{R}$, $0 \leq \alpha \leq 1$,

$$f(\alpha x_1 + (1 - \alpha)x^*) \geq \alpha f(x_1) + (1 - \alpha)f(x^*). \tag{7.15}$$

By taking α sufficiently close to 0, i.e., $0 < \alpha < \delta/(|x^* - x_1|)$, the point $(\alpha x_1 + (1 - \alpha)x^*)$ will satisfy (7.13), that is,

$$f(x^*) \geq f(\alpha x_1 + (1 - \alpha)x^*). \tag{7.16}$$

Then by (7.15), we have

$$f(x^*) \geq \alpha f(x_1) + (1 - \alpha)f(x^*),$$

which contradicts (7.14). Thus no point $x_1 \in I$ can be found for which (7.14) holds. Thus

$$f(x^*) \geq f(x) \quad \text{for all } x \in I.$$

That is, f has a global maximum at x^*. □

One can prove an analogous theorem for convex functions, as follows:

Theorem 7.4. *If f is convex on closed interval, I with a local minimum at $x^* \in I$ then f must have a global minimum at x^*.*

We leave the proof of Theorem 7.4 as an exercise for the reader. We now prove a theorem which leads to Theorem 7.7, the main result of this section.

Theorem 7.5. *If f is concave on a closed interval I and there exists a neighbourhood, $N(x_1)$ of an interior point $x_1 \in I$ such that f' is continuous in $N(x_1)$, then*

$$f(x) \leq f(x_1) + f'(x_1)[x - x_1] \quad \text{for all } x \in I.$$

PROOF. Let x be a point in I. Then as f is concave on I, by (7.11) we have

$$f(\alpha x + (1 - \alpha)x_1) \geq \alpha f(x) + (1 - \alpha)f(x_1) \quad \text{for all } \alpha \in \mathbf{R}, 0 \leq \alpha \leq 1.$$

On rearranging, we obtain

$$f(x_1 + \alpha(x - x_1)) - f(x_1) \geq \alpha(f(x) - f(x_1)). \tag{7.17}$$

By Taylor's theorem (see Section 9.2 in the Appendix), with

$$h = \alpha(x - x_1),$$

we have

$$f(x_1 + \alpha(x - x_1)) - f(x_1)$$
$$= \alpha f'(x_1 + \theta\alpha(x - x_1))(x - x_1), \quad \text{for some } \theta, 0 < \theta < 1.$$

$$f'(x_1 + \theta\alpha(x - x_1))\alpha(x - x_1) \geq \alpha(f(x) - f(x_1)),$$

hence

$$f'(x_1 + \theta\alpha(x - x_1))(x - x_1) \geq f(x) - f(x_1), \quad \text{for } \alpha > 0.$$

Therefore,

$$\lim_{\alpha \to 0} f'(x_1 + \theta\alpha(x - x_1))(x - x_1) = f'(x_1)(x - x_1) \geq f(x) - f(x_1),$$

and

$$f(x_1) + f'(x_1)(x - x_1) \geq f(x),$$

as required.   □

The corresponding theorem for convex functions is left for the reader to prove.

Theorem 7.6. *If f is convex on a closed interval I and is differentiable at a point x_1 in I, then*

$$f(x) \geq f(x_1) + f'(x_1)(x - x_1) \quad \text{for all } x \in I.$$

We leave the proof of Theorem 7.6 as an exercise for the reader. We are now in a position to prove the following theorem:

Theorem 7.7. *If f is concave on a closed interval I and there exists a point $x^* \in I$ such that*

$$f'(x^*) = 0,$$

then f has a global maximum at x^*.

PROOF. Let N be a neighbourhood of x^* contained in I. Let x be any point point in N. Then, by Theorem 7.5, we have

$$f(x) \leq f(x^*) + f'(x^*)(x - x^*)$$

which reduces to

$$f(x) \le f(x^*).$$

Thus f has a local maximum at x^*. Hence by Theorem 7.3 f has a global maximum at x^*. $\square$

The sequel for convex functions is

Theorem 7.8. *If f is convex on a closed interval I and there exists a point x^* in I such that*

$$f'(x^*) = 0,$$

then f has a global minimum at x^*.

We now turn our attention to the *minimization* of a concave function. Of course no minima may exist for such a function defined on interval I where $I = \mathbf{R}$ or where I is bounded but open. However, if I is bounded the global minimum will occur at an endpoint. This is now stated and proven formally.

Theorem 7.9. *If f is concave on a closed interval $I = [a, b]$, then f will have a global minimum at a or b or both.*

PROOF. This proof is by contradiction. Suppose that there exists a point x^* in the interior of I which is a global minimum and a and b are not global minima, i.e.,

$$f(x^*) < f(a), \qquad f(x^*) < f(b). \tag{7.18}$$

Then, as $a < x^* < b$, there exists $\alpha \in \mathbf{R}, 0 < \alpha < 1$ such that

$$x^* = \alpha a + (1 - \alpha)b.$$

Now from Definition 7.10, with $x_1 = a$ and $x_2 = b$, we have

$$f(x^*) \ge \alpha f(a) + (1 - \alpha)f(b),$$

and, by (7.18),

$$f(x^*) > \alpha f(x^*) + (1 - \alpha)f(x^*),$$

which is a contradiction. $\square$

For completeness we state the equivalent result for convex functions. We leave the proof as an exercise for the reader.

Theorem 7.10. *If f is convex on a closed interval $I = [a, b]$, then f will have a global maximum at a or b or both.*

7.3 Optimization of Unconstrained Functions of Several Variables

Of course many models of real-world problems involve functions of many variables. In this section we study the classical mathematics required for the optimization of such functions and generalize the results obtained in the earlier sections of this chapter.

7.3.1 Background

As in earlier chapters we denote a vector of n variables by $X = (x_1, x_2, \ldots, x_n)^T$. Consider a function, $f : S \to \mathbf{R}$ where S is a region in n-dimensional Euclidean space. Then Definitions 7.1–7.11 hold for multidimensional functions with X replacing x.

7.3.2 A Necessary Condition for Local Extrema

As in the single-variable, case we develop a necessary condition for the existence of local extrema.

Theorem 7.11. *If $\partial f(X)/\partial x_j$ exists for all $X \in S$ and for all $j = 1, 2, \ldots, n$, and if f has a local extremum at X^* in the interior of S, then*

$$\frac{\partial f(X^*)}{\partial x_j} = 0, \qquad j = 1, 2, \ldots, n.$$

PROOF. Suppose that the conditions of the theorem hold. Let

$$X^* = (x_1^*, x_2^*, \ldots, x_n^*)^T.$$

Consider the points in S which are generated when all the variables x_i except for x_j are held fixed at $x_i = x_i^*$, $i = 1, 2, \ldots, j-1, j+1, \ldots, n$ for some j, $1 \le j \le n$. Now define

$$\bar{f}(x_j) = f(x_1^*, x_2^*, \ldots, x_{j-1}^*, x_j, x_{j+1}^*, \ldots, x_m^*).$$

As f has a local extremum at X^*, $\bar{f}$ must have a local extremum at x_j. Thus, by Theorem (7.1), we have

$$\bar{f}'(x_j) = 0.$$

But as

$$f'(x_j) = \frac{\partial f(X^*)}{\partial x_j}.$$

we have

$$\frac{\partial f(X^*)}{\partial x_j} = 0.$$

Thus, as j was chosen arbitrarily, we have

$$\frac{\partial f(X^*)}{\partial x_j} = 0, \qquad j = 1, 2, \ldots, n. \qquad \square$$

7.3.3 A Sufficient Condition for Local Extrema

As with the single-variable case, it is desirable to develop sufficient conditions for a local extrema to exist. Then any points identified by an application of the result of Theorem 7.11 can be examined to see if they are local extrema. Theorem 7.12 provides the desired conditions. We assume that the second partial derivatives not only exist but are continuous in some neighbourhood of any point X^* for which the condition of Theorem 7.11 holds.

Theorem 7.12. *If*

$$\frac{\partial f(X^*)}{\partial x_j} = 0, \qquad j = 1, 2, \ldots, n,$$

for some X^* in the interior of S, and if $H(X^*)$, the Hessian matrix of f evaluated at X^*, is negative definite, then f has a local maximum at X^*.

PROOF. Consider the Taylor series expansion of f about X^* (see Section 9.2 of the Appendix), where it is assumed that the first and second derivatives of f exist in S:

$$f(X^* + h) = f(X^*) + Vf(X^*)^T h$$
$$+ \tfrac{1}{2} h^T H(\theta X^* + (1 - \theta)(X^* + h)) h, \quad \text{for some } \theta, 0 < \theta < 1. \quad (7.19)$$

In view of the hypothesis, we have

$$Vf(X^*) = \left(\frac{\partial f(X^*)}{\partial x_1}, \frac{\partial f(X^*)}{\partial x_2}, \ldots, \frac{\partial f(X^*)}{\partial x_n} \right)^T = 0.$$

Hence (7.19) can be rearranged to become

$$f(X^* + h) - f(X^*) = \tfrac{1}{2} h^T H(\theta X^* + (1 - \theta)(X^* + h)) h, \qquad 0 < \theta < 1. \quad (7.20)$$

Let us now consider the sign of the right-hand side of (7.20). As the second partial derivatives of f are continuous in some neighbourhood of X^*, for h sufficiently small, the entries of $H(\theta X^* + (1 - \theta)(X^* + h))$ will have the same sign as the corresponding entries of $H(X^*)$. Now, as $H(X^*)$ is negative definite, $H(\theta X^* + (1 - \theta)(X^* + h))$ will be negative definite. Thus $h^T H(\theta X^* + (1 - \theta)(X + h))h$ will be negative (see Section 9.2 on quadratic forms). $\square$

We have shown, using (7.20), that, for all $X^* + h$ in a neighbourhood of X^*,

$$f(X^* + h) - f(X^*) < 0,$$

i.e., f has a local maximum at X^*. One can prove the following theorem in an analogous fashion:

Theorem 7.13. *If*

$$\frac{\partial f(X^*)}{\partial x_j} = 0, \qquad j = 1, 2, \ldots, n,$$

for some X^* in the interior of S and $H(X^*)$ is positive definite then f has a local minimum at X^*.

We leave the proof of Theorem 7.13 as an exercise for the reader.

7.3.4 Illustrative Examples

Find the extreme points of

$$f(X) = -x_1^2 - 6x_2^2 - 4x_1 + 8x_2 + 143, \qquad X \in \mathbf{R}^2.$$

Using the result of Theorem 7.11, we have

$$\frac{\partial f}{\partial x_1} = -2x_1 - 4 = 0 \Rightarrow x_1 = -2$$

$$\frac{\partial f}{\partial x_2} = -12x_2 + 8 = 0 \Rightarrow x_2 = \tfrac{2}{3},$$

Thus $X_0 = (-2, \tfrac{2}{3})$ is the only candidate for an extreme point. Now

$$H(X) = \begin{pmatrix} -2 & 0 \\ 0 & -12 \end{pmatrix},$$

which is negative definite. Thus, by Theorem 7.12, X_0 is a local maximum. As X_0 is the only maximum point, f has a global maximum at X_0, with value

$$f(X_0) = 149\tfrac{2}{3}.$$

Let us now consider an example that is a little more challenging. Find the extreme points of

$$f(X) = x_1^2 + x_2^2 + x_3^2 + x_1 x_2 + x_1 x_3 + x_2 x_3 - 7x_1 - 8x_2 - 9x_3 + 101, \qquad X \in \mathbf{R}^3.$$

Using the result of Theorem 7.11, we have

$$\frac{\partial f}{\partial x_1} = 2x_1 + x_2 + x_3 - 7 = 0$$

$$\frac{\partial f}{\partial x_2} = x_1 + 2x_2 + x_3 - 8 = 0$$

$$\frac{\partial f}{\partial x_3} = x_1 + x_2 + 2x_3 - 9 = 0.$$

Solving these three equations simultaneously yields the unique solution $X_0 = (1, 2, 3)$, which is thus the only candidate for an extreme point. Now

$$H(X) = \begin{pmatrix} 2 & 1 & 1 \\ 1 & 2 & 1 \\ 1 & 1 & 2 \end{pmatrix},$$

which is positive definite. Thus, by Theorem 7.12, X_0 is a local minimum. As X_0 is the only minimum point, f has a global minimum at X_0, with value

$$f(X_0) = 76.$$

7.3.5 Discussion

Lest the reader begin to believe that the above procedure is always as straight-forward as in analyzing the examples of Section 7.3.4, a few words of caution are in order. First, a system of equations derived from

$$\frac{\partial f(X)}{\partial x_j} = 0, \qquad j = 1, 2, \dots, n \tag{7.21}$$

must be solved in order to find the stationary points. The system of equations will be nonlinear if f has terms of cubic or higher powers. This can sometimes be achieved using what is known as *Newton's method for systems*. However, this usually requires a great deal of computational effort, and unless there is some information available about the likely location of roots, the method may fail to converge. The reader is referred to Henrici (1964) for a more full discussion of this problem. Of course it is possible that the system (7.21) may be inconsistent in the sense that it has no solutions. In this case f has no extreme points.

Even if it is possible to locate the possible candidates for extrema by finding all solutions X_0 to (7.21), one still has to establish the definiteness of $H(X_0)$. For nontrivial systems this is often a difficult task. In fact, for systems arising from most real-world problems it is usually far more efficient to try and establish the nature of X_0 by examining the behaviour of f in the neighbourhood of X_0 directly.

As has been seen in Theorems 7.12 and 7.13, if (7.21) holds and $H(X^*)$ is negative (positive) definite then f has a local maximum (minimum) at X^*. However, if $h^T H(X^*)h$ changes sign for different h, then X^* is *not* a local extremum. The reader will note that nothing has been said about the cases where $H(X^*)$ is negative semidefinite or positive semidefinite. These are equivalent to the single-variable situations covered in Theorem 7.2. The multivariable situation, however, is complicated and will not be examined here. The reader is referred to Hancock (1960) for a detailed treatment.

In the previous paragraph we alluded to the single-variable case. It is easily seen that this is merely a special case of the theory developed in Section 7.3.

7.3.6 Global Extrema

In the examples in Section 7.3.4 S, the domain of f, was defined to be $\mathbf{R}^2$. When S is thus unrestricted, the point satisfying (7.21) which yields the highest (lowest) value of f will be the global maximum (minimum). As in Section 7.2.6, if the domain of f is closed, the possibility exists that the global extrema occur on the boundary. This will certainly happen if no interior points of S satisfy (7.21).

For example, consider the first function given in Section 7.3.4, with S redefined as the rectangle:

$$S = \{(x_1, x_2): -3 \le x_1 \le 0, 0 \le x_2 \le 1\}.$$

Then, as $X_0 = (-2, \frac{2}{3})$, the only candidate for a local extremum, belongs to S, it is still the global maximum. However, if we redefine S as the square:

$$S = \{(x_1, x_2): 0 \le x_1 \le 1, 0 \le x_2 \le 1\},$$

then $X_0 \notin S$. Thus we must examine the boundaries of S, which are

$$B_1 = \{\alpha(0,0) + (1-\alpha)(0,1): 0 \le \alpha \le 1\}$$
$$B_2 = \{\alpha(0,1) + (1-\alpha)(1,1): 0 \le \alpha \le 1\}$$
$$B_3 = \{\alpha(1,1) + (1-\alpha)(1,0): 0 \le \alpha \le 1\}$$
$$B_4 = \{\alpha(1,0) + (1-\alpha)(0,0): 0 \le \alpha \le 1\}.$$

We now find the extrema of f on each boundary B_i, $= i = 1, 2, \ldots, n$. The value of f at any point on a B_i can be expressed as a function of one variable in α. Starting with B_1, let

$$g(\alpha) = f(0, 1 - \alpha) = -6(1 - \alpha)^2 + 8(1 - \alpha) + 143.$$

Using the methods of Section 7.2, we obtain

$$g'(\alpha) = -12(1 - \alpha)(-1) - 8,$$

which has a unique zero at

$$\alpha^* = \tfrac{1}{3}.$$

Also, we have

$$g''(\alpha^*) < 0,$$

indicating that α^* corresponds to a maximum. As the interval over which g is defined is closed we must also check the endponts:

$$g(0) = 145$$
$$g(1) = 143$$
$$g(\tfrac{1}{3}) = 145\tfrac{2}{3}.$$

Thus the maximum value of f on B_1 occurs at $(0, \frac{1}{3})$, with value $145\frac{2}{3}$, and the minimum at $(0, 1)$ with value 143.

A complete display of this analysis for all four boundaries is given in Figure 7.3, with values of f given. It can be seen that f has a global maximum at $(0, \frac{1}{3})$ with value $145\frac{2}{3}$ and a global minimum at $(1, 0)$ with value 138.

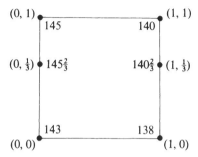

Figure 7.3. Examining boundaries for global extrema.

7.3.7 Concave and Convex Functions

The definitions for concave and convex functions of a single variable can be generalized for functions of several variables.

Definition 7.12. A function f defined on S, a simply connected region in n-dimensional Euclidean space, is said to be *concave on S* if for all $\alpha \in \mathbf{R}$, $0 \leq \alpha \leq 1$, and for all $X_1, X_2 \in S$,

$$f(\alpha X_1 + (1 - \alpha)X_2) \geq \alpha f(X_1) + (1 - \alpha)f(X_2). \qquad (7.22)$$

Definition 7.13. A function f defined on S, a simply connected region in n-dimensional Euclidean space, is said to be *convex on S* if $-f$ is concave on S.

As with the one-dimensional case, we can prove far stronger results for concave and convex functions than for more general functions. We begin by generalizing Theorem 7.3. The proof of Theorem 7.14 follows along the lines of that for theorem 7.3.

Theorem 7.14. *If f is concave on a simply connected region $S \subseteq \mathbf{R}^n$ with a local maximum $X^* \in S$ then f has a global maximum at X^*.*

PROOF. As f has a local maximum at X^* there exists a δ-neighbourhood about X^* such that f at X^* is no less than at any other point in the neighbourhood. That is, there exists $\delta \in \mathbf{R}^+$ such that, for all $X \in S$ such that $\|X^* - X\| < \delta$, we have

$$f(X^*) \geq f(X). \tag{7.23}$$

Hence if we can show for all $X \in S$ that (7.23) holds we have shown that X^* is a global maximum. This is done by contradiction. Suppose (7.23) does not hold for all $X \in S$, i.e., there exists $X_1 \in S$ such that

$$f(X^*) < f(X_1). \tag{7.24}$$

Now, as f is concave, we have

$$f(\alpha X_1 + (1 - \alpha)X^*) \geq \alpha f(X_1) + (1 - \alpha)f(X^*) \quad \text{for all } \alpha, 0 \leq \alpha \leq 1. \tag{7.25}$$

By taking α to within $\delta/(\|X^* - X_1\|)$ of 0, i.e.,

$$0 < \alpha < \delta/(\|X^* - X\|), \tag{7.26}$$

the point $\alpha X_1 + (1 - \alpha)X^*$ will satisfy (7.23); that is,

$$f(X^*) \geq f(\alpha X_1 + (1 - \alpha)X^*). \tag{7.27}$$

Then, by (7.25), we have

$$f(X^*) \geq \alpha f(X_1) + (1 - \alpha)f(x^*),$$

which contradicts (7.24). Thus

$$f(X^*) \geq f(X) \quad \text{for all } X \in S.$$

That is, f has a global maximum at X^*. $\qquad\square$

The proof of the analogous theorem for convex functions is left to the reader.

Theorem 7.15. *If f is concave on a simply connected region $S \subseteq \mathbf{R}^n$ with a local minimum $X^* \in S$, then f has a global minimum at X^*.*

7.4 Optimization of Constrained Functions of Several Variables

Feasible solutions to many realistic optimization problems are constrained to be within a subset of n-dimensional Euclidean space. As examples, a company may not be able to invest more funds than it possesses, time allocated to a machine must be nonnegative, and pollution laws may require values of a certain variables to be less than a given level. When this is the case, as it is in nearly all real-world problems, one must maximize the objective function subject to a number of constraints on its variables. These constraints are usually expressed in the form of equations or inequalities. We begin with the case where all the constraints are equations.

7.4.1 Multidimensional Optimization with Equality Constraints

This problem can be stated in general as follows:

$$\text{Maximize:} \quad f(X) \tag{7.28}$$
$$\text{subject to:} \quad g_j(X) = 0, \quad j = 1, 2, \ldots, m \tag{7.29}$$

where

$$X = (x_1, x_2, \ldots, x_n)^T.$$

A typical numerical example of such a problem is given in the next section. One obvious approach is to use the equations to eliminate some of the variables from the problem. For example consider the following problem:

$$\text{Maximize:} \quad f(X) = x_1^2 + 2(x_2 - 4)^2 + 8$$
$$\text{subject to:} \quad x_1 - x_2^2 + 4 = 0.$$

As

$$x_1 = x_2^2 - 4,$$

we are left with the following unconstrained problem in one dimension:

$$\text{Maximize:} \quad f(x_1) = (x_2^2 - 4)^2 + 2(x_2 - 4)^2 + 8,$$

which is easier to solve. Of course, this approach of elimination will be successful in reducing the number of variables in the problem only if it is possible to express a solution for one or more of the variables explicitly. Often, however, this cannot be done.

It can be shown that when the variables of the objective function must satisfy constraints which are equations, the optimal point must lie on the boundary of the feasible region F. There are a number of methods available for locating optima which lie in the interior of F. We now present the Jacobian method and Lagrange's method, which both transform such a problem into one with its optima all lying in the interior of F.

7.4.1.1 The Jacobian Method

We now present a method which solves the problem (7.28), (7.29). It is assumed that f and $g_j, j = 1, 2, \ldots, m$ have continuous second derivatives. The strategy is to find a suitable expression for the first derivatives of f at all points which satisfy (7.29). The feasible stationary points of f are the ones among these for which

$$\frac{\partial f}{\partial x_i} = 0, \quad i = 1, 2, \ldots, n. \tag{7.30}$$

The maximum points are identified among those satisfying (7.30) by using Theorem 7.12.

These ideas are now placed on a firm mathematical basis. Consider any point X which satisfies (7.29). In any neighbourhood of X there will exist at least one point $X + h$ which satisfies (7.29), because X is on the boundary of the region defined by (7.29). Expanding f and $g_j, j = 1, 2, \ldots, m$, in a Taylor series about X, we get

$$f(X+h) = f(X) + \nabla f(X)^T h + \tfrac{1}{2} h H_f(\theta X + (1-\theta)(X+h))h,$$
$$g_j(X+h) = g_j(X) + \nabla g_j(X)^T h + \tfrac{1}{2} h H_{g_j}(\theta X + (1-\theta)(X+h))h, \qquad j = 1, 2, \ldots, m,$$

for some θ, $0 < \theta < 1$. As $X + h$ approaches X, we get

$$f(X + h) \approx f(X) + \nabla f(X)^T h$$

$$g_j(X + h) \approx g_j(X) + \nabla g_j(X)^T h, \qquad j = 1, 2, \ldots, m.$$

Therefore

$$\partial f(X) \approx \nabla f(X)^T \partial X$$
$$\partial g_j(X) \approx \nabla g_j(X)^T \partial X, \qquad j = 1, 2, \ldots, m.$$

Using (7.29), we get

$$\partial g_j(X) = 0, \qquad j = 1, 2, \ldots, m.$$

Thus we can state, to within a first order approximation,

$$\nabla g_j(X)^T \partial X = 0, \qquad j = 1, 2, \ldots, m. \tag{7.31}$$

Now as $\nabla f(X)$ and $\nabla g_j(X)$, $j = 1, 2, \ldots, m$, consist of known constants, (7.31) constitutes a set of $(m + 1)$ linear equations in $(n + 1)$ unknowns, ∂x_1, $\partial x_2, \ldots, \partial x_n, \partial f(X)$. If the equations are linearly dependent one discards the smallest number whose removal leaves an independent set. Hence we can assume that there are no more equations than variables, i.e.,

$$m \le n.$$

Now

$$m = n$$

leads to the unique solution

$$\partial X = 0,$$

which implies that there are no feasible points other than X in any neighbourhood of X. That is, the set of feasible points is discrete. Hence we can assume that

$$m < n.$$

We redefine $X = (x_1, x_2, \ldots, x_n)^T$ as

$$X = (w_1, w_2, \ldots, w_m, y_1, y_2, \ldots, y_{n-m})^T. \tag{7.32}$$

The variables w_i, $i = 1, 2, \ldots, m$ are called *state variables* and the variables y_i, $i = 1, 2, \ldots, (n - m)$ are called *decision variables*. Now (7.31) can be rewritten using (7.32), as follows:

$$\sum_{i=1}^{m} \frac{\partial f(X)}{\partial w_i} \partial w_i + \sum_{i=1}^{n-m} \frac{\partial f(X)}{\partial y_i} \partial y_i = \partial f(X) \tag{7.33}$$

$$\sum_{i=1}^{m} \frac{\partial g_j(X)}{\partial w_i} \partial w_i + \sum_{i=1}^{n-m} \frac{\partial g_j(X)}{\partial y_i} \partial y_i = 0, \qquad j = 1, 2, \ldots, m. \tag{7.34}$$

Suppose now that the ∂y_i, $i = 1, 2, \ldots, (n - m)$ are given arbitrary values. When these are substituted into (7.34) unique values for the ∂w_i, $i = 1, 2, \ldots, m$ can be found which keep $X + h$ inside the feasible region. One can then use all these values in (7.33) to see if

$$\partial f(X) > 0,$$

i.e., the new point $X + h$ is an improvement over X.

We now state the explicit steps needed to carry this out using vector notation. The matrix

$$\begin{pmatrix} \dfrac{\partial g_1}{\partial w_1} & \dfrac{\partial g_1}{\partial w_2} & \cdots & \dfrac{\partial g_1}{\partial w_m} \\[2mm] \dfrac{\partial g_2}{\partial w_1} & \dfrac{\partial g_2}{\partial w_2} & \cdots & \dfrac{\partial g_2}{\partial w_m} \\[2mm] \vdots & \vdots & & \vdots \\[2mm] \dfrac{\partial g_m}{\partial w_1} & \dfrac{\partial g_m}{\partial w_2} & \cdots & \dfrac{\partial g_m}{\partial w_m} \end{pmatrix}$$

is called the *Jacobian matrix*, and the matrix

$$C = \begin{pmatrix} \dfrac{\partial g_1}{\partial y_1} & \dfrac{\partial g_1}{\partial y_2} & \cdots & \dfrac{\partial g_1}{\partial y_{n-m}} \\[2mm] \dfrac{\partial g_2}{\partial y_1} & \dfrac{\partial g_2}{\partial y_2} & \cdots & \dfrac{\partial g_2}{\partial y_{n-m}} \\[2mm] \vdots & \vdots & & \vdots \\[2mm] \dfrac{\partial g_m}{\partial y_1} & \dfrac{\partial g_m}{\partial y_2} & \cdots & \dfrac{\partial g_m}{\partial y_{n-m}} \end{pmatrix}$$

is called the *control matrix*. It is important in defining the state and decision variables that the left-hand sums in (7.33) and (7.34) be linearly independent. It is always possible to make a choice of which x_i's become state variables, so this happens because we have assumed that the equations in (7.31) are linearly independent. The implication of this is that J is nonsingular. Now let

$$W = (w_1, w_2, \ldots, w_m)^T$$
$$Y = (y_1, y_2, \ldots, y_{n-m})^T.$$

Then (7.33) and (7.34) become

$$\nabla_w f^T \partial W + \nabla_y f^T \partial Y = \partial f(W, Y) \tag{7.35}$$

and

$$J \partial W + C \partial Y = 0, \tag{7.36}$$

respectively. As J is nonsingular, we can multiply (7.36) by J^{-1}:

$$\partial W = -J^{-1} C \partial Y. \tag{7.37}$$

It can be seen, as was stated earlier, that if the elements in ∂Y are given values, ∂W can be calculated using (7.37). Substituting this into (7.35) yields

$$\partial f(W, Y) = (\nabla_y f^T - \nabla_w f^T J^{-1} C) \partial Y. \tag{7.38}$$

From (7.38) we can form what is known as the *constrained gradient* of f with respect to Y, which is

$$\nabla_y^c f = \frac{\partial^c f(w, y)}{\partial^c y} = \nabla_y f^T - \nabla_w f^T J^{-1} C. \tag{7.39}$$

Each element of $\nabla_y^c f$, namely $\partial^c f / \partial^c y_i$, $i = 1, 2, \ldots, (n - m)$, is called a *constrained derivative*. It represents the rate of change of f resulting from perturbing x_i from y_i (all other x_i's being held constant) to *feasible* points.

When constrained derivatives are used one can show that Theorem 7.11 is applicable, i.e., if X^* is a feasible maximum it is necessary that

$$\nabla_y^c f(X^*) = \mathbf{0}. \tag{7.40}$$

Equation (7.40) can be used to identify all the stationary points; it remains to find which one is the global maximum. To do this we use Theorem 7.12, with the modification that H is the matrix of *constrained* second derivatives with respect to the independent variables $y_1, y_2, \ldots, y_{n-m}$ only, and not $w_1, w_2, \ldots, w_m$. The complete method will be illustrated with a numerical example.

7.4.1.2 *Numerical Example*

Consider the following problem:

Maximize: $f(X) = f((x_1 x_2 x_3)) = -2x_1^2 - x_2^2 - 3x_3^2$
subject to: $g_1(X) = x_1 + 2x_2 + x_3 - 1 = 0$
              $g_2(X) = 4x_1 + 3x_2 + 2x_3 - 2 = 0.$

Here $m = 2$, $n = 3$, and we define

$$W = (w_1, w_2)^T = (x_1, x_2)^T$$
$$Y = (y_1) = (x_3)$$
$$\nabla_w f = (-4x_1, -2x_2)^T$$
$$\nabla_y f = (-6x_3)$$

$$J = \begin{pmatrix} 1 & 2 \\ 4 & 3 \end{pmatrix}$$

$$J^{-1} = \begin{pmatrix} -\frac{3}{5} & \frac{2}{5} \\ \frac{4}{5} & -\frac{1}{5} \end{pmatrix}$$

$$C = \begin{pmatrix} 1 \\ 2 \end{pmatrix}.$$

Now, by (7.39), we have

$$V_y^c f(X) = V_y f - V_w f^T J^{-1} C$$

$$= -6x_3 - (-4x_1, -2x_2) \begin{pmatrix} -\frac{3}{5} & \frac{2}{5} \\ \frac{4}{5} & -\frac{1}{5} \end{pmatrix} \begin{pmatrix} 1 \\ 2 \end{pmatrix}$$

$$= -6x_3 + \tfrac{4}{5}x_1 + \tfrac{4}{5}x_2$$

$$= 0, \quad \text{by (7.40)}.$$

Combining this equation with the two original constraints, we have

$$\tfrac{4}{5}x_1 + \tfrac{4}{5}x_2 - 6x_3 = 0$$
$$x_1 + 2x_2 + x_3 = 1$$
$$4x_1 + 3x_2 + 2x_3 = 2,$$

which have a unique solution:

$$X^* = (\tfrac{5}{27}, \tfrac{10}{27}, \tfrac{2}{27})^T,$$

which is a stationary point. It is now determined whether this is a maximum point by using Theorem 7.12:

$$V_y^c = \tfrac{4}{5}x_1 + \tfrac{4}{5}x_2 - 6x_3$$

Therefore

$$\frac{\partial^{c2} f}{\partial y_1^2} = \frac{4}{5}\frac{dx_1}{dx_3} + \frac{4}{5}\frac{dx_2}{dx_3} - 6. \tag{7.41}$$

Now, from (7.37), we have

$$\begin{pmatrix} \dfrac{dx_1}{dx_3} \\ \dfrac{dx_2}{dx_3} \end{pmatrix} = \frac{\partial w}{\partial y} = -J^{-1}C$$

$$= -\begin{pmatrix} -\frac{3}{5} & \frac{2}{5} \\ \frac{4}{5} & -\frac{1}{5} \end{pmatrix}\begin{pmatrix} 1 \\ 2 \end{pmatrix}$$

$$= \begin{pmatrix} -\frac{1}{5} \\ -\frac{2}{5} \end{pmatrix}.$$

Substituting these values into (7.41) yields

$$\frac{\partial^2 cf}{\partial y_1^2} = (\tfrac{4}{5})(-\tfrac{1}{5}) + (\tfrac{4}{5})(-\tfrac{2}{5}) - 6 < 0.$$

Thus X^* is indeed a maximum point.

7.4.1.3 *The Method of Lagrange*

The following method was developed by Lagrange in 1761. From the argument that was used to develop (7.31) it can be said that

$$\partial f(W, Y) = \nabla_w f^T \partial W + \nabla_y f^T \partial Y, \tag{7.42}$$

$$\partial g = J \partial W + C \partial Y, \tag{7.43}$$

where $g = (g_1, g_2, \ldots, g_m)^T$, the vector of constraint functions. Eliminating ∂W from (7.42) and (7.43) produces

$$J \partial f(W, Y) - \nabla_w f^T \partial g = \nabla_y f^T J \partial Y - \nabla_w f^T C \partial Y.$$

Therefore

$$\partial f(W, Y) = \nabla_w f^T J^{-1} \partial g + \nabla_y f^T \partial Y - \nabla_w f^T J^{-1} C \partial Y,$$

and from (7.39),

$$\partial f(W, Y) = \nabla_w f^T J^{-1} \partial g + \nabla_y^c f^T \partial Y. \tag{7.44}$$

Now, if $X^{*T} = (W^{*T}, Y^{*T})$ is a local maximum, then

$$\nabla_c^y f(X^*) = 0.$$

So, from (7.44), we have

$$\partial f(W^*, Y^*) = \nabla_w f^T J^{-1} \partial g,$$

hence

$$\frac{\partial f(W^*, Y^*)}{\partial g} = \nabla_w f^T J^{-1}. \tag{7.45}$$

Equation (7.45) is useful in allowing one to analyze the rate at which $f(W^*, Y^*)$, the optimal value, changes when g is perturbed. The individual components of this vector are called *sensitivity coefficients* for this reason.

These sensitivity coefficients (which are constant, as can be seen from (7.45)) are now introduced into (7.29). Let

$$\lambda = (\lambda_1, \lambda_2, \ldots, \lambda_m) = \frac{\partial f(W^*, Y^*)}{\partial g} = \nabla_w f^T J^{-1}. \tag{7.46}$$

From (7.46) we have

$$\partial f(W^*, Y^*) = \lambda \partial g. \tag{7.47}$$

Define F, the *Lagrangian*, as

$$F(X, \lambda) = f(X) - \lambda g.$$

Then the system of equations (7.29) and (7.47) correspond to

$$\frac{\partial F}{\partial \lambda_j} = 0, \qquad j = 1, 2, \ldots, m \qquad (7.48)$$

$$\frac{\partial F}{\partial x_i} = 0, \qquad i = 1, 2, \ldots, n. \qquad (7.49)$$

It is necessary that any stationary point satisfies (7.48) and (7.49), which constitute a system of $m + n$ equations in $m + n$ unknowns, $\lambda_1, \lambda_2, \ldots, \lambda_m$, $x_1, x_2, \ldots, x_n$. Any stationary point will produce a unique set of values for the elements of λ, as long as (7.48) and (7.49) are independent. Hence these values are independent of which members of X are assigned to W and which to Y. We now illustrate these ideas by using the method to solve the previous numerical example.

7.4.1.4 Numerical Example

$$\begin{aligned}
\text{Maximize:} \quad & f(X) = -2x_1^2 - x_2^2 - 3x_3^2 \\
\text{subject to:} \quad & g_1(X) = x_1 + 2x_2 + x_3 - 1 = 0 \\
& g_2(X) = 4x_1 + 3x_2 + 2x_3 - 2 = 0.
\end{aligned}$$

Now

$$\begin{aligned}
F(X, \lambda) &= F(x_1, x_2, x_3, \lambda_1, \lambda_2) \\
&= -2x_1^2 - x_2^2 - 3x_3^2 - \lambda_1(x_1 + 2x_2 + x_3 - 1) \\
&\quad - \lambda_2(4x_1 + 3x_2 + 2x_3 - 2)
\end{aligned}$$

$$\frac{\partial F}{\partial x_1} = -4x_1 - \lambda_1 - 4\lambda_2 = 0$$

$$\frac{\partial F}{\partial x_2} = -2x_2 - 2\lambda_1 - 3\lambda_2 = 0$$

$$\frac{\partial F}{\partial x_3} = -6x_3 - \lambda_1 - 2\lambda_2 = 0$$

$$\frac{\partial F}{\partial \lambda_1} = -(x_1 + 2x_2 + x_3 - 1) = 0$$

$$\frac{\partial F}{\partial \lambda_2} = -(4x_1 + 3x_2 + 2x_3 - 2) = 0.$$

This yields

$$\begin{aligned}
x_1^* &= \tfrac{5}{27} \\
x_2^* &= \tfrac{10}{27} \\
x_3^* &= \tfrac{2}{27} \\
\lambda_1^* &= -\tfrac{4}{27} \\
\lambda_2^* &= -\tfrac{4}{27},
\end{aligned}$$

which is of course the same optimal solution as that produced by the Jacobian method.

7.4.2 Multidimensional Optimization with Inequality Constraints

This problem can be stated in general as follows:

$$\text{Maximize:} \quad f(X) \tag{7.50}$$

$$\text{subject to:} \quad g_j(X) \leq 0, \quad j = 1, 2, \ldots, m. \tag{7.51}$$

Necessary (and in some cases, sufficient) conditions for X to be a stationary point for (7.50), (7.51) were developed by Kuhn and Tucker (1951). Many of the algorithms for solving the above problem are based on these conditions, and termination criteria concerned with recognizing when a stationary point has been reached are derived from them. Wilde and Beightler (1967) presented a development of the conditions based on constrained derivatives. The nonrigorous formulation given here uses the Lagrangian.

7.4.2.1 *The Kuhn–Tucker Conditions*

Consider problem (7.50), (7.51). Adding nonnegative slack variables s_j to the left-hand side of (7.51) produces the following equations:

$$g_j(X) + s_j = 0, \quad j = 1, 2, \ldots, m \tag{7.52}$$

$$s_j \geq 0, \quad j = 1, 2, \ldots, m. \tag{7.53}$$

Apart from (7.53), we are now confronted with a problem in equality constraints and can use one of the methods of Section 7.4.1. In particular, we use the method of Lagrange and form the Lagrangian:

$$F(X, \lambda) = f(X) - \sum_{j=1}^{m} \lambda_j(g_j(X) + s_j). \tag{7.54}$$

From (7.46), we have

$$\lambda_j = \frac{\partial f}{\partial g_j}, \quad j = 1, 2, \ldots, m.$$

That is, λ_j is the rate of change of f with respect to g_j, the jth constraint function. Now, if

$$g_j(X) \leq 0$$

becomes

$$g_j(X) \leq \varepsilon,$$

where ε is a relatively small positive number, the set of feasible solutions for the original problem is no smaller. Hence $f(X^*)$, the optimal solution value, will be no less than what it is for (7.50), (7.51). Hence

$$\lambda_j \geq 0. \tag{7.55}$$

Following the method of Lagrange, for any local maximum X^*, with corresponding value $s = s^*$ and $\lambda = \lambda^*$, we have

$$\frac{\partial F(X^*)}{\partial X} = 0,$$

$$\frac{\partial F(X^*)}{\partial \lambda} = 0,$$

and, by (7.55),

$$\frac{\partial F(X^*)}{\partial s_j} = -\lambda_j \leq 0, \qquad j = 1, 2, \ldots, m$$

that is,

$$\nabla f(X^*) - \sum_{j=1}^{m} \lambda_j \nabla g_j(X^*) = 0$$

$$g(X^*) + S^* = 0$$

$$\lambda^* \geq 0,$$

where S is the vector $(s_1, s_2, \ldots, s_m)^T$.

Now if $\lambda_j > 0$, by (7.46) we have

$$\frac{\partial f}{\partial g_j} > 0,$$

that is, the jth constraint is tight and $s_j = 0$. Hence

$$g_j(X^*) = 0.$$

However, if $g_j(X^*) < 0$, then $s_j > 0$ and the jth constraint is not tight. In this case

$$\frac{\partial f}{\partial g_j} = 0.$$

Therefore

$$\lambda_j = 0.$$

Recapitulating, we have

$$\lambda_j > 0 \Rightarrow g_j(X^*) = 0$$

$$g_j(X^*) < 0 \Rightarrow \lambda_j = 0.$$

Therefore

$$\lambda_j g_j(X^*) = 0, \qquad j = 1, 2, \ldots, m.$$

We have given an intuitive, nonrigorous outline of the following theorem:

Theorem 7.16. *If f has a local maximum X^* in the feasible region R of the problem:*

$$\text{maximize:} \qquad f(X)$$

$$\text{subject to:} \qquad g_j(X) \leq 0, \qquad j = 1, 2, \ldots, m,$$

where f and g_j, $j = 1, 2, \ldots, m$ have continuous first derivatives, and R is well-behaved at its boundary, then it is necessary that

$$\nabla f(X^*) - \sum_{j=1}^{m} \lambda_j^* \nabla g_j(X^*) = 0 \tag{7.56}$$

$$g(X^*) \le 0 \tag{7.57}$$

$$\lambda_j^* g_j(X^*) = 0, \qquad j = 1, 2, \ldots, m \tag{7.58}$$

$$\lambda_j^* \ge 0, \qquad j = 1, 2, \ldots, m \tag{7.59}$$

for some set of real numbers $\lambda^* = (\lambda_1^*, \lambda_2^*, \ldots, \lambda_m^*)$.

The phrase "R is well-behaved at its boundary" needs some explanation. When Kuhn and Tucker developed the above theory they found that for certain feasible regions their theorem did not hold. This occurred when it was possible to find a point X in such a region R with the following property. There does not exist a continuous, differentiable curve C beginning at X such that one can travel along C from X for a positive distance and remain in R. This situation is thankfully rare in practice and occurs when X is at the vertex of a cusp in R. Kuhn and Tucker therefore qualified the conditions of their theorem by stating that it must be possible to find such a curve C for any point X in R. This is called the *constraint qualification*.

The values $\lambda^* = (\lambda_1^*, \lambda_2^*, \ldots, \lambda_m^*)$ are called *generalized Lagrange multipliers* for obvious reasons. Of course (7.57) represents the fact that X^* must be a feasible point for the original set of constraints. (7.56), (7.57), (7.59) are termed the *Kuhn–Tucker conditions*.

For completeness we state the analogous theorem for the problem with a minimization objective. We leave the proof as an exercise for the reader.

Theorem 7.17. *If f has a local minimum X^* in the feasible region R of the problem:*

> Minimize: $f(X)$
>
> subject to: $g_j(X) \le 0, \qquad j = 1, 2, \ldots, m,$

where f and g_j, $j = 1, 2, \ldots, m$ have continuous first derivatives and the constraint qualification is satisfied, then it is necessary that

$$\nabla f(X^*) - \sum_{j=1}^{m} \lambda_j^* \nabla g_j(X^*) = 0 \tag{7.60}$$

$$g_j(X^*) \le 0 \tag{7.61}$$

$$\lambda_j^* g_j(X^*) = 0, \qquad j = 1, 2, \ldots, m \tag{7.62}$$

$$\lambda_j^* \le 0, \qquad j = 1, 2, \ldots, m \tag{7.63}$$

for some set of real numbers $\lambda^* = (\lambda_1^*, \lambda_2^*, \ldots, \lambda_m^*)$.

Note that in (7.63) the inequality signs have the opposite sense to those in (7.59).

7.4.2.1.1 *When the Kuhn–Tucker Conditions are Sufficient.* In previous sections in this chapter we have shown that the necessary conditions for f to have a local maximum (minimum) at X^* are sufficient if f is concave (convex). This is also true for Theorem 7.16 if the feasible region R defined by (7.51) is convex. When will R be convex? A sufficient condition is given by Theorem 7.18.

Theorem 7.18 *The region R defined by*

$$g_j(X) \le 0, \qquad j = 1, 2, \ldots, m$$

will be convex if g_j is convex for all $j = 1, 2, \ldots, m$.

PROOF. Consider two distinct points X_1, $X_2 \in R$. Then for all j, $j = 1$, $2, \ldots, m$,

$$g_j(X_1) \le 0$$
$$g_j(X_2) \le 0.$$

Also for all $\alpha \in R$, $0 \le \alpha \le 1$,

$$g_j(\alpha X_1 + (1 - \alpha)X_2) \le \alpha g_j(X_1) + (1 - \alpha)g_j(X_2)$$
$$\le \alpha 0 + (1 - \alpha)0.$$

Hence

$$g_j(\alpha X_1 + (1 - \alpha)X_2) \le 0.$$

Therefore

$$\alpha X_1 + (1 - \alpha)X_2 \in R. \qquad \square$$

Before stating the main result of this section we first prove two lemmas which are needed in the proof of Theorem 7.2.1. The lemmas (Theorem 7.19 and 7.20) are an n-dimensional generalization of Theorem 7.5.

Theorem 7.19. *If f is convex on a convex region $R \subset \mathbf{R}^n$ with continuous first partial derivatives within R, then for any two points X, $X + h \in R$,*

$$f(X + h) - f(X) \ge \nabla f(X)^T h.$$

PROOF. As f is convex, we have

$$f(\alpha(X + h) + (1 - \alpha)X) \le \alpha f(X + h) + (1 - \alpha)f(X), \quad \text{for all } \alpha \in R, 0 \le \alpha \le 1.$$

On rearranging, we obtain

$$f(X + \alpha h) - f(X) \le \alpha(f(X + h) - f(X)). \tag{7.64}$$

Making a first-order expansion of the Taylor series of the left-hand side of (7.64), we obtain

$$Vf(X + \theta\alpha h)^T\alpha h \leq \alpha(f(X + h) - f(X)), \quad \text{for some } \theta, 0 < \theta < 1$$
$$Vf(X + \theta\alpha h)^T h \leq f(x + h) - f(X), \quad \text{for } \alpha > 0$$

and

$$\lim_{\alpha \to 0} Vf(X + \theta\alpha h)^T h = Vf(X)^T h \leq f(X + h) - f(X). \qquad \square$$

We also need the analogous result for concave functions:

Theorem 7.20. *If f is concave on a convex region R with continuous partial derivatives within R, then for any two points X, X + h ∈ R,*

$$f(X + h) - f(X) \leq Vf(X)^T h.$$

We leave the proof of Theorem 7.20 as an exercise for the reader.

We come now to the main result of this section: the sufficiency of Kuhn–Tucker conditions for concave functions.

Theorem 7.21. *If, in the problem:*

$$\text{Maximize:} \quad f(X),$$
$$\text{subject to:} \quad g_j(X) \leq 0, \quad j = 1, 2, \ldots, m,$$

f is concave and g_j is convex for $j = 1, 2, \ldots, m$ and there exist X^* and $\lambda^* = (\lambda_1^*, \lambda_2^*, \ldots, \lambda_m^*)$ which satisfy (7.56)–(7.59), then f has a global maximum at X^*.

PROOF. As f is concave, by Theorem 7.19 we have

$$Vf(X^*)^T h \geq f(X^* + h) - f(X^*),$$

and by (7.56) we have

$$\sum_{j=1}^{m} \lambda_j^* Vg_j(X^*)^T h \geq f(X^* + h) - f(X^*). \tag{7.65}$$

Now using the result of Theorem 7.19 in the left-hand side of (7.65) as g_j is convex and (7.59), we obtain

$$\sum_{j=1}^{m} \lambda_j^* [g_j(X^* + h) - g_j(X^*)] \geq f(X^* + h) - f(X^*),$$

which by (7.58) becomes

$$\sum_{j=1}^{m} \lambda_j^* g_j(X^* + h) \geq f(X^* + h) - f(X^*). \tag{7.66}$$

Now if $(X^* + h)$ is a feasible solution to the problem, then

$$g_j(X^* + h) \leq 0, \quad j = 1, 2, \ldots, m.$$

Hence, by (7.59), we have

$$\lambda_j g_j(X^* + h) \le 0 \qquad j = 1, 2, \ldots, m.$$

Therefore, from (7.66), we have

$$0 \ge f(X^* + h) - f(X^*),$$

for all feasible $X^* + h$. That is, f has a global maximum at X^*.   □

We state the analogous result for the problem with a minimization objective; we leave the proof as an exercise for the reader.

Theorem 7.22. *If, in the problem*

Minimize: $f(X)$

subject to: $g_j(X) \le 0, \qquad j = 1, 2, \ldots, m,$

f and $g_j, j = 1, 2, \ldots, m$ are convex and there exist X^* and $\lambda^* = (\lambda_1^*, \lambda_2^*, \ldots, \lambda_m^*)$ which satisfy (7.60)–(7.63), then f has a global minimum at X^*.

7.4.2.1.2 *Numerical Example.* Let us return to the numerical example of Section 7.4.1.2 and relax the equality constraints so that the problem becomes

Maximize: $f(X) = 2x_1^2 - x_2^2 - 3x_3^2$

subject to: $g_1(X) = \quad x_1 + 2x_2 + \quad x_3 - 1 \le 0$

                   $g_2(X) = 4x_1 + 3x_2 + 2x_3 - 2 \le 0.$

It can be shown that the feasible region defined by $g_1(X)$ and $g_2(X)$ obeys the constraint qualification and so we can apply the result of theorem 7.16. Let $X^* = (x_1^*, x_2^*, x_3^*)$ be a local maximum, then:

$$(-4x_1^*, -2x_2^*, -6x_3^*) - \lambda_1^*(1, 2, 1) - \lambda_2^*(4, 3, 2) = 0 \qquad (7.56)'$$

$$\begin{aligned} x_1^* + 2x_2^* + \quad x_3^* - 1 \le 0 \\ 4x_1^* + 3x_2^* + 2x_3^* - 2 \le 0 \end{aligned} \qquad (7.57)'$$

$$\begin{aligned} \lambda_1^*(\quad x_1^* + 2x_2^* + \quad x_3^* - 1) = 0 \\ \lambda_2^*(4x_1^* + 3x_2^* + 2x_3^* - 2) = 0 \end{aligned} \qquad (7.58)'$$

$$\lambda_1^* \ge 0, \lambda_2^* \ge 0. \qquad (7.59)'$$

From (7.56)' and (7.58)' we have the following system of five equations in five unknowns:

$$-4x_1^* - \lambda_1^* - 4\lambda_2^* = 0$$

$$-2x_2^* - 2\lambda_1^* - 3\lambda_2^* = 0$$

$$-6x_3^* - \lambda_1^* - 2\lambda_2^* = 0$$

$$\lambda_1^* x_1^* + 2\lambda_1^* x_2^* + \lambda_1^* x_3^* = \lambda_1^*$$

$$4\lambda_2^* x_1^* + 3\lambda_2^* x_2^* + 2\lambda_2^* x_3^* = 2\lambda_2^*,$$

which has the following solution:

$$(x_1^*, x_2^*, x_3^*) = (0, 0, 0),$$
$$(\lambda_1^*, \lambda_2^*) = (0, 0).$$

Hence $X^* = \mathbf{0}$ is a local maximum. However, as f is concave and g_1, and g_2 are convex, we can apply Theorem 7.21 and need look no further for other local maxima. $X^* = \mathbf{0}$ is a unique local maximum and hence is the global maximum.

7.5 The Calculus of Variations

The calculus of variations is the branch of mathematics which is concerned with the optimization of functionals. A functional is a special kind of function which has as its domain a set of functions and as its range the set of real numbers. The calculus of variations has applications in many areas: astronautics, economics, business management, the physical sciences, engineering, and others. As will be seen in the next section, some of the problems of this subject have been studied since the dawn of mathematics.

7.5.1 Historical Background

One of the earliest recorded problems on this topic is concerned with the finding of a curve of fixed length which encloses the greatest area with a given straight line. It is said that this problem was solved intuitively by the Phoenician queen Dido in approximately 850 B.C. According to Virgil she persuaded a North African chieftain to allow her to have as much of his land as she could enclose within the hide of a bull. She apparently had the hide cut up into very thin strips which were joined together to form a single length. This she laid out in semicircle with the Mediterranean coast as diameter. The piece of land enclosed, which has the maximum possible area for the given length, was used to found the city of Carthage.

The calculus of variations received a large impetus in the seventeenth and eighteenth centuries when some of the great mathematicians of those times studied some of its problems. Many tasks were undertaken, such as finding the shape of an object which caused least resistance when propelled at constant velocity through a fluid. One of the most famous problems has already been discussed in Chapter 1—the brachistochrone. Newton also considered a related problem: that of finding the shape of a tunnel through the earth joining two points on the surface which would cause a bead on a frictionless wire in the tunnel to travel between the two points in minimum time when falling under gravity. Contrary to the intuitive feeling of some people, the solution turns out to be not a straight line joining the two points, but a hypocycloid.

150 years later the German mathematician Zermelo solved the following problem. Find the path of a boat crossing a river in minimum time from a given point on one bank to a given point on the other. The river current is known at all points and it is assumed that the boat has constant power.

In 1962 an isoperimetric problem similar to Queen Dido's was solved by the Soviet mathematician Chaplygin. The problem was to find the course of an aeroplane which encloses the greatest area in a given time while a constant wind blows. It is assumed that the aeroplane has constant power. The solution is an ellipse, which tends to a circle as the wind velocity tends to zero.

7.5.2 Modern Applications

As was mentioned earlier there are numerous applications of the calculus of variations to diverse areas, a few of which will be detailed now. One of the main applications is concerned with problems in rocket control. For example, designers often wish to find the minimum amount of fuel required for a rocket of given specifications to achieve a given height above the earth's surface while it experiences atmospheric resistance; or a designer may wish to find the minimum time required for a rocket to reach the height when it has only a given amount of fuel. Other applications occur in the financial planning of both companies and individuals. For instance, a manager may wish to discover how to maximize the production of certain commodities within a fixed budget where costs are due to storage, machine set up, production runs, and inflation.

7.5.3 A Simple Variational Problem

In this section we introduce a simple general problem of the calculus of variations. Unfortunately problems of this type cannot be solved by the methods of elementary calculus. Hence we extend the theory so that such problems can be tackled.

Definition 7.14. A *functional J* is a function:

$$J: D \to \mathbf{R},$$

where D is a set of real-valued functions each of which is defined on a real interval.

Note that in all the optimization problems studied so far in this book we have wished to optimize a function f whose domain is some subset S

of $\mathbf{R}^n$ ($n \geq 1$). That is, we have searched for a vector $X = (x_1, x_2, \ldots, x_n)^T$, such that $f(X)$ is a maximum or minimum among all vectors in S. In this section we consider the optimization of a functional rather than a function. That is, we search for a function f (rather than a vector) such that $J(f)$ is a maximum or minimum among all functions in D. The mathematics necessary to optimize functionals is known as the *calculus of variations*.

Let D be defined by:

$$D = \{f : f(x) = \sin nx, \; x_0 \leq x \leq x_1, \; n = 0, 1, 2, \ldots\}$$

where x_0, x_1 are two given real numbers. Let J be defined by

$$J(f) = \min_{x_0 \leq x \leq x_1} \{f(x)\}.$$

Then a typical variational problem is to find $f_* \in D$ such that

$$J(f_*) = \max_{f \in D} \{J(f)\}.$$

7.5.3.1 *Necessary and Sufficient Conditions for a Local Optimum*

As can be seen from the description of the historical problems given earlier, often in the calculus of variations one wishes to optimize some functional of time, distance, area, or volume. In many such cases J is of the form:

$$J(f) = \int_{x_0}^{x_1} F(x, f, f') \, dx, \qquad f \in D,$$

where D is the set of continuous bounded functions defined on $[x_0, x_1]$ with continuous second derivatives. D is called the *set of admissable curves*. F is a continuous three-variable function with continuous partial derivatives. We assume J is of this form throughout this section, and we further assume that any $f \in S$ obeys *boundary conditions*. That is, there exists $y_0, y_1 \in \mathbf{R}$ such that

$$\left. \begin{array}{l} f(x_0) = y_0 \\ f(x_1) = y_1 \end{array} \right\} \quad \text{for all } f \in D.$$

What we wish to do is to find an $f_* \in D$ which optimizes J. In some variational problems,

$$J(f_*) = \max_{f \in D} \{J(f)\},$$

while in others,

$$J(f_*) = \min_{f \in D} \{J(f)\}.$$

J is said to have a *local maximum* of $f_* \in D$ if there exists a positive real number β such that

$$J(f_*) \geq J(f)$$

for all $f \in D$, such that

$$|f_*(x) - f(x)| < \beta, \qquad x_0 \leq x \leq x_1.$$

Since all $f \in D$ are bounded there exist $m, M \in \mathbf{R}$ such that

$$m \le f_*(x) \le M, \qquad x_0 \le x \le x_1.$$

We now address ourselves to the task of developing necessary conditions for J to have a local maximum at f_*. It is possible to find a positive real number δ such that

$$m \le f_*(x) - \delta \le f_*(x) + \delta \le M, \qquad x_0 \le x \le x_1.$$

Let $f \in D$ be a function within the δ-neighbourhood of f_*, i.e.,

$$|f_*(x) - f(x)| < \delta, \qquad x_0 \le x \le x_1. \tag{7.67}$$

Then we can represent f as follows.

Let α be an arbitrary real-valued function with domain $[x_0, x_1]$ and continuous second derivative such that

$$\alpha(x_0) = \alpha(x_1) = 0. \tag{7.68}$$

Then it is possible to find a small number ε such that

$$f(x) = f_*(x) + \varepsilon\alpha(x), \qquad x_0 \le x \le x_1. \tag{7.69}$$

The function $\varepsilon\alpha$ is called the *variation of* f_*. We now define the *variation of J* as follows:

$$\Delta J = \int_{x_0}^{x_1} F(x, f, f')\,dx - \int_{x_0}^{x_1} F(x, f_*, f'_*)\,dx. \tag{7.70}$$

As it has been assumed that J has a local maximum at f_*, for all $f \in D$ satisfying (7.67) we have

$$\Delta J \le 0$$

for sufficiently small δ. Substituting (7.69) in (7.70), we get

$$\Delta J = \int_{x_0}^{x_1} F(x, f_* + \varepsilon\alpha, f'_* + \varepsilon\alpha')\,dx - \int_{x_0}^{x_1} F(x, f_*, f'_*)\,dx$$

$$= \int_{x_0}^{x_1} \{F(x, f_* + \varepsilon\alpha, f'_* + \varepsilon\alpha') - F(x, f_*, f'_*)\}\,dx. \tag{7.71}$$

We can form the Taylor series of the integrand of (7.71) about (x, f_*, f'_*) to obtain.

$$\Delta J = \int_{x_0}^{x_1} \left\{ \frac{\partial F}{\partial f_*} \varepsilon\alpha + \frac{\partial F}{\partial f'_*} \varepsilon\alpha' + \frac{1}{2} \frac{\partial^2 F}{\partial f_*^2} \varepsilon^2\alpha^2 + \frac{\partial^2 F}{\partial f_* \partial f'_*} \varepsilon\alpha\varepsilon\alpha' \right.$$

$$\left. + \frac{1}{2} \frac{\partial^2 F}{\partial f'_*} (\alpha')^2\varepsilon^2 + \cdots \right\} dx$$

$$= \varepsilon \int_{x_0}^{x_1} \left\{ \frac{\partial F}{\partial f_*} \alpha + \frac{\partial F}{\partial f'_*} \alpha' \right\} dx + \frac{\varepsilon^2}{2} \int_{x_0}^{x_1} \left\{ \frac{\partial^2 F}{\partial f_*^2} \alpha^2 + 2 \frac{\partial^2 F}{\partial F_* \partial f'} \alpha\alpha' \right.$$

$$\left. + \frac{\partial^2 F}{\partial f'_*} (\alpha')^2 \right\} dx + O(\varepsilon^3). \tag{7.72}$$

The expression to the right of the first integral in (7.72), $O(\varepsilon^3)$, can be neglected if ε is small in magnitude. Thus we have

$$\Delta J = \varepsilon \int_{x_0}^{x_1} \left\{ \frac{\partial F}{\partial f_*} \alpha + \frac{\partial F}{\partial f'_*} \alpha' \right\} dx.$$

Now there is no restriction on the sign of ε, so that

$$\left(\int_{x_0}^{x_1} \left\{ \frac{\partial F}{\partial f_*} \alpha + \frac{\partial F}{\partial f'_*} \alpha' \right\} dx \geq 0 \quad \text{and} \quad \varepsilon > 0 \right) \Rightarrow \Delta J \geq 0$$

$$\left(\int_{x_0}^{x_1} \left\{ \frac{\partial F}{\partial f_*} \alpha + \frac{\partial F}{\partial f'_*} \alpha' \right\} dx \geq 0 \quad \text{and} \quad \varepsilon < 0 \right) \Rightarrow \Delta J \leq 0$$

and

$$\int_{x_0}^{x_1} \left\{ \frac{\partial F}{\partial f_*} \alpha + \frac{\partial F}{\partial f'_*} \alpha' \right\} dx \leq 0 \Rightarrow \Delta J \geq 0.$$

But we must have

$$\Delta J \leq 0.$$

Hence

$$\int_{x_0}^{x_1} \left\{ \frac{\partial F}{\partial f_*} \alpha + \frac{\partial F}{\partial f'_*} \alpha' \right\} dx = 0.$$

Therefore

$$\int_{x_0}^{x_1} \frac{\partial F}{\partial f_*} \alpha \, dx + \int_{x_0}^{x_1} \frac{\partial F}{\partial f'_*} \alpha' \, dx = 0,$$

which becomes, on integrating the second expression by parts,

$$\int_{x_0}^{x_1} \frac{\partial F}{\partial f_*} \alpha \, dx + \left[\frac{\partial F}{\partial f'_*} \alpha \right]_{x_0}^{x_1} - \int_{x_0}^{x_1} \alpha \frac{d}{dx} \frac{\partial F}{\partial f'_*} dx = 0.$$

On rearranging, we obtain

$$\int_{x_0}^{x_1} \alpha \left(\frac{\partial F}{\partial f_*} - \frac{d}{dx} \frac{\partial F}{\partial f'_*} \right) dx + \left[\frac{\partial F}{\partial f'_*} \alpha \right]_{x_0}^{x_1} = 0.$$

Because of (7.68) the term to the right of the integral vanishes. Hence

$$\int_{x_0}^{x_1} \alpha \left(\frac{\partial F}{\partial f_*} - \frac{d}{dx} \frac{\partial F}{\partial f'_*} \right) dx = 0. \tag{7.73}$$

Now suppose there exists at least one $\bar{x} \in [x_0, x_1]$ for which

$$\frac{\partial F}{\partial f_*} - \frac{d}{dx} \frac{\partial F}{\partial f'_*} \tag{7.74}$$

is nonzero. As α is arbitrary, it is possible to define $\alpha(\bar{x})$ to have the same sign as (7.74). Thus

$$\alpha \left\{ \frac{\partial F}{\partial f_*} - \frac{d}{dx} \frac{\partial F}{\partial f'_*} \right\}$$

will be positive everywhere it is nonzero, and there is at least one point $\bar{x}$ where this occurs. This contradicts (7.73). Thus

$$\frac{\partial F}{\partial f_*} - \frac{d}{dx}\frac{\partial F}{\partial f'_*} = 0$$

is a necessary condition for J to have a local maximum at f_*. The proof for the case of a local minimum is analogous. We have proven a result known as the Euler–Lagrange lemma, which is now stated formally:

Theorem 7.23. (The Euler–Lagrange lemma.) *If J has a local extremum at f_* it is necessary that*

$$\frac{\partial F}{\partial f_*} - \frac{d}{dx}\frac{\partial F}{\partial f'_*} = 0. \tag{7.75}$$

Of course, the result in Theorem 7.23 is only necessary and not sufficient. One strategy that may be considered to identify the global extremum is to find all local extrema using the lemma (if there are not too many) and then choose the best. A sufficient condition for the existence of an extremum has been provided by Elsgolc (1961): If J has a local extremum at f_* a sufficient condition for f_* to be a local maximum (minimum) is

$$\frac{\partial^2 F}{\partial (f'_*)^2} \leq 0\,(\geq 0).$$

The reader will have noticed the strong similarity between the results on the optimization of functions in the calculus of variations and the optimization of functions in elementary calculus. We now apply the result of theorem 7.23 to some examples.

7.5.3.2 Applications of the Euler–Lagrange Lemma

(*i*) *The Shortest Length Problem.* Consider the problem of joining two given points (x_0, y_0), $(x_1, y_1) \in \mathbf{R}^2$ with the curve of shortest length. A curve f joining the points has arc length

$$\int_{x_0}^{x_1} \sqrt{1 + (f'(x))^2}\, dx,$$

where

$$f(x_0) = y_0, \qquad f(x_1) = y_1. \tag{7.76}$$

In the context of the general problem, we have

$$F(x, f, f') = \sqrt{1 + (f'(x))^2}$$

and

$$J(f) = \int_{x_0}^{x_1} \sqrt{1 + (f'(x))^2}\, dx.$$

Applying the Euler–Lagrange lemma, we obtain

$$\frac{\partial F}{\partial f} = 0$$

$$\frac{\partial F}{\partial f'} = f'(1 + (f')^2)^{-1/2}$$

$$\frac{d}{dx}\left(\frac{\partial F}{\partial f'}\right) = f''(1 + (f')^2)^{-1/2} - (f')^2 f''(1 + (f')^2)^{-3/2}.$$

Substituting these results into (7.75) produces

$$f''_*(1 + (f'_*)^2)^{-1/2} - (f'_*)^2 f''_*(1 + (f'_*)^2)^{-3/2} = 0.$$

Hence

$$f''_*(1 + (f'_*)^2) = (f'_*)^2 f''_*,$$

and therefore

$$f''_*(x) = 0$$
$$f'_*(x) = a, \quad \text{a constant}$$
$$f_*(x) = ax + b, \quad b \text{ a constant.}$$

This is the curve of a straight line with (7.76) uniquely determining a and b. Hence we have shown that the shortest distance between two points is a straight line.

(ii) *The Problem of Least Surface Area of Rotation.* Consider once again a curve f joining two points $(x_0, y_0), (x_1, y_1) \in \mathbf{R}^2$. Suppose now that f is rotated about the x-axis. The surface described by this rotation has area:

$$J(f) = 2\pi \int_{x_0}^{x_1} f(x)\sqrt{1 + (f'(x))^2}\, dx.$$

The problem is to find the curve f which describes least surface area, that is, minimizes $J(f)$. Now

$$\frac{\partial F}{\partial f} = (1 + (f')^2)^{1/2}$$

$$\frac{\partial F}{\partial f'} = ff'(1 + (f')^2)^{-1/2}$$

$$\frac{d}{dx}\frac{\partial F}{\partial f'} = (f')^2(1 + (f')^2)^{-1/2} + ff''(1 + (f')^2)^{-1/2} - f(f')^2 f''(1 + (f')^2).$$

Applying the Euler–Lagrange lemma, we obtain

$$(1 + (f'_*)^2)^{1/2} - \{(f'_*)^2 + f_* f''_*\}(1 + (f'_*)^2)^{-1/2} + f_*(f'_*)^2 f''_*(1 + (f'_*)^2)^{-3/2} = 0,$$

hence

$$(1 + (f'_*)^2)^2 - \{(f'_*)^2 + f_* f''_*\}(1 + (f'_*)^2) + f_*(f'_*)^2 f'' = 0$$

and therefore

$$1 + 2(f'_*)^2 + (f'_*)^4 - (f'_*)^2 - f_* f''_* - (f'_*)^4 - f_*(f'_*)^2 f''_* + f_*(f'_*)^2 f''_* = 0$$

or

$$1 + (f'_*)^2 - f_* f''_* = 0.$$

The solution of this differential equation is a curve f_* called a *catenary*:

$$f_*(x) = a \cosh \frac{x+b}{a},$$

where a and b can be determined uniquely by (7.76).

(*iii*) *The Brachistochrone*. This problem was described in Chapter 1. It involves finding a curve f joining points $(x_0, y_0), (x_1, y_1) \in \mathbf{R}^2$ which, if made of frictionless wire, would cause a bead to slide under gravity from one point to the other in least time. Thus the problem is to find the curve f_* which minimizes

$$J(f) = \int_{x_0}^{x_1} \sqrt{\frac{1 + (f'(x))^2}{2gf(x)}} \, dx,$$

where g is the acceleration due to gravity. In order to solve this problem the following theorem is useful.

Theorem 7.24. *If F does not depend upon x, then f_*, the solution to* (7.75), *obeys.*

$$F(f_*, f'_*) - f'_* F_{f'_*}(f_*, f'_*) = c,$$

where c is a constant.

PROOF. Consider the expression

$$F(f, f') - f' F_{f'}(f, f').$$

Upon differentiation, this gives

$$\frac{d}{dx}\{F(f,f') - f'F_{f'}(f,f')\} = F_f(f,f')f' + F_{f'}(f,f')f''$$

$$- f''F_{f'}(f,f') - f'\frac{d}{dx}F_{f'}(f,f')$$

$$= f'\left\{F_f(f,f') - \frac{d}{dx}F_{f'}(f,f')\right\}.$$

Now, setting

$$f = f_*,$$

and using the Euler–Lagrange lemma, we get

$$\frac{d}{dx}\{F(f_*, f'_*) - f'_* F_{f'_*}(f_*, f'_*)\} = f'_* \cdot 0 = 0.$$

Therefore

$$F(f_*, f') - f_*' F_{f_*'}(f_*, f_*') = c, \text{ a constant.} \qquad \square$$

Returning to the brachistochrone, F is defined by:

$$F(x, f, f') = F(f, f') = \sqrt{\frac{\{1 + (f')^2\}}{2gf}},$$

which is not directly dependent upon x. Hence, on using Theorem 2.24, we obtain

$$(1 + (f')^2)^{1/2}(2gf)^{-1/2} - f'(1 + (f')^2)^{-1/2}f'(2gf)^{-1/2} = c.$$

Hence

$$\frac{1 + (f')^2 - (f')^2}{f^{1/2}(1 + (f')^2)^{1/2}} = (\sqrt{2g})c,$$

and therefore

$$f(1 + (f')^2) = a,$$

where

$$a = (2gc^2)^{-1}.$$

Therefore

$$f'(x) = \sqrt{\frac{a - f(x)}{f(x)}}.$$

The solution to this differential equation can be expressed in parametric form as follows:

$$x = x_0 + \frac{a}{2}(t - \sin t)$$

$$f(x) = -\frac{a}{2}(1 - \cos t) \qquad t_0 \le t \le t_1, \qquad (7.77)$$

where it has been assumed that $f(x_0) = 0$, and t_0, t_1 correspond to the end-points of the wire, $(x_0, y_0), (x_1, y_1)$. (7.77) describes a curve known as a *cycloid*.

7.5.4 The Relationship Between C.V. and D.P.

The calculus of variations (C.V.) and dynamic programming (D.P.) (introduced in the previous chapter) have a great deal in common as branches of mathematics. We now present one instance of how D.P. can be used to solve a simple, general variational problem.

Suppose it is wished to find the curve f_* satisfying (7.76) which minimizes some functional J, where

$$J(f) = \int_{x_0}^{x_1} F(x, f, f') \, dx.$$

Consider now an intermediate point (x', y') on f_*. Then, as f_* is optimal, the part of the curve from (x', y') to (x_1, y_1) must also be optimal for the problem:

$$\text{Minimize:} \qquad \int_{x'}^{x_1} F(x, f, f')\, dx.$$

The reasoning behind this statement is embodied in the principle of optimality stated in Chapter 6. We can look upon the problem as one of D.P. in which there is an infinite number of stages–points along the x-axis from x_0 to x_1, each point x' corresponding to a state $(x', f(x'))$.

Suppose f_* is the optimal curve for $[x_0, x_1 - \Delta x]$ and is arbitrary on $[x_1 - \Delta x, x_1]$, except that

$$f_*(x_1) = y_1.$$

Then

$$\int_{x_0}^{x_1} F(x, f_*, f'_*)\, dx = \int_{x_0}^{x_1 - \Delta x} F(x, f_*, f'_*)\, dx + \int_{x_1 - \Delta x}^{x_1} F(x, f_*, f'_*)\, dx.$$

Now define a two-variable function dependent upon (x, y) by

$$S(x, y) = \min_{f \in D} \int_{x_0}^{x} F(x, f, f')\, dx.$$

As f_* is taken to be optimal from x_0 to $x_1 - \Delta x$,

$$S(x_1 - \Delta x, f_*(x_1 - \Delta x)) = \int_{x_0}^{x_1 - \Delta x} F(x, f_*, f'_*)\, dx.$$

Expanding $f_*(x_1 - \Delta x)$ in a Taylor series, we obtain

$$S(x_1 - \Delta x, f_*(x_1) - f'_*(x_1)\Delta x + O(\Delta x^2)) = \int_{x_0}^{x_1 - \Delta x} F(x, f_*, f'_*)\, dx.$$

Also, it can be shown that

$$\int_{x_0}^{x_1 - \Delta x} F(x, f, f')\, dx = F(x, f, f')\Delta x + O(\Delta x^2).$$

Putting these results together, we have

$$\min_{f \in D} \int_{x_0}^{x_1} F(x, f, f')\, dx = \int_{x_0}^{x_1 - \Delta x} F(x, f_*, f'_*)\, dx + \min_{f \in D} \int_{x_1 - \Delta x}^{x_1} F(x, f_*, f'_*)\, dx$$

$$= S(x_1 - \Delta x, f_*(x_1) - f'_*(x_1)\Delta x + O(\Delta x^2))$$
$$\quad + \min_{f \in D} \{F(x, f, f')\Delta x + O(\Delta x^2)\}$$

$$= S(x_1, f_*(x_1)) - \frac{\partial S}{\partial x_1}\Delta x - \frac{\partial S}{\partial f_*} f'_*(x_1)\Delta x + O(\Delta x^2)$$
$$\quad + \min_{f \in D} \{F(x, f, f')\Delta x + O(\Delta x^2)\}.$$

Now, as $\Delta x \to 0$,

$$0 = \min_{f \in D} \left\{ F(x, f, f') - \frac{\partial S}{\partial x} - f'(x_1)\frac{\partial S}{\partial x} \right\}.$$

This is the basic partial differential equation of D.P.

However, the use of D.P. to solve C.V. problems which have simple analytical solutions is like using a sledge hammer to crack a peanut. The approach is most appropriate when f is so complicated that it has to be approximated by numerical methods.

7.5.5 Further Horizons of C.V.

As the reader has no doubt gathered, the material presented so far in this chapter represents only a mere glimpse at the most elementary theory of the calculus of variations. While a detailed analysis of the more advanced ideas is beyond the scope of this book, we present a brief outline of the scope of the topic.

7.5.5.1 *Multivariable Functions*

Until now we have assumed that D, the domain of the functionals under consideration, comprises functions of a single variable. It is desirable to generalize this to functions of many variables as applications of this generalization arise in many areas. So now we are considering functionals J where

$$J: D \to \mathbf{R},$$

and D is a set of functions such that if $f \in D$:

$$f: \mathbf{R}^n \to \mathbf{R}, \qquad n > 1.$$

In this case the variational problem becomes

$$\text{Optimize:}_{f \in D} \quad J(f) = \int_{x_0^1}^{x_1^1} \int_{x_0^2}^{x_1^2} \cdots \int_{x_0^n}^{x_1^n} F(x_1, x_2, \ldots, x_n, f, f') \, dx \cdots dx_n,$$

where each $f \in D$ must satisfy appropriate boundary conditions. If certain conditions are met Theorem 7.23 can be generalized as follows:

Theorem 7.25. *If f has continuous second partial derivatives and if J has a local extremum at f_* it is necessary that*

$$\frac{\partial F}{\partial f_*} - \sum_{i=1}^{n} \frac{\partial}{\partial x_i} \frac{\partial F}{\partial f_{*x_i}} = 0.$$

PROOF. See Gelfand and Fomin (1963).

7.5.5.2 *Multivariable Functionals*

We can also make a different generalization to the case in which the functional J is multivariable, that is, J depends upon, say n functions:

$$J: D^n \to \mathbf{R}.$$

In this case the variational problem becomes

$$\text{optimize:} \quad J(f) = \int_{x_0}^{x_1} F(x, f_1, f_2, \ldots, f_n, f'_1, f'_2, \ldots, f'_n)\, dx,$$

where the functions f_i, $i = 1, 2, \ldots, n$, are assumed to satisfy appropriate boundary conditions. Once again if certain conditions are met Theorem 7.23 can be generalized:

Theorem 7.26. *If each f_i, $i = 1, 2, \ldots, n$, has continuous second partial derivatives and if J has a local extremum at $(f_{1*}, f_{2*}, \ldots, f_{n*})$ then it is necessary that*

$$\frac{\partial F}{\partial f_{i*}} - \frac{d}{dx}\frac{\partial F}{\partial f'_{i*}} = 0.$$

PROOF. See Gelfand and Fomin (1963).

7.5.5.3 Parametric Form

The solution to the brachistochrone was expressed in parametric form in Section 7.5.3.2. Indeed, it is often convenient to express the curves of certain variational problems in parametric form. Consider the simple variational problem given at the beginning of Section 7.5.3.1 and suppose that x and f depend upon the parameter t. If

$$x(t_0) = x_0 \quad \text{and} \quad x(t_1) = x_1,$$

then J becomes

$$J(f) = \int_{t_0}^{t_1} F\left(x(t), f(t), \frac{df}{dt}\bigg/\frac{dx}{dt}\right)\frac{dx}{dt}\, dt$$

$$= \int_{t_0}^{t_1} G(x, f, \dot{f}, \dot{x}, t)\, dt,$$

where $\dot{f}$ and $\dot{x}$ denote, respectively, the derivative of f and x with respect to t and G is the appropriate five-variable function.

The following theorem provides a necessary condition for a local extremum for J.

Theorem 7.27. *If J has a local extremum at f_* it is necessary that*

$$\frac{\partial G}{\partial x} - \frac{d}{dt}\left(\frac{\partial G}{\partial \dot{x}}\right) = 0 \tag{7.78}$$

$$\frac{\partial G}{\partial f} - \frac{d}{dt}\left(\frac{\partial G}{\partial \dot{f}}\right) = 0. \tag{7.79}$$

PROOF. See Gelfand and Fomin (1963).

Equations (7.78) and (7.79) are not independent and are equivalent to (7.75).

7.5.5.4 *Constrained Variational Problems*

In the simple variational problem of Section 7.5.3 the curves in D had to obey very few conditions. Namely, any such curve had to be bounded, have continuous second derivatives and obey boundary conditions. However it is necessary in any applications that the curves also obey additional constraints. These are usually of three types: integral, differential, or algebraic equations or inequalities.

7.5.5.4.1 *Integral Constraints.* As an example of a problem with integral constraints we introduce the *isoperimetric problem*:

$$\text{Optimize:} \qquad J(f) = \int_{x_0}^{x_1} F(x, f, f') \, dx, \qquad f \in D$$

$$\text{subject to:} \qquad K(f) = \int_{x_0}^{x_1} G(x, f, f') \, dx - q = 0, \tag{7.80}$$

where F and G have continuous second derivatives and q is a given real constant. Applying the ideas of Section 7.4.1.3 we form the Lagrangian:

$$\begin{aligned} J + \lambda K &= \int_{x_0}^{x_1} \left\{ F(x, f, f') \, dx + \lambda \int_{x_0}^{x_1} G(x, f, f') - q \right\} dx \\ &= \int_{x_0}^{x_1} \{ F(x, f, f') + \lambda G(x, f, f') \} \, dx - \lambda q (x_1 - x_0), \end{aligned}$$

which will have the same optimum as

$$\int_{x_0}^{x_1} \{ F(x, f, f') + \lambda G(x, f, f') \, dx.$$

Applying Theorem 7.23 to this last expression produces the following necessary condition for J to have a local extremum at f_*:

$$\frac{\partial F}{\partial f} - \frac{d}{dx} \frac{\partial F}{\partial f'} + \lambda \left(\frac{\partial G}{\partial f} - \frac{d}{dx} \frac{\partial G}{\partial f'} \right) = 0. \tag{7.81}$$

Equations (7.80) and (7.81) can be solved to find f_* and λ.

7.5.5.4.2 *Differential Constraints.* Let us now consider problems involving differential constraints. Consider the following problem of two functions, f_1 and f_2:

$$\text{Optimize:} \qquad J(f_1, f_2) = \int_{x_0}^{x_1} F(x, f_1, f'_1, f_2, f'_2) \, dx, \qquad f_1, f_2 \in D \tag{7.82}$$

$$\text{subject to:} \qquad K(f_1, f_2) = G(x, f_1, f'_1, f_2, f'_2) = 0,$$

where once again it is assumed that F and G have continuous second derivatives. Here we do not form the Lagrangian, but instead the integral:

$$\begin{aligned} I(f_1, f_2) &= J + \lambda(x) \int_{x_0}^{x_1} K \, dx \\ &= \int_{x_0}^{x_1} \{ F(x, f_1, f'_1, f_2, f'_2) + \lambda(x) G(x, f_1, f'_1, f_2, f'_2) \} \, dx. \end{aligned}$$

It can be shown that if I has a local extremum at f_{1*} and f_{2*} then it is necessary that

$$\frac{d}{dx}\left(\frac{\partial F}{\partial f'_{i*}} + \lambda(x)\frac{\partial G}{\partial f'_{i*}}\right) = \frac{\partial F}{\partial f_{i*}} + \lambda(x)\frac{\partial G}{\partial f_{i*}}, \qquad i = 1, 2. \qquad (7.83)$$

It can further be shown (see Gottfried and Weisman (1973)) that the application of necessary conditions for the extremization of I is equivalent to the application of them for the original constrained problem (7.82). Hence (7.83) constitutes a set of necessary conditions for (7.82).

7.5.5.4.3 *Algebraic Constraints.* The constraint in (7.82) involved f'_1 and f'_2 and hence was called a differential constraint. If these functions are not present, we are left with a variational problem with a solely algebraic constraint:

Optimize: $J(f_1, f_2) = \int_{x_0}^{x_1} F(x, f_1, f_2, f'_1, f'_2)\, dx, \qquad f_1, f_2 \in D$

subject to: $K(f_1, f_2) = G(x, f_1, f_2) = 0.$

Once again one can use the integral

$$I(f_1, f_2) = J + \lambda(x) \int_{x_0}^{x_1} K\, dx$$

to develop necessary conditions for the existence of a local extremum. These are presented in the following theorem.

Theorem 7.28. *If J has a local extremum at (f_{1*}, f_{2*}) it is necessary that*

$$\frac{\partial F}{\partial f_1} + \lambda(x)\frac{\partial G}{\partial f_1} - \frac{d}{dx}\frac{\partial F}{\partial f'_1} = 0$$

and

$$\frac{\partial F}{\partial f_2} + \lambda(x)\frac{\partial G}{\partial f_2} - \frac{d}{dx}\frac{\partial F}{\partial f'_2} = 0.$$

PROOF. See Gelfand and Fomin (1963).

7.5.5.5 *The Maximum Principle*

The topic of control theory or optimal control is concerned with the finding of a policy for the efficient operation of a physical system. Sometimes the state of the system can be described by a real vector $x = (x_1, x_2, \ldots, x_m)$, the elements of which vary with time as follows:

$$\frac{dx_i}{dt} = F_i(x_1, x_2, \ldots, x_n, f_1, f_2, \ldots, f_m), \qquad i = 1, 2, \ldots, n. \qquad (7.84)$$

Here $f_1, f_2, \ldots, f_m$ are bounded, piecewise continuous real functions, dependent on time, forming a vector $f = (f_1, f_2, \ldots, f_m)$, and the F_i are also continuous.

Now the x_i, $i = 1, 2, \ldots, n$ also depend upon t and it is assumed that a set of initial boundary conditions:

$$x_i(t_0) = a_i, \qquad i = 1, 2, \ldots, n, \qquad (7.85)$$

are satisfied for the beginning t_0 of the time span $[t_0, t_1]$ under consideration, where the a_i are given constants. Consider now some measurement $F_0(x, f)$ (a differentiable function) of the performance of the system. Then for any solution $f_1, f_2, \ldots, f_m$ to (7.84) we can calculate a real number $J(f)$ where

$$J(f) = \int_{t_0}^{t_1} F_0(x, f) \, dt.$$

Let $D = \{ f = (f_1, f_2, \ldots, f_m) : f_i, \ i = 1, 2, \ldots, m$ are continuous real functions defined on $[t_0, t_1]$, satisfying (7.84)$\}$. D is called the set of *admissable* processes and sometimes has further restrictions placed upon it. Then J is said to have a local minimum at $f^* \in D$ if

$$J(f^*) \le J(f), \quad \text{for all } f \in D.$$

We now examine what conditions are necessary for J to have a local minimum at f^*. To this end we introduce a new variable x_0, where

$$\frac{dx_0}{dt} = F_0(x, f), \qquad (7.86)$$

and define

$$x_0(t_0) = a_0 = 0. \qquad (7.87)$$

Integrating (7.86), we obtain

$$\int_{t_0}^{t_1} \frac{dx_0}{dt} = \int_{t_0}^{t_1} (x, f) \, dt = J(f).$$

Therefore   -

$$J(f) = [x_0(t)]_{t_0}^{t_1} = x_0(t_1) - x_0(t_0)$$
$$= x_0(t_1), \qquad \text{by (7.87)}.$$

Hence the problem can be restated as follows:

$$\text{Minimize:} \quad x_0(t_1)$$
$$\scriptstyle f \in D$$

$$\text{subject to:} \quad \frac{dx_i}{dt} = F_i(x, f), \qquad i = 0, 1, \ldots, n \qquad (7.88)$$

$$x_i(t_0) = a_i, \qquad i = 0, 1, \ldots, n. \qquad (7.89)$$

Applying the necessary conditions of Section 7.5.5.4.2 to this problem, it is easy to show that, on ignoring the constraints in (7.88) and (7.89) corresponding to $i = 0$, we obtain

$$\frac{\partial F_0}{\partial x_i} - \sum_{k=1}^{n} \lambda_k(t) \frac{\partial F_k}{\partial x_i} - \frac{d\lambda_i}{dt} = 0, \qquad i = 1, 2, \ldots, n \qquad (7.90)$$

and

$$\frac{\partial F_0}{\partial f_j} - \sum_{k=1}^{m} \lambda_k(t) \frac{\partial F_k}{\partial f_j} = 0, \qquad j = 1, 2, \ldots, m. \qquad (7.91)$$

We now construct what is known as the *Hamiltonian function H* where

$$H(x, f, \lambda) = \sum_{i=0}^{n} \lambda_i(t) F_i(x, f). \qquad (7.92)$$

Then (7.88) can be expressed as follows:

$$\frac{\partial H}{\partial \lambda_i} = \frac{dx_i}{dt}, \qquad i = 0, 2, \ldots, n. \qquad (7.93)$$

Taking the partial derivatives of H with respect to x_i, we obtain

$$\frac{\partial H}{\partial \lambda_i} = \sum_{k=0}^{n} \lambda_k(t) \frac{\partial F_k}{\partial x_i},$$

which, by (7.90), yields

$$\frac{\partial H}{\partial \lambda_i} = \lambda_0(t) \frac{\partial F_0}{\partial x_i} + \frac{\partial F_0}{\partial x_i} - \frac{d\lambda_i}{dt}.$$

Now, as (7.86), which is the first constraint in the family (7.88), is artificial, we can assign $\lambda_0(t)$ an arbitrary constant value for all $t \in [t_0, t_1]$. Thus, let

$$\lambda_0(t) = -1, \qquad t \in [t_0, t_1]. \qquad (7.94)$$

Then we have

$$\frac{\partial H}{\partial x_i} = \frac{d\lambda_i}{dt}, \qquad i = 0, 1, 2, \ldots, n. \qquad (7.95)$$

Taking the partial derivative of H with respect to f_j, we obtain

$$\frac{\partial H}{\partial f_j} = \sum_{k=0}^{} \lambda_k(t) \frac{\partial F_k}{\partial f_j}$$

$$= \frac{\partial F_0}{\partial f_j} \qquad \text{by (7.91)}$$

$$= \frac{\partial}{\partial f_j} \frac{dx_0}{dt} \qquad \text{by (7.86)}$$

$$= 0$$

$$\frac{\partial H}{\partial f_j} = 0, \qquad j = 1, 2, \ldots, m. \qquad (7.96)$$

Thus we can replace the necessary conditions (7.90) and (7.91) by (7.93), (7.95), and (7.96), and state the following theorem:

Theorem 7.29. (The maximum principle) *If* $f^* = (f_1^*, f_2^*, \ldots, f_m^*)$ *is optimal and* $x = (x_1, x_2, \ldots, x_n)$ *obeys* (7.84) *and* (7.85) *then there exists* $\lambda(t) = (\lambda_0(t),$ $\lambda_1(t), \ldots, \lambda_n(t))$ *such that* (7.93), (7.91), (7.95), *and* (7.96) *are satisfied for H as defined in* (7.92).

These results have been known for many years. However, recently Pontryagin et al. (1962) have extended this theory to cover the case when the functions $f_1, f_2, \ldots, f_m$ must also obey a family of inequality constraints. Their results have come to be known as the *maximum principle*. It is identical to Theorem 7.29 when the optimal vector f is in the interior of the region defined by the inequality constraints.

7.6 Exercises

1. Locate all extrema of the following functions and identify the nature of each, where $x \in \mathbf{R}$.
   (a) $f(x) = x^3 + \frac{3}{2}x^2 - 18x + 19$
   (b) $f(x) = 6x^4 + 3x^2 + 42$
   (c) $f(x) = x^2 + 4x - 8$
   (d) $f(x) = 6x^2 + \sqrt{3x} - 9$
   (e) $f(x) = x^{12} - 14x^{11} + x^{10} + 90x^9 + 8x^8 + 6$.

2. Given that $x \in [-\frac{5}{2}, \frac{5}{2}]$, find the global extrema of each function in Exercise 1.

3. Prove that a function $f : I \to \mathbf{R}$ is convex if and only if, for all $\alpha \in R, 0 \leq \alpha \leq 1$ and for all $x_1, x_2 \in I$,

$$f(\alpha x_1 + (1 - \alpha)x_2) \leq \alpha f(x_1) + (1 - \alpha)f(x_2).$$

4. Prove Theorem 7.4.

5. Prove Theorem 7.6.

6. Prove Theorem 7.8.

7. Prove Theorem 7.10.

8. Use the results of Section 7.2.7 to find the global extrema of the following functions, which are either concave, convex, or both.
   (a) $f(x) = \sin x, 0 \leq x \leq \pi$.
   (b) $f(x) = \cos x, -\pi/2 \leq x \leq \pi$.
   (c) $f(x) = 4x - 2, -6 \leq x \leq 9$.
   (d) $f(x) = 3x^2 - 18x + 2, -3 \leq x \leq 20$.

9. Prove Theorem 7.13.

10. Locate all extrema of the following functions and identify the nature of each, where $X \in \mathbf{R}^2$.
    (a) $f(x_1, x_2) = x_1^2 - x_1 + 3x_2^2 + 18x_2 + 14$.
    (b) $f(x_1, x_2) = 3x_1^2 + 4x_2^2 - 6x_1 - 7x_2 + 13x_1 x_2 + 1$.
    (c) $f(x_1, x_2) = x_1^2 - 6x_1 + x_2^2 - 16x_2 + 25$.

11. Given that

$$-1 \le x_1 \le 5$$
$$-2 \le x_2 \le 6$$

    find the global extrema of each function in Exercise 10.

12. Prove that a function $f: S \to \mathbf{R}$, $S \subset \mathbf{R}^n$ is convex on S if and only if, for all $\alpha \in \mathbf{R}$, $0 \le \alpha \le 1$ and for all $X_1, X_2 \in S$,

$$f(\alpha X_1 + (1 - \alpha)X_2) \le \alpha f(X_1) + (1 - \alpha)f(X_2).$$

13. Prove Theorem 7.15.

14. Solve the following problems using the Jacobian method.

    (a) Maximize: $f(x_1, x_2, x_3, x_4) = -4x_1^2 - 3x_2^2 - 6x_3^2 - x_4^2$
        subject to: $x_1 + x_2 + x_3 + x_4 - 2 = 0$
                       $3x_1 + 2x_2 + 4x_3 + x_4 - 3 = 0$
                       $x_1 + 4x_2 + 3x_3 + x_4 - 1 = 0$
                       $[X^* = (0.5752, -0.3856, 0.0784, 1.732)]$.

    (b) Maximize: $f(x_1, x_2) = 6x_1^2 + 3x_2^2 + 4x_1 x_2$
        subject to: $x_1 x_2 = 7$.

    (c) Maximize: $f(x_1, x_2) = 2x_1^2 + x_2^2 + 3x_1 + 4x_2 + 9$
        subject to: $x_1^2 + x_2 + 3x_1 x_2 = 11$
                       $x_1 + x_2^2 + 4x_1 x_2 = 12$.

    (d) Maximize: $f(x_1, x_2, x_3, x_4) = -4x_1^2 - 2x_2^2 - x_3^2 - 2x_4^2$
        subject to: $2x_1 + x_2 + x_3 + x_4 - 2 = 0$
                       $x_1 + 2x_2 + 2x_3 + x_4 - 1 = 0$
                       $3x_1 + 3x_2 + x_3 + x_4 \quad = 0$
                       $[X^* = (\frac{16}{18}, \frac{10}{18}, -\frac{12}{18}, \frac{6}{18})]$.

    (e) Maximize: $f(x_1, x_2, x_3, x_4) = -x_1^2 - 2x_x^2 - 3x_3^2 - 4x_4^2 + 5$
        subject to: $x_1 + x_2 - x_3 + x_4 + 1 = 0$
                       $2x_1 + 3x_2 - x_3 + 2x_4 - 2 = 0$
                       $2x_1 + x_2 + x_3 + 3x_4 - 1 = 0$
                       $[X^* = (\frac{36}{13}, \frac{22}{13}, \frac{32}{13}, -\frac{37}{13})]$.

    (f) Minimize: $f(x_1, x_2) = x_1^2 + x_2^2$
        subject to: $x_1 x_2 = 8$.

(g) Maximize: $f(x_1, x_2, x_3, x_4) = -x_1^2 + 2x_2^2 + 4x_3^2 - 3x_4^2$

    subject to: $x_1 + 3x_2 + 4x_3 - 2x_4 \quad = 0$

    $x_1 + \quad x_2 + \quad x_3 + \quad x_4 \quad = 0$

    $4x_1 + 3x_2 + 2x_3 + \quad x_4 - 1 = 0$

    $[X^* - (366, -168, -43, -155)].$

(h) Maximize: $f(x_1, x_2, x_3, x_4) = -x_1^2 - 2x_2^2 - 3x_3^2 - x_4^2$

    subject to: $x_1 + \quad x_2 + \quad x_3 + \quad x_4 = 4$

    $x_1 - \quad x_2 + 2x_3 - \quad x_4 = 5$

    $3x_1 + 2x_x - \quad x_3 - 2x_4 = 3$

    $[X^* = (\frac{240}{53}, -\frac{154}{53}, -\frac{1}{53}, \frac{127}{53})].$

(i) Maximize: $f(x_1, x_2, x_3, x_4) = -2x_1^2 - 3x_2^2 - x_3^2 - 3x_4^2$

    subject to: $2x_1 - x_4 = 0$

    $x_2 + \quad x_3 \quad\quad + 1 = 0$

    $x_2 + 2x_3 + x_4 + 6 = 0$

    $[X^* = (-\frac{17}{15}, \frac{26}{15}, -\frac{41}{15}, -\frac{34}{15})].$

(j) Maximize: $f(x_1, x_2, x_3, x_4) = 4x_1 - x_1^2 - x_2^2 - 2x_3^2 - 3x_4^2$

    subject to: $x_1 - x_2 \quad\quad\quad - 4 = 0$

    $x_2 + 2x_3 + x_4 + 2 = 0$

    $x_1 \quad\quad + \quad x_3 - x_4 - 3 = 0$

    $[X^* = (\frac{14}{5}, -\frac{6}{5}, -\frac{1}{5}, -\frac{2}{5})].$

(k) Maximize: $f(x_1, x_2, x_3, x_4) = -3x_1^2 - x_2^2 - 9x_3^2 - 6x_4^2$

    subject to: $x_1 + 3x_2 + \quad x_3 + 3x_4 - 1 = 0$

    $3x_2 + 4x_3 + 2x_4 - 2 = 0$

    $x_1 + 6x_2 + 4x_3 + 3x_4 - 1 = 0$

    $[X^* = (-\frac{1}{4}, -\frac{1}{2}, \frac{1}{2}, \frac{3}{4})].$

(l) Maximize: $f(x_1, x_2, x_3, x_4) = -x_1^2 - x_2^2 - 3x_3^2 - 2x_4^2$

    subject to: $2x_1 + 3x_2 + 4x_3 + \quad x_4 = 5$

    $3x_1 + 4x_2 + \quad x_3 + 2x_4 = 3$

    $x_1 + \quad x_2 + \quad x_3 + \quad x_4 = 1$

    $[X^* = (-\frac{3}{28}, \frac{13}{14}, \frac{3}{4}, -\frac{8}{14})].$

(m) Maximize: $f(x_1, x_2, x_3, x_4) = -3x_1^2 - 4x_2^2 - x_3^2 - 2x_4^2$

    subject to: $x_1 + x_2 + \quad x_3 + \quad x_4 - 3 = 0$

    $2x_1 + x_2 + 3x_3 + \quad x_4 - 5 = 0$

    $4x_1 + x_2 + \quad x_3 + 3x_4 - 4 = 0$

    $[X^* = (-0.34, 1.16, 1.17, 1.04)].$

15. Solve the linear programming Problem 2.1 of Chapter 2 by the Jacobian method.

16. Solve the linear programming Problem 2.1 of Chapter 2 by the method of Lagrange.

17. Solve the problems in Exercise 14 by the method of Lagrange.

18. Suppose now that the right-hand side constants of each of the constraints in the problems in Exercise 17 are increased by 0.01. Use the sensitivity coefficient of Section 7.4.1.3 to calculate the increase in the value of the optimal solution to each problem.

19. Prove Theorem 7.17.

20. Prove Theorem 7.20.

21. Prove Theorem 7.22

22. Develop the Kuhn–Tucker conditions for the following problem:

    Minimize: $f(X)$

    subject to: $g_j(X) = 0, \qquad j = 1, 2, \ldots, m.$

23. Replace each of the "=" signs by "≤" signs in each of the problems in Exercise 17 and present the Kuhn–Tucker conditions for each of the problems.

24. Minimize: $\int_0^1 (1 + x)\sqrt{1 + (f'(x))^2}\, dX$
    $f \in D$

    where $D = \{f: [0,1] \to \mathbf{R} \mid f$ has continuous derivatives,

    is bounded, $f(0) = 0, f(1) = 1\}.$

25. Find the curve joining points $(0,0)$ and $(4,4)$ whose arc length is 6, the area under which is a maximum.

26. Prove that if

$$f(x) = f(x_1, x_2, \ldots, x_n) = \qquad f_i(x_i),$$

where each $f_i, i = 1, 2, \ldots, n$ is concave (convex), then f is concave (convex).

Chapter 8

Nonlinear Programming

8.1 Introduction

This chapter is concerned with presenting algorithms for finding the optimal points of a continuous function. As was pointed out in the previous chapter, there exists a body of knowledge called classical optimization which provides an underlying theory for the solution of such problems. We now use that theory to develop methods which are designed to solve the large nonlinear optimization problems which occur in real-world applications.

The general nonlinear programming problem (N.P.P.) is

$$\text{Maximize:} \quad f(X) = x_0 \tag{8.1}$$

$$\text{subject to:} \quad g_j(X) = 0, \quad j = 1, 2, \ldots, m \tag{8.2}$$

$$h_j(X) \leq 0, \quad j = 1, 2, \ldots, k \tag{8.3}$$

where $X = (x_1, x_2, \ldots, x_n)^T$ is an n-dimensional real vector, and f; $g_j, j = 1, 2, \ldots, m$; $h_j, j = 1, 2, \ldots, k$, are real valued functions defined on $\mathbf{R}^n$.

Before dealing with the specific techniques we classify some of the special cases of the N.P.P. If f is quadratic, the g_j's are all linear and $h_j(X) = -x_j$, $j = 1, 2, \ldots, k$, then the N.P.P. is said to be a *quadratic programming* problem. In this case the problem can be expressed as follows

$$\text{Maximize:} \quad x_0 = C^T X + X^T D X$$

$$\text{subject to:} \quad A X = B$$

$$X \geq 0.$$

If D is symmetric negative definite, x_0 is concave, which guarantees that an optimum exists.

If there are no equality constraints and x_0 and the h_j's are all convex then the N.P.P. is said to be a *convex programming problem*, which can be handled by Zoutendijk's method of feasible directions (see Section 8.3.1).

If x_0 can be expressed as

$$x_0 = f_1(x_1) + f_2(x_2) + \cdots + f_n(x_n),$$

where the f_i's are all continuous functions of one variable, then the N.P.P. is said to be a *separable programming problem*. Unconstrained problems with this type of objective function can be attacked using pattern search (see Section 8.2.4.1). For constrained problems in which each constraint function is also separable, an approximate solution can be found by making a linear approximation of each function (including x_0) and using linear programming (see Section 8.3.4.2).

If x_0 and the constraint functions are of the form:

$$\sum_{j=1}^{P} \left(c_j \prod_{i=1}^{n} x_i^{a_{ij}} \right),$$

where

$$c_j > 0, \qquad j = 1, 2, \ldots, p,$$

then the N.P.P. is said to be a *geometric programming problem*. Problems of this type have been solved by a recently developed technique due to Duffin, Petersen, and Zener (1967; see Section 8.3.6).

Of course many N.P.P.'s belong to more than one of the above groups. Unfortunately, some N.P.P.'s belong to none. This chapter develops some of the more popular techniques for various nonlinear problems. Before beginning with the unconstrained case, we mention two simple but important concepts, *resolution* and *distinguishability*.

It may often happen when using the methods outlined in this chapter that the limit of precision to which numbers are calculated is exceeded. For example, a computer with 6 decimal place precision will not distinguish between the numbers 6.8913425 and 6.8913427, and the last digit is arbitrarily chopped. This phenomenon may occur when an objective function f is being evaluated, and in this case it would be said that the distinguishability of f is 10^{-6}. Formally:

Definition 8.1. The *distinguishability* of f is the minimum postive number γ such that for all X_1, X_2 in the domain of f, if $|f(X_1) - f(X_2)| \geq \gamma$, then it can be concluded that $f(X_1)$ and $f(X_2)$ are unequal.

Hence if f is of distinguishability γ and

$$|f(X_1) - f(X_2)| < \gamma,$$

one cannot conclude that $f(X_1)$ and $f(X_2)$ are unequal.

In the application of any numerical optimization technique there will be a practical limit on the accuracy with which one can deal with the values of $x_1, x_2, \ldots, x_n$. This accuracy may be governed by the conditions of an experiment or task which produces values of f. Readings on a gauge, the availability of only certain units of quantity of a commodity (for example, drugs with which to dose rats may be available only in 5 cc lots), or one's eyesight in reading a slide rule are examples. Hence if one is working with four-figure logarithm tables it may make little sense to attempt to consider a value of x_i of 4.0693. In this case we say the resolution of x_i is 0.001. Formally:

Definition 8.2. The *resolution* of a variable x_i is the smallest positive number ε_i such that, for all pairs x_i^1, x_i^2 of x_i, if $|x_i^1 - x_i^2| \geq \varepsilon_i$, then it can be concluded that x_i^1 and x_i^2 are unequal.

8.2 Unconstrained Optimization

In this case there are no constraints of the form of (8.2) or (8.3) and one is confronted solely with maximizing a real-valued function with domain $\mathbf{R}^n$. When such problems arise in practice first or second derivatives of the function are often difficult or impossible to compute and hence classical methods are usually unsuitable. Whether derivatives are available or not, the usual strategy is first to select a point in $\mathbf{R}^n$ which is thought to be the most likely place where the maximum exists. If there is no information available on which to base such a selection, a point is chosen at random. From this first point an attempt is made to construct a sequence of points, each of which yields an improved objective function value over its predecessor. The next point to be added to the sequence is chosen by analyzing the behaviour of the function at the previous points. This construction continues until some termination criterion is met. Methods based upon this strategy are called *ascent methods*.

Thus ascent methods are ways to construct a sequence: $X_1, X_2, X_3, \ldots$, of n-dimensional real vectors, where $f(X_1) < f(X_2) < f(X_3), \ldots$. In generating a new point X_{j+1} from the previous points $X_1, X_2, \ldots, X_j$, it is usual to express X_{j+1} as some function of X_j. Hence it must be decided (i) in what direction X_{j+1} lies from X_j and (ii) how far (in terms of the Euclidean metric) X_{j+1} is from X_j. So X_{j+1} can be expressed as follows:

$$X_{j+1} = X_j + s_j D_j.$$

The vector D_j is called the jth *direction vector*, and the magnitude $|s_j|$ of the scalar s_j is called the jth *step size*. Thus we find the new point X_{j+1} by "moving" s_j from X_j a distance in the direction D_j.

There are a host of methods which arise from using the information gained about the behaviour of f at the previous points $X_1, X_2, \ldots, X_j$ to specify D_j and s_j. Of course, in order to generate a new point $X_j + s_j D_j$

which will satisfy

$$f(X_j + s_j D_j) > f(X_j) \tag{8.4}$$

it is usually necessary to have to consider only certain s_j, D_j pairs. Indeed, some methods consider only D_j's for which (8.4) holds for a small value of s_j, that is, f must yield improved values near X_j.

Ascent methods can be classified according to the information about the behaviour of f that is required. *Direct methods* require only that the function be evaluated at each point. *Gradient methods* require the evaluation of first derivatives of f. *Hessian methods* require the evaluation of second derivatives. Although Hessian methods usually require the least number of points to be generated in order to locate a local maximum (which is all that any ascent method aims to produce), these methods are not always the most efficient in terms of computational effort. In fact, there is no superior method for all problems, the efficiency of a method being very much dependent upon the function to be maximized.

8.2.1 Univariate Search

Many search methods for unconstrained problems require searches for the maximal point of f in a specified direction. Suppose it is necessary to find the maximal point of f along a direction d_i from a point X_i. The feasible points can be expressed as

$$X_i + s_i D_i, \qquad s_i \in \mathbf{R}.$$

(Negative values of s_i represent the possibility that the maximal point may lie in the $-D_i$ direction from X_i.) Thus the problem is to maximize a function α of s_i, where

$$\alpha(s_i) = f(X_i + s_i D_i), \qquad s_i \in \mathbf{R},$$

with X_i and D_i fixed. Because this type of problem has to be solved repeatedly in many direct, gradient, and Hessian search methods, it is important that these one-dimensional searches be performed efficiently.

One crude technique is to first somehow find an interval I of the line $X_i + s_i D_i$ in which the maximum is known to lie. One then evaluates α at equally spaced points along I. Then I is replaced by a smaller interval I' which includes the best point found so far. The procedure is then repeated with I' replacing I. It is not hard to construct simple examples for which this technique performs rather poorly. It is usually better to make just one function evaluation each time and to decide where to make the next on the basis of the outcome. This approach is still inefficient unless it is assumed that α belongs to a restrictive class of functions of one variable called unimodal functions, which are described next.

It will be assumed in this section that the global maximum of α is known to lie in a closed interval and that within this interval the maximum occurs at a unique point. Thus α must strictly increase in value as s (we shall drop the subscript i) increases until the maximum is attained. Then α strictly decreases as s assumes values greater than the maximum. A function satisfying these properties is said to be *unimodal*. Hence, if α is unimodal and

$$s_0 < s_1 < s^* \quad \text{or} \quad s_0 > s_1 > s^*,$$

then

$$\alpha(s_0) < \alpha(s_1) < \alpha(s^*),$$

(8.5)

where s^* is the maximum of α.

If a function α is unimodal and its unique maximum is known to lie within a closed interval $[a, b]$, then when α is evaluated at any pair of points s_1, s_2 where $s_1 > s_2$, such that either

$$[s_2, s_1] \subseteq [a, b) \quad \text{or} \quad [s_2, s_1] \subseteq (a, b]$$

upon comparison of $\alpha(s_1)$ and $\alpha(s_2)$, the interval in which the maximum s^* lies can be reduced in length from $b - a$. This is because one of three events must occur: either

$$\alpha(s_1) > \alpha(s_2)$$

(8.6)

or

$$\alpha(s_1) < \alpha(s_2)$$

(8.7)

or

$$\alpha(s_1) = \alpha(s_2),$$

(8.8)

so that, by (8.5), we have

$$(8.6) \Rightarrow s^* \in (s_2, b]$$
$$(8.7) \Rightarrow s^* \in [a, s_1)$$
$$(8.8) \Rightarrow s^* \in (s_2, s_1).$$

(8.9)

A one-dimensional search procedure is termed *adaptive* if it uses the information gained about the behaviour of α at the previous point to decide where to evaluate α next. There are many adaptive procedures available which take advantage of (8.9).

The above concepts will be illustrated by some examples. Consider the functions shown in Figure 8.1. It can readily be seen that α is unimodal. It can be seen that

$$s^* = \tfrac{1}{2}.$$

Now,

$$\tfrac{1}{8} = s_0 < s_1 = \tfrac{1}{4} < s^*,$$

so that

$$\alpha(s_0) < \alpha(s_1) < \alpha(s^*),$$

as can be seen from Figure 8.1. Also, if

$$s^* < \tfrac{3}{4} = s_2 < s_3 = \tfrac{7}{8}$$

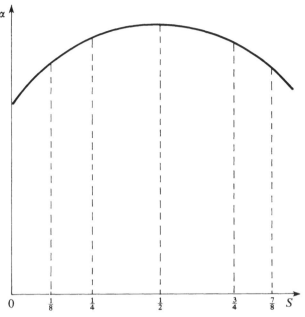

Figure 8.1.   A unimodal function.

then

$$\alpha(s^*) > \alpha(s_2) > \alpha(s_3)$$

as can be seen from Figure 8.1.

Now suppose that this diagram is unavailable and that information can be gained about the location of s^* only by evaluating α at selected points. The results of (8.9) will now be illustrated. Suppose $s_0 = \frac{1}{8}$ and $s_1 = \frac{1}{4}$ are evaluated. Then, as

$$\alpha(s_0) < \alpha(s_1),$$

the interval $\left[0, \frac{1}{8}\right]$ can be eliminated as shown in Figure 8.2(a). The same elimination could have occurred if any s_i had been chosen instead of s_1, as long as

$$\alpha(s_0) < \alpha(s_i).$$

Suppose $s_1 = \frac{1}{4}$ and $s_3 = \frac{7}{8}$ are evaluated instead. Then, as

$$\alpha(s_1) > \alpha(s_3),$$

the interval $\left[\frac{7}{8}, 1\right]$ can be eliminated, as shown in Figure 8.2(b). The same elimination could have occurred if any s_i had been chosen instead of s_1, as long as

$$\alpha(s_i) > \alpha(s_3).$$

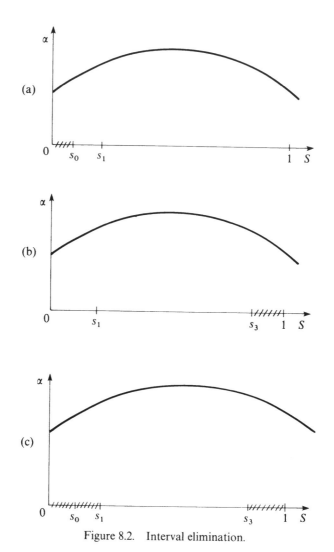

Figure 8.2. Interval elimination.

Suppose $s_1 = \frac{1}{4}$ and $s_2 = \frac{3}{4}$ are evaluated instead. Then, as

$$\alpha(s_1) = \alpha(s_2),$$

the intervals $[0, \frac{1}{4}]$ and $[\frac{3}{4}, 1]$ can be eliminated, as shown in Figure 8.2(c).

So far it has been assumed that the optimal point lies within a known closed interval $[a, b]$. There are many ways by which this initial interval can be found. One method is carried out as follows. Let the most likely location of the optimal point be a_1. If no information is known about the likely whereabouts of the optimum the point a_1 is chosen at random along the line. Next a positive real number β is chosen. The function is then evaluated at a_1 and $a_1 + \beta$. Three cases must be examined.

CASE I. $\alpha(a_1) < \alpha(a_1 + \beta)$. α is evaluated at $a_1 + 2\beta$, $a_1 + 4\beta, \ldots$, until a decrease occurs in the value of α at, say, $a_1 + 2^n\beta$. Then set

$$[a, b] = [a_1 + 2^{n-2}\beta, a_1 + 2^n\beta].$$

CASE II. $\alpha(a_1) > \alpha(a_1 + \beta)$. α is evaluated at $a_1 - \beta$, $a_1 - 2\beta$, $a_1 - 4\beta, \ldots$, until no increase occurs in the value of α at, say, $a_1 - 2^m\beta$. Then set

$$[a, b] = [a_1 - 2^m\beta, a_1 - 2^{m-2}\beta].$$

CASE III. $\alpha(a_1) = \alpha(a_1 + \beta)$. Set

$$[a, b] = [a_1, a_1 + \beta].$$

Of course if two points a', b' are found such that

$$\alpha(a') = \alpha(b'),$$

then set

$$[a, b] = [a', b'].$$

For example, let

$$a_1 = 4, \qquad \beta = 1.$$

Now if

$$\alpha(a_1) = \alpha(4) = -6$$
$$\alpha(a_1 + \beta) = \alpha(5) = -14,$$

then we have case II. Suppose, then, that

$$\alpha(a_1 - \beta) = \alpha(3) = 0$$
$$\alpha(a_1 - 2\beta) = \alpha(2) = 4$$
$$\alpha(a_1 - 4\beta) = \alpha(0) = 6$$
$$\alpha(a_1 - 8\beta) = \alpha(-4) = -14.$$

Then we have

$$[a, b] = [-4, 2].$$

This is shown in Figure 8.3.

One of the most efficient adaptive one-dimensional search procedures is called *Fibonacci serach*. It is described next.

8.2.1.1 *Fibonacci Search*

Fibonacci search depends upon the Fibonacci numbers $A_0, A_1, A_2, \ldots$, defined as follows:

$$A_0 = 0$$
$$A_1 = 1$$
$$A_i = A_{i-1} + A_{i-2}, \qquad i = 2, 3, 4, \ldots.$$

The procedure is used to reduce the interval of uncertainty of a unimodal function α. Suppose the initial interval is $[a_1, b_1]$. After a number of iterations

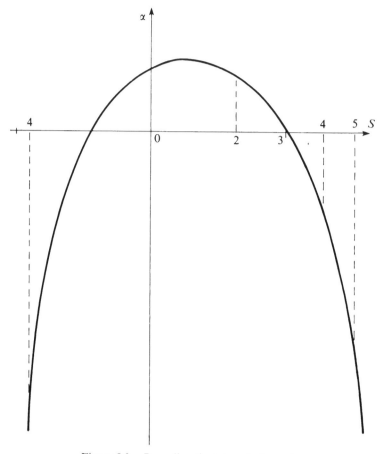

Figure 8.3. Bounding the interval of search.

the interval is reduced to $[a_i, b_i]$. In order to make a further reduction two points s_i and $\bar{s}_i$ are generated by the following formula

$$\left.\begin{array}{l} s_i = a_i + (b_i - a_i)A_{n-i}/A_{n+2-i} \\ \bar{s}_i = a_i + (b_i - a_i)A_{n+1-i}/A_{n+2-i} \end{array}\right\} \quad i = 1, 2, \ldots, n - 1. \quad (8.10)$$

(Note that s_i and $\bar{s}_i$ are placed symmetrically within $[a_i, b_i]$.) Here n is the number of function evaluations which must be made in order to achieve the desired interval reduction.

Now α is evaluated at s_i and $\bar{s}_i$. If

$$\alpha(s_i) > \alpha(\bar{s}_i),$$

then the remaining interval $[a_{i+1}, b_{i+1}]$ is defined as $[a_i, \bar{s}_i]$. If

$$\alpha(s_i) < \alpha(\bar{s}_i),$$

then the remaining interval is defined as $[s_i, b_i]$. If

$$\alpha(s_i) = \alpha(\bar{s}_i),$$

then the remaining interval is defined as $[s_i, \bar{s}_i]$. In this case only, the search is begun all over again, starting with this new interval $[s_i, \bar{s}_i]$, and a new number n of evaluations must be calculated.

The last two points generated by the procedure as it stands would be placed at

$$s_{n-1} = a_{n-1} + (b_{n-1} - a_{n-1})A_1/A_3 = \tfrac{1}{2}(b_{n-1} + a_{n-1})$$
$$\bar{s}_{n-1}' = a_{n-1} + (b_{n-1} - a_{n-1})A_2/A_3 = \tfrac{1}{2}(b_{n-1} + a_{n-1}).$$

This means that both points would be placed at the same spot, which would be of no advantage. Hence s_{n-1} is to be placed in the position as defined above and $\bar{s}_{n-1}$ is placed as close as possible to the right of s_{n-1} so as to guarantee that the points s_{n-1} and $\bar{s}_{n-1}$ are distinguishable. This minimum distance of distinguishability is the resolution ε:

$$s_{n-1} = \tfrac{1}{2}(b_{n-1} + a_{n-1})$$
$$\bar{s}_{n-1} = \tfrac{1}{2}(b_{n-1} + a_{n-1}) + \varepsilon.$$

Then the interval $[a_{n-1}, b_{n-1}]$ is reduced as before.

(i) If

$$\alpha(s_{n-1}) > \alpha(\bar{s}_{n-1}),$$

set

$$[a_n, b_n] = [a_{n-1}, \bar{s}_{n-1}].$$

(ii) If

$$\alpha(s_{n-1}) < \alpha(\bar{s}_{n-1}),$$

set

$$[a_n, b_n] = [s_{n-1}, b_n].$$

(iii) If

$$\alpha(s_{n-1}) = \alpha(\bar{s}_{n-1})$$

set

$$[a_n, b_n] = [s_{n-1}, \bar{s}_{n-1}].$$

The final interval will be of maximum length when (i) occurs. This maximum length is

$$b_n - a_n = \bar{s}_{n-1} - a_{n-1}$$
$$= \tfrac{1}{2}(b_{n-1} + a_{n-1}) + \varepsilon - a_{n-1}$$
$$= \tfrac{1}{2}(b_{n-1} - a_{n-1}) + \varepsilon.$$

Hence

$$2[(b_n - a_n) - \varepsilon] = b_{n-1} - a_{n-1},$$

and therefore

$$2[(b_n - a_n) - \varepsilon] = (b_1 - a_1)/(\tfrac{3}{2}A_{n-2} + A_{n-3})$$

by Theorem 8.1 (below). Thus we have

$$\frac{b_n - a_n}{b_1 - a_1} = \frac{\varepsilon}{b_1 - a_1} + \frac{1}{3A_{n-2} + 2A_{n-3}}.$$

In order to be certain of a reduction of at least the fraction r, that is,

$$r \geq \frac{b_n - a_n}{b_1 - a_1},$$

in the least number of function evaluations, n must be the minimum integer satisfying

$$r \geq \frac{\varepsilon}{b_1 - a_1} + \frac{1}{3A_{n-2} + 2A_{n-3}}. \tag{8.11}$$

Thus the number A_n can be determined by (8.11). After the first iteration only one point, either s_i or $\bar{s}_i$, needs to be calculated, as the other is already present. Kiefer (1957) has shown that for a given number of function evaluations, Fibonacci search minimizes the maximum interval of uncertainty and in that sense is optimal.

We now prove Theorem 8.1, which was used in the derivation of (8.11).

Theorem 8.1

$$b_1 - a_1 = (\tfrac{3}{2}A_{n-2} + A_{n-3})(b_{n-1} - a_{n-1}).$$

PROOF. We present the proof of this theorem in outline only. Let

$$I_i = b_i - a_i, \qquad i = 1, 2, \ldots, n - 1.$$

In particular,

$$I_1 = b_1 - a_1,$$

the length of the initial interval. It can be shown that the lengths of successive intervals are related by:

$$I_i = I_{i+1} + I_{i+2}, \qquad i = 1, 2, \ldots, n - 3.$$

Therefore

$$\begin{aligned}
I_1 &= I_2 + I_3 \\
&= (I_3 + I_4) + I_3 \\
&= 2I_3 + I_4 \\
&= 2(I_4 + I_5) + I_4 \\
&= 3I_4 + 2I_5 \\
&\vdots \\
&= A_{n-2}I_{n-2} + A_{n-3}I_{n-1} \\
&= A_{n-2}(\tfrac{3}{2}I_{n-1}) + A_{n-3}I_{n-1}.
\end{aligned}$$

Hence

$$I_1 = (\tfrac{3}{2}A_{n-2} + A_{n-3})I_{n-1}. \qquad \square$$

As an example of Fibonacci search, consider the reduction of the interval $[-10, 10]$ to at most 10% of its present length. Let

$$\varepsilon = \tfrac{1}{8}.$$

Then the number of evaluations can be found by (8.11):

$$\frac{2}{20} \geq \frac{\tfrac{1}{8}}{10 - (-10)} + \frac{1}{3A_{n-2} + 2A_{n-3}}.$$

Hence

$$\frac{3}{32} \geq \frac{1}{3A_{n-2} + 2A_{n-3}}, \quad \text{for minimum } n.$$

Therefore

$$n = 6.$$

Thus 6 evaluations will be necessary. The first two points are placed at

$$s_1 = -10 + (10 - (-10))\tfrac{5}{13} = -\tfrac{30}{13}$$

and

$$\bar{s}_1 = -10 + (10 - (-10))\tfrac{8}{13} = \tfrac{30}{13}.$$

Suppose

$$f(s_1) < f(\bar{s}_1).$$

Then

$$[a_2, b_2] = [-\tfrac{30}{13}, 10]$$

$$s_2 = \bar{s}_1$$

$$\bar{s}_2 = -\tfrac{30}{13} + (10 - (-\tfrac{30}{13}))\tfrac{5}{8} = \tfrac{70}{13}.$$

Suppose

$$f(s_2) > f(\bar{s}_2).$$

Then

$$[a_3, b_3] = [-\tfrac{30}{13}, \tfrac{70}{13}]$$

and therefore

$$s_3 = -\tfrac{30}{13} + (\tfrac{70}{13} - (-\tfrac{30}{13}))\tfrac{2}{5} = \tfrac{10}{13}$$

$$\bar{s}_3 = s_2.$$

Suppose

$$f(s_3) > f(\bar{s}_3).$$

Then

$$[a_4, b_4] = [-\tfrac{30}{13}, \tfrac{30}{13}].$$

Hence,

$$s_4 = -\tfrac{30}{13} + (\tfrac{30}{13} - (-\tfrac{30}{13} - (-\tfrac{30}{13}))\tfrac{1}{3}.$$

Therefore

$$s_4 = -\tfrac{10}{13}$$

and

$$s_4 = s_3.$$

Suppose

$$f(s_4) < f(\bar{s}_4).$$

Then

$$[a_5, b_5] = [-\tfrac{10}{13}, \tfrac{30}{13}].$$

It can be seen that the point remaining in the interval $[a_5, b_5]$ is at the centre of $[a_5, b_5]$. Also

$$s_5 = \bar{s}_4.$$

Thus in order to place $\bar{s}_5$ symmetrically it must coincide with s_5, which is of no advantage. Hence $\bar{s}_5$ is placed ε to the right of s_5:

$$\bar{s}_5 = \tfrac{1}{2}(\tfrac{30}{13} - \tfrac{10}{13}) + \varepsilon$$
$$= \tfrac{93}{104}.$$

Suppose

$$f(s_5) > f(\bar{s}_5).$$

The final interval is:

$$[a_6, b_6] = [-\tfrac{10}{33}, \tfrac{93}{104}],$$

which is only 8.3% as long as the original interval.

8.2.1.2 Golden Section Search

The Fibonacci search technique, although most efficient, requires that one know in advance how many points are going to be evaluated. Golden section search, although not quite as efficient, does not make such a requirement. Recall that we must know n, the number of evaluations in order to calculate the ratios

$$\frac{A_{n-i-1}}{A_{n+1-i}}, \qquad \frac{A_{n-i}}{A_{n+1-i}}$$

in (8.10) in order to find s_i and $\bar{s}_i$ at each iteration. Golden section search overcomes this problem by using an approximation of these ratios based on

$$\lim_{n \to \infty} \frac{A_{n-1}}{A_{n+1}} = \frac{3 - \sqrt{5}}{2}$$

$$\lim_{n \to \infty} \frac{A_n}{A_{n+1}} = \frac{\sqrt{5} - 1}{2} = 1 - \frac{3 - \sqrt{5}}{2}.$$

Using these results at each step, (8.10) becomes

$$s_i = a_i + (b_i - a_i) \frac{3 - \sqrt{5}}{2}, \qquad i = 0, 1, 2, \ldots$$

$$\bar{s}_i = a_i + (b_i - a_i) \frac{\sqrt{5} - 1}{2}, \qquad i = 0, 1, 2, \ldots .$$

With this strategy it can be shown that the ratio of the lengths of successive intervals found is a constant and

$$\frac{b_i - a_i}{b_{i+1} - a_{i+1}} = \frac{b_{i-1} - a_{i-1}}{b_i - a_i} = \frac{1 + \sqrt{5}}{2} = \left(\frac{\sqrt{5} - 1}{2}\right)^{-1}.$$

The method proceeds as in the previous section. Two initial evaluation points s_i and $\bar{s}_i$ are found, then at each successive step there will be one point present in the remaining interval, and the new point is placed symmetrically with respect to it. The procedure is therefore very similar to that of Fibonacci search, except that the initial points s_1 and $\bar{s}_1$ would most likely differ. Hence all the remaining points s_i and $\bar{s}_i$ are likely to differ in the two procedures for the same problem. Also, golden section does not have an automatic stopping point as does Fibonacci search. The search proceeds until some termination criterion is met: the interval is sufficiently reduced, or the next point is to be placed within the resolution distance of the last.

The performance of golden section search on the problem of Section 8.2.1.2 will be compared with that of Fibonacci search. The problem is now solved by golden section search. The first two points are placed at

$$s_0 = -10 + (10 - (-10))(2 - \tau)$$
$$\bar{s}_0 = -10 + (10 - (-10))(\tau - 1),$$

where

$$\tau = \frac{1 + \sqrt{5}}{2}$$

is the ratio of the *golden section* of Greek geometry (hence the name of the method). Hence

$$s_0 = 10(2 - \sqrt{5})$$
$$\bar{s}_0 = 10(\sqrt{5} - 2).$$

Now if

$$\alpha(s_0) < \alpha(\bar{s}_0),$$

then the new interval becomes

$$[a_1, b_1] = [10(2 - \sqrt{5}), 10]$$

and

$$s_1 = \bar{s}_0$$
$$\bar{s}_1 = 10(2 - \sqrt{5}) + [10 - 10(2 - \sqrt{5})]$$
$$= 50 - 20\sqrt{5}.$$

Now if

$$\alpha(s_1) > \alpha(\bar{s}_1),$$

then the new interval becomes

$$[a_2, b_2] = [10(2 - \sqrt{5}), 10(5 - 2\sqrt{5})]$$

and

$$s_2 = 10(2 - \sqrt{5}) + (10(5 - 2\sqrt{5}) - 10(2 - \sqrt{5}))(2 - \tau)$$
$$= 10(9 - 4\sqrt{5})$$
$$\bar{s}_2 = s_1.$$

Now if

$$\alpha(s_2) > \alpha(\bar{s}_2),$$

then the new interval becomes

$$[a_3, b_3] = [10(2 - \sqrt{5}), 10(\sqrt{5} - 2)]$$

and

$$s_3 = 10(2 - \sqrt{5}) + (10(\sqrt{5} - 2) - 10(2 - \sqrt{5}))(2 - \tau)$$
$$= 10(4\sqrt{5} - 9)$$

$$\bar{s}_3 = s_2.$$

Now if

$$\alpha(s_3) < \alpha(\bar{s}_3),$$

then the new interval becomes

$$[a_4, b_4] = [10(4\sqrt{5} - 9), 10(\sqrt{5} - 2)]$$

and

$$s_4 = \bar{s}_3$$
$$\bar{s}_4 = 10(4\sqrt{5} - 9) + (10(\sqrt{5} - 2) - 10(4\sqrt{5} - 9))(\tau - 1)$$
$$= 10(9\sqrt{5} - 20).$$

Now if

$$\alpha(s_4) > \alpha(\bar{s}_4),$$

the final interval is

$$[a_5, b_5] = [10(4\sqrt{5} - 9), 10(9\sqrt{5} - 20)],$$

which has length $10(5\sqrt{5} - 11)$. This interval is a little over 9% of the original interval in length. This comparison is typical, and in general golden section search is not quite as efficient as Fibonacci search.

8.2.1.3 *The Method of Bolzano*

If first derivatives of the objective function are available, then the Bolzano technique for finding the root of a decreasing function in numerical analysis can be profitably modified. In using Bolzano's method (also called the method of successive bisection) one successively evaluates the function in the middle of the current interval of uncertainty. The right-hand or left-hand half of the interval is eliminated depending upon whether the derivative is negative or positive, respectively.

In attempting to find the maximum of the objective function α one is trying to find the unique root of the first derivative of α. The root is unique because α is assumed unimodal. So the Bolzano technique can be applied to α' in order to find the maximum of α. The modified technique will now be described in precise terms. Assume that the maximum is bounded by an initial interval $[a_0, b_0]$. Then

$$\alpha'\left(\frac{a_0 + b_0}{2}\right) > 0 \Rightarrow [a_1, b_1] = \left[\frac{a_0 + b_0}{2}, b_0\right]$$

and

$$\alpha'\left(\frac{a_0 + b_0}{2}\right) < 0 \Rightarrow [a_1, b_1] = \left[a_0, \frac{a_0 + b_0}{2}\right].$$

The general step is

$$\alpha'\left(\frac{a_i + b_i}{2}\right) > 0 \Rightarrow [a_i, b_i] = \left[\frac{a_i + b_i}{2}, b_i\right]$$

and

$$\alpha'\left(\frac{a_i + b_i}{2}\right) < 0 \Rightarrow [a_i, b_i] = \left[a_i, \frac{a_i + b_i}{2}\right].$$

Of course, if

$$\alpha'\left(\frac{a_i + b_i}{2}\right) = 0,$$

the maximum has been found.

It can readily be seen that at each step the remaining interval is halved. Thus after n steps,

$$\frac{b_n - a_n}{b_0 - a_0} = \frac{1}{2^n}.$$

Thus the number of derivative evaluations required to achieve a specified reduction ratio is the minimum integer n satisfying:

$$\frac{b_n - a_n}{b_0 - a_0} \geq \frac{1}{2^n}.$$

Bolzano's method will also be tried on the example of Section 8.2.1.1. Recall that

$$[a_0, b_0] = [-10, 10].$$

Let

$$s_i = \frac{a_i + b_i}{2}.$$

If

$$\alpha'(s_0) > 0,$$

then

$$[a_1, b_1] = [0, 10].$$

If

$$\alpha'(s_1) < 0,$$

then

$$[a_2, b_2] = [0, 5].$$

If

$$\alpha'(s_2) < 0,$$

then

$$[a_3, b_3] = [0, 2.5].$$

If

$$\alpha'(s_3) < 0,$$

then

$$[a_4, b_4] = [0, 1.25].$$

If

$$\alpha'(s_4) < 0,$$

then
$$[a_5, b_5] = [0, 0.625].$$
If
$$\alpha'(s_5) > 0,$$
then
$$[a_6, b_6] = [0.3125, 0.625].$$

Hence after only six iterations the interval has been reduced to one of length 0.3125, or just 1.56% of the original length. This rapid decrease compared with the previous two procedures comes at the cost of calculating derivatives, which may be no easy task, if not impossible.

8.2.1.4 *Even Block Search*

A simplified version of the general even block search method will be presented in this section. When derivatives are unavailable, it is still possible to simulate the Bolzano method in the following way. The sign of the derivative of a function can be approximated at a point by making two distinct evaluations, each as close to the point as the resolution ε will allow. The points about which these evaluations are made are the same as those that would be used in the normal Bolzano method.

Suppose that the first derivative of α is unavailable. Let $[a_0, b_0]$ be the initial interval, bracketing the maximum. Thus the first evaluation would have been of α' at $(a_0 + b_0)/2$. Instead, we approximate the sign of $\alpha'(s_0)$, denoted by $\sigma(\alpha'(s_0))$, by

$$\sigma(\alpha'(s_0)) \approx (\alpha(s_0 + \varepsilon) - \alpha(s_0))/\varepsilon.$$

In general, we have

$$\alpha(s_i + \delta) - \alpha(s_i) > 0 \Rightarrow [a_{i+1}, b_{i+1}] = [s_i, b_i]$$
$$\alpha(s_i + \delta) - \alpha(s_i) < 0 \Rightarrow [a_{i+1}, b_{i+1}] = [a_i, s_i + \varepsilon]$$
$$\alpha(s_i + \delta) - \alpha(s_i) = 0 \Rightarrow [a_{i+1}, b_{i+1}] = [s_i, s_i + \varepsilon].$$

In the last case the procedure must be terminated, as no further observations can be made in the remaining interval.

Neglecting resolution, this simple even block method will require twice as many evaluations as Bolzano's method. However, because it usually takes far less effort to evaluate a function than to calculate and evaluate its derivative, even block search is often more efficient.

8.2.2 Hessian Methods

Recall from the initial remarks of Section 8.2 that ascent methods generate a new point X_{i+1} by a calculation of the form:

$$X_{i+1} = X_i + s_i D_i.$$

In the case of gradient methods and Hessian methods this equation has the special form:

$$X_{i+1} = X_i + s_i B_i \nabla f(X_i).$$

The matrix B_i may be a constant matrix or may vary according to previous calculations. In the description of the methods of this section, B_i is a function of Hessian matrix $H(X_i)$.

8.2.2.1 The Method of Newton and Raphson

The following is a "classical" method and should be related to Section 7.3.3. In attempting to optimize an n-dimensional function we are attempting to find a root of

$$\nabla f(X) = 0. \tag{8.12}$$

In what follows it is necessary to assume that the Hessian matrix evaluated at each point X_i is nonsingular, i.e., $H^{-1}(X_i)$ exists. Now suppose we have found an estimate X_{i+1}; the Taylor series of $\nabla f(X_{i+1})$ is expanded about X_i as follows:

$$\nabla f(X_{i+1}) = \nabla f(X_i) + H(X_i)(X_{i+1} - X_i).$$

Now if X_{i+1} is an estimate of a root of (8.12) it is hoped that

$$\nabla f(X_{i+1}) \approx 0.$$

Hence

$$0 = \nabla f(X_i) + H(X_i)(X_{i+1} - X_i)$$

and we have found an iterative method for generating $X_{i+1}, X_{i+2}, \ldots,$ namely

$$X_{i+1} = X_i - H^{-1}(X_i)\nabla f(X_i), \qquad i = 1, 2, \ldots.$$

8.2.2.2 Variable Metric Method

The variable metric method does not require that the Hessian matrix of the function be calculated and inverted, as does the method of Newton and Raphson. Instead, the inverse of the Hessian matrix is estimated more and more accurately until the optimum is found. This means that the method is often the most efficient currently available when the gradient is available and when the Hessian matrix is not available, is expensive to calculate, or must be found by numerical methods. Apart from the initial step, the one-dimensional searches performed in pursuit of the optimum are not usually in the direction of the gradient. They are carried out in a direction $E_i \nabla f(X_i)$, where X_i is the current estimate of the optimum and E_i is a negative definite matrix. Thus the direction from each point X_i is "deflected" away from the gradient by matrix E_i.

The method will be outlined with view to maximizing the following quadratic f, in which H is assumed negative definite:

$$f(X) = C^T X + \tfrac{1}{2} X^T H X. \tag{8.13}$$

For a complete description of how the method maximizes a general function see Davidon (1959), or Fletcher and Powell (1963). Suppose it is desired to find the optimum X^* to (8.13) from a present estimate X_1. Note that

$$\nabla f(X) = C + HX,$$

hence

$$\nabla f(X_1) = C + HX_1.$$

Therefore

$$X_1 = H^{-1}(\nabla f(X_1) - C)$$

and

$$X^* = H^{-1}(\nabla f(X^*) - C).$$

But

$$\nabla f(X^*) = 0.$$

Hence

$$X^* = -H^{-1}C.$$

Thus

$$X^* = X_1 - H^{-1}\nabla f(X_1). \tag{8.14}$$

Equation (8.14) shows why it is worthwhile to search along a direction which is different from the gradient direction. Thus when H is known, the optimum to (8.13) can be found in one step by using (8.14). Problems arise when, for one reason or another, H^{-1} is not readily at hand.

The method proceeds by calculating X_{i+1} from X_i by using the relation:

$$X_{i+1} = X_i - s_i E_i \nabla f(X_i),$$

where E_i is a negative definite matrix and s_i is the step size taken in the $E_i \nabla f(X_i)$ direction. If f contains n variables, then

$$E_{n+1} = H^{-1}. \tag{8.15}$$

The method generates the estimates of $X^*(X_2, X_3, \ldots, X_{n+1})$ in such a way that

$$(X_{i+1} - X_i)^T \nabla f(X_{n+1}) = 0, \qquad i = 1, 2, \ldots, n. \tag{8.16}$$

Now, as

$$(X_{i+1} - X_i), \qquad i = 1, 2, \ldots, n$$

are constructed to be linearly independent, from (8.16) it must be that

$$\nabla f(X_{n+1}) = 0.$$

Hence the optimum is found after n iterations if f is quadratic.

The method begins by setting

$$E_1 = I,$$

where I is the identity matrix (for simplicity), unless the analyst has further information and can choose E_1 such that $E_1 \nabla f(X_1)$ is a more promising direction than $\nabla f(X_1)$. This first step just turns out to be a basic gradient search, which will be explained in Section 8.2.3.

In general E_i is computed by the relation:

$$E_i = E_{i-1} + F_i + G_i, \qquad (8.17)$$

where the matrices F_i and G_i are chosen so that

$$\sum_{i=1}^{n} F_i = H^{-1}$$

and

$$\sum_{i=1}^{n} G_i = -E_1.$$

Thus usually

$$\sum_{i=1}^{n} G_i = -I.$$

Now at any iteration of the method F_i and G_i must be found from previous information, namely

$$\nabla f(X_i), \nabla f(X_{i-1}), \ldots, X_i, X_{i-1}, \ldots, E_{i-1}, E_{i-2}, \ldots .$$

One possible choice for F_i and G_i is

$$F_i = \frac{(X_i - X_{i-1})(X_i - X_{i-1})^T}{(X_i - X_{i-1})^T(\nabla f(X_i) - \nabla f(X_{i-1}))}$$

and

$$G_i = -\frac{E_{i-1}(\nabla f(X_i) - \nabla f(X_{i-1}))^T(\nabla f(X_{i-1}))E_{i-1}}{(\nabla f(X_i) - \nabla f(X_{i-1}))^T E_{i-1}(\nabla f(X_i) - \nabla f(X_{i-1}))}.$$

This is the well-known D.F.P. formula (Davidon (1959), Fletcher and Rowell (1963)). However, recent numerical evidence supports the complementary D.F.P. formula, labelled B.F.G.S.. For this see Broyden (1970), Fletcher (1970), Goldfarb (1970), and Shanno (1970).

8.2.3 Gradient Methods

Recall that

$$\nabla f = \left(\frac{\partial f}{\partial x_1}, \frac{\partial f}{\partial x_2}, \ldots, \frac{\partial f}{\partial x_n} \right)^T,$$

a vector of first partial derivatives. Because this vector points in the direction of greatest slope of the function at any point, it is called the gradient. (For a proof of this fact see Theorem 9.7 in the Appendix.)

Gradient methods for seeking a maximum for f involve evaluating the gradient at an initial point, moving along the gradient direction for a calculable distance, and repeating this process until the maximum is found.

One of the problems of gradient methods is that they require ∇f and hence the first partial derivatives to be calculated. In many problems the mathematical form of f is unknown and hence it is impossible to find derivatives. In such cases gradients may be approximated by numerical procedures. This introduces errors, which make the methods less attractive. Throughout this section it will be assumed that the necessary gradients are available either by direct computation or by approximation.

In the basic gradient method first an initial point X_1 is selected. The gradient $\nabla f(X_1)$ of f at X_1 is computed. A line is then drawn through X_1 in the gradient direction $\nabla f(X_1)$. The point on this line X_2 is then selected which yields the greatest value for f of all points on the line. Suppose the distance from X_1 to X_2 is s_1. Then

$$X_2 = X_1 + \nabla f(X_1)s_1 \tag{8.18}$$

and

$$f(X_2) = \max_{s_1 \in \mathbf{R}} \{f(X): X = X_1 + \nabla f(X_1)s_1\}$$

$$= \max_{s_1 \in \mathbf{R}} \{f(X_1 + D_1 s_1)\},$$

where

$$D_1 = \nabla f(X_1).$$

The obvious task is to decide where along the line the best point for f lies. That is, s_1 must be found. There are two ways of going about this. Whenever derivatives can be evaluated and f is well behaved, the best method is to substitute (8.18) into the equation for f and differentiate with respect to s. One can solve for the maximizing value of s, say s_1, by setting the derivative equal to zero. The second method is to use one of the one-dimensional search methods explained in Section 8.2.1.

Having found s_1 and thus X_2, the procedure is repeated with X_2 replacing X_1. The process continues until no improvement can be made. There are a number of variations on this process. One such variation, which requires considerably less effort, is to use a fixed step size s_i at each step. This has the disadvantage of there being no way to predict a satisfactory step size for a given f. A relatively small step size will usually produce an improvement at each step but require a large number of steps. A relatively large step size may sometimes produce a *decrease* in objective function value from one step to another.

Gradient methods were first introduced by Cauchy (1847) and were later used by Box and Wilson (1951) on problems in industrial statistics.

8.2.3.1 *Gradient Partan*

A specialized version of the gradient method will now be presented. It might have occurred to the reader that the maximization of a function of two variables has some aspects in common with climbing a mountain, the maximum being the peak. As everyone who has looked at a mountain knows, most mountains have ridges. Quite often these ridges lead to the summit. This geological fact can sometimes be used to advantage in unconstrained optimization.

The preceding idea was used by Forsythe and Motzkin (1951) to find the maximum of certain two-dimensional functions. Consider the objective function in Figure 8.4, which has ellipsoidal concentric contours. Unless the initial search point lies on an axis of the ellipses, the normal gradient search will proceed to the maximum X^* as shown. It can be seen that the points in the search, $X_1, X_2, \ldots$, are bounded by two "ridges" which both pass through the maximum X^*. Thus a short-cut could be made after three search points have been identified. When X_1, X_2, and X_3 have been found the next search should be made in the direction of the line through X_1 and X_3. If the contours of f are concentric ellipses then the maximum will be found immediately.

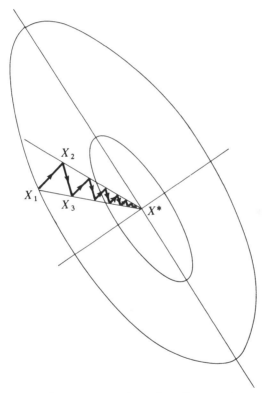

Figure 8.4. Accelerated gradient search.

This method can be generalized to maximize a function f of n variables. The method is most efficient when f is a negative definite quadratic (see the Appendix). Suppose that the initial starting point is X_1. Let the best point found in the direction of the X_1 gradient be X_3 (rather than X_2). The next point found by the gradient method is X_4. The point X_5 is found by maximizing along the line through X_1 and X_4. For $n > 2$ it is unlikely that X_5 will be a maximum.

The process can be described in general as follows. Once the process has "warmed up" and $i > 3$, X_i is found by gradient search from X_{i-1} for i odd, and X_i is found by an accelerated step by maximizing over the line through X_{i-1} and X_{i-3}. It can be shown that the global maximum of a negative definite quadratic function of n variables can be found after $2n - 1$ steps using this procedure.

The above method requires the calculation of first derivatives. A similar method, due to Shah, Buehler, and Kempthorne (1964), does not involve such a restriction. Consider a function f of two variables whose contours are negative definite quadratics. It can be shown that the contours are concentric ellipses. Let X^* be the global maximum of f and X_1 and X_2 two points lying on an arbitrary line through X^*. It can be shown that the tangents to the contours at X_1 and X_2 are parallel. Conversely, it can be shown that if the situation of the previous sentence holds then X_1 and X_2 are colinear with X^*. It is also true that X_1 and X_2 represent optima for f along the lines corresponding to the respective tangents.

The foregoing facts suggest an interesting method for locating X^*. One begins by arbitrarily selecting two parallel lines and finds the maximum of f along each line. These two maxima are denoted by X_1 and X_2. A line is then passed through these two points and a search is carried out along this line. The resulting point is X^*.

The method can be generalized to maximize ellipsoidal functions of any number of variables. The reader may very well have spotted that *partan* is an acronym for *parallel tangents*.

8.2.3.2 *Conjugate Directions*

This section and the next deal with methods which represent improvements on the basic gradient method of Section 8.2.3. The reader will recall from his study of linear algebra that vectors X and Y from n-dimensional Euclidean space are *orthogonal* if
$$X^T Y = 0.$$
Now if the identity matrix with n rows is inserted into the expression, the equation still holds, i.e.,
$$X^T I Y = 0.$$
Let us now replace I by any n-square matrix H. If
$$X^T H Y = 0,$$
X and Y are said to be *H-conjugate*. Thus orthogonality is a special case of conjugacy, in the sense that orthogonal vectors are *I-conjugate*.

The *methods of conjugate directions* can be used to produce a sequence of points $X_1, X_2, X_3, \ldots$, which each yield improving values in maximizing a quadratic

$$f(X) = AX + \tfrac{1}{2}X^T H X,$$

where H is the Hessian matrix of f. The directions of search D_i and D_k all obey the relationship

$$D_i^T H D_k = 0, \quad \text{for all } i, k, i \neq k.$$

The general method of conjugate directions can be implemented as follows. A point X_1 is chosen initially as the point most likely to be optimal. (A random choice is made if no relevant information is available.) A one-dimensional search is carried out in the direction D_1, the first conjugate direction. This produces a new point X_2. Another one-dimensional search is carried out along D_2, the next search direction, where

$$D_1^T H D_2 = 0.$$

Here H is the Hessian matrix of f. In general, when point X_{i-1} is found, X_i is found by a one-dimensional search along D_i from X_i and

$$D_j^T H D_k = 0, \quad j, k \leq i, j \neq k.$$

The maximum will be located in at most n steps.

The trouble is that the sequence of conjugate directions $D_1, D_2, \ldots$ is not known in advance. The usual method is to generate each new direction D_i as it is needed. One way of doing this is to find the new conjugate direction D_i from point X_i, using $\nabla f(X_i)$. Such directions are called conjugate gradient directions and are discussed in the next section.

8.2.3.3 Conjugate Gradients

The *conjugate gradient method* generates each new conjugate direction from the gradient at the point concerned in such a way that the direction is conjugate to all those previously generated. First a starting point, X_1 is chosen. A one-dimensional search is then performed in the gradient direction from X_1. That is, to start the ball rolling the first direction is

$$D_1 = \nabla f(X_1).$$

The maximum point X_2 on this line is found. In general the direction D_i is constructed from $\nabla f(X_i)$ so as to make it conjugate to D_{i-1}. For this purpose define a scalar a_{i-1} such that

$$D_i = \nabla f(X_i) + a_{i-1} D_{i-1} \qquad (8.19)$$

and

$$D_i^T H D_{i-1} = 0. \qquad (8.20)$$

Here Q still represents the Hessian matrix of f. Now from (8.19) and (8.20) we have

$$(\nabla f(X_i) + a_{i-1} D_{i-1})^T H D_{i-1} = 0.$$

Hence

$$(\nabla f(X_i))^T HD_{i-1} + a_{i-1} D_{i-1}^T HD_{i-1} = 0$$

and therefore

$$a_{i-1} = -\frac{(\nabla f(X_i))^T HD_{i-1}}{D_{i-1}^T HD_{i-1}}.$$

It can be shown by induction that the directions D_i are mutually conjugate.

When it comes to actual implementation the D_i's can be calculated by a simple recurrence relation and only a few vectors and no matrices need be stored. In fact, storage requirements vary with dimensionality n rather than n^2 as for the quasi-Newton methods (Fletcher and Reeves 1964).

8.2.4 Direct Methods

If the mathematical expression defining f is difficult to manipulate, its gradient vector and Hessian matrix are likely to be complicated to calculate and evaluate. If the expression is unknown and hence unavailable these evaluations cannot be performed. When these situations arise, or when f has many local maxima, the following direct search approach is appropriate. It assumes that f can be evaluated at any point $X = (x_1, x_2, \ldots, x_n)^T \in \mathbf{R}$ by performing some task or experiment, or employing some algorithm with input the specific values of $x_1, x_2, \ldots, x_n$.

8.2.4.1 *Pattern Search*

The first direct search method to be examined is called *pattern search* and was developed by Hooke and Jeeves (1961). As with most direct search methods, pattern search begins by evaluating f at the point X_1 most likely to be maximal, or at a random point if all points are initially equally likely to be maximal. Exploration about X_1 now begins in order to find the best direction for improvement. A pre-defined step is taken in the increasing direction of the first variable. To make this more clear let $X_a = X_1 = (x'_1, x'_2, \ldots, x'_n)$ and let the initial perturbation size for each variable, x_i be a given positive real number, e_i. Then f is first evaluated at X_a and then at $(x'_1 + e_1, x'_2, \ldots, x'_n)$ and the two values thus obtained are compared. If the second value is greater than the first, (an improvement over the original point has been found) the next variable, x_2 is increased from the new point. If no improvement was found x_1 is decreased by e_1, i.e. f is evaluated at $(x'_1 - e_1, x'_2, \ldots, x'_n)$. If this second value is greater than $f(X_a)$ then x_2 is increased from the new point. Otherwise x_2 is increased from X_a. Note that if $(x'_1 + e_1, x'_2, \ldots, x'_n)$ corresponds to an improvement in f over X_a we don't bother to evaluate f at $(x'_1 - e_1, x'_2, \ldots, x'_n)$. The result of all this is that we end up with a point which is the best for f found so far, call it $(\bar{x}'_1, x'_2, \ldots, x'_n)$ where $\bar{x}'_1$ is either x'_1 or $x'_1 + e_1$ or $x'_1 - e_1$. We now perturb x_2 about this best point. We evaluate f at $(\bar{x}'_1, x'_2 + e_2, x'_3, \ldots, x'_n)$ and if this represents an improvement we perturb x_3 about it. If not we evaluate f about $(\bar{x}'_1, x'_2 - e_2, x'_3, \ldots, x'_n)$ and so on. This strategy is continued, perturbing (increasing, and if necessary decreasing, each variable) about the best point found so far. When all the variables have been perturbed in this manner the final best point X_b is identified. The first direction vector D_1 is defined as

$$D_1 = (X_b - X_a).$$

The first step size s_1 is defined to be twice the Euclidean distance between X_b and X_a. Thus the new point X_2 is derived as

$$X_2 = X_1 + s_1 D_1 = X_a + 2(X_b - X_a)$$
$$= 2X_b - X_a.$$

The process is now repeated with X_2 replacing X_1. Indeed each successive step size s_{j+1} is twice its predecessor s_j unless no perturbations about X_j bring about an improvement. If this happens the pre-defined perturbation sizes e_i are successively halved until the final best point represents an improvement over X_j. If no improvement in f is found before an e_i becomes less than the corresponding resolution ε_i for x_i then the process is terminated and X_j is declared the best point that could be found by the method.

It can be seen that the search seeks a general trend or direction of improvement. Hooke and Jeeves call this a *pattern*. If exploration about a point is unfruitful the perturbation size by which each variable is decreased and the process is repeated about this point. The method seems to reflect the saying: "nothing succeeds like success", for at each successful iteration the step size is doubled. However when the exploration process fails to yield further improvement the method begins over again and slowly builds up increasingly large steps once more.

Pattern search often has excellent success because of its ability to follow a ridge of the geometric mountain it is trying to climb. Before the reader begins to regard pattern search as the ultimate in direct search methods there is a shortcoming. The method may fail to yield any further improvement in a function with tightly curved ridges or sharp-cornered contours while still far from a local maximum. However the technique is often successful on real applications and is easily programmed which can be seen by looking at the following algorithmic statement.

Pattern Search Algorithm. Let

e_i = the initial perturbation size for x_i, $i = 1, 2, \ldots, n$,

X_a = the current point about which perturbations are being made,

X_b = the current best point found so far while perturbations are being made,

X_c = the best point found once all variables have been perturbed,

K_i = the vector $(0, 0, \ldots, 0, 1, 0, \ldots, 0)$ consisting of all zero entries except for a unit entry in the ith position, $i = 1, 2, \ldots, n$.

1. Initialization: set $X_a = X_1$, the initial point of the search, and set $j = 1$.
2. An exploration is carried out about X_a.
   (a) Set $X_b = X_a$ and set $i = 1$.
   (b) If $f(X_b + e_i K_i) > f(X_b)$, set X_b to become $X_b + e_i K_i$, go to step 2(c), otherwise continue. If $f(X_b - e_i K_i) > f(X_b)$, set X_b to become $X_b - e_i K_i$, otherwise continue.
   (c) Set i to become $i + 1$. If $i < n$ go to step 2(b), otherwise continue.
3. If any termination criterion is met go to step 6, otherwise continue.
4. If $f(X_b) \leq f(X_c)[f(X_b) = f(X_a)$ in the first iteration] go to step 5, otherwise continue. An extrapolation is made in the new search direction: $(X_b - X_c)$. Set X_a to become $X_c + 2^j(X_b - X_c)$ and j to become $j + 1$ $(X_a + 2(X_b - X_a)$ in the first iteration). Set $X_c = X_b$. Go to step 2.
5. Set e_i to become $e_i/2$ for $i = 1, 2, \ldots, n$ and $j = 1$. If there exists some i, $1 \leq i \leq n$ such that $e_i < \varepsilon_i$, then set $X_b = X_c$ (except for first iteration) and go to step 6. Otherwise, set $X_a = X_c$ (except for first iteration) and go to step 2.
6. The best point found was X_b.

Example. The following is an account of the use of pattern search in the maximization of a function $f(X) = f(x_1, x_2)$, $x_1, x_2 \in \mathbf{R}$, using pattern search, where

$$X_1 = (0, 0)$$
$$e_1 = e_2 = 0.1$$
$$\varepsilon_1 = \varepsilon_2 = 0.03.$$

The pattern the search takes is shown in Figure 8.5. Following the algorithm,

$$X_a = (0, 0).$$

Now we have
$$f((0, 0) + 0.1(1, 0) < f(0, 0).$$

But
$$f((0, 0) - 0.1(1, 0)) > f(0, 0).$$

Hence
$$X_b = (-0.1, 0).$$

Further,
$$f((-0.1, 0) + 0.1(0, 1)) > f(-0.1, 0).$$

So
$$X_b = (-0.1, 0.1).$$

Proceeding to step 4, as this is the first iteration, X_a becomes
$$X_a + 2(X_b - X_a) = (0, 0) + 2((-0.1, 0.1) - (0, 0))$$
$$= (-0.2, 0.2) \qquad (= X_2)$$
$$X_c = (-0.1, 0.1).$$

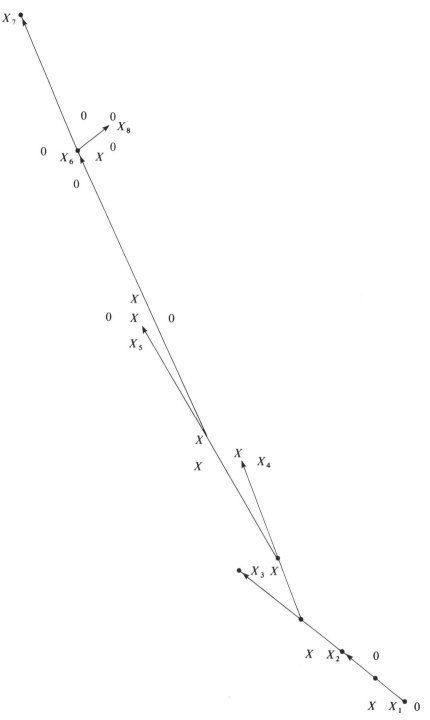

Figure 8.5.   An example of pattern search.

Returning to step 2, we now explore about X_2:

$$f((-0.2, 0.2) + 0.1(1, 0)) < f(-0.2, 0.2).$$

But

$$f((-0.2, 0.2) - 0.1(1, 0)) > f(-0.2, 0.2).$$

Hence

$$X_b = (-0.3, 0.2).$$

Further,

$$f((-0.3, 0.2) + 0.1(0, 1)) > f(-0.3, 0.2).$$

So

$$X_b = (-0.3, 0.3).$$

Proceeding to step 4, X_a becomes

$$\begin{aligned} X_c + 2(X_b - X_c) &= (-0.1, 0.1) + 2((-0.3, 0.3) - (-0.1, 0.1)) \\ &= (-0.5, 0.5) \qquad (= X_3) \\ X_c &= (-0.3, 0.3). \end{aligned}$$

Returning to step 2, we now explore about X_3:

$$f((-0.5, 0.5) + 0.1(1, 0)) > f(-0.5, 0.5).$$

So

$$X_b = (-0.4, 0.5).$$

Further,

$$f((-0.4, 0.5) + 0.1(0, 1)) > f(-0.4, 0.5).$$

Hence

$$X_b = (-0.4, 0.6).$$

Proceeding to step 4, X_a becomes

$$\begin{aligned} X_c + 2(X_b - X_c) &= (-0.3, 0.3) + 2((-0.4, 0.6) - (-0.3, 0.3)) \\ &= (-0.5, 0.9) \qquad (= X_4) \\ X_c &= (-0.4, 0.6). \end{aligned}$$

At this point it is interesting to note that the direction of search has changed slightly, as seen in Figure 8.5.

Returning to step 2, we now explore about X_4:

$$f((-0.5, 0.9) + 0.1(1, 0)) < f(-0.5, 0.9).$$

But

$$f((-0.5, 0.9) - 0.1(1, 0)) > f(-0.5, 0.9).$$

Hence

$$X_b = (-0.6, 0.9).$$

Further,

$$f((-0.6, 0.9) + 0.1(0, 1)) > f(-0.6, 0.9).$$

So

$$X_b = (-0.6, 1.0).$$

Proceeding to step 4, X_a becomes

$$X_c + 2(X_b - X_c) = (-0.4, 0.6) + 2((-0.6, 1.0) - (-0.4, 0.6))$$
$$= (-0.8, 1.4) \qquad (=X_5)$$
$$X_c = (-0.6, 1.0).$$

Returning to step 2 we now explore about X_5:

$$f((-0.8, 1.4) + 0.1(1, 0)) < f(-0.8, 1.4)$$

and

$$f((-0.8, 1.4) - 0.1(1, 0)) < f(-0.8, 1.4).$$

Thus X_b remains at $(-0.8, 1.4)$. However,

$$f((-0.8, 1.4) + 0.1(1, 0)) > f(-0.8, 1.4).$$

Hence

$$X_b = (-0.8, 1.5).$$

Proceeding to step 4, X_a becomes

$$X_c + 2(X_b - X_c) = (-0.6, 1.0) + 2((-0.8, 1.5) - (-0.6, 1.0))$$
$$= (-1.0, 2.0) \qquad (=X_6)$$
$$X_c = (-0.8, 1.5).$$

Returning to step 2, we now explore about X_6:

$$f((-1.0, 2.0) + 0.1(1, 0)) < f(-1.0, 2.0)$$

and

$$f((-1.0, 2.0) - 0.1(1, 0)) < f(-1.0, 2.0).$$

Thus X_b remains at $(-1.0, 2.0)$. Further,

$$f((-1.0, 2.0) + 0.1(0, 1)) < f(-1.0, 2.0)$$

and

$$f((-1.0, 2.0) - 0.1(0, 1)) < f(-1.0, 2.0).$$

Thus X_b still remains at $(-1.0, 2.0)$. Proceeding to step 4,

$$f(-1.0, 2.0) > f(-0.8, 1.5).$$

So X_a becomes

$$X_c + 2(X_b - X_c) = (-0.8, 1.5) + 2((-1.0, 2.0) - (-0.8, 1.5))$$
$$= (-1.2, 2.5) \qquad (=X_7)$$
$$X_c = (-1.0, 2.0).$$

Returning to step 2, we now explore about X_7:

$$f((-1.2, 2.5) + 0.1(1, 0)) < f(-1.2, 2.5)$$

and

$$f((-1.2, 2.5) - 0.1(1, 0)) < f(-1.2, 2.5).$$

Thus X_b remains at $(-1.2, 2.5)$. Further,

$$f((-1.2, 2.5) + 0.1(0, 1)) < f(-1.2, 2.5)$$

and

$$f((-1.2, 2.5) - 0.1(0, 1)) < f(-1.2, 2.5).$$

Thus X_b still remains at $(-1.2, 2.5)$. Proceeding to step 4,

$$f(-1.0, 2.0) > f(-1.2, 2.5).$$

The extrapolation to X_7 has failed to yield an improvement over X_6 and the pattern is destroyed. Proceeding to step 6, an attempt is made to find a new pattern by casting about near X_6:

$$e_1 = e_2 = 0.05.$$

As this exceeds the resolution, setting

$$X_a = X_c = (-1.0, 2.0)$$

and proceeding to step 2, we obtain

$$f((-1.0, 2.0) + .05(1, 0)) > f(-1.0, 2.0).$$

Hence

$$X_b = (-0.95, 2.0).$$

Further,

$$f((-0.95, 2.0) + 0.05(0, 1)) > f(-0.95, 2.0).$$

So

$$X_b = (-0.95, 2.05).$$

Proceeding to step 4, as

$$f(-0.95, 2.05) > f(-1.0, 2.0),$$

a new search direction has been formed. X_a becomes

$$X_c + 2(X_b - X_c) = (-1.0, 2.0) + 2((-0.95, 2.05) - (-1.0, 2.0))$$
$$= (-0.9, 2.1) \qquad (= X_8)$$
$$X_c = (-0.95, 2.05).$$

As can be seen from Figure 8.5, the search took a new turn at X_7 and proceeded in the $(0.1, 0.1)$ direction. Returning to step 2, we now explore about X_8:

$$f((-0.9, 2.1) + 0.05(1, 0)) < f(-0.9, 2.1)$$

and

$$f((-0.9, 2.1) - 0.05(1, 0)) < f(-0.9, 2.1).$$

Thus X_b remains at $(-0.9, 2.1)$. Further

$$f((-0.9, 2.1) + 0.05(1, 0)) < f(-0.9, 2.1)$$

and

$$f((-0.9, 2.1) - 0.05(0, 1)) < f(-0.9, 2.1).$$

Thus X_b still remains at $(-0.9, 2.1)$. Proceeding to step 4,

$$f(-0.9, 2.1) < f(-0.95, 2.05).$$

The extrapolation to X_8 has failed to yield an improvement over X_7 and the pattern is destroyed. Proceeding to step 6 an attempt is made to find a new pattern by casting about near X_7,

$$e_1 = e_2 = .025.$$

As these values do not exceed resolution, the search is terminated. Let $X_b = X_c = (-0.95, 2.05)$, which is the best point found.

8.2.4.2 One-at-a-Time Search

One-at-a-time search, or *sectioning*, as it is often called, is a classical method of direct search. Given an initial estimate,

$$X_1 = (x_1', x_2', \ldots, x_n')^T,$$

one first searches in the direction of the first variable x_1. Suppose the maximum of f in this direction from X_1, lies at a distance δ_1 from X_i. This maximum point can be found by one of the one-dimensional search methods of Section 8.2.1. This yields a new estimate X_2, where

$$X_2 = (x_1' + \delta_1, x_2', \ldots, x_n')^T.$$

Next one searches for a maximum in the x_2 direction from X_2. This yields X_3 where

$$X_3 = (x_1' + \delta_1, x_2' + \delta_2, x_3', \ldots, x_n')^T.$$

Eventually X_{n+1} is found, where

$$X_{n+1} = (x_1' + \delta_1, x_2' + \delta_2, \ldots, x_n' + \delta_n)^T.$$

The process is then repeated with X_{n+1} replacing X_1. The above steps are carried out until the steps δ_i, $i = 1, 2, \ldots, n$ become less than the resolution.

The rate of convergence of this method is usually painfully slow. Indeed, if the function has ridges which are far from parallel to any coordinate axis the method will probably grind to a halt far from the optimum.

8.2.4.3 The Method of Rosenbrock

One of the problems of the previously mentioned direct search strategies is that they often fail if they encounter a ridge. Rosenbrock (1960) tried to overcome this by developing a method which attempts to identify a ridge and then searches in the direction of the ridge.

The method begins by making a first attempt at finding a ridge direction by using one-at-a-time search. Each variable direction is searched for the optimum about that direction, as in Section 8.2.4.2. Thus if the initial estimate is

$$X_1 = (x_1, x_2, \ldots, x_n)^T,$$

let the result of this search be X_2, where

$$X_2 = (x_1 + \delta_1, x_2 + \delta_2, \ldots, x_n + \delta_n)^T.$$

Now instead of repeating the process for X_2, the method searches in the direction

$$(\delta_1, \delta_2, \ldots, \delta_n)^T$$

from X_2. Thus we replace the original point X_1 by a new estimate X_2.

The direction from X_1 to X_2 is given by $(\delta_1, \delta_2, \ldots, \delta_n)^T$, hence it seems a promising direction in which to search. This acceleration step is similar in idea to that mentioned in connection with the gradient method of Section 8.2.3.

Once the optimum is found in this direction the remaining search directions are chosen so that they are orthogonal to all previous directions. These remaining directions can be generated by Gram–Schmidt orthogonalization (see Section 9.1 in the Appendix). Once each of these directions has been searched, an acceleration step is made in the direction of the line from the first point to the last point corresponding to these directions. This furnishes the first direction for the next orthogonalization process. This cycle of acceleration and orthogonalization is repeated until no significant improvement can be found, or some termination criterion is satisfied. Improved procedures, requiring computational time of dimension n^2 rather than n^3 for the Gram-Schmidt process have been given by Powell (1968).

8.2.4.4 *The Method of Powell*

This method was developed by M. J. D. Powell (1964). It is similar to the method of conjugate directions of Section 8.2.3.2, except that derivatives are not required. The method is also similar to Rosenbrock's method of the previous section, except that each search is carried out along a conjugate direction. The directions become conjugate with respect to an approximation of the Hessian matrix. An algorithmic statement of the method is given next.

Let X_0 be the initial estimate of the maximum point. Let $d_1, d_2, \ldots, d_n$ be the search directions.

1. Set $d_1, d_2, \ldots, d_n$ to be equal to the coordinate directions.
2. Set $i = 1$.
3. Find the maximum of f in the d_i direction from X_{i-1}, at, say, $X_{i-1} + s_i d_i$.
4. Let

$$X_i = X_{i-1} + s_i d_i.$$

If $i < n$, let i become $i + 1$, and go to step 3. If $i = n$, go to step 5.
5. Let

$$d_i = d_{i+1}, \qquad i = 1, 2, \ldots, n-1$$
$$d_n = X_n - X_0.$$

6. Find the maximum of f in the d_n direction from X_n, at, say, $X_n + s_n d_n$. Let

$$X_0 = X_n + s_n d_n.$$

7. Return to (2) unless some termination criterion is met.

It can be seen that the initial coordinate directions are gradually replaced by new directions, one per iteration. When the method is applied to a quadratic function these new directions are usually mutually conjugate. This means that the method is likely to terminate after at most n iterations for a quadratic objective function.

8.2.4.5 Brent's Praxis Method

There is a problem that may occur in the implementation of Powell's method, described in the previous section. That is, that even if f is quadratic, it may happen that

$$s_i = 0, \qquad \text{for some } i, 1 \leq i \leq n.$$

That is, having some estimate X_{i-1} of the optimum, the next estimate is calculated to be the same point. This occurs when the directions $d_1, d_2, \ldots,$ d_n become linearly dependent. When the method is implemented on a digital computer it is unlikely that the step size will ever become exactly zero (because of roundoff). However, it can come alarmingly close, and to avoid this problem a new direction $X_n - X_0$ should replace one of the d_i's so as to make the set linearly independent. While this modification has been found to be quite successful (Fletcher 1965; Box 1966), there is no longer any guarantee that for a quadratic function the set of directions will be mutually conjugate. This means that the method may not produce fast convergence to the optimum.

Brent (1973) has suggested a different approach to the problem of avoiding linear dependence among the search directions. His modification of Powell's method is to periodically reset the search directions to be a set of orthogonal directions based on the original conjugate directions which are replaced. This results in faster convergence than would occur if the search directions were reset as the coordinate directions, as in that case information built up about the function is thrown away at each reset.

The new set of normalized search directions $\bar{d}_1, \bar{d}_2, \ldots, \bar{d}_n$ are built up by assuming that the objective function f is a quadratic. If f is indeed quadratic, the $\bar{d}_1, \bar{d}_2, \ldots, \bar{d}_n$ will be mutually conjugate. They can be assembled as column vectors into a matrix

$$\bar{D} = (\bar{d}_1, \bar{d}_2, \ldots, \bar{d}_m).$$

Then

$$H = (\bar{D}\bar{D}^T)^{-1}$$

will be the Hessian of f evaluated at the optimum. $\bar{D}$ is then replaced by an orthogonal matrix satisfying this last equation. This ensures fast convergence.

As the directions $\bar{d}_1, \bar{d}_2, \ldots, \bar{d}_n$ are calculated to be orthogonal, they span $\mathbf{R}^n$ and thus no potential optimum is overlooked. The computational details are given in Brent (1973), p. 129.

8.2.4.6 *The Method of Stewart*

As can be seen, the variable metric method of Section 8.2.2.2 requires that the partial derivatives of f be evaluated at successive points of the search. If it is difficult to perform these evaluations a natural question arises: Is it possible to estimate successfully the required values of the derivatives at the necessary points? Although the answer is often yes, difficulties may arise due to rounding and approximation errors in the computation of the estimates.

Stewart (1957) followed this approach by estimating the value of each partial derivative by a difference quotient, i.e.,

$$\frac{\partial f(X)}{\partial x_i} \approx \frac{f(X + \Delta_i K_i) - f(X)}{\Delta_i}, \qquad i = 1, 2, \ldots, n, \qquad (8.21)$$

where Δ_i, $i = 1, 2, \ldots, n$ is a small positive number (some suggestions for the choosing of which are given in Stewart's paper) and K_i, $i = 1, 2, \ldots, n$ is a vector with all zero entries except for a unit entry in the ith position.

If the Δ_i are relatively small, rounding error will be relatively high and the error in (8.17) will be unacceptable. However, if the Δ_i are relatively large (8.21) does not provide a very accurate approximation. Stewart attempts to steer a middle course by choosing the Δ_i according to the curvature at X (requiring estimates of second derivatives). For greater accuracy, rather than using the forward difference formula (8.21) he uses a central difference formula:

$$\frac{\partial f(X)}{\partial x_i} \approx \frac{f(X + \frac{1}{2}\Delta_i K_i) - f(X - \frac{1}{2}\Delta_i K_i)}{\Delta_i}, \qquad i = 1, 2, \ldots, n.$$

Despite these precautions, it sometimes happens that the method fails to yield an improvement in f after reaching a particular point X_i. In this case, E_i, the ith approximation of H^{-1}, is reset to I, the identity matrix, i.e., the search once again sets off initially in the direction $\nabla f(X_i)$.

8.2.5 A Comparison of Unconstrained Optimization Methods

Having presented a large number of unconstrained optimization methods we must make some attempt to compare them. As would be expected, Hessian methods usually take *fewer steps* to converge to the maximum of a given problem than gradient methods. Also, the latter usually take *fewer steps* than direct search methods. However, because derivatives are often expensive to compute, it is not always true that direct search methods require more overall computational effort than the more sophisticated methods.

Whenever second derivatives are available a modified version of the Newton–Raphson method of Section 8.2.2.1 should be used. The modification consists of using an acceleration step $X_n - X_0$ as a direction of search after X_n has been found after n steps. (Further discussion is given in Jacoby, Kowalik, and Pizzo (1972).) When the Hessian matrix is unavailable, variable metric methods are generally the most powerful. (See especially the comment at the end of Section 8.2.2.2.) However when the objective function is known to be quadratic or near quadratic, the method of conjugate gradients (Section 8.2.3.3) is just as efficient. When derivatives are unavailable, Powell's method (Section 8.2.4.4) seems best on problems with a small number of variables. With problems with a larger number of variables, Stewart's method (Section 8.2.4.5) is often appropriate. Lill (1970) has modified the linear search in Stewart's algorithm to yield good results. However Himmelblau (1972) found Stewart's method with golden section search inferior to a modification of Powell's method (the modification being to include the method of Davies, Swann, and Campey, mentioned by Swann (1964), in both the linear search and iterative quadratic fitting.)

8.3 Constrained Optimization

Feasible solutions to many realistic optimization problems are constrained to lie within a subset of n-dimensional Euclidean space. These constraints are usually expressed as equalities or inequalities involving the decision variables, as outlined in Section 8.1.

8.3.1 The Method of Zoutendijk

The first method to be described for solving nonlinear optimization problems with inequality constraints is called the *method of feasible directions*, presented by Zoutendijk (1960).

The method starts with a point X_1 which satisfies the inequalities (8.3). A *feasible* direction in which to move from X_1 is now determined. A feasible direction is one in which small steps produce points which are both improvements over X_1 (in magnitude of f) and satisfy (8.3). An obvious candidate would be the gradient direction, as this would yield the biggest improvement in f. If this direction is feasible it is chosen. However, suppose it is not feasible. Suppose also that X_1 lies on the boundary of the feasible region (otherwise an unconstrained optimization strategy can be used until a constraint is met). The problem now is to find a direction which both increases the objective function and leads to points within the feasible region.

The feasible direction chosen is the one which in general makes the smallest angle θ with the gradient direction. There may be pitfalls however. If the "active" constraint (the constraint which forms the part of the boundary on which X_1 lies) is linear, everything is satisfactory and one of the two directions defined by this constraint is chosen. However, if the active constraint is nonlinear, it is possible that this procedure will produce a direction leading out of the feasible region. Having made such a step, one would then have to "jump" back into the feasible region. But there is no guarantee that such pairs of steps would not be performed repeatedly, causing an inefficient zig zag. To avoid these and other traps, it becomes increasingly obvious that we must choose a direction which moves decisively away from the boundary of the feasible region while also increasing the value of f. For this purpose the desirable direction d is found by solving the following program:

Maximize: E

subject to: $(\nabla h_i(X_1))^T d + t_i E \leq 0,$ for all i for which $h_i(X_1) = 0,$

$$0 \leq t_i \leq 1$$
$$(\nabla f(X_1))^T d \geq E$$
$$d^T d = 1.$$

The direction d^* which is the solution to this problem is the most desirable direction to use. One proceeds in this direction as far as possible from X_1 until the function begins to decrease or a boundary of the feasible region is met. The process is then repeated until the maximum value for E is non-positive. The process is then terminated. If all the functions in (8.3) are concave the global maximum will have been found. The method often performs well on maximization problems when concavity is not present.

8.3.2 The Gradient Projection Method

Let X^* be the optimal solution to the problem (8.1), (8.3). It is very likely that at this point some of the m constraints in (8.3) will be active. These active constraints form a subspace of the original feasible region. Thus if this subspace is examined using unconstrained optimization techniques, the optimum will be found. The main problem is to identify the correct subspace from among the multitude of subspaces defined by combinations of the constraints. Rosen (1960) developed the gradient projection method which solves this problem efficiently if all the constraints are linear. Unlike Zoutendijk's method, Rosen's method does not require the solution of a linear programming problem each time a new search direction is to be found. This decrease in computational effort has a price. At each iteration, the method does not search in the feasible direction which brings about the greatest objective function increase. Instead it chooses a direction which both increases the function value and ensures that a small step in this direction does not lead to an infeasible point. This direction is defined as the projection of the gradient vector onto the intersection of the hyperplanes associated with the active constraints. If there are no active constraints, the gradient direction is taken as the direction of search.

In outlining the method here it will be assumed that all the functions h_i, $i = 1, 2, \ldots, m$ in (8.3) are linear. Specifically, assume that the method begins with an estimate X_1 of X^*. Let $X_k = X_1$.

1. First calculate $\nabla f(X_k)$.
2. Let P_k be the set of hyperplanes corresponding to active constraints at X_k.
3. Find the projection of $\nabla f(X_k)$ onto the intersection of the hyperplanes in P_k. (If there are no active constraints, X_k is an interior point and P_k is empty. In this case the projection is $\nabla f(X_k)$.)
4. Maximize along the direction of the projection, taking care to remain within the feasible region.
5. This produces a new point X_{k+1}.
6. (a) If P_k was not empty in step 2, replace X_k by X_{k+1} and return to step 1.
   (b) If P_k was empty in step 2, then

$$\frac{\partial f}{\partial x_i} + \sum_{i=1}^{q} \lambda_i \frac{\partial h_i}{\partial x_i} = 0,$$

where $h_1, h_2, \ldots, h_q$ are the functions which correspond to the hyperplanes in P_k. If

$$\lambda_i \leq 0, \qquad i = 1, 2, \ldots, q,$$

X_k satisfies the Kuhn–Tucker conditions (see Chapter 7). Hence X_k is a maximum. If at least one λ_i is such that

$$\lambda_i > 0, \tag{8.22}$$

a plane corresponding to a function h_i for which (8.22) holds is removed from P_k. Return to step 2.

The method is unlikely to be as efficient on problems with nonlinear constraints. In these cases the projections are made onto hyperplanes which are tangent to the constraint surfaces. Steps taken in these hyperplanes may very well move out of the feasible region. Thus jumps back into the feasible region are likely to be necessary.

8.3.3 A Penalty Function Method

Carroll (1961) presented a method for solving (8.1), (8.3) which generates a sequence $X_1, X_2, \ldots$, of successively better estimates of X^*, each of which is feasible. Fiacco and McCormick (1968) refined the method and call their modification SUMT (Sequential Unconstrained Minimization Technique). The method can just as easily accommodate maximization problems.

The SUMT method creates a new objective function:

$$F(X,q) = f(X) - q \sum_{i=1}^{k} \frac{1}{h_i(X)},$$

where q is negative as the objective is maximization. First an initial feasible starting point X_1 must be chosen. Then an initial search is carried out for the maximum of F having chosen a large value of q, say q_0. The methods of Powell (Section 8.2.4.4) or Davidon (Section 8.2.2.2) appear to be two of the most appropriate for this search. Note that the maximum X_2 for F will not lie on the boundary of the feasible region, as, if

$$h_i(X_2) = 0,$$

for any i, $i = 1, 2, \ldots, m$, then F will become arbitrarily small. Hence X_2 will not be the maximum for the original problem if X^* lies on the boundary. A more accurate approximation of X^* is found by maximizing F by searching from X_2 after reducing the value of q. The above series of steps are repeated, with q being successively reduced at each iteration. The sequence of feasible points found approaches an optimum if certain assumptions are met.

8.3.3.1 The Generalized Reduced Gradient Method

Consider the nonlinear programming problem:

Minimize: $f(X)$
subject to: $g_j(X) = 0,$ $j = 1, 2, \ldots, m$
               $L_i \leq x_i \leq U_i,$ $i = 1, 2, \ldots, n,$

where $X = (x_1, x_2, \ldots, x_n)$ and the L_i, U_i, $i = 1, 2, \ldots, n$ are given constants. We now outline a method due to Abadie and Carpentier (described in Fletcher (1969)) which solves this problem. Inequality constraints can be handled by this method by introducing nonnegative slack variables to force equality. To make this clear, suppose a problem contained the constraint:

$$h_j(X) \geq 0.$$

We introduce a new (slack) variable s_j, forcing equality:

$$h_j(X) - s_j = 0.$$

We set

$$L_j = 0$$
$$U_j = \infty,$$

and thus the problem is now in the desired form.

As in the simplex method (assuming nondegeneracy) one partitions the set of variables $\{x_1, x_2, \ldots, x_n\}$ into two subsets: X_B, comprising m *basic* variables (one for each constraint), and X_{NB} comprising $n - m$ *nonbasic* variables. Now

$$df(X) = \nabla_{X_{NB}} f(X)^T dX_{NB} + \nabla_{X_B} f(X)^T dX_B$$

and

$$\frac{df(X)}{dX_{NB}} = \nabla_{X_{NB}} f(X) + \nabla_{X_{NB}} f(X)^T \frac{dX_B}{dx_{NB}},$$

where

$$\nabla_{X_B} f(X) = \left(\frac{\partial f(X)}{\partial x_B^1}, \frac{\partial f(X)}{dX_B^2}, \ldots, \frac{\partial f(X)}{\partial x_B^m} \right)^T$$

and

$$\nabla_{X_{NB}} f(X) = \left(\frac{\partial f(X)}{\partial x_{NB}}, \frac{\partial f(X)}{\partial x_{NB}^2}, \ldots, \frac{\partial f(X)}{\partial X_{NB}^{n-m}} \right)^T.$$

Now, as

$$g_j(X) = 0, \qquad j = 1, 2, \ldots, m,$$

$$dg_j(X) = \nabla_{X_{NB}} g_j(X)^T dX_{NB} + \nabla_{X_B} g_j(X)^T dX_B = 0, \qquad j = 1, 2, \ldots, m$$

we have

$$\frac{dg(X)}{dX_{NB}} = \frac{\partial g(X)}{\partial X_{NB}} + \left(\frac{\partial g(X)}{\partial X_B} \right)^T \frac{dX_B}{dX_{NB}} = 0,$$

where

$$\frac{dg}{dX_{NB}} = \left(\frac{dg_1}{dX_{NB}}, \frac{dg_2}{dX_{NB}}, \ldots, \frac{dg_m}{dX_{NB}} \right)^T.$$

Thus

$$\frac{dX_B}{dX_{NB}} = -\left(\frac{\partial g}{\partial X_B} \right)^{-1} \frac{\partial g}{\partial X_{NB}}.$$

On substitution, we obtain

$$\frac{df(X)}{dX_{NB}} = \nabla_{X_{NB}} f(X) - \nabla_{X_B} f(X)^T \left(\frac{\partial g}{\partial X_B} \right)^{-1} \frac{\partial g}{\partial X_{NB}}.$$

This last expression is called the *generalized reduced gradient*, and permits a reduction in the dimensionality of the problem.

Now if f has a local minimum at X^*, it is necessary that

$$\frac{df(X^*)}{dX_{NB}} = 0.$$

The search for X^* begins at point X_0 on the boundary of the feasible region. One then searches from X_0 along the boundary until

$$\frac{df(X)}{dX_{NB}} = 0,$$

at which point a local minimum has been found.

8.3.4 Linear Approximation

It must have been obvious to the reader who studied Chapter 2 that linear programming is a very powerful tool. Hence it seems fruitful to consider the possibility of converting nonlinear optimization problems into linear ones so that L.P. theory can be applied.

One of the best known linearization methods is due to Wolfe (Abadie 1967). The feasible region defined by the constraints in (8.3) is approximated by selecting a number of points called *grid points* and forming their convex hull. The r grid points $X_1, X_2, \ldots, X_r$ are chosen by methods described by Wolfe. The function f is approximated between a pair of grid points X_{j-1}, and X_j by linear interpolation; i.e., if

$$X = \alpha X_{j-1} + (1 - \alpha)X_j, \qquad j = 2, 3, \ldots, r, \quad 0 \le \alpha \le 1,$$

then

$$f(X) = f(\alpha X_{j-1} + (1 - \alpha)X_j)$$

$$\approx \sum_{j=1}^{r} \alpha_j f(X_j), \qquad \alpha_j \ge 0, \quad j = 1, 2, \ldots, r,$$

where

$$\sum_{j=1}^{r} \alpha_j = 1.$$

The grid points must be carefully selected so that only a small number of the α_i are nonzero for the representation of any point within the convex hull.

The constraint functions in (8.3) can also be approximated by linear interpolation:

$$h_i(X) = h_i\left(\sum_{j=1}^{r} \alpha_j X_j \right)$$

$$\approx \sum_{j=1}^{r} \alpha_j h_i(X_j), \qquad = 1, 2, \ldots, k.$$

Thus the original problem can now be replaced by an approximate linear programming problem:

$$\text{Maximize:} \quad f = \sum_{j=1}^{r} \alpha_j f(X_j)$$

$$\text{subject to:} \quad \sum_{j=1}^{r} \alpha_j h_i(X) \le 0, i = 1, 2, \ldots, k$$

$$\sum_{j=1}^{r} \alpha_j = 1$$

$$\alpha_j \ge 0, \qquad j = 1, 2, \ldots, r.$$

The expressions:

$$f(X_j), h_i(X), \qquad i = 1, 2, \ldots, k; \qquad j = 1, 2, \ldots, r$$

are known constants and the decision variables in the L.P. are the α_j's.

Of course many grid points and hence many α_j are needed for a good approximation of the nonlinear functions in (8.3). This is accomplished by allowing the simplex method to choose which new grid points are best by way of solving certain subproblems. Thus the approximation of (8.3) becomes increasingly more accurate as the optimum is approached.

8.3.4.1 The Method of Griffith and Stewart

Griffith and Stewart (1961) presented a linearization method which attacks a nonlinear programming problem by starting with a feasible solution and reducing each nonlinear function to a linear one by a Taylor series approximation. This produces a linear programming problem which is then solved to yield another solution to the original problem. The cycle continues and a sequence of L.P. problems are solved.

Consider the following problem:

$$\begin{aligned} \text{Maximize:} \qquad & f(X) \\ \text{subject to:} \qquad & g_i(X) = 0, \qquad j = 1, 2, \ldots, m \\ & 0 \le x_i \le U_i, \qquad i = 1, 2, \ldots, n. \end{aligned}$$

As in the generalized reduced gradient method, any inequality constraints can be converted into equations. It is assumed that all of the above functions have continuous first partial derivatives. Assume now that X_0 is a feasible solution. Then the first-order Taylor approximations of the above functions are:

$$f(X_0 + h) = f(X_0) + \nabla f(X_0)^T h$$
$$g_j(X_0 + h) = g_j(X_0) + \nabla g_j(X_0)^T h = 0, \qquad j = 1, 2, \ldots, m.$$

Now we attempt to find $h = (h_1, h_2, \ldots, h_m)$ such that $X_0 + h$ represents an improvement over X_0, i.e.,

$$f(X_0 + h) > f(X_0).$$

Ignoring the constant $f(X_0)$ we

$$\text{Maximize:} \qquad \nabla f(X_0)^T h = \sum_{i=1}^{n} \frac{\partial f(X_0)}{\partial x_i} h_i$$

$$\text{subject to:} \qquad \sum_{i=1}^{n} \frac{\partial g_j(X_0)}{\partial x_i} h_i = -g_j(X_0), \qquad j = 1, 2, \ldots, m,$$

which is a linear programming problem with variables $h_1, h_2, \ldots, h_n$. Of course, these are unrestricted in sign, so the technique for converting the problem to one with all nonnegative variables (given in Chapter 2) must be used.

The solution to the above L.P. may produce a new point $X_0 + h$ which is outside the feasible region of the original problem. Thus we must place restrictions on the magnitude that the h_i's can attain in the above L.P. to ensure that this does not happen. Hence we add to the above L.P. the following constraints:

$$-m_i \le h_i \le m_i, \qquad i = 1, 2, \ldots, n.$$

Thus the method proceeds by establishing a feasible point X_0, constructing an L.P. based on the Taylor approximations and the bounds on the h_i's and then solving this L.P. to produce an improved point $X_0 + h$. Once this new point is found, the process is repeated until the improvement in f from one iteration to the next falls below some given level or two successive solutions are sufficiently close together. The success of the method depends upon choosing efficient m_i's at each iteration. If the m_i values are too large, the method may produce an infeasible solution. However, relatively small m_i's lead to a large number of steps.

8.3.4.2 *Separable Programming*

Consider a special case of (8.01), (8.02), (8.03) in which

$$f(X) = \sum_{i=1}^{n} f_i(x_i)$$

$$g_j(X) = \sum_{i=1}^{n} g_{ij}(x_i), \qquad j = 1, 2, \ldots, m$$

$$h_i(X) = -x_i, \qquad i = 1, 2, \ldots, k = n$$
$$(\text{i.e., } x_i \ge 0, \qquad i = 1, 2, \ldots, n),$$

where

$$X = (x_1, x_2, \ldots, x_n)^T.$$

That is, each function in the problem can be expressed as the sum of a number of functions of one variable. Such a problem is called a *separable programming* problem.

In this section we develop a technique for approximating the above problem by a linear programming formulation. We then show that when f is concave and the constraint functions are all convex the approximating technique can be made more efficient.

We begin by approximating each function by a piece-wise linear function as follows. First, we construct such an approximating function, $\hat{f}_i$ for each f_i. Suppose that for a particular i, where $1 \le i \le n$, f_i can be represented by the graph in Figure 8.06. Suppose

$$0 \le x_i \le u_i$$

for any feasible solution. The values u_i must be calculated by examining the constraints. Suppose the interval over which x_i is defined, $[0, u_i]$ is divided into p_i subintervals not necessarily of equal length:

$$[x_{i0}, x_{i1}], [x_{i1}, x_{i2}], [x_{i2}, x_{i3}], \ldots, [x_{ip_i-1}, x_{ip_i}],$$

where

$$x_{i0} < x_{i1} < x_{i2} < x_{i3} < \cdots < x_{ip_i-1} < x_{ip_i}$$

and

$$x_{i0} = 0,$$
$$x_{ip_i} = u_i.$$

$\hat{f}_i$ is defined over each subinterval $[x_{ik-1}, x_{ik}]$, $k = 1, 2, \ldots, p_i$ by the line segment joining $(x_{ik-1}, f_i(x_{ik-1}))$ and $(x_{ik}, f_i(x_{ik}))$. That is $\hat{f}_i$ is shown in Figure 8.6 by the straight line segments approximating f_i. Formally:

$$\hat{f}_i(x_i) = f_i(x_{ik-1}) + [f_i(x_{ik}) - f_i(x_{ik-1})] \frac{(x_i - x_{ik-1})}{(x_{ik} - x_{ik-1})},$$

$$x_i \in [x_{ik-1}, x_{ik}]. \tag{8.23}$$

Now, as $x_i \in [x_{ik-1}, x_{ik}]$, x_i can be expressed as

$$x_i = \alpha_{ik-1} x_{ik-1} + \alpha_{ik} x_{ik}, \tag{8.24}$$

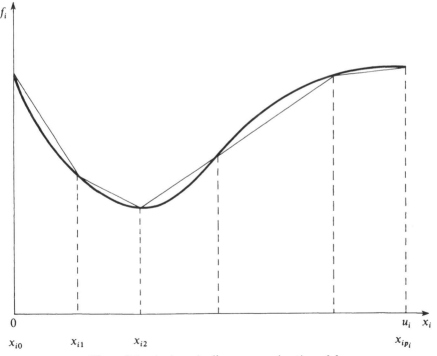

Figure 8.6. A piecewise linear approximation of f_i.

where
$$0 \leq \alpha_{ik-1}, 0 \leq \alpha_{ik}$$

and
$$\alpha_{ik-1} + \alpha_{ik} = 1. \tag{8.25}$$

By (8.24)
$$(x_i - x_{ik-1})/(x_{ik} - x_{ik-1}) = (\alpha_{ik-1}x_{ik-1} + \alpha_{ik}x_{ik} - x_{ik-1})/(x_{ik} - x_{ik-1}).$$

By (8.25)
$$= \alpha_{ik}(x_{ik} - x_{ik-1})/(x_{ik} - x_{ik-1})$$
$$= \alpha_{ik}.$$

Thus (8.23) becomes
$$\hat{f}_i(x_i) = f_i(x_{ik-1}) + [f_i(x_{ik}) - f_i(x_{ik-1})]\alpha_{ik}.$$

By (8.25)
$$\hat{f}_i(x_i) = \alpha_{ik-1}f_i(x_{ik-1}) + \alpha_{ik}f_i(x_{ik}). \tag{8.26}$$

In general
$$\hat{f}_i(x_i) = \alpha_{ik-1}f_i(x_{ik-1}) + \alpha_{ik}f_i(x_{ik}), \quad x_i \in [x_{ik-1}, x_{ik}], \tag{8.27}$$

where
$$\alpha_{ik-1} + \alpha_{ik} = 1$$

and
$$k = 1, 2, \ldots, p_i.$$

The family of p_i equations given by (8.27) can be combined as
$$\hat{f}_i(x_i) = \alpha_{i0}f_i(x_{i0}) + \alpha_{i1}f_i(x_{i1}) + \cdots + \alpha_{ip_i}f(x_{ip_i}),$$

where
$$\alpha_{i0} + \alpha_{i1} + \cdots + \alpha_{ip_i} = 1$$
$$\alpha_{ik} \geq 0, k = 0, 1, \ldots, p_i.$$

And for each i, $1 \leq i \leq n$, at most two α_{ik}'s can be positive, and if two are positive they must be adjacent. That is, if
$$\alpha_{ik} > 0$$

for some $k \in \{0, 1, \ldots, p_i\}$, then at most one of
$$\alpha_{ik-1} > 0, \alpha_{ik+1} > 0 \tag{8.27a}$$

holds and all other α_{ik}'s are zero.

We define the approximating functions for all of the functions in the problem in this way:

$$\hat{f}_i(x_i) = \sum_{k=0}^{p_i} \alpha_{ik}f_i(x_{ik}), \qquad i = 1, 2, \ldots, n, \tag{8.28}$$

$$\hat{g}_i(x_i) = \sum_{k=0}^{p_i} \alpha_{ik}g_{ij}(x_{ik}), \qquad \begin{matrix} i = 1, 2, \ldots, n, \\ j = 1, 2, \ldots, m. \end{matrix} \tag{8.29}$$

We can now formulate an approximation of the original problem by substituting $\hat{f}_i$ for f_i and $\hat{g}_{ij}$ for g_{ij} using (8.28) and (8.29):

$$\text{Maximize:} \quad \sum_{i=1}^{n} \sum_{k=0}^{p_i} \alpha_{ik} f_{ij}(x_{ik}) \qquad j = 1, 2, \ldots, m,$$

$$\text{subject to:} \quad \sum_{i=1}^{n} \sum_{k=0}^{p_i} \alpha_{ik} g_{ij}(x_{ik}) = 0, \qquad i = 1, 2, \ldots, n,$$

$$x_i \geq 0, \qquad i = 1, 2, \ldots, n,$$

$$\sum_{k=0}^{p_i} \alpha_{ik} = 1, \qquad i = 1, 2, \ldots, n,$$

$$\alpha_{ik} \geq 0, \qquad k = 0, 1, \ldots, p_i,$$
$$i = 1, 2, \ldots, n.$$

And constraint (8.27a) holds.

This is a linear programming problem in which the $f_i(x_{ik})$'s and the $g_{ij}(x_{ik})$'s are constants and the α_{ik}'s are the decision variables. (8.27a) poses a minor problem in that it imposes restrictions on which variables can enter the basis if the simplex method is used to solve it. This means that a check must be made at each iteration to ensure that 8.27a is not violated when a new variable is brought into the basis. If the selection criterion specifies that the variable to enter is such that 8.27a is violated, then that variable is ignored and the criterion is applied anew. This type of strategy will also appear in the quadratic programming technique of Section 8.3.5.

Unfortunately there is no guarantee that the solution to the L.P. problem will be even feasible let alone optimal for the original problem. This is because the approximate feasible region may yield an optimal point which lies outside the original feasible region. Even if the optimal point for the L.P. is feasible it may be only a local maximum.

When each f_i is concave, f, being the sum of a number of convex functions is also concave. (The proof of this fact is left as an exercise.) Similarly, when each g_{ij} is convex, each g_j is convex. Now the necessary conditions for a point to be a global maximum are sufficient when f is concave and the feasible region is convex. As the feasible region is defined by a set of convex functions it is convex. Thus if each f_i is concave and g_{ij} is convex, any stationary point will be a global maximum.

In this case the optimal solution to the approximating problem will be optimal for the original problem. To see why this is so we formulate a new approximating problem. Let

$$x_i = y_{i1} + y_{i2} + \cdots + y_{ip_i},$$

where

$$y_{ih} \begin{cases} = x_{ih} - x_{ih-1} & h < k \\ = x_i - x_{ih-1} & h = k \\ = 0 & h > k \end{cases} \qquad (8.30)$$

when

$$x_i \in \left[x_{ik-1}, x_{ik} \right].$$

As an example, if

$$x_{iq} = 2q, q = 0, 1, 2, \ldots, 5,$$

as in Figure 8.7:

$$p_i = 5$$

and

$$x_i = 5.5,$$

then

$$k = 3,$$

i.e., x_i lies in the 3rd interval. In this case

$$y_{i1} = x_{i1} - x_{i0} = 2 - 0 = 2$$
$$y_{i2} = x_{i2} - x_{i1} = 4 - 2 = 2$$
$$y_{i3} = x_i - x_{i3} = 5.5 - 4 = 1.5$$
$$y_{i4} = y_{i5} = 0.$$

Thus

$$x_i = y_{i1} + y_{i2} + y_{i3} + y_{i4} + y_{i5}$$
$$= 2 + 2 + 1.5 + 0 + 0 = 5.5.$$

Consider $\hat{f}_i(x_i)$ where

$$x_i \in \left[x_{ik-1}, x_{ik} \right].$$

By (8.23)

$$f_i(x_i) = f_i(x_{ik-1}) + \left[f_i(x_{ik}) - f_i(x_{ik-1}) \right] \frac{y_{ik}}{(x_{ik} - x_{ik-1})}$$

$$= f_i(x_{i0}) + (f_i(x_{i1}) - f_i(x_{i0})) + (f_i(x_{i2}) - f_i(x_{i1})) + \cdots + f_i(x_{ik-1})$$

$$+ f_i(x_{ik-2}) + \left[f_i(x_{ik}) - f_i(x_{ik-1}) \right] \frac{y_{ik}}{(x_{ik} - x_{ik-1})}.$$

Since

$$y_{ik} \begin{cases} = x_{ih} - x_{ih-1}, & h < k, \\ = 0, & h > 0, \end{cases}$$

$$\hat{f}_i(x_i) = f_i(x_{i0}) + \sum_{r=1}^{p_i} \left[f_i(x_{ir}) - f_i(x_{ir-1}) \right] \frac{y_{ir}}{(x_{ir} - x_{ir-1})}.$$

The expression

$$\frac{f_i(x_{ir}) - f_i(x_{ir-1})}{(x_{ir} - x_{ir-1})}$$

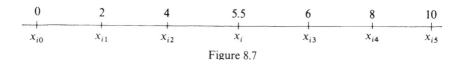

Figure 8.7

is an approximation of the slope of f_i over the interval $[x_{ir-1}, x_{ir}]$, which we denote by $\hat{f}'_{ir}$. Therefore

$$\hat{f}_i(x_i) = f_i(x_{i0}) + \sum_{r=1}^{p_i} \hat{f}'_{ir} y_{ir}.$$

Similar approximations can be constructed for each g_{ij}:

$$\hat{g}_{ij}(x_i) = g_{ij}(x_{i0}) + \sum_{r=1}^{p_i} \hat{g}_{ijr} y_{ir},$$

where

$$\hat{g}_{ijr} = \frac{g_{ij}(x_{ir}) - g_{ij}(x_{ir-1})}{(x_{ir} - x_{ir-1})}.$$

Summing all expressions of each type we have

$$\hat{f}(x) = \sum_{i=1}^{n} f_i(x_{i0}) + \sum_{i=1}^{n} \sum_{r=1}^{p_i} \hat{f}'_{ir} y_{ir}$$

$$\hat{g}_j(x_i) = \sum_{i=1}^{n} g_{ij}(x_{i0}) + \sum_{i=1}^{n} \sum_{r=1}^{p_i} \hat{g}_{ijr} y_{ir}.$$

These expressions can be used to formulate a new linear programming problem:

Maximize:
$$\sum_{i=1}^{n} \sum_{r=1}^{p_i} \hat{f}'_{ir} y_{ir}$$

subject to:
$$\sum_{i=1}^{n} \sum_{r=1}^{p_i} \hat{g}_{ijr} y_{ir} \leq - \sum_{i=1}^{n} g_{ij}(x_{i0}), \qquad j = 1, 2, \ldots, m.$$

$$0 \leq y_{ir} \leq x_{ir} - x_{ir-1}, \qquad j = 1, 2, \ldots, n$$
$$r = 0, 1, \ldots, p_i.$$

Note that the expression

$$\sum_{i=1}^{n} f_i(x_{i0})$$

has been omitted as it is a constant. The terms: $\hat{f}'_{ir}, \hat{g}_{ijr}, g_{ij}(x_{i0}), x_{ir}$ are all constants and the y_{ir}'s are the decision variables. Conditions (8.30) have not been included in the above formulation as they are implicitly satisfied when f is concave and the g_j's are convex.

This is because for any i and r:

$$\hat{f}'_{ir} \geq \hat{f}'_{is}, \quad \text{for all } s < r, \tag{8.31}$$

as f_i is concave. That is, the slope of $f_i(x_i)$ decreases as x_i increases by virtue of the nature of concavity. Therefore the objective function coefficients $\hat{f}'_{ir}$ are automatically assembled in nonincreasing order for each given i. However, by analogous reasoning, for any pair i and j

$$\hat{g}_{ijr} \leq \hat{g}_{irs} \quad \text{for all } s < r. \tag{8.32}$$

Thus the technological coefficients, $\hat{g}'_{ijr}$ are automatically assembled in nondecreasing order for each pair i and j.

Equations (8.31) and (8.32) together imply that for a given i, the order preference of the variables is: $y_{i1}, y_{i2}, \ldots, y_{ip_i}$. Hence conditions (8.30) will be satisfied automatically. The very last family of constraints can be ignored, as f is concave and the g_j's are convex.

8.3.5 Quadratic Programming

There exists a special class of nonlinear optimization problems called *quadratic programming* (Q.P.), for which a considerable amount of theory has been developed. The maximization version of the Q.P. problem is given below:

$$\text{Maximize:} \quad f(X) = C^T X + X^T D X$$
$$\text{subject to:} \quad AX \leq B$$
$$X \geq 0.$$

$X = (x_1, x_2, \ldots, x_n)^T$ is the vector of decision variables, C is an $n \times 1$ vector and B an $m \times 1$ vector, both of given real numbers. A and D are real matrices of appropriate dimensions, and D is assumed symmetric negative definite. This last assumption means that f is strictly concave (see Section 9.1.6 in the Appendix). As can be seen, f is a quadratic function subject to only linear constraints. The solution to this problem can be obtained by applying the Kuhn–Tucker conditions of Section 7.4.2.1. As f is strictly concave and the constraint set defines a convex region, satisfaction of the Kuhn–Tucker necessary conditions guarantees a global optimum. The constraints, including the nonnegativity conditions, can be rewritten in a form compatible with Section 7.4.2.1:

$$AX - B \leq 0 \tag{8.33a}$$
$$-X \leq 0. \tag{8.33b}$$

Let $\lambda_1, \lambda_2, \ldots, \lambda_m$ be the Lagrange multipliers associated with (8.33a), where A is $m \times n$. Let $\delta_1, \delta_2, \ldots, \delta_n$ be the Lagrange multipliers associated with (8.33b). Then the Kuhn–Tucker conditions yield (on dropping the *'s):

$$C + 2X^T D - \lambda A + \delta = 0$$
$$AX - B \leq 0$$
$$-X \leq 0$$
$$\lambda(AX - B) = 0$$
$$\delta X = 0$$
$$\lambda \geq 0$$
$$\delta \geq 0,$$

where

$$\lambda = (\lambda_1, \lambda_2, \ldots, \lambda_m)$$
$$\delta = (\delta_1, \delta_2, \ldots, \delta_n).$$

Introducing slack variable $s_1, s_2, \ldots, s_m$, where

$$S = (s_1, s_2, \ldots, s_m)^T$$

we obtain

$$AX + S = B.$$

The conditions can be rearranged as follows:

$$-2X^TD + \lambda A - \delta = C \tag{8.34}$$

$$AX + S = B \tag{8.35}$$

$$\delta X = 0 \tag{8.36}$$

$$\lambda S = 0 \tag{8.37}$$

$$\lambda \geq 0, \qquad \delta \geq 0, \qquad S \geq 0, \qquad X \geq 0.$$

The problem is now to solve (8.34) and (8.35) while also satisfying (8.36) and (8.37). Because f is strictly concave and the feasible region is convex, the solution found must be optimal for the original problem. Thus it is enough to find a feasible solution to the system (8.34), (8.35) viewed as the constraint set of an L.P. problem. The only restrictions are (8.36) and (8.37), which imply that δ_j and x_j or λ_i and s_i cannot both be simultaneously positive for any i or j. Restrictions of this type occurred in the separable programming method of Section 8.3.4.2.

The solution is found by using phase I of the two-phase method of Section 2.5.4, making sure that (8.36) and (8.37) are never violated. In practice, this means that, if one of δ_i and x_i or λ_j and S_j are in the basis (assuming no degeneracy), then the other cannot enter the basis. When phase I has been completed, the optimal solution (if it exists) will have been found. As described in Chapter 2, if all the artificial variables are zero on termination of phase I, the problem has a feasible solution; otherwise it has no feasible solution.

8.3.5.1 Examples of Q.P.

Consider the following problem:

Maximize: $f(X) = f(x_1, x_2) = 3x_1 + 2x_2 - x_1^2 - x_1x_2 - x_2^2,$

subject to: $4x_1 + 5x_2 \leq 20$

$$x_1, x_2 \geq 0.$$

Written in matrix form, the problem is:

Maximize: $f(X) = (3, 2)\begin{pmatrix} x_1 \\ x_2 \end{pmatrix} + (x_1, x_2)\begin{pmatrix} -1 & -\frac{1}{2} \\ -\frac{1}{2} & -1 \end{pmatrix}\begin{pmatrix} x_1 \\ x_2 \end{pmatrix}.$

Thus

$$C^T = (3, 2)$$

$$D = \begin{pmatrix} -1 & -\frac{1}{2} \\ -\frac{1}{2} & -1 \end{pmatrix}$$

$$A = (4, 5)$$

$$B = 20.$$

We need one Lagrange multiplier λ_1 to be associated with the constraint and δ_1 and δ_2 to be associated with the nonnegativity conditions. Substituting all this into the rearranged Kuhn–Tucker conditions yields:

$$-2(x_1, x_2)\begin{bmatrix} -1 & -\frac{1}{2} \\ -\frac{1}{2} & -1 \end{bmatrix} + \lambda_1(4, 5) - (\delta_1, \delta_2) = (3, 2),$$

$$(4, 5)\begin{pmatrix} x_1 \\ x_2 \end{pmatrix} + s_3 = 20,$$

$$(\delta_1, \delta_2)\begin{pmatrix} x_1 \\ x_2 \end{pmatrix} = \begin{pmatrix} 0 \\ 0 \end{pmatrix},$$

$$\lambda_1 s_3 = 0,$$

$$\lambda_1, \delta_1, \delta_2, s_3, x_1, x_2 \geq 0,$$

where the vector of slack variables S consists of a single variable, say s_3. Writing out the first two equations and introducing artificial variables s_1 and s_2 in the first, we have:

$$2x_1 + x_2 + 4\lambda_1 - \delta_1 + s_1 = 3,$$
$$x_1 + 2x_2 + 5\lambda_1 - \delta_2 + s_2 = 2,$$
$$4x_1 + 5x_2 + s_3 = 20.$$

Recall that in the two-phase method the objective is to remove the artificial variables from the basis by minimizing their sum, i.e.,

$$\text{Minimize } s_0 = s_1 + s_2.$$

In tabular form the system is:

	x_1	x_2	λ_1	δ_1	δ_2	s_1	s_2	s_3	r.h.s.
	2	1	4	−1	0	1	0	0	3
	1	2	5	0	−1	0	1	0	2
	4	5	0	0	0	0	0	1	20
s_0	0	0	0	0	0	1	1	0	0

In canonical form:

	x_1	x_2	λ_1	δ_1	δ_2	s_1	s_2	s_3	r.h.s.
	2	1	4	-1	0	1	0	0	3
	1	2	5	0	-1	0	1	0	2
	4	5	0	0	0	0	0	1	20
s_0	-3	-3	-9	1	1	0	0	0	-5

We apply phase I to this tableau, taking care that none of the pairs: δ_1, x_1; δ_2, x_2; λ_1, s_3 are simultaneously positive. λ_1 cannot enter the basis as $s_3 \neq 0$. But x_1 can enter the basis as $\delta_1 = 0$:

	x_1	x_2	λ_1	δ_1	δ_2	s_1	s_2	s_3	r.h.s.
	1	$\frac{1}{2}$	2	$-\frac{1}{2}$	0	$\frac{1}{2}$	0	0	$\frac{3}{2}$
	0	$\frac{3}{2}$	3	$\frac{1}{2}$	-1	$-\frac{1}{2}$	1	0	$\frac{1}{2}$
	0	3	-8	2	0	-2	0	1	14
s_0	0	$-\frac{3}{2}$	-3	$-\frac{1}{2}$	1	$\frac{3}{2}$	0	0	$-\frac{1}{2}$

Now x_2 can enter the basis as $\delta_2 = 0$:

	x_1	x_2	λ_1	δ_1	δ_2	s_1	s_2	s_3	r.h.s.
	1	0	1	$-\frac{2}{3}$	$\frac{1}{3}$	$\frac{2}{3}$	$-\frac{1}{3}$	0	$\frac{4}{3}$
	0	1	2	$\frac{1}{3}$	$-\frac{2}{3}$	$-\frac{1}{3}$	$\frac{2}{3}$	0	$\frac{1}{3}$
	0	0	-14	1	2	-1	-2	1	13
s_0	0	0	0	0	0	1	1	0	0

Phase I has now been completed with $s_0 = 0$. Thus the original problem does have a feasible solution. Indeed the optimal solution to the original problem can be found from this tableau with

$$x_1^* = \tfrac{4}{3},$$
$$x_2^* = \tfrac{1}{3},$$

and

$$f(x_1^*, x_2^*) = \tfrac{7}{3}.$$

8.3.6 Geometric Programming

Geometric programming is a technique developed in the late 1960's for solving a certain class of nonlinear programming problems. Although this class includes certain kinds of problems which have constraints, the only constraints we allow here are that the decision variables must all be strictly positive. For the general techniques the reader should refer to the book written by the inventors of the subject, Duffin, Peterson and Zener (1967) and to Beightler and Phillips (1976) for applications.

Geometric programming is concerned with the optimization of the class of functions of the form:

$$\text{Minimize:} \qquad f(X) = \sum_{j=1}^{m} \left(c_j \prod_{i=1}^{n} x_i^{a_{ij}} \right),$$

$$\text{subject to:} \qquad x_i > 0, \qquad i = 1, 2, \ldots, n,$$

where

$$X = (x_1, x_2, \ldots, x_n),$$

$$c_j > 0, \qquad j = 1, 2, \ldots, m,$$

and $a_{ij}, i = 1, 2, \ldots, n; j = 1, 2, \ldots, m$ are arbitrary real numbers. Note that $f(X)$ is not in general a polynomial as the a_{ij}'s may possibly be negative. As the coefficients, c_j of the terms of $f(X)$ must be positive, Duffin, Peterson, and Zener call $f(X)$ a *posynomial*.

Let X^* be the optimal solution to the above problem. Then if we define

$$p_j(X) = \prod_{i=1}^{n} x_i^{a_{ij}}, \qquad j = 1, 2, \ldots, m,$$

we can express f as

$$f(X) = \sum_{j=1}^{m} c_j p_j(X).$$

As each x_i is constrained to be positive, each term $c_j p_j(X)$ must be positive when $X = X^*$. Thus once we know the value of $f(X^*)$ we can calculate the fractional contribution w_j made to it by the ith term, that is

$$w_j = \frac{c_j p_j(X^*)}{f(X^*)}, \qquad j = 1, 2, \ldots, m.$$

Of course

$$\sum_{j=1}^{m} w_j = 1. \tag{8.38}$$

(8.38) is called the *normality condition*. Also

$$0 \le w_j \le 1, \qquad j = 1, 2, \ldots, m.$$

The fractions w_j are called *weights*.

Applying the necessary conditions for f to have a minimum at X^* we have

$$\frac{\partial f(X^*)}{\partial x_k} = \sum_{j=1}^{m} c_j a_{kj} (x_k^*)^{(a_{kj}-1)} \prod_{i \ne k}^{n} (x_i^*)^{a_{ij}} = 0, \qquad k = 1, 2, \ldots, n,$$

where

$$X^* = (x_1^*, x_2^*, \ldots, x_n^*).$$

Multiplying the kth equation by x_k^* (>0) these conditions reduce to

$$\sum_{j=1}^{m} c_j a_{kj} \prod_{i=1}^{n} (x_i^*)^{a_{ij}} = 0, \qquad k = 1, 2, \ldots, n$$

or

$$\sum_{j=1}^{m} c_j a_{kj} p_j(X^*) = 0, \qquad k = 1, 2, \ldots, n.$$

Now as $f(X^*) > 0$,

$$\sum_{j=1}^{m} \left\{ \frac{c_j a_{kj} p_j(X^*)}{f(X^*)} \right\} = 0, \qquad k = 1, 2, \ldots, n,$$

or

$$\sum_{j=1}^{m} a_{kj} w_j = 0, \qquad k = 1, 2, \ldots, n. \qquad (8.39)$$

(8.39) is called the set of *orthogonality conditions*.

Now

$$f(X^*) = \{f(X^*)\}^1 = \{f(X^*)\}^{\sum_{j=1}^{m} w_j} = \prod_{j=1}^{m} \{f(X^*)\}^{w_j}$$

$$= \prod_{j=1}^{m} \left\{ \frac{c_j p_j(X^*)}{w_j} \right\}^{w_j}$$

$$= \prod_{j=1}^{m} \left\{ \frac{c_j \prod_{i=1}^{n} (x_i^*)^{a_{ij}}}{w_j} \right\}^{w_j}$$

$$= \left\{ \prod_{j=1}^{m} \left(\frac{c_j}{w_j} \right)^{w_j} \right\} \left\{ \prod_{j=1}^{m} \left[\prod_{i=1}^{n} (x_i^*)^{a_{ij}} \right]^{w_j} \right\}$$

$$= \left\{ \prod_{j=1}^{m} \left(\frac{c_j}{w_j} \right)^{w_j} \right\} \left\{ \prod_{j=1}^{m} \prod_{i=1}^{n} (x_i^*)^{a_{ij} w_j} \right\}$$

$$= \left\{ \prod_{j=1}^{m} \left(\frac{c_j}{w_j} \right)^{w_j} \right\} \left\{ \prod_{i=1}^{n} (x_i^*)^{\sum_{j=1}^{m} a_{ij} w_j} \right\}$$

$$= \left\{ \prod_{j=1}^{m} \left(\frac{c_j}{w_j} \right)^{w_j} \right\} \left\{ \prod_{i=1}^{n} (x_i^*)^0 \right\}$$

$$= \prod_{j=1}^{m} \left(\frac{c_j}{w_j} \right)^{w_j}.$$

As the c_j's are given constants, once the weights are found we can compute $f(X^*)$. The w_j's are found using (8.38) and (8.39) which represent a system of $(n + 1)$ linear equations in m unknowns. When $n + 1 = m$ the system can be solved by conventional methods (such as the Newton–Raphson method referenced earlier). When m exceeds $(n + 1)$ special techniques must be employed to find the optimal weights. Indeed the more m exceeds $(n + 1)$, the harder the problem. This has led to the quantity $m - (n + 1)$ being called the *degree of difficulty* of the problem.

We now solve a problem with degree of difficulty zero:

Minimize: $\quad f(X) = f(x_1, x_2, x_3)$
$$= 2x_1 x_2^{-1} + 3x_2 x_3^{-2} + 2x_1^{-2} x_2 x_3 + x_1 x_2$$

subject to: $\quad x_1, x_2, x_3 > 0.$

Now
$$C = (c_1, c_2, c_3, c_4) = (2, 3, 2, 1),$$
$$p_1(X) = x_1 x_2^{-1} x_3^0,$$
$$p_2(X) = x_1^0 x_2 x_3^{-2},$$
$$p_3(X) = x_1^{-2} x_2 x_3,$$
$$p_4(X) = x_1 x_2 x_3^0.$$

Thus
$$(a_{ij})_{3 \times 4} = \begin{pmatrix} 1 & 0 & -2 & 1 \\ -1 & 1 & 1 & 1 \\ 0 & -2 & 1 & 0 \end{pmatrix}.$$

The orthogonality and normality conditions are
$$\begin{pmatrix} 1 & 0 & -2 & 1 \\ -1 & 1 & 1 & 1 \\ 0 & -2 & 1 & 0 \\ 1 & 1 & 1 & 1 \end{pmatrix} \begin{pmatrix} w_1 \\ w_2 \\ w_3 \\ w_4 \end{pmatrix} = \begin{pmatrix} 0 \\ 0 \\ 0 \\ 1 \end{pmatrix}.$$

Thus we have a system of four linear equations in four unknowns with zero degree of difficulty. This system has a unique solution:
$$(w_1^*, w_2^*, w_3^*, w_4^*) = (\tfrac{7}{14}, \tfrac{2}{14}, \tfrac{4}{14}, \tfrac{1}{14}).$$

Thus
$$f(X^*) = \left(\frac{2}{\frac{7}{14}}\right)^{7/14} \left(\frac{3}{\frac{2}{14}}\right)^{2/14} \left(\frac{2}{\frac{4}{14}}\right)^{4/14} \left(\frac{1}{\frac{1}{14}}\right)^{1/14}$$
$$= 6.50491068$$

and
$$\left.\begin{array}{l} 2x_1^*(x_2^*)^{-1} = \tfrac{7}{14}(6.504) \\ 3x_2^*(x_3^*)^{-2} = \tfrac{2}{14}(6.504) \\ 2(x_1^*)^{-2} x_2^* x_3^* = \tfrac{4}{14}(6.504) \end{array}\right\} \Rightarrow \begin{array}{l} x_1^* = 0.869255252 \\ x_2^* = 0.534522483 \\ x_3^* = 1.213626700. \end{array}$$

Note that the final set of equations solved are nonlinear. However one can linearize these by taking logarithms. Further, it is interesting to note that the above derivation of (8.38) and (8.39) does not rely on the c_j's. Thus the x_i^*'s are independent of these values. Hence for the above problem, the solution found is optimal for any set of (positive) c_j values. Of course the optimal solution value $f(X^*)$ will change as the c_j's change. We end this section with a brief guide of how to solve problems with positive degree of difficulty by solving a numerical example.

Consider the following problem:

$$\text{Minimize:} \quad f(X) = f(x_1, x_2)$$
$$= 2x_1 x_2^{-1} + 3x_2 + 2x_1^{-2} + x_1 x_2^{-3}$$
$$\text{subject to:} \quad x_1, x_2 > 0.$$

The orthogonality and normality conditions are:

$$\begin{pmatrix} 1 & 0 & -2 & 1 \\ -1 & 1 & 0 & -3 \\ 1 & 1 & 1 & 1 \end{pmatrix} \begin{pmatrix} w_1 \\ w_2 \\ w_3 \\ w_4 \end{pmatrix} = \begin{pmatrix} 0 \\ 0 \\ 1 \end{pmatrix}.$$

As the number of unknowns is one more than the number of constraints, the degree of difficulty of the problem is one. Solving for the first three weights in terms of the fourth, we obtain

$$w_1 = \tfrac{2}{5}(1 - \tfrac{9}{2}w_4)$$
$$w_2 = \tfrac{2}{5}(1 + 3w_4)$$
$$w_3 = \tfrac{1}{5}(1 - 2w_4).$$

There is an infinite number of solutions to this system, and we now set about selecting the optimal one. Recall that

$$f(X^*) = \prod_{j=1}^{m} \left(\frac{c_j}{w_j} \right)^{w_j}.$$

Thus in this case

$$f(X^*) = \left(\frac{2}{\tfrac{2}{5}(1 - \tfrac{9}{2}w_4)} \right)^{2/5(1 + 9/2w_4)} \left(\frac{3}{\tfrac{2}{5}(1 + 3w_4)} \right)^{2/5(1 + 3w_4)}$$
$$\times \left(\frac{2}{\tfrac{1}{5}(1 - 2w_4)} \right)^{1/5(1 - 2w_4)} \left(\frac{1}{w_4} \right)^{w_4}.$$

Now $\ln f(X^*)$ will be minimal when $f(X^*)$ is minimal. Taking logarithms of both sides, we get

$$\ln f(X^*) = \tfrac{2}{5}(1 - \tfrac{9}{2}w_4)[\ln 10 - \ln(2 - 9w_4)] + \tfrac{2}{5}(1 + 3w_4)[\ln 15 - \ln(2 + 6w_4)]$$
$$+ \tfrac{2}{5}(1 - 2w_4)[\ln 10 - \ln(1 - 2w_4)] + w_4[\ln 1 - \ln w_4].$$

Employing the necessary conditions for $\ln f$ to have a minimum at X^*, we get

$$\frac{\partial \ln f(X^*)}{\partial w_4} = -\tfrac{9}{5}[\ln 10 - \ln(2 - 9w_4)] + \tfrac{9}{5} + \tfrac{6}{5}[\ln 15 - \ln(2 + 6w_4)]$$
$$- \tfrac{6}{5} - \tfrac{2}{5}[\ln 10 - \ln(1 - 2w_4)] + \tfrac{2}{5} + [\ln 1 - \ln w_4] - 1$$
$$= 0.$$

Hence

$$\tfrac{9}{5}\ln(2 - 9w_4) - \tfrac{6}{5}\ln(2 + 6w_4) + \tfrac{2}{5}\ln(1 - 2w_4) - \ln w_4 = 1.81602693,$$

so that

$$w_4^* \approx 0.07980696$$
$$w_1^* = 0.25634747$$
$$w_2^* = 0.49576835$$
$$w_3^* = 0.16807722.$$

Also

$$f(X^*) = \left(\frac{2}{w_1^*}\right)^{w_1^*}\left(\frac{3}{w_2^*}\right)^{w_2^*}\left(\frac{2}{w_3^*}\right)^{w_3^*}\left(\frac{1}{w_4^*}\right)^{w_4^*}$$

$$= (1.69322)(2.44125)(1.51625)(1.22356)$$

$$= 7.66869727.$$

Further

$$w_3^* = \frac{2(x_1^*)^{-2}}{f(X^*)}$$

$$w_2^* = \frac{3x_2^*}{f(X^*)},$$

so that

$$x_1^* = 1.24566074$$
$$x_2^* = 1.26729914.$$

8.4 Exercises

(I) Computational

1. Suppose it is wished to locate the maximum value of the following functions within the given interval I. Reduce the interval to within 10% of its original length using using Fibonacci search.

   (a) $\alpha(S) = -2S^2 + S + 4$, $I = (-5, 5)$, $\varepsilon = \tfrac{1}{9}$
   $$[I^* = (-\tfrac{5}{13}, \tfrac{58}{117})].$$

   (b) $\alpha(S) = -4S^2 + 2S + 2$, $I = (-6, 6)$, $\varepsilon = \tfrac{1}{10}$
   $$[I^* = (-0.46, 0.56)].$$

   (c) $\alpha(S) = S^3 + 6S^2 + 5S - 12$, $I = (-5, -2)$, $\varepsilon = \tfrac{1}{10}$
   $$[I^* = (-3.615, -3.384)].$$

   (d) $\alpha(S) = 2S - S^2$, $I = (0, 3)$, $\varepsilon = \tfrac{1}{100}$
   $$[I^* = (\tfrac{12}{13}, \tfrac{15}{13})].$$

   (e) $\alpha(S) = S^2 - S - 10$, $I = (-10, 10)$, $\varepsilon = \tfrac{1}{5}$
   $$[I^* = (-\tfrac{1}{2}, \tfrac{29}{20})].$$

   (f) $\alpha(S) = -(S + 6)^2 + 4$, $I = (-10, 10)$, $\varepsilon = \tfrac{1}{8}$
   $$[I^* = (-6.924, -5.386)].$$

(g) $\alpha(S) = 3S^2 + 2S + 1$, $I = (-5, 5)$, $\varepsilon = \frac{1}{12}$
$$[I^* = (-0.383, 0.467)].$$

(h) $\alpha(S) = -S^2 + 4S + 4$, $I = (-5, 5)$, $\varepsilon = \frac{1}{10}$
$$[I^* = (\tfrac{237}{130}, \tfrac{35}{13})].$$

(i) $\alpha(S) = S^3 - 2S^2 + S - 4$, $I = (-\frac{5}{2}, \frac{5}{2})$, $\varepsilon = \frac{1}{12}$
$$[I^* = (\tfrac{1}{4}, \tfrac{1}{2})].$$

(j) $\alpha(S) = -S^2 + 4S - 3$, $I = (0, 5)$, $\varepsilon = \frac{1}{4}$
$$[I^* = (\tfrac{15}{8}, \tfrac{17}{8})].$$

(k) $\alpha(S) = S - S^2 + 4$, $I = (0, 2)$, $\varepsilon = \frac{1}{20}$
$$[I^* = (\tfrac{8}{17}, \tfrac{11}{20})].$$

(l) $\alpha(S) = S^2 + 2S - 3$, $I = (-4, 6)$, $\varepsilon = \frac{1}{4}$
$$[I^* = (\tfrac{76}{13}, 6)].$$

(m) $\alpha(S) = -2S^2 + 3S + 6$, $I = (-3, 5)$, $\varepsilon = \frac{1}{8}$
$$[I^* = (0, \tfrac{9}{8})].$$

2. Repeat Exercise 1 using golden section search.

3. Repeat Exercise 1 using Bolzano's method.

4. Repeat Exercise 1 using even block search.

5. Attempt to maximize the following unconstrained functions using pattern search, given initial starting point X_0, $e_1 = e_2 = 0.1$, and resolution $\varepsilon_1 = \varepsilon_2 = 0.001$.
   (a) $f(X) = -x_1^2 + x_1 x_2 - x_2^2$, $X_0 = (2, 3)$
   $$[X^* = (0, 0)].$$

   (b) $f(X) = 4x_1 - x_1^2 - + 2x_1 x_1 - 2x_2^2$, $X_0 = (1\,'\,1)$
   $$[X^* = (2, 4)].$$

   (c) $f(X) = 3x_1 - x_1^2 - 2x_2^2 + 4x_2 + 15$, $X_0 = (0, 0)$
   $$[X^* = (\tfrac{3}{2}, 1)].$$

   (d) $f(X) = -(2x_2 - x_1)^2 - 4(x_1 + 3)^2$, $X_0 = (0, 0)$
   $$[X^* = (-3, -\tfrac{3}{2})].$$

   (e) $f(X) = -x_1^2 - x_1 x_2 - x_2^2$, $X_0 = (5, 5)$
   $$[X^* = (0, 0)].$$

   (f) $f(X) = -x_1^2 - x_2^2$, $X_0 = (3, -4)$
   $$[X^* = (0, 0)].$$

   (g) $f(X) = -(x_1 - 1)^2 - (x_2 + 2)^2$, $X_0 = (0, 0)$
   $$[X^* = (1, -2)].$$

   (h) $f(X) = -(x_1 - 2)^2 - (x_2 - 3)^2$, $X_0 = (0, 0)$
   $$[X^* = (2, 3)].$$

6. Repeat Exercise 5 using one-at-a-time search.

7. Repeat Exercise 5 using Rosenbrock's method.

8. Repeat Exercise 5 using Powell's method.

9. Maximize the following unconstrained functions using the gradient method of Section 8.2.3, with $X_0 = (0, 0, 0)$.
   (a) $f(X) = -3(x_1 - 2)^2 - 4(x_2 - 3)^2 - 2(x_3 + 5)^2$
   $$[X^* = (2, 3, -5)].$$

(b) $f(X) = -2x_1(x_1 - 4) - x_2(x_2 - 2)$
$$[X^* = (2, 1)].$$

(c) $f(X) = x_1x_2 - 2x_3^2 - x_1x_3$
[unbounded].

(d) $f(X) = x_1x_2 - 2x_1^2 - 2x_1x_3$
$$[X^* = (0, \tfrac{6}{5}, \tfrac{3}{5})].$$

(e) $f(X) = (x_1 - 3)^2 - 4(x_2 - 2)^2 - x_3^2$
$$[X^* = (3, 2, 0)].$$

(f) $f(X) = -x_1^2 - (x_2 - 2)^2 - 2(x_3 - 3)^2$
$$[X^* = (0, 2, 3)].$$

(g) $f(X) = x_1x_2 - x_1^2 - x_2^2 - x_3^2$
$$[X^* = (0, 0, 0)].$$

(h) $f(X) = 5x_1^2 + x_2^2 + x_1^2 - 4x_1x_2 - 2x_1 - 6x_3$
$$[X^* = (1, 2, 3)].$$

10. Solve the above problems by using the gradient partan method of Section 8.2.3.1

11. Solve the above problems by using the conjugate gradient method of Section 8.2.3.3.

12. Attempt to maximize the following functions using the method of Newton and Raphson of Section 8.2.2.1.

(a) $f(X) = -(x_1 - 3)^3 - (x_2 - 4)^2 + 1$
$$[X^* = (3, 4)].$$

(b) $f(X) = -(x_1 + 2)^2 - (x_2 - 1)^2$
$$[X^* = (-2, 1)].$$

(c) $f(X) = (x_1 + 1)^2 - x_1x_2 - 2x_3^2$
$$[X^* = (0, 2, 0)].$$

(d) $f(X) = -(x_1 - 3)^2 - 4(x_2 - 6)^3$.

(e) $f(X) = -5(x_1x_2 - 3)^2 - 4x_1x_2 - 2(x_1x_2 - 1)^3$.

(f) $f(X) = (x_1 - 2)^2 + (x_2 - 5)^2 - 4x_1^2x_2$.

(g) $f(X) = x_1^3 - 3x_1x_2^2$.

(h) $f(X) = 4x_1 - (2x_2 - 2x_1)^2 + 4x_1x_2 + 3x_2$.

(i) $f(X) = 5x_1^2 + x_2^2 + x_3^2 - 4x_1x_2 - 2x_1 - 6x_3$
$$[X^* = (1, 2, 3)].$$

13. Repeat Exercise 8.1.2 using the variable metric method of Section 8.2.2.2.

14. Minimize the following functions using geometric programming.

(a) $f(X) = 3x_1^3x_2x_3 + x_1^2x_2^2x_3 + 4x_1^2x_2x_3 + 6x_1^{-5}x_2^{-5/2}x_3^{-2}$.

(b) $f(X) = 3x_1x_2^{-3}x_3^2 + 2x_1^{-2}x_2 + x_2^3x_3^{-1}$
$$[X^* = (1.094, 1.077, 1.041)].$$

(c) $f(X) = 4x_1^4 + 4x_1^{-2}x_2^2 + 5x_2^{-4}x_3^2 + x_3^{-3}$
$$[X^* = (0.9073, 1.0561, 0.8211)].$$

(d) $f(X) = x_1^3x_2x_3^{-5} + 4x_1^2x_2^{1/2}x_3^{-3} + 4x_1x_2^{-2}x_3^3$
$$[X^* = (2(1/4)^{1/4}, (1/4)^{1/14}, (1/4)^{-1/28})].$$

(e) $f(X) = 3x_1^2 x_2 + 5x_2 x_3^{-2} + x_1^{-1} x_2^{-1} + 2x_1^{-1} x_3^2$
$$[X^* = (1.82, 0.234, 1/\sqrt{2})].$$

(f) $f(X) = 8x_1 x_2 + 2x_1 x_3 + 4x_1^{-2} x_2^{-1} x_3^{-2} + x_1^{-1}$
$$[X^* = (1.5004, 0.4992, 3.9936)].$$

(g) $f(X) = x_1^{1/2} x_2 + 2x_1^2 x_2^3 x_3^{-2} + x_1^{-1} x_3^{-1} + 4x_2^{-1} x_2$
$$[X^* = (0.2878, 2.3438, 1.842)].$$

(h) $f(X) = 3x_2^3 x_3^{-1} + 6x_1^2 x_2^{-2} + 2x_1^{-1} + x_1^{-1} x_2^{-2} x_3^2$
$$[X^* = (1.03, 0.872, -1.05)].$$

(II) Theoretical

15. Show that the directions D_i, $i = 1, 2, \ldots$ generated in Section 8.2.3.3 are mutually conjugate.

16. Compare the performance of the methods used in Exercises 1–4 with the method outlined in Section 8.2.1 by using it to solve the problems in Exercise 1.

17. Prove that the global maximum of a negative definite quadratic function of n variables can be found after $2n - 1$ steps by the gradient partan method.

18. Show that if a function of two variables has contours which are negative definite quadratics that these contours are concentric ellipses.

19. If X^* is a global maximum for the function described in Exercise 18 and X_1, $X_2 \in \mathbf{R}^2$, prove tangents T_1 and T_2 to the contours of f at X_1 and X_2 are parallel if and only if X_1 and X_2 are collinear with X^*.

20. Prove that X_1 in Exercise 19 is the maximum point for f along T_1.

21. Justify the formulae for F_i and G_i in Section 8.2.2.2.

22. Apply the Kuhn–Tucker conditions to the quadratic programming problem of Section 8.3.5.

Chapter 9

Appendix

This appendix comprises two parts: an introduction to both linear algebra and basic calculus. We begin with linear algebra.

9.1 Linear Algebra

9.1.1 Matrices

Definition. A *matrix* is a rectangular array of elements (often real numbers).

If A is a matrix, we write

$$A = \begin{pmatrix} a_{11} & a_{12} & \cdots & a_{1n} \\ a_{21} & a_{22} & \cdots & a_{2n} \\ \vdots & \vdots & & \vdots \\ a_{m1} & a_{m2} & \cdots & a_{mn} \end{pmatrix}.$$

Here A is said to have m *rows* and n *columns* and the element in the ith row and jth column, $1 \le i \le m$, $1 \le j \le n$ is called the i, j element, or a_{ij}. A is also denoted by $(a_{ij})_{m \times n}$.

Definition. A matrix is termed *square* if $m = n$.

Definition. The *identity* matrix is a square matrix in which

$$a_{ij} = \begin{cases} 0, & \text{if } i \ne j \\ 1, & \text{otherwise.} \end{cases}$$

The identity matrix with n columns (and rows) is denoted by I_n, or simply I if no confusion arises; it is of the form:

$$
I_n = \overbrace{\begin{pmatrix} 1 & 0 & 0 & \cdots & 0 & 0 \\ 0 & 1 & 0 & \cdots & 0 & 0 \\ 0 & 0 & 1 & \cdots & 0 & 0 \\ \vdots & \vdots & \vdots & & \vdots & \vdots \\ 0 & 0 & 0 & \cdots & 1 & 0 \\ 0 & 0 & 0 & \cdots & 0 & 1 \end{pmatrix}}^{n \text{ columns}} \left.\vphantom{\begin{pmatrix} 1 \\ 0 \\ 0 \\ \vdots \\ 0 \\ 0 \end{pmatrix}}\right\} n \text{ rows.}
$$

Definition. A matrix is termed a *zero matrix* if all its elements are zero, i.e.

$$a_{ij} = 0, \qquad 1 \le i \le m, 1 \le j \le n.$$

The zero matrix with m rows and n columns is denoted by $\mathbf{0}_{m \times n}$, or simply $\mathbf{0}$ if no confusion arises, and is of the form:

$$
\mathbf{0}_{m \times n} = \overbrace{\begin{pmatrix} 0 & 0 & \cdots & 0 \\ 0 & 0 & \cdots & 0 \\ \vdots & \vdots & & \vdots \\ 0 & 0 & \cdots & 0 \end{pmatrix}}^{n \text{ columns}} \left.\vphantom{\begin{pmatrix} 0 \\ 0 \\ \vdots \\ 0 \end{pmatrix}}\right\} m \text{ rows.}
$$

Definition. The *transpose* of a matrix $A = (a_{ij})_{m \times n}$ is a matrix with n rows and m columns with its i, j element $\bar{a}_{ij}$ defined by

$$\bar{a}_{ij} = a_{ji}, \qquad 1 \le i \le n, 1 \le j \le m.$$

The transpose of A is denoted by A^T and is obtained from A by making the ith row in A the jth column in A^T, $1 \le i \le m$. Note that for any matrix A,

$$(A^T)^T = A.$$

Definition. Two matrices $A = (a_{ij})_{m \times n}$ and $B = (b_{ij})_{m \times n}$ are termed *equal* if, and only if,

$$a_{ij} = b_{ij}, \qquad 1 \le i \le m, 1 \le j \le n.$$

Note that equality is not defined if A and B do not have the same number of rows and of columns.

9.1.2 Vectors

Definition. A *vector* is a matrix which has either:
a. exactly one row ($m = 1$), or
b. exactly one column ($n = 1$).

It is usual to drop the first (second) subscript in case a (b). In case a the vector is often called a *row vector* and is denoted by:

$$(a_{11}, a_{12}, a_{13}, \ldots, a_{1n}) = (a_1, a_2, a_3, \ldots, a_n).$$

In case b the vector is often called a *column vector* and is denoted by

$$\begin{pmatrix} a_{11} \\ a_{21} \\ a_{31} \\ \vdots \\ a_{m1} \end{pmatrix} = \begin{pmatrix} a_1 \\ a_2 \\ \vdots \\ a_m \end{pmatrix}.$$

A vector (either row or column) with n entries is called an *n-vector*.

Definition. A finite set $\{X_1, X_2, \ldots, X_q\}$, of n-vectors is said to be *linearly independent* if and only if

$$\sum_{i=1}^{q} \alpha_i X_i = \mathbf{0},$$

for α_i, $i = 1, 2, \ldots, q$ real numbers, implies that

$$\alpha_i = 0, \qquad i = 1, 2, \ldots, m.$$

Definition. A finite set of n-vectors is said to be *linearly dependent* if it is not linearly independent.

For example, the set of 3-vectors $\{X_1, X_2, X_3\}$ given by

$$X_1 = (1, 0, 0), \qquad X_2 = (0, 1, 0), \qquad X_3 = (0, 0, 1)$$

is linearly independent. However, if $X_3 = (1, 1, 0)$ the set is linearly dependent.

9.1.3 Arithmetical Operations on Vectors and Matrices

Unless otherwise mentioned, the following operations are defined for vectors, which can be thought of as simply a special type of matrix.

(i) Scalar Multiplication

The *scalar product* of a matrix $A = (a_{ij})_{m \times n}$ and a real number α is defined as a new matrix, denoted by αA. The i, j element of αA is defined by

$$\alpha a_{ij}, \qquad 1 \leq i \leq m, 1 \leq j \leq n.$$

(ii) Addition

Two matrices $A = (a_{ij})_{m \times n}$ and $B = (b_{ij})_{m \times n}$ can be added together to form a new matrix, called the *sum* of A and B, denoted by $A + B$. The i, j element of $A + B$ is defined by

$$a_{ij} + b_{ij}, \qquad 1 \leq i \leq m, 1 \leq j \leq n.$$

Note that the sum of two matrices is not defined if they do not have both the same number of rows and of columns.

(iii) Subtraction

Two matrices $A = (a_{ij})_{m \times n}$ and $B = (b_{ij})_{m \times n}$ can be subtracted to form a new matrix, denoted by $A - B$. $A - B$ is formed by forming the sum of A and the scalar product of -1 and B. Thus the i, j element of $A - B$ is defined by

$$a_{ij} - b_{ij}, \qquad 1 \le i \le m, 1 \le j \le n.$$

Note that $A - B$ is not defined if A and B do not have both the same number of rows and of columns.

(iv) Multiplication

Two matrices $A = (a_{ij})_{m \times q}$ and $B = (b_{ij})_{q \times n}$ can be multiplied to form a new matrix, $C = (c_{ij})_{m \times n}$, called the *product* of A and B. The i, j element of C is defined by

$$c_{ij} = \sum_{k=1}^{q} a_{ik} b_{kj}, \qquad 1 \le i \le m, 1 \le j \le n.$$

C is denoted by AB. Note that the product AB is not defined unless A has the same number of columns as B has rows. Thus, although AB may be defined, for a given pair of matrices A and B, BA may not necessarily be defined, and even if it is, it is not necessarily so that

$$AB = BA.$$

Examples of this multiplication are given below:

$$\begin{pmatrix} 1 & 2 \\ 3 & 4 \end{pmatrix}\begin{pmatrix} 5 & 6 \\ 7 & 8 \end{pmatrix} = \begin{pmatrix} 1 \times 5 + 2 \times 7 & 1 \times 6 + 2 \times 8 \\ 3 \times 5 + 4 \times 7 & 3 \times 6 + 4 \times 8 \end{pmatrix} = \begin{pmatrix} 19 & 22 \\ 43 & 48 \end{pmatrix}$$

$$\begin{pmatrix} 1 & 2 & 3 \\ 4 & 5 & 6 \end{pmatrix}\begin{pmatrix} 7 & 10 \\ 8 & 11 \\ 9 & 12 \end{pmatrix} = \begin{pmatrix} 1 \times 7 + 2 \times 8 + 3 \times 9 & 1 \times 10 + 2 \times 11 + 3 \times 12 \\ 4 \times 7 + 5 \times 8 + 6 \times 9 & 4 \times 10 + 5 \times 11 + 6 \times 12 \end{pmatrix}$$

$$= \begin{pmatrix} 50 & 68 \\ 122 & 167 \end{pmatrix}$$

$$(1, 2)\begin{pmatrix} 3 & 4 \\ 5 & 6 \end{pmatrix} = (1 \times 3 + 2 \times 5 \quad 1 \times 4 + 2 \times 6) = (13, 16)$$

$$\begin{pmatrix} 3 & 4 \\ 5 & 6 \end{pmatrix}\begin{pmatrix} 1 \\ 2 \end{pmatrix} = \begin{pmatrix} 3 \times 1 + 4 \times 2 \\ 5 \times 1 + 6 \times 2 \end{pmatrix} = \begin{pmatrix} 11 \\ 17 \end{pmatrix}$$

$$(1, 2)\begin{pmatrix} 3 \\ 4 \end{pmatrix} = (1 \times 3 + 2 \times 4) = (11).$$

Note in the last example that the product of two vectors yields a scalar.

It is not difficult to prove some simple properties of the above operations. For all matrices A, B, C, D, and E with m rows and n columns, and all real numbers α:

$$A + 0 = 0 + A = A$$
$$A + B = B + A$$
$$A + (B + C) = (A + B) + C$$
$$A - (B - C) = (A - B) - C$$
$$(A + B)^T = A^T + B^T$$
$$(A - B)^T = A^T - B^T$$
$$\alpha(A + B) = \alpha A + \alpha B,$$

and if all the necessary multiplication is compatible:

$$IA = AI = A$$
$$A(DE) = (AD)E$$
$$A(D + E) = AD + AE$$
$$(D + E)A = DA + EA$$
$$\alpha(AB) = (\alpha A)B = A(\alpha B).$$

(v) *Vector Distance*

Definition. The *distance* between two n-dimensional vectors, $X = (x_1, x_2, \ldots, x_m)^T$ and $Y = (y_1, y_2, \ldots, y_n)^T$ is $\|X - Y\|$, defined as follows:

$$\|X - Y\| = \sqrt{\sum(x_i - y_i)^2}.$$

9.1.4 Determinants

Any square matrix A whose entries are real numbers has associated with it a unique real number called its *determinant*, denoted by $|A|$ or det A. Rather than define $|A|$ explicitly, we will outline a method for calculating $|A|$ for any A. First we introduce some basic concepts. Associated with each element a_{ij} of A is a number which is the determinant of the matrix arrived at by deleting the ith row and jth column of A. This determinant is called the i, j *minor* and is denoted by M_{ij}:

$$M_{ij} = \begin{vmatrix}
a_{11} & a_{12} & \cdots & a_{1j-1} & a_{1j+1} & \cdots & a_{1n} \\
a_{21} & a_{22} & \cdots & a_{2j-1} & a_{2j+1} & \cdots & a_{2n} \\
\vdots & \vdots & & \vdots & \vdots & & \vdots \\
a_{i-11} & a_{i-12} & \cdots & a_{i-1j-1} & a_{i-1j+1} & \cdots & a_{i-1n} \\
a_{i+11} & a_{i+12} & \cdots & a_{i+1j-1} & a_{i+1j+1} & \cdots & a_{i+1n} \\
\vdots & \vdots & & \vdots & \vdots & & \vdots \\
a_{m1} & a_{m2} & \cdots & a_{mj-1} & a_{mj+1} & \cdots & a_{mn}
\end{vmatrix}.$$

For example, if

$$A = \begin{pmatrix} 6 & 9 & 7 & 5 \\ 8 & 3 & 9 & 5 \\ 4 & 1 & 11 & 13 \\ 2 & 6 & 12 & 20 \end{pmatrix}$$

$$M_{23} = \begin{pmatrix} 6 & 9 & 5 \\ 4 & 1 & 13 \\ 2 & 6 & 20 \end{pmatrix},$$

where the inside parentheses are omitted by common convention. Associated with each minor M_{ij} is a *cofactor* C_{ij} defined by

$$C_{ij} = (-1)^{i+j} M_{ij}. \tag{9.1}$$

That is, each cofactor is either $+1$ or -1 times the determinant of the associated minor.

We can now begin to calculate $|A|$. Let $A = (a_{ij})_{n \times n}$ and r be such that $1 \le r \le n$. Then

$$|A| = \sum_{j=1}^{n} a_{rj} C_{rj}. \tag{9.2}$$

Thus to calculate $|A|$ we choose any row, say row r of A:

$$a_{r1} \quad a_{r2} \quad a_{r3} \quad \cdots \quad a_{rn}.$$

We multiply each of these elements a_{rj} by the cofactor C_{rj} and sum up all the products. Thus (9.2) reduces the problem of finding a determinant of an $n \times n$ matrix to n problems of calculating the determinant of an $(n-1) \times (n-1)$ matrix. Substituting (9.1) into (9.2) produces

$$|A| = \sum_{r=1}^{n} a_{rj} (-1)^{r+j} M_{rj}. \tag{9.3}$$

We can now use (9.2) to reduce the problem of finding M_{rj} from that of finding determinants of $(n-1) \times (n-1)$ matrices to that of finding $(n-2) \times (n-2)$ determinants. Eventually the problem is reduced to finding the determinants of 2×2 matrices. We use the following definition in this case:

$$|A| = |(a_{ij})_{2 \times 2}| = \begin{vmatrix} a_{11} a_{12} \\ a_{21} a_{22} \end{vmatrix} = a_{11} a_{22} - a_{12} a_{21}. \tag{9.4}$$

We now illustrate this approach by finding the determinant of the matrix A given earlier in this section. Let $r = 1$ in (9.2) throughout the rest of this example. Then, by (9.3), we have

$$\begin{aligned} |A| &= a_{11}(-1)^{1+1} M_{11} + a_{12}(-1)^{1+2} M_{12} \\ &\quad + a_{13}(-1)^{1+3} M_{13} + a_{14}(-1)^{1+4} M_{14} \\ &= 6M_{11} - 9M_{12} + 7M_{13} - 5M_{14}. \end{aligned} \tag{9.5}$$

Now

$$M_{11} = \begin{vmatrix} 3 & 9 & 5 \\ 1 & 11 & 13 \\ 6 & 12 & 20 \end{vmatrix},$$

which, by (9.3), becomes

$$M_{11} = 3(-1)^{1+1}\begin{vmatrix} 11 & 13 \\ 12 & 20 \end{vmatrix} + 9(-1)^{1+2}\begin{vmatrix} 1 & 13 \\ 6 & 20 \end{vmatrix} + 5(-1)^{1+3}\begin{vmatrix} 1 & 11 \\ 6 & 12 \end{vmatrix},$$

$$= 3(11 \times 20 - 13 \times 12) - 9(1 \times 20 - 6 \times 13) + 5(1 \times 12 - 6 \times 11), \quad \text{by (9.4)}$$

$$= 474.$$

Similarly,

$$M_{12} = 8(-1)^{1+1}\begin{vmatrix} 11 & 13 \\ 12 & 20 \end{vmatrix} + 9(-1)^{1+2}\begin{vmatrix} 4 & 13 \\ 2 & 20 \end{vmatrix} + 5(-1)^{1+3}\begin{vmatrix} 4 & 11 \\ 2 & 12 \end{vmatrix}$$

$$= 8(220 - 156) - 9(80 - 26) + 5(48 - 22)$$

$$= 156$$

$$M_{13} = 8(-1)^{1+1}\begin{vmatrix} 1 & 13 \\ 6 & 20 \end{vmatrix} + 3(-1)^{1+2}\begin{vmatrix} 4 & 13 \\ 2 & 20 \end{vmatrix} + 5(-1)^{1+3}\begin{vmatrix} 4 & 1 \\ 2 & 6 \end{vmatrix}$$

$$= 8(20 - 78) - 3(80 - 26) + 5(24 - 2)$$

$$= -512$$

$$M_{14} = 8(-a)^{1+1}\begin{vmatrix} 1 & 11 \\ 6 & 12 \end{vmatrix} + 3(-1)^{1+2}\begin{vmatrix} 4 & 11 \\ 2 & 12 \end{vmatrix} + 9(-1)^{1+3}\begin{vmatrix} 4 & 1 \\ 2 & 6 \end{vmatrix}$$

$$= 8(12 - 66) - 3(48 - 22) + 9(24 - 2),$$

$$= -312.$$

Substituting all this information into (9.5) yields

$$|A| = 6 \times 474 - 9 \times 156 + 7 \times (-512) - 5 \times (-312)$$

$$= -584.$$

The reader may like to verify that the following properties hold for some numerical examples and then prove them true in general.

1. If A and B are two matrices which have the property that one can be obtained from the other by the interchange of two rows (or columns) then

$$|A| = -|B|.$$

2. $|A| = |A^T|$ for all matrices A.
3. If a matrix A has a row (or column) of zeros, then

$$|A| = 0.$$

4. If B is a matrix which is obtained by adding a scalar multiple of a row (or column) to another row (or column) of another matrix A, then

$$|B| = |A|.$$

From this and property 3 it follows that if two rows (or columns) of a matrix A are identical then

$$|A| = 0.$$

5. (9.2) can be used to show that if a matrix B is obtained by multiplying by a scalar α all the elements of a row (or column) of another matrix A, then

$$|B| = \alpha |A|.$$

6. If $A = (a_{ij})_{n \times n}$ and $B = (b_{ij})_{n \times n}$, then the determinant of their product equals the product of their determinants, i.e.,

$$|AB| = |A| |B|.$$

Definition. The *cofactor matrix* $\bar{A} = (\bar{a}_{ij})_{n \times n}$ of a matrix, $A = (a_{ij})_{n \times n}$ is a matrix defined by

$$\bar{a}_{ij} = C_{ij}, \qquad 1 \le i \le n, 1 \le j \le n$$

where C_{ij} is the cofactor defined in (9.1).

Definition. The *adjoint matrix*, A_{adj} of a matrix, $A = (a_{ij})_{n \times n}$ is a matrix defined by $A_{adj} = \bar{A}^T$, the transpose of the cofactor matrix. So

$$A_{adj} = \begin{pmatrix} C_{11} & C_{21} & \cdots & C_{n1} \\ C_{12} & C_{22} & \cdots & C_{n2} \\ \vdots & \vdots & & \vdots \\ C_{1n} & C_{2n} & \cdots & C_{nn} \end{pmatrix}.$$

9.1.5 The Matrix Inverse

Definition. The *inverse* of a square matrix, $A = (a_{ij})_{n \times n}$ is a matrix $B = (b_{ij})_{n \times n}$ with the property that

$$AB = I_n.$$

It is usual to denote B by A^{-1}.

Definition. A matrix A is termed *nonsingular* if

$$|A| \ne 0.$$

The reader is encouraged to attempt to prove the following properties of the inverse.

1. If A is a nonsingular square matrix, A^{-1} is unique.
2. If A and B are both nonsingular square matrices with the same number of rows then

$$(AB)^{-1} = B^{-1}A^{-1}.$$

3. $A^{-1}A = I$ for all square, nonsingular matrices.
4. If A is nonsingular and the given multiplications are defined, then $AB = AC$ implies $B = C$. The inverse A^{-1} of a square, nonsingular matrix A may be computed using the following formula:

$$A^{-1} = \frac{1}{|A|}(A_{\text{adj}}). \qquad (9.6)$$

As an example, let us take the inverse of the matrix, A where

$$A = \begin{pmatrix} 1 & 2 & 3 \\ 4 & 3 & 6 \\ 1 & 1 & 1 \end{pmatrix}$$

we have

$$\bar{A} = \begin{pmatrix} -3 & 2 & 1 \\ 1 & -2 & 1 \\ 3 & -6 & -5 \end{pmatrix}$$

and

$$A_{\text{adj}} = \begin{pmatrix} -3 & 1 & 3 \\ 2 & -2 & -6 \\ 1 & 1 & -5 \end{pmatrix}.$$

Further,

$$|A| = 4.$$

Hence

$$A^{-1} = \frac{1}{|A|}(A_{\text{adj}}) = \tfrac{1}{4}\begin{pmatrix} -3 & 1 & 3 \\ 2 & -2 & -6 \\ 1 & 1 & 5 \end{pmatrix} = \begin{pmatrix} -\frac{3}{4} & \frac{1}{4} & \frac{3}{4} \\ \frac{1}{2} & -\frac{1}{2} & -\frac{3}{2} \\ \frac{1}{4} & \frac{1}{4} & \frac{5}{4} \end{pmatrix}.$$

This can be verified as follows:

$$AA^{-1} = \begin{pmatrix} 1 & 2 & 3 \\ 4 & 3 & 6 \\ 1 & 1 & 1 \end{pmatrix}\begin{pmatrix} -\frac{3}{4} & \frac{1}{4} & \frac{3}{4} \\ \frac{1}{2} & -\frac{1}{2} & -\frac{3}{2} \\ \frac{1}{4} & \frac{1}{4} & \frac{5}{4} \end{pmatrix} = \begin{pmatrix} 1 & 0 & 0 \\ 0 & 1 & 0 \\ 0 & 0 & 1 \end{pmatrix} = I.$$

A second way in which the inverse of a matrix can be calculated is by *Gauss–Jordan elimination*. Suppose it is desired to calculate the inverse of the square, nonsingular matrix A. We first form an appended matrix:

$$B = [A \mid I],$$

where I is the identity matrix with as many rows as A. If the left-hand part of B is now transformed into I by adding scalar multiples of the rows of B to other rows, the right-hand part is transformed into A^{-1}. As an example, we once again calculate the inverse of the matrix just displayed:

$$B = \begin{pmatrix} 1 & 2 & 3 & \vdots & 1 & 0 & 0 \\ 4 & 3 & 6 & \vdots & 0 & 1 & 0 \\ 1 & 1 & 1 & \vdots & 0 & 0 & 1 \end{pmatrix} \begin{matrix} R_1 \\ R_2 \\ R_3 \end{matrix}$$

becomes

$$\begin{pmatrix} 1 & 2 & 3 & \vdots & 1 & 0 & 0 \\ 0 & -5 & -6 & \vdots & -4 & 1 & 0 \\ 0 & -1 & -2 & \vdots & -1 & 0 & 1 \end{pmatrix} \begin{matrix} R_1 \\ R_2 - 4R_1 \\ R_3 - R_1 \end{matrix}$$

$$\begin{pmatrix} 1 & 2 & 3 & \vdots & 1 & 0 & 0 \\ 0 & 1 & \frac{6}{5} & \vdots & \frac{4}{5} & -\frac{1}{5} & 0 \\ 0 & -1 & -2 & \vdots & -1 & 0 & 1 \end{pmatrix} \begin{matrix} R_1 \\ R_2/(-5) \\ R_3 \end{matrix}$$

$$\begin{pmatrix} 1 & 0 & \frac{3}{5} & \vdots & -\frac{3}{5} & \frac{2}{5} & 0 \\ 0 & 1 & \frac{6}{5} & \vdots & \frac{4}{5} & -\frac{1}{5} & 0 \\ 0 & 0 & -\frac{4}{5} & \vdots & -\frac{1}{5} & -\frac{1}{5} & 1 \end{pmatrix} \begin{matrix} R_1 - 2R_2 \\ R_2 \\ R_3 + R_2 \end{matrix}$$

$$\begin{pmatrix} 1 & 0 & \frac{3}{5} & \vdots & -\frac{3}{5} & \frac{2}{5} & 0 \\ 0 & 1 & \frac{6}{5} & \vdots & \frac{4}{5} & -\frac{1}{5} & 0 \\ 0 & 0 & 1 & \vdots & \frac{1}{4} & \frac{1}{4} & \frac{5}{4} \end{pmatrix} \begin{matrix} R_1 \\ R_2 \\ R_3/(-\frac{4}{5}) \end{matrix}$$

$$\begin{pmatrix} 1 & 0 & 0 & \vdots & -\frac{3}{4} & \frac{1}{4} & -\frac{3}{4} \\ 0 & 1 & 0 & \vdots & \frac{1}{2} & -\frac{1}{2} & -\frac{3}{2} \\ 0 & 0 & 1 & \vdots & \frac{1}{4} & \frac{1}{4} & \frac{5}{4} \end{pmatrix} \begin{matrix} R_1 - \frac{3}{5}R_3 \\ R_2 - \frac{6}{5}R_3 \\ R_3. \end{matrix}$$

Thus

$$A^{-1} = \text{right-hand part of } B$$

$$= \begin{pmatrix} -\frac{3}{4} & \frac{1}{4} & -\frac{3}{4} \\ \frac{1}{2} & -\frac{1}{2} & -\frac{3}{2} \\ \frac{1}{4} & \frac{1}{4} & \frac{5}{4} \end{pmatrix},$$

which is the same result as that obtained by the previous method.

9.1.6 Quadratic Forms, Definiteness, and the Hessian

A quadratic form is an expression of the type:

$$f(X) = X^T H X,$$

where X is an n-dimensional vector and H is an $n \times n$ matrix. A quadratic form $f(X)$ and its associated matrix H is said to be:

negative definite if

$$f(X) < 0, \quad \text{for all } X \neq 0$$

negative semidefinite if

$$f(X) \leq 0, \quad \text{for all } X \neq 0$$
$$f(X) = 0, \quad \text{for some } X \neq 0$$

positive definite if

$$f(X) > 0, \quad \text{for all } X \neq 0$$

positive semidefinite if

$$f(X) \geq 0, \quad \text{for all } X \neq 0$$
$$f(X) = 0, \quad \text{for some } X \neq 0$$

indefinite if

$$f(X) > 0, \quad \text{for some } X$$
$$f(X) < 0, \quad \text{for some } X.$$

The following rules can be invoked in order to determine the definiteness of any matrix H.

1. H is negative (positive) definite if and only if all the eigenvalues of H are negative (positive). (See Section 9.1.7 for a discussion of eigenvalues.)
2. H is negative (positive) semidefinite if and only if all the eigenvalues of H are nonpositive (negative) and at least one is zero.
3. H is indefinite if and only if H has some positive and some negative eigenvalues.

Thus the definiteness of $H(X)$ can be discovered by examining the eigenvalues of $H(X)$. This is often no trivial task. However the rules 1–3 require that only the signs of the eigenvalues be known, not the values themselves. These signs can be found by using Descarte's rule of signs.

Let us now examine the behavior of f at a critical point X^*. That is,

$$\nabla f(X^*) = 0.$$

If:

(a) $H(X^*)$ is negative definite, X^* is a local maximum for f.
(b) $H(X^*)$ is positive definite, X^* is a local minimum for f.
(c) $H(X^*)$ is indefinite, X^* is a saddle point for f.
(d) $H(X^*)$ is either positive or negative semidefinite, nothing can be said about X^*.

9.1.7 Eigenvalues and Eigenvectors

Given any $n \times n$ matrix H, the scalars $\lambda_1, \lambda_2, \ldots, \lambda_n$ which are the zeros of the *characteristic equation*

$$\det(H - \lambda I) = 0$$

are called the *eigenvalues* of H. Here I is the $n \times n$ identity matrix.

Corresponding to the n eigenvalues there are n *eigenvectors* $X_1, X_2, \ldots, X_n$, which satisfy

$$HX_i - \lambda_i X_i = 0, \qquad X_i \neq 0, i = 1, 2, \ldots, n.$$

It can be shown that all the eigenvectors of a real symmetric matrix are real. (Note that the Hessian matrix of a multivariable function with continuous second partial derivatives is symmetric.) Further, the eigenvectors which correspond to distinct eigenvalues are orthogonal, i.e.,

$$\lambda_i \neq \lambda_j \Rightarrow X_i^T X_j = 0.$$

The relationship between the Hessian, the eigenvectors and eigenvalues can be expressed as

$$H = E^T \wedge E, \tag{9.7}$$

where

$$E = \begin{pmatrix} X_1^T \\ X_2^T \\ \vdots \\ X_n^T \end{pmatrix}$$

and

$$\lambda = \begin{pmatrix} \lambda_1 & & & 0 \\ & \lambda_2 & & \\ & & \ddots & \\ 0 & & & \lambda_n \end{pmatrix}$$

if the eigenvectors $X_1, X_2, \ldots, X_n$ are replaced by unit vectors in the same direction. Now it can be shown that

$$X_i^T H X_i = X_i^T E^T \wedge E X_i, \quad i = 1, 2, \ldots, n, \text{ by (9.7),}$$
$$= \lambda_i, \qquad\qquad i = 1, 2, \ldots, n.$$

Hence it can be shown that a necessary and sufficient condition for H to be negative semidefinite is

$$\lambda_i \leq 0, \qquad i = 1, 2, \ldots, n.$$

9.1.8 Gram–Schmidt Orthogonalization

Given n linearly independent vectors $d_1, d_2, \ldots, d_n$, the Gram–Schmidt orthogonalization process can be used to construct from them n *orthonormal* vectors $e_1, e_2, \ldots, e_n$. A set of vectors $\{e_1, e_2, \ldots, e_n\}$ is termed orthonormal if

$$e_i e_j = \begin{cases} 1, & \text{if } i = j \\ 0, & \text{otherwise.} \end{cases} \tag{9.8}$$

The process begins by setting,

$$e_1 = \frac{d_1}{|d_1|}$$

so that (9.8) is satisfied for $i = j = 1$. Next choose

$$e_2 = \gamma_1 d_1 + \gamma_2 d_2$$
$$= \delta_1 e_1 + \delta_2 d_2.$$

Now

$$e_1 e_2 = 0.$$

Hence

$$e_1(\delta_1 e_1 + \delta_2 d_2) = \delta_1 e_1 e_1 + \delta_2 e_1 d_2$$
$$= \delta_1 + \delta_2 e_1 d_2$$
$$= 0$$

and

$$\delta_1 = -\delta_2 e_1 d_2.$$

Therefore

$$e_2 = \delta_1 e_1 + \delta_2 d_2$$
$$= -\delta_2(e_1 d_2)e_1 + \delta_2 d_2$$
$$= \delta_2[d_2 - (e_1 d_2)e_1].$$

Let

$$g_2 = d_2 - (e_1 d_2)e_1$$

hence

$$e_2 = \delta_2 g_2.$$

Now

$$e_2 e_2 = 1.$$

Therefore

$$e_2 = \frac{g_2}{|g_2|}.$$

Next choose

$$e_3 = \sigma_1 d_1 + \sigma_2 d_2 + \sigma_3 d_3$$
$$= w_1 e_1 + w_2 e_2 + w_3 d_3.$$

Now

$$e_1 e_3 = 0$$

and

$$e_2 e_3 = 0.$$

Therefore

$$0 = e_1(w_1 e_1 + w_2 e_2 + w_3 d_3) = w_1 e_1 e_1 + w_2 e_1 e_2 + w_3 e_1 d_3$$
$$0 = e_2(w_1 e_1 + w_2 e_2 + w_3 d_3) = w_1 e_2 e_1 + w_2 e_2 e_2 + w_3 e_2 d_3.$$

Hence

$$0 = w_1 + w_3 e_1 d_3 \Rightarrow w_1 = -w_3 e_1 d_3$$
$$0 = w_2 + w_3 e_2 d_3 \Rightarrow w_2 = -w_3 e_2 d_3.$$

Therefore

$$e_3 = w_1 e_1 + w_2 e_2 + w_3 d_3$$
$$= -w_3(e_1 d_3)e_1 - w_3(e_2 d_3)e_2 + w_3 d_3$$
$$= w_3[d_3 - (e_1 d_3)e_1 - (e_2 d_3)e_2].$$

Let

$$g_3 = d_3 - (e_1 d_3)e_1 - (e_2 d_3)e_2.$$

Hence

$$e_3 = w_3 g_3.$$

Now

$$e_3 e_3 = 1.$$

Therefore

$$e_3 = \frac{g_3}{|g_3|}.$$

The process continues in this manner until e_n is constructed.

As an example, consider the orthogonalization of the following (row) vectors

$$\{d_1, d_2, d_3, d_4\} = \{(0, 2, 0, 0), (2, 0, 1, 0), (0, 0, 1, 1), (1, 0, 0, 2)\}.$$

Now

$$e_1 = \frac{d_1}{|d_1|}$$
$$= (0, 1, 0, 0)$$

$$g_2 = d_2 - (e_1 d_2)e_1$$
$$= (2, 0, 1, 0) - [(0, 1, 0, 0)(2, 0, 1, 0)](0, 1, 0, 0)$$
$$= (2, 0, 1, 0).$$

Therefore

$$e_2 = \frac{g_2}{|g_2|}$$

$$= \frac{1}{\sqrt{5}}(2,0,1,0).$$

Now

$$g_3 = d_3 - (e_1 d_3)e_1 - (e_2 d_3)e_2$$

$$= (0,0,1,1) - [(0,1,0,0)(0,0,1,1)](0,1,0,0)$$

$$- \left[\frac{1}{\sqrt{5}}(2,0,1,0)(0,0,1,1)\right]\left(\frac{1}{\sqrt{5}}(2,9,1,0)\right)$$

$$= (\tfrac{3}{5},0,\tfrac{4}{5},1).$$

Therefore

$$e_3 = \frac{g_3}{|g_3|}$$

$$= \frac{1}{\sqrt{2}}(\tfrac{3}{5},0,\tfrac{4}{5},1)$$

$$g_4 = d_4 - (e_1 d_4)e_1 - (e_2 d_4)e_2 - (e_3 d_4)e_3$$

$$= (1,0,0,2) - [(0,1,0,0)(1,0,0,2)](0,1,0,0)$$

$$- \left[\frac{1}{\sqrt{5}}(2,0,1,0)(1,0,0,2)\right]\frac{1}{\sqrt{5}}(2,0,1,0)$$

$$- \left[\frac{1}{\sqrt{2}}(\tfrac{3}{5},0,\tfrac{4}{5},1)(1,0,0,2)\right]\frac{1}{\sqrt{2}}(\tfrac{3}{5},0,\tfrac{4}{5},1)$$

$$= (\tfrac{29}{50},0,\tfrac{62}{50},\tfrac{7}{10}).$$

Therefore

$$e_4 = \frac{1}{3\sqrt{390}}(-29,0,-62,35).$$

9.2 Basic Calculus

9.2.1 Functions of One Variable

Definition. A real-valued *function of one real variable f* comprises a set D together with a rule for associating exactly one real number $f(x)$ with each element x of D.

The set D is called the *domain* of f and the set $\{f(x): x \in D\}$ is called the *range* of f. In this book the range is assumed to be a subset of **R** the set of real numbers, as is the domain (expect for the functionals of Chapter 7).

We now turn to the concept of a limit of a function.

Definition. The *limiting value of* f as $x \in D$ tends to b is said to be d if and only if for all $\varepsilon > 0$ there exists a $\delta > 0$ such that

$$0 < |x - b| < \delta \Rightarrow |f(x) - d| < \varepsilon.$$

The fact that f has a limiting value of d as x tends to b is denoted by:

$$\lim_{x \to b} f(x) = d.$$

Sometimes a function f is not defined for values of x that are either greater than a given value b or less than b. In these cases the above definition of a limit is invalid and we are lead to the concept of one-sided limits:

Definition. The limiting value of f as $x \in D$ tends to b from the right (left) is said to be d and only if for all $\varepsilon > 0$ there exists a $\delta > 0$ such that

$$0 < x - b < \delta \Rightarrow |f(x) - d| < \varepsilon$$
$$(0 < b - x < \delta \Rightarrow |f(x) - d| < \varepsilon).$$

These limits are denoted by;

$$\text{from the right:} \qquad \lim_{x \to b^+} f(x) = d$$

and

$$\text{from the left:} \qquad \lim_{x \to b^-} f(x) = d.$$

We come now to the concept of continuity of a function of one variable.

Definition. A function f is said to be *continuous* at a point $d \in D$ if and only if for all $\varepsilon > 0$, there exists a $\delta > 0$ such that

$$|x - d| < \delta \Rightarrow |f(x) - f(d)| < \varepsilon.$$

Note that, unlike the definition of a limit, the above definition is such that there is no necessity that the left-hand quantity $|x - d|$ be positive.

Definition. A function f is said to be *differentiable* at $d \in D$ if and only if the limit:

$$\lim_{h \to 0} \frac{f(d + h) - f(d)}{h}$$

exists.

If this limit does exist it is denoted by $f'(d)$, and is called the *derivative of f at d*.

Definition. If f is differentiable at x for all $x \in D$, f is said to be *differentiable*.

The proof of the following theorem is left as an exercise for the reader:

Theorem 9.1. *If a function, f is differentiable at $d \in D$, then f is continuous at d.*

One can attempt to find derivatives of f'; if f' is differentiable then the derivative of f' at a point $d \in D$ is denoted by $f''(d)$. In general, when the process is repated k times, the final derivative is denoted by $f^{(k)}(d)$.

9.2.2 Some Differential Theorems of Calculus

Theorem 9.2 (Weierstrass' Theorem). *If f is a continuous function on $[x_1, x_1 + h] \subseteq D$, then f attains both a maximum and a minimum value on $[x_1, x_i + h]$.*

We omit proof of Theorem 9.2. Stated in other words, this theorem means that there exist $\bar{x}^*, x^* \in [x_1, x_1 + h]$ such that

$$f(x) \le f(\bar{x}^*) \text{ for all } x \in [x_1, x_1 + h],$$
$$f(x) \ge f(x^*) \text{ for all } x \in [x_1, x_1 + h].$$

Theorem 9.3 (Rolle's Theorem). *If f is differentiable on $(x_1, x_1 + h) \subseteq D$ and continuous on $[x_1, x_1 + h]$ and*

$$f(x_1) = f(x_1 + h) = 0,$$

then there exists $\theta, 0 < \theta < 1$, such that

$$f'(\theta x_1 + (1 - \theta)(x_1 + h)) = 0.$$

PROOF. If

$$f(x) = 0, \quad \text{for all } x \in [x_1, x_1 + h],$$

the result is true. If not, there exists $x_2 \in [x_1, x_1 + h]$ such that

$$f(x_2) \ne 0.$$

Assume

$$f(x_2) > 0. \tag{9.9}$$

(If f is negative at this point an analogous proof follows.) Since f is continuous we can invoke Weierstrass' theorem and state that f attains a maximum value on $[x_1, x_1 + h]$. Thus there exists $\bar{x}^* \in [x_1, x_1 + h]$ such that

$$f(x) \le f(\bar{x}^*), \quad \text{for all } x \in [x_1, x_1 + h]. \tag{9.10}$$

By (9.9), we have

$$f(\bar{x}^*) > 0.$$

But as

$$f(x_1) = f(x_1 + h) = 0,$$

we have

$$\bar{x}^* \in (x_1, x_1 + h).$$

Hence $f'(\bar{x}^*)$ exists, by assumption. Assume

$$f'(\bar{x}^*) > 0.$$

Then there exists $\delta > 0$ such that

$$f(\bar{x}^* - \delta) < f(\bar{x}^*) < f(\bar{x}^* + \delta).$$

(Prove this.) But this contradicts (9.10). Assuming

$$f'(\bar{x}^*) < 0$$

leads to a similar contradiction. Thus

$$f'(\bar{x}^*) = 0.$$

Define θ to be such that

$$\bar{x}^* = \theta x_1 + (1 - \theta)(x_1 + h).$$

Then $0 < \theta < 1$ and the theorem is proved. $\qquad\square$

Theorem 9.4 (First Mean Value Theorem). *If f is differentiable on $(x_1, x_1 + h)$ and continuous on $[x_1, x_1 + h] \subseteq D$ then there exists θ, $0 < \theta < 1$ such that*

$$f(x_1 + h) = f(x_1) + hf'(\theta x_1 + (1 - \theta)(x_1 + h)).$$

PROOF. Set

$$g(x) = \left\{ \frac{x - x_1}{h} (f(x_1 + h) - f(x_1)) \right\} + f(x_1) - f(x).$$

Now

$$g(x_1) = g(x_1 + h) = 0,$$

and g is differentiable on $(x_1, x_1 + h)$ as f is. Thus, by Rolle's theorem there exists θ, $0 < \theta < 1$, such that

$$g'(\theta x_1 + (1 - \theta)(x_1 + h)) = 0.$$

Hence

$$0 = \frac{(f(x_1 + h) - f(x_1))}{h} - f'(\theta x_1 + (1 - \theta)(x_1 + h)).$$

Hence the result. $\qquad\square$

9.2.3 Taylor's Theorem

Let a function f with domain D be differentiable at $x \in [x_1, x_1 + h]$ in D. The first mean value theorem states that

$$f(x_1 + h) = f(x_1) + hf'(\theta x_1 + (1 - \theta)(x_1 + h)), \quad \text{for some } \theta, 0 < \theta < 1.$$

If $f^{(k)}$ is continuous on $[x_1, x_1 + h]$ and $f^{(k+1)}(x)$ exists for all $x \in (x_1, x_1 + h)$ we can generalize this result as follows:

Theorem 9.5 (Taylor's Theorem). *If $f^{(k)}$ is continuous on $[x_1, x_1 + h]$ and $f^{(k+1)}(x)$ exists for all $x \in (x_1, x_1 + h)$ then*

$$f(x_1 + h) = f(x_1) + hf'(x_1) + \frac{h^2}{2} f''(x_1) + \cdots$$

$$+ \frac{h^{k+1}}{(k + 1)!} f^{(k+1)}(\theta x_1 + (1 - \theta)(x_1 + h)), \quad \text{for some } \theta, 0 < \theta < 1.$$

PROOF (By induction). For $k = 1$, the above hypothesis is the first mean value theorem. Assume the result holds for k. Let

$$g(x) = f(x_1) + (x - x_1)f'(x_1) + \frac{(x - x_1)^2}{2} f''(x_1) + \cdots + \frac{(x - x_1)^k}{k!} f^{(k)}(x_1)$$

$$+ \frac{(x - x_1)^{k+1}}{(k + 1)!} R,$$

where R is such that

$$g(x_1 + h) = f(x_1 + h).$$

Now we wish to show that

$$R = f^{(k+1)}(\theta x_1 + (1 - \theta)(x_1 + h)), \quad \text{for some } \theta, 0 < \theta < 1.$$

Let

$$m(x) = f(x) - g(x), \quad x \in [x_1, x_1 + h].$$

As

$$m(x_1) = m(x_1 + h) = 0,$$

by Rolle's theorem we have

$$m'(\theta_1 x_1 + (1 - \theta_1)(x_1 + h)) = 0, \quad \text{for some } \theta_1, 0 < \theta < 1.$$

Therefore

$$f'(\eta) = f'(x_1) + (\eta - x_1)f''(x_1) + \cdots$$

$$+ \frac{(\eta - x_1)^k}{k!} f^{(k+1)}(\theta_2 x_1 + (1 - \theta_2)\eta), \quad \text{for some } \theta_2, 0 < \theta_2 < 1.$$

Thus

$$R = f^{(k+1)}(\theta_2 x_1 + (1 - \theta_2)\eta).$$

Define θ such that

$$\theta x_1 + (1 - \theta)(x_1 + h) = \theta_2 x_1 + (1 - \theta_2)\eta$$

and the theorem is proven. ☐

9.2.4 Functions of Severable Variables

Many of the results of the previous sections can be generalized to functions of several variables. The definition of a function can be so generalized by simply defining D, the domain of f, to be a set of n-dimensional real vectors.

The definition of the limiting value of f can be amended as follows:

Definition. The *limiting value of f* as $X \in D$ tends to b is said to be d if and only if for all $\varepsilon > 0$, there exists $\delta > 0$ such that

$$0 < \|X - b\| < \delta \Rightarrow |f(X) - d| < \varepsilon.$$

The definition of continuity follows analogously:

Definition. A function, f is said to be *continuous* at $X \in D$ if and only if for all $\varepsilon > 0$ there exists $\delta > 0$ such that

$$\|X - d\| < \delta \Rightarrow |f(X) - f(d)| < \varepsilon.$$

Things are a little more complicated when it comes to generalizing differentiation:

Let $X_0 = (x_1, x_2, \ldots, x_n)$ and

$$X_1 = (x_1, x_2, \ldots, x_{i-1}, x_i + h, x_{i+1}, \ldots, x_n)$$

be two points in the domain D of a function f. Then f is said to be *differentiable with respect to x_i* if and only if the limit:

$$\lim_{h \to 0} \frac{f(X_1) - f(X_0)}{h}$$

exists.

If this limit does exist, it is denoted by $\partial f(X_0)/\partial x_i$ and is called the *first partial derivative of f with respect to x_i*.

Assuming that all the partial derivatives $\partial f(X)/\partial x_i$, $i = 1, 2, \ldots, n$, exist for all $X \in D$, each can be thought of as a function on D. Each of these functions may have partial derivatives, which are termed second partial derivatives of f. Thus if $\partial f/\partial x_i$ has a partial derivative with respect to x_j at X, the derivative is denoted by $\partial^2 f/\partial x_j x_i$. This process can of course be repeated if the necessary limits exist.

The first partial derivatives of f at X can be assembled into a vector:

$$\left(\frac{\partial f(X)}{\partial x_1}, \frac{\partial(f(X))}{\partial x_2}, \ldots, \frac{\partial f(X)}{\partial x_n}\right)^T,$$

called the *gradient vector of f at X*, denoted by $Vf(X)$. The set of second partial derivatives of f at X can be assembled into a matrix:

$$\begin{bmatrix} \dfrac{\partial^2 f}{\partial x_1 \partial x_1} & \dfrac{\partial^2 f}{\partial x_1 \partial x_2} & \cdots & \dfrac{\partial^2 f}{\partial x_1 \partial x_n} \\[2ex] \dfrac{\partial^2 f}{\partial x_2 \partial x_1} & \dfrac{\partial^2 f}{\partial x_2 \partial x_2} & \cdots & \dfrac{\partial^2 f}{\partial x_2 \partial x_n} \\[1ex] \vdots & \vdots & & \vdots \\[1ex] \dfrac{\partial^2 f}{\partial x_n \partial x_1} & \dfrac{\partial^2 f}{\partial x_n \partial x_2} & \cdots & \dfrac{\partial^2 f}{\partial x_n \partial x_n} \end{bmatrix},$$

called the *Hessian matrix of f at X*, denoted by $H(X)$ in hour of the German mathematician who discovered it, Hesse.

Taylor's theorem can be extended to functions of several variables:

Theorem 9.6 (Taylor's Theorem for Functions of Several Variables). *If the second partial derivatives of f are continuous and X and $X + h$ are two points in the domain of f, then*

$$f(X + h) = f(X) + Vf(X)^T h + \tfrac{1}{2} h^T H(\theta X + (1 - \theta)(X + h))h, \quad \text{for some } \theta,$$
$$0 < \theta < 1.$$

Another theorem involving the gradient is of some importance in optimization:

Theorem 9.7 (Gradient Direction Theorem). *The gradient*

$$Vf = \left(\frac{\partial f}{\partial x_1}, \frac{\partial f}{\partial x_2}, \ldots, \frac{\partial f}{\partial x_n}\right)^T$$

points in the direction of steepest slope of the hypersurface of f.

PROOF. Construct an n-dimensional hypersphere of radius r about an arbitrary point $X \in D$. Let points on the sphere be of form:

$$C + \Delta X,$$

where

$$\Delta X = (\Delta x_1, \Delta x_2, \ldots, \Delta x_n)^T.$$

Then

$$\Delta x_1^2 + \Delta x_2^2 + \cdots + \Delta x_n^2 = r^2.$$

Now by using a first-order Taylor series approximation it can be shown that

$$\Delta f = f(X + \Delta X) - f(X) = \nabla f(X)^T \Delta X.$$

Let us now attempt to find the point on the hypersphere for which Δf is a maximum. We must form the Lagrangian:

$$L(\Delta C) = \nabla f^T \Delta X - \lambda[(\Delta X)^T(\Delta X) - r^2]$$

$$\frac{\partial L}{\partial \Delta X} = \nabla f - 2\lambda \Delta X$$

$$= 0$$

for a maximum. Therefore

$$\Delta X^* = \frac{1}{2\lambda} \nabla f.$$

Hence the vector ΔX^* yielding the greatest improvement in f has the same direction as the gradient, ∇f (as $1/2\lambda$ is a scalar).   □

Theorem 9.8. *If* $x_1, x_2, \ldots, x_n$ *are nonnegative numbers and* $\lambda_1, \lambda_2, \ldots, \lambda_n$ *are positive numbers such that*

$$\sum_{i=1}^{n} \lambda_i = 1,$$

then

$$\sum_{j=1}^{n} \lambda_j x_j \geq \prod_{j=1}^{n} x_j^{\lambda_j}. \tag{9.11}$$

If

$$\lambda_j = \frac{1}{n}, \quad j = 1, 2, \ldots, n,$$

then the left- and right-hand sides of (9.11) are the arithmetic mean and the geometric mean of $x_1, x_2, \ldots, x_n$, *respectively.*

9.3 Further Reading

The purpose of this section is to outline some of the texts that are available to the reader who wishes to pursue some of the topics of this book to a deeper level. We begin with general books on optimization and then cover, in order, linear programming, integer programming, network analysis, dynamic programming, and finally nonlinear programming. In addition to those listed here, the reader should be aware of books in the fields of operations research, management science, industrial engineering, and computer science which sometimes contain substantial content of an optimization nature.

One of the most important general books on optimization is by Beightler, Phillips, and Wilde (1979) covering classical optimization; linear, integer, nonlinear, and dynamic programming; and optimal control, all at an advanced level. Some of the topics are also covered at a more elementary level by Wilde (1964). Another more elementary text covering linear programming and a little nonlinear programming is Claycombe and Sullivan (1975). Mital (1976) covers most of the topics of this book, together with a chapter on game theory at an elementary level, as do Cooper and Steinberg (1970). Husain and Gangiah (1976) cover nonlinear programming and variational methods, with special emphasis on the techniques required for certain problems in chemical engineering. Sivazlian and Stanfel (1975) have written an undergraduate text covering the optimization techniques needed to solve many of the deterministic models of operations research. Gottfried and Weisman (1973) cover most of the optimization topics at a more advanced level than Sivazlian and Stanfel, and include a chapter on optimization under uncertainty and risk. Finally Geoffrion (1972) has edited a collection of expository papers covering a wide range of optimization topics at an advanced level.

Dantzig (1963) is the most important early reference on linear programming at an advanced level. Coverages at an elementary level include: Claycombe and Sullivan (1975), Daellenbach and Bell (1970) (with good sections on the formulation of L.P.'s and a computer code), Driebeek (1969) (with a good coverage of real-world applications), Campbell (1965) (covering the linear algebra underlying L.P.), and Fryer (1978). Intermediate level texts include: Hadley (1962), Spivey and Thrall (1970), Garvin (1960) (with some good applications of L.P.), Smythe and Johnson (1966), and Bazaraa and Jarvis (1977). Of the advanced texts on linear programming we mention: Gass (1969) and Simmonard (1966) (both requiring a mathematical background). Also Gal (1978) has written an advanced text covering postoptimal analysis and parametric programming.

Recent publications in integer programming include Salkin (1974) (advanced level) and Taha (1978) (very readable). Of the earlier works, Plane and McMillan (1971) is easily accessible to those with little mathematical background, Garfinkel and Nemhauser (1972) is more advanced and Greenberg (1971) is an intermediate text with interesting I.P. applications. A recent survey of integer programming articles published between 1976 and 1978 was completed by Hausmann (1978). An earlier survey was published by Geoffrion and Marsten (1972).

The best early reference on network analysis is Ford and Fulkerson (1962). Since then Busacker and Saaty (1965) have covered some aspects of network flow in a mathematically sophisticated fashion and Hu (1969) has given network flow problems a thorough examination at an advanced level. Geoffrion (1972), mentioned earlier, contains articles on optimization in networks. Plane and McMillan (1971), already mentioned, contains a chapter covering most of the material in Chapter 5 which is accessible to those with little background. Finally, Bazaraa and Jarvis (1977) cover the network flow and shortest path problems in a book that is very easy to read.

As was mentioned in Chapter 6, Bellman (1957) wrote the first book on dynamic programming. It is an advanced-level treatise. Since then Bellman and Dreyfus (1962), Hadley (1964), Nemhauser (1966) and White (1969) have written books which are also somewhat advanced in level. Bellman and Dreyfus present many applications. Hadley contains, apart from two chapters on D.P., a great deal of useful material on classical optimization, stochastic, integer, and nonlinear programming. Nemhauser's book is difficult to read, while White concentrates on the mathematical aspects of D.P. For the reader with limited mathematics background, Dreyfus and Law (1977) is recommended. While Dreyfus and Law state their book is graduate-level, it concentrates on applications and numerical examples of D.P.

Of the many books which specialize in classical optimization we mention Panik (1974). This book is intermediate in level and contains a great deal of mathematical background before covering classical optimization and many of its extensions. Panik does not cover variational problems and hence we cite the following, which cover the calculus of variations in increasing depth: Arthurs (1975), Craggs (1973), Young (1969), Smith (1974), Pars (1962), Ewing (1969). Well worth special mention are Hestenes (1966) and Gelfand and Fomin (1963). Hestenes covers introductory variational theory and also optimal control theory in some detail. Gelfand and Fomin slant their approach toward physical applications, and the proofs of many of the theorems of Chapter 7 of the present book can be found there. Finally, Blatt and Gray (1977) have provided an elementary derivation of Pontryagin's maximum principle. Many of the books mentioned in the next paragraph also contain sections on classical optimization.

There is an enormous amount of literature on nonlinear programming and we mention only a relatively small number of references here. Among the general references, Wilde and Beightler (1967) was mentioned earlier; Abadie (1967) surveys many of the areas of nonlinear programming in a collection of expository papers; Luenberger (1969) contains an advanced coverage of the mathematical aspects of nonlinear programming; Zangwill (1969) has become something of a classic and represents one of the first attempts at unifying nonlinear programming theory; Pierre (1969) covers classical optimization, the calculus of variations, linear and dynamic programming, the maximum principle, as well as many nonlinear programming techniques at the graduate level; Beveridge and Schechter (1970) constitutes a comprehensive treatment of most of the theory of nonlinear programming as it was in 1970 with, a valuable section on optimization in practice, at the senior/graduate level; Aoki (1971) is an undergraduate level text written for an audience interested in the applications rather than the mathematical theory of N.L.P. and contains some applications to engineering; Martos (1975) sets out to give a systematic treatment of the most important aspects of N.L.P. with many numberical examples; and finally Simmons (1975) covers some classical optimization but emphasises solution algorithms that have shown themselves to be of continuing importance and practical utility. Worthy of special mention is Himmelblau (1972) and Adby and Dempster (1974), which describe and compare in simple terms many of the N.L.P. methods which have proven to be effective.

Nonlinear programming books of a more specialized nature include: Zoutendijk (1960), the original work on feasible directions; Hadley (1964), which includes advanced-level material on separable problems, the Kuhn–Tucker conditions, quadratic programming, and gradient methods; Kunzi, Tzschach, and Zehnder (1968), which contains a list of FORTRAN and ALGOL computer codes for some N.L.P. algorithms; Kowalik and Osborne (1968), Fiacco and McCormick (1968), Murray (1972), (covering methods of computing optima of unconstrained problems); and Duffin, Peterson, and Zener (1967), the fathers of geometric programming, present the mathematical theory of G.P. and its application to problems in engineering design. Also there is an excellent survey of unconstrained optimization methods by Powell in Geoffrion (1972). Though dated, the survey gives a good introduction to the area.

References*

Abadie, J., ed., (1967) *Nonlinear Programming*. North-Holland. [350, 393]

Adby, P. R., and Dempster, M. A. H. (1974) *Introduction to Optimization Methods*. Chapman and Hall. [394]

Aoki, M. (1971) *Introduction to Optimization Techniques*. Macmillan. [394]

Arthurs, A. M. (1975) *Calculus of Variations*. Routledge and Kegan Paul. [393]

Balas, E. (1965) An additive algorithm for solving linear programs with zero–one variables. *Operations Research*, **13**: 517–546. [159]

Balinski, M. L., and Quandt, R. E. (1964) On an integer program for a delivery problem. *Operations Research* **12**: 300–304. [174]

Bazaraa, M. S. and Jarvis, J. J. (1977) *Linear Programming and Network Flows*. Wiley. [392, 393]

Beightler, C. S., and Phillips, D. T. (1976) *Applied Geometric Programming*. Wiley. [361]

Beightler, C. S., Phillips, D. T., and Wilde, D. J. (1979) *Foundations of Optimization*, 2nd ed. Prentice-Hall. [392]

Bellman, R. (1975) *Dynamic Programming*. Univ. Press. [3, 235, 393]

Bellman, R., and S. E. Dreyfus (1962) *Applied Dynamic Programming*. Princeton University Press. [235, 253, 393]

Bellmore, M., and Malone, J. C. (1971) Pathology of travelling salesmen subtour elimination algorithms. *Operations Research* **19**: 278–307. [174]

Beltrami, E., and Bodin, L. (1974) Networks and vechicle routing for municipal waste collection. *Networks* **1**: 65–94. [174]

Beveridge, G. S., and Schechter, R. S. (1970) *Optimization: Theory and Practice*. McGraw-Hill. [394]

Blatt, J. M., and Gray, J. D. (1977) An elementary derivation of Pontryagin's maximum principle of optimal control theory. *J. Aust. Math Soc.*, **20B**: 142–6. [393]

Box, G. E. P., and Wilson, K. B. (1951) On the experimental attainment of optimum conditions. *J. Roy. Stat Soc.* **B13**: 1. [330]

Box, M. J. (1966) A comparison of several current optimization methods and the use of transformation in constrained problems. *Comp. J.*, **9**: 67–77. [343]

Brent, R. P., (1973) *Algorithms for Minimization Without Derivatives*. Prentice-Hall. [343]

---

* References to page numbers in the text are given in square parentheses.

Broyden, C. G. (1971). The Convergence of an algorithm for solving sparse nonlinear systems. *Math. Comp*: **25**: 285–294. [329]

Busacker, R. G., and Gowan, P. J. (1961) A procedure for determining a family of minimal-cost network flow patterns. ORO Technical Rept. 15, Operations Research Office, Johns Hopkins University. [206]

Busacker, R. G., and Saaty, T. L. (1965) *Finite Graphs and Networks*. McGraw-Hill. [188, 392]

Campbell, H. G. (1965) *Matrices, Vectors and Linear Programming*. Appleton-Century-Crofts. [392]

Carroll, G. W. (1961) The created response surface technique for optimizing nonlinear restrained systems. *Operations Research* **9**: 169–184. [347]

Cauchy, A. (1847) Méthode générale pur la résolution des systèmes d' équations simultanées. *Compt. Rend. Acad. Sci.* (Paris) **25**: 536–38. [330]

Clarke, G., and Wright, S. W. (1964) Scheduling of vehicles from a central depot to a number of delivery points. *Operations Research* **12**: 568–681. [177]

Claycombe, W. W., and Sullivan, W. G. (1975) *Foundations of Mathematical Programming*. Reston. [392]

Conte, S. D., and de Boor, C. (1965) *Elementary Numerical Analysis*, 2nd ed. McGraw-Hill. [264]

Cooper, L., and Steinberg, D. (1970) *Introduction to Methods of Optimization*. Saunders. [392]

Craggs, J. W. (1973) *Calculus of Variations*. Allen and Unwin. [393]

Daellenbach, H. G., and Bell, E. J. (1970) *User's Guide to Linear Programming*. Prentice-Hall. [392]

Dakin, R. J. (1965) A tree search algorithm for mixed integer programming problems. *Computer J.* **8**: 250–255. [155]

Dantzig, G. B. (1963) *Linear Programming and Extensions*. Princeton Univ. Press. [2, 122, 392]

Dantzig, G. B., and, Wolfe, P. (1960) Decomposition principle for linear programs. *Operations Research*, **8**: 101–111. [122]

Davidon, W. C. (1959) Variable metric methods for minimization. Argonne Laboratory Rep. ANL-5990, Revised. [328, 329, 348]

Deo, N., (1974) *Graph Theory with Applications*. Prentice-Hall. [188]

Dijkstra, E. W. (1959) A note on two problems in connection with graphs. *Numerische Mathematik* **1**: 269 [194]

Dreyfus, S. E., and Law, A. M. (1977) *The Art and Theory of Dynamic Programming*. Academic Press. [393]

Driebeek, N. J. (1969) *Applied Linear Programming*. Addison-Wesley. [392]

Duffin, R. J., Peterson, E., and Zener, C. (1967) *Geometric Programming: Theory and Applications*. Wiley. [311, 361, 393]

Eastman, W. L. (1958) Linear programming with pattern constraints. Ph.D. Diss., Harvard University. [174]

Elsgolc, L. E. (1961) *Calculus of Variations*. Pergamon Press Ltd. [295]

Ewing, G. M. (1969) *Calculus of Variations with Applications*. Norton. [393]

Fiacco, A. V., and McCormick, G. P. (1968) *Nonlinear Programming: Sequential Unconstrained Minimization Techniques*. Wiley. [347, 393]

Fletcher, R. (1965) Function minimization without evaluating derivations—a review. *Comp. J.* **8**: 33–41. [343]

Fletcher, R., ed. (1969) *Optimization*. Academic Press. [348]

Fletcher, R. (1970) A new approach to variable metric algorithms. *Comp. J.* **13**: 317–322. [329]

Fletcher, R., and Powell, M. J. D. (1963) A rapidly convergent descent method for minimization. *Comp. J.* **6**: 163–8. [328, 329]

Fletcher, R., and Reeves, C. M. (1964) Function minimization by conjugate gradients. *Comp. J.* **7**: 149. [334]

Floyd, R. W. (1962) Algorithm 97—Shortest Path, *Comm. ACM* **5**: 345. [194]

Ford, L. R., and Fulkerson, D. R. (1962) *Flows in Networks*. Princeton Univ. Press. [201, 209, 392]

Forsythe, G. E., and Motzkin, T. S. (1951) Acceleration of the optimum gradient method. *Bull. Amer. Math. Soc.* **57**: 304–305. [331]

Foster, B. A., and Ryan, D. M. (1976) An integer programming approach to the vehicle scheduling problem. *Opnal. Res. Quart.* **27**: 367–384. [177]

Foulds, L. R., Robinson, D. F., and Read, E. G. (1977a) A manual procedure for the school bus routing problem. *Australian Road Research* **7**: 21–25. [174]

Foulds, L. R., O'Brien, L. E., and Pun, T. J. (1977b) Computer-based milk tanker scheduling. *NZ J. Dairy Sci. and Tech.* **12**: 141–145. [174]

Fryer, M. J., (1978) *An Introduction to Linear Programming and Game Theory*. Arnold. [392]

Gal, T. (1978) *Postoptimal Analysis. Parametric Programming and Related Topics*, McGraw-Hill. [392]

Gale, D. (1960) *The Theory of Linear Economic Models*. McGraw-Hill. [56]

Garfinkel, R. S., and Nemhauser, G. L. (1970) Optimal political districting by implicit enumeration techniques. *Mgmt. Sci.* **16**: 495–508. [179]

Garfinkel, R. S., and Nemhauser, G. L. (1972) *Integer Programming*. Wiley. [174, 392]

Garvin W. W. (1960) *Introduction to Linear Programming*. McGraw-Hill. [392]

Garvin, W. W., Crandall, H. W., John J. B., and Spellman, R. A. (1957) Applications of linear programming in the oil industry. *Mgmt. Sci.* **3**: 407. [174, 175]

Gass, S. I. (1969) *Linear Programming, Methods and Applications*, 3rd ed. McGraw-Hill. [35, 38, 392]

Gelfand, I. M., and Fomin, S. V. (1963) *Calculus of Variations*. Prentice-Hall. [300, 301, 393]

Geoffrion, A. M., ed. (1972) *Perspectives on Optimization*. Addison-Wesley. [392, 393]

Geoffrion, A., and Marsten, R. (1972) Integer programming: A framework and state-of-the-art survey. *Mgmt. Sci.* **18**: 465–91. [392]

Golden, B. Magnanti, T., and Nguyen, H. (1975) Implementing vehicle routing algorithms. M.I.T. Operations Research Center Technical Rept. No. 115. [174]

Goldfarb, D. (1970) A family of variable-metric methods derived by variational means. *Math. Comp.* **24**: 23–26. [329]

Gomory, R. E. (1958) Outline of an algorithm for integer solutions to linear programs. *Bull. Amer. Math. Soc.* **64**: 275–278. [2, 162]

Gottfried, B. S., and Weisman, J. (1973) *Introduction to Optimization Theory*. Prentice-Hall. [303, 392]

Greenberg, N. (1971) *Integer Programming*. Academic Press. [392]

Griffith, R. E., and Stewart, R. A. (1961) A nonlinear programming technique for the optimization of continuous processing systems. *Mgt. Sci.* **7**: 379–392. [351]

Hadley, G. H. (1962) *Linear Programming*. Addison-Wesley. [122, 392]

Hadley, G. (1964) *Nonlinear and Dynamic Programming*. Addison-Wesley. [235, 393]

Hancock, H. (1960) *Theory of Maxima and Minima*. Dover Publications. [273]

Harary, F. (1969) *Graph Theory*. Addison-Wesley. [188]

Hausmann, D., ed. (1978) *Integer Programming and Related Areas*. Springer-Verlag. [392]

Henrici, P. (1964) *Elements of Numerical Analysis*. Wiley. [273]

Hess, S., Weaver, J., Siegfeldt, H., Whelan, J., and Zitlau, P. (1965) Nonpartisan political districting by computer. *Operations Research* **13**: 998–1006. [179]

Hestenes, M. R. (1966) *Calculus of Variations and Optimal Control Theory*. Wiley. [393]

Himmelblau, D. M. (1972) *Applied Nonlinear Programming*. McGraw-Hill. [265, 345, 394]

Hooke, K., and Jeeves, T. A. (1961) Direct search solution of numerical and statistical problems. *J. ACM* **8**: 212–229. [334]

Hu, T. C. (1969) *Integer Programming and Network Flows*. Addison-Wesley. [22, 392]

Husain, A., and Gangiah, K. (1976) *Optimization Techniques*. Macmillan India. [392]

Jacoby, S. L. S., Kowalik, J. S., and Pizzo, J. T. (1972) *Iterative Methods for Nonlinear Optimization Problems*. Prentice-Hall. [345]

Kiefer, J. (1957) Sequential minimax search for a maximum. *Proc. Amer. Math. Soc.* **4**: 502–506. [320]

Kowalik, J., and Osborne, M. R. (1968) *Methods for Unconstrained Optimization Problems*. Elsevier. [393]

Kruskal, J. B. (1956) On the shortest spanning subtree of a graph and the traveling salesman problem. *Proc. Amer. Math. Soc.* **7**: 48. [194]

Kuhn, H. W., and Tucker, A. W. (1951) Nonlinear programming. Proc. 2nd Berkeley Symp. on Math. Stat. Prob., Univ. Calif. Press., (1951) p. 481–492. [3, 284]

Kunzi, H. P., Tzschach, H. G., and Zehnder, C. A. (1968) *Numerical Methods of Mathematical Optimization*. Academic Press. [393]

Land, A., and Doig, A. (1960) An automatic method of solving discrete programming problems. *Econometrica* **28**: 497–520. [155]

Lill, S. A. (1970) A modified Davidon method for finding the minimum of a function using difference approximations for derivatives. *Computer J.* **13**: 111–113. [345]

Little, J. D. C., Murty, K. G., Sweeney, P. W., and Karel, C. (1963) An algorithm for the traveling salesman problem. *Operations Research* **11**: 979–989. [174]

Luenberger, D. G. (1969) *Optimization by Vector Space Methods*. Wiley. [393]

Martos, B. (1975) *Nonlinear Programming*. North-Holland. [394]

Mital K. V. (1976) *Optimization Methods*. Wiley Eastern. [392]

Murchland, J. D. (1967) The once-through method of finding all shortest distances in a graph from a single origin. London Graduate School of Business Studies, Rept. LBS-TNT-56. [194]

Murray, W., ed. (1972) *Numerical Methods for Unconstrained Optimization*. Academic Press. [393]

Nemhauser, G. L., (1966) *Introduction to Dynamic Programming*. Wiley. [235, 393]

Panik, M. J. (1976) *Classical Optimization*. North-Holland. [393]

Pars, L. A. (1962) *An Introduction to the Calculus of Variations*. Heineman. [393]

Pierre, D. A. (1969) *Optimization Theory with Applications*. Wiley. [394]

Plane, D. R., and McMillan, C. (1971) *Discrete Optimization*. Prentice-Hall. [209, 392]

Pontryagin, L. S., Boltyanskii, V. G., Gamkrelidze, R. V., and Mishchenko, E. F. (1962) *The Mathematical Theory of Optimal Processes*. Wiley–Interscience. [5, 306]

Powell, M. J. D. (1964) An efficient method for finding the minimum of a function of several variables without calculating derivatives. *Comp. J.* **7**: 155–62. [342, 345, 348]

Powell, M. J. D. (1968) On the calculation of orthogonal vectors. *Comp. J.* **11**: 302–304. [342]

Prim, R. C. (1957) Shortest connection networks and some generalizations. *Bell Syst. Tech. J.* **36**: 1389. [196]

Rosen, J. B. (1960) The gradient projection method for nonlinear programming. Part 1: Linear constraints. *SIAM J. Appl. Math.* **8**: 181–217. [346]

Rosenbrock, H. H. (1960) An automatic method for finding the greatest or least value of a function. *Comp. J.* **3**: 175–184. [341]

Salkin, H. M. (1974) *Integer Programming*. Addison-Wesley. [392]

Shah, B. V., Buehler, R. J., and, Kempthorne O. (1964) Some algorithms for minimizing a function of several variables. *SIAM J. Appl. Math.* **12**: 74–92. [332]

Shanno, D. F. (1970) Conditioning of Quasi-Newton methods for function minimization *Math. Comp.* **24**:647–656. [329]

Simmonard, M. (1966) *Linear Programming.* Prentice-Hall. [392]

Simmons, D. M. (1975) *Nonlinear Programming for Operations Research.* Prentice-Hall. [394]

Sivazlian, B. D., and Stanfel, L. E. (1975) *Optimization Techniques in Operations Research.* Prentice-Hall. [392]

Smith, D. R. (1974) *Variational Methods in Optimization.* Prentice-Hall. [393]

Smith, R. G., Foulds L. R., and Read, E. G. (1976) A political redistricting problem. *New Zealand Operational Research* **4**:37–52. [179]

Smythe, W. R., and Johnson, L. A. (1966) *Introduction to Linear Programming with Applications.* Prentice-Hall. [392]

Spivey, W. A., and Thrall R. M. (1970) *Linear Optimization.* Holt, Rinehart and Winston. [392]

Stewart, G. W. (1957) A modification of Davidon's minimization method to accept difference approximations of derivatives. *J. A.C.M.* **14**:72–83. [344, 345]

Swann, W. H. (1964) Report on the development of a new direct search method for optimization. I.C.I. C.I. Lab. Res. Note. 6413. [345]

Taha, H. A. (1976) *Operations Research*, 2nd ed. Macmillan. [221]

Taha, H. A. (1978) *Integer Programming.* Macmillan. [392]

Turner, W. C., Ghare, P. M., and Foulds, L. R. (1974) Transportation routing problem: A survey. *AIIE Transactions* **6**:288–301. [174]

Wagner, J. (1968) An application of integer programming to legislative redistricting. 34th National Meeting of the Operations Research Society of America. [179]

Watson-Gandy, C. and Foulds, L. R. (1981) The Vehicle Scheduling Problem: A survey. *New Zealand Operational Research* **9** no. 2:73–92. [174]

White, D. J. (1969) *Dynamic Programming.* Oliver and Boyd. [235, 393]

Wilde, D. J. (1964) *Optimum Seeking Methods.* Prentice-Hall. [392]

Wilde, D. J., and Beightler, C. S. (1967) *Foundations of Optimization.* Prentice-Hall. [284, 393]

Young, L. C. (1969) *Calculus of Variations and Optimal Control Theory.* Saunders. [393]

Zangwill, W. I. (1969) *Nonlinear Programming.* Prentice-Hall. [394]

Zoutendijk, K. G. (1960) *Method of Feasible Directions.* Elsevier. [345, 393]

Solutions to Selected Exercises

Chapter 2

Section 2.8

1(a). Let

x_1 = the number of chocolate cakes

x_2 = the number of banana cakes.

Then the problem is to:

$$\text{Maximize:} \quad 75x_1 + 60x_2$$
$$\text{subject to:} \quad 4x_1 + 6x_2 \le 96$$
$$2x_1 + x_2 \le 24$$
$$x_1, x_2 \ge 0.$$

The optimum point can be found graphically (see Figure S.1) or by solving the two equations:

$$4x_1 + 6x_2 = 96$$
$$\underline{2x_1 + x_2 = 24}$$
$$\Rightarrow \quad 4x_2 = 48.$$

Thus the optimal solution is

$$x_1^* = 6, \qquad x_2^* = 12.$$

The value is

$$6 \times 0.75 + 12 \times 0.60 = 11.7.$$

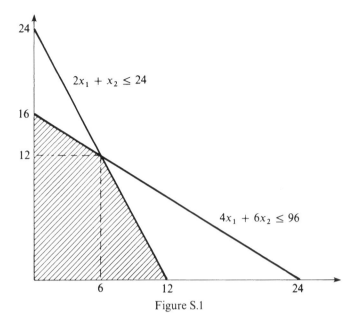

Figure S.1

Hence the best profit the baker can hope to make is $11.70 by baking six chocolate cakes and twelve banana cakes a day.

2(a). Let

x_1 = the number of trucks manufactured

x_2 = the number of automobiles manufactured

x_3 = the number of vans manufactured.

Then the problem is (on dividing the profits by 1000) to:

$$\begin{aligned}
\text{Maximize:} \quad & 6x_1 + 4x_2 + 3x_3 \\
\text{subject to:} \quad & 4x_1 + 5x_2 + 3x_3 \leq 12 \\
& 3x_1 + 4x_2 + 2x_3 \leq 10 \\
& 4x_1 + 2x_2 + \ x_3 \leq 8 \\
& x_1, x_2, x_3 \geq 0.
\end{aligned}$$

On introducing slack variables, the problem becomes

$$\begin{aligned}
\text{Maximize:} \quad & 6x_1 + 4x_2 + 3x_2 \\
\text{subject to:} \quad & 4x_1 + 5x_2 + 3x_3 + x_4 \qquad\qquad\quad = 12 \\
& 3x_1 + 4x_2 + 2x_3 \qquad + x_5 \qquad\quad = 10 \\
& 4x_1 + 2x_2 + \ x_3 \qquad\qquad + x_6 = 8 \\
& x_i \geq 0, \qquad i = 1, 2, 3, 4, 5, 6.
\end{aligned}$$

The problem can now be solved using the simplex method.

1

	x_1	x_2	x_3	x_4	x_5	x_6	r.h.s	Ratio
	4	5	3	1	0	0	12	$\frac{12}{4}$
	3	4	2	0	1	0	10	$\frac{10}{3}$
	④	2	1	0	0	1	8	$\frac{8}{4}$
x_0	-6	-4	-3	0	0	0	0	

2

	x_1	x_2	x_3	x_4	x_5	x_6	r.h.s	Ratio
	0	3	2	1	0	-1	4	$\frac{4}{2}$
	0	$\frac{5}{2}$	$\frac{5}{4}$	0	1	$-\frac{3}{4}$	4	$\frac{16}{5}$
	1	$\frac{1}{2}$	$\frac{1}{4}$	0	0	$\frac{1}{4}$	2	8
x_0	0	-1	$-\frac{3}{2}$	0	0	$\frac{3}{2}$	12	

3

	x_1	x_2	x_3	x_4	x_5	x_6	r.h.s.
	0	$\frac{3}{2}$	1	$\frac{1}{2}$	0	$-\frac{1}{2}$	2
	0	$\frac{5}{8}$	0	$-\frac{5}{8}$	1	$-\frac{1}{8}$	$\frac{3}{2}$
	1	$\frac{1}{8}$	0	$-\frac{1}{8}$	0	$\frac{3}{8}$	$\frac{3}{2}$
x_0	0	$\frac{5}{4}$	0	$\frac{3}{4}$	0	$\frac{3}{4}$	15

Tableau 3 yields the optimal solution:

$$x_1^* = \tfrac{3}{2}, \qquad x_3^* = 2, \qquad x_5^* = \tfrac{3}{2},$$
$$x_2^* = x_4^* = x_6^* = 0 \qquad x_0^* = 15{,}000.$$

3(a). Let

x_1 = the number of pounds of chutney produced per week

x_2 = the number of pounds of sauce produce per week.

Then, with the introduction of the slack variables x_3, x_4, and x_5, and the artificial variable x_6, the problem is

$$
\begin{array}{llr}
\text{Maximize:} & x_0 = 4x_1 + 5x_2 - Mx_6 & = x_0 \\
\text{subject to:} & 3x_1 + 5x_2 + x_3 & = 24 \quad (1) \\
& 4x_1 + 2x_2 + x_4 & = 16 \quad (2) \\
& x_1 + x_2 - x_5 + x_6 & = 3 \quad (3) \\
& x_j \geq 0, \quad j = 1, 2, \ldots, 6.
\end{array}
$$

This formulation will be solved by the big M method. The feasible region for the problem is shown in Figure S.2.

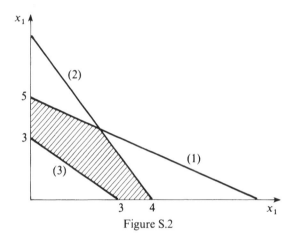

Figure S.2

The initial tableau for the problem is:

1

Constraints	x_1	x_2	x_3	x_4	x_5	x_6	r.h.s.
(1)	3	5	1	0	0	0	24
(2)	4	2	0	1	0	0	16
(3)	1	①	0	0	-1	1	3
x_0	-4	-5	0	0	0	M	0

The initial basis is (x_3, x_4, x_6). However, because the objective function coefficient of the basic variable x_6 is nonzero, the tableau is not yet in canonical form. This is remedied by replacing the x_0 row by the sum of the x_0 row and M times (3). This gives:

2

Constraints	x_1	x_2	x_3	x_4	x_5	x_6	r.h.s.	Ratio
(1)	3	5	1	0	0	0	24	$\frac{24}{5}$
(2)	4	2	0	1	0	0	16	$\frac{16}{2}$
(3)	1	1	0	0	-1	1	3	$\frac{3}{1}$
x_0	$-(M+4)$	$-(M+5)$	0	0	M	0	$-3M$	

The simplex iterations required to reach the optimal solutions are

3

Constraints	x_1	x_2	x_3	x_4	x_5	x_6	r.h.s.	Ratio
(1)	-2	0	1	0	$\circled{5}$	-5	9	$\frac{9}{5}$
(2)	2	0	0	1	2	-2	10	$\frac{10}{2}$
(3)	1	1	0	0	-1	1	3	—
x_0	1	0	0	0	-5	$(M+5)$	15	

4

Constraints	x_1	x_2	x_3	x_4	x_5	x_6	r.h.s.
(1)	$-\frac{2}{5}$	0	$\frac{1}{5}$	0	1	-1	$\frac{9}{5}$
(2)	$2\frac{4}{5}$	0	$-\frac{2}{5}$	1	0	0	$6\frac{2}{5}$
(3)	$\frac{3}{5}$	1	$\frac{1}{5}$	0	0	0	$4\frac{4}{5}$
x_0	-1	0	1	0	0	M	24

5

Constraints	x_1	x_2	x_3	x_4	x_5	x_6	r.h.s.
(1)	0	0	$\frac{1}{7}$	$\frac{1}{7}$	1	-1	$2\frac{5}{7}$
(2)	1	0	$-\frac{1}{7}$	$\frac{5}{14}$	0	0	$2\frac{4}{7}$
(3)	0	1	$\frac{2}{7}$	$-\frac{3}{14}$	0	0	$3\frac{3}{7}$
x_0	0	0	$\frac{6}{7}$	$\frac{5}{14}$	0	M	$26\frac{2}{7}$

The optimal solution is

$$x_1^* = 2\tfrac{2}{7}, \qquad x_2^* = 3\tfrac{3}{7}, \qquad x_5^* = 2\tfrac{5}{7},$$
$$x_3^* = x_4^* = 0, \qquad x_0^* = 26\tfrac{2}{7}.$$

Thus the housewife should make $2\tfrac{2}{7}$ lbs chutney and $3\tfrac{3}{7}$ lbs sauce to obtain a maximum profit of $2.63.

4(a). This problem can be expressed in mathematical terms. The variables are defined as follows. Let

$x_1 = $ the number of classrooms constructed

$x_2 = $ the number of houses constructed.

The problem can now be stated:

$$
\begin{array}{lll}
\text{Maximize:} & 4x_1 + 5x_2 & \\
\text{subject to:} & 4x_1 + 5x_2 \le 32 & (1) \\
& 4x_1 + 3x_2 \le 24 & (2) \\
& 3x_1 + 2x_2 \le 20 & (3) \\
& 2x_1 + x_2 \le 16 & (4)
\end{array}
$$

which converted to standard form is

$$
\begin{array}{ll}
\text{Maximize:} & 4x_1 + 5x_2 \\
\text{subject to:} & 4x_1 + 5x_2 + x_3 = 32 \\
& 4x_1 + 3x_2 + x_4 = 24 \\
& 3x_1 + 2x_2 + x_5 = 20 \\
& 2x_1 + x_2 + x_6 = 16.
\end{array}
$$

This problem can now be solved using the simplex method.

1

Constraints	x_1	x_2	x_3	x_4	x_5	x_6	r.h.s.	Ratio
(1)	4	⑤	1	0	0	0	32	$\tfrac{35}{2}$
(2)	4	3	0	1	0	0	24	$\tfrac{24}{3}$
(3)	3	2	0	0	1	0	20	$\tfrac{20}{2}$
(4)	2	1	0	0	0	1	16	$\tfrac{16}{1}$
x_0	-4	-5	0	0	0	0	0	

2

Constraints	x_1	x_2	x_3	x_4	x_5	x_6	r.h.s.	Ratio
(1)	$\frac{4}{5}$	1	$\frac{1}{5}$	0	0	0	$\frac{32}{5}$	8
(2)	$\frac{8}{5}$	0	$-\frac{3}{5}$	1	0	0	$\frac{24}{5}$	3
(3)	$\frac{7}{5}$	0	$-\frac{2}{5}$	0	1	0	$\frac{36}{5}$	$5\frac{1}{5}$
(4)	$\frac{6}{5}$	0	$-\frac{1}{5}$	0	0	1	$\frac{48}{5}$	8
x_0	0	0	1	0	0	0	32	

The last tableau yields the optimal solution:

$$x_2^* = \tfrac{32}{5}, \qquad x_4^* = \tfrac{24}{5}, \qquad x_5^* = \tfrac{36}{5},$$
$$x_6^* = \tfrac{48}{5}, \qquad x_1 = x_3^* = 0, \qquad x_0^* = 32.$$

However, the nonbasic variable x_1, has a zero x_0-row coefficient, indicating that the objective function value would remain unchanged if x_1 was brought into the basis:

3

Constraints	x_1	x_2	x_3	x_4	x_5	x_6	r.h.s.
(1)	0	1	$\frac{1}{2}$	$-\frac{1}{2}$	0	0	4
(2)	1	0	$-\frac{3}{8}$	$\frac{5}{8}$	0	0	3
(3)	0	0	$\frac{1}{8}$	$-\frac{7}{8}$	1	0	3
(4)	0	0	$\frac{1}{4}$	$-\frac{6}{8}$	0	1	6
x_0	0	0	1	0	0	0	32

This tableau yields the optimal solution:

$$x_1^* = 3, \qquad x_2^* = 4, \qquad x_5^* = 3, \qquad x_6^* = 6, \qquad x_3^* = x_4^* = 0, \qquad x_0^* = 32.$$

Thus the builder should build 3 classrooms and 4 houses and maximize his profit at $32,000.

The x_0 row value of x_4 is zero indicating that x_4 could replace x_1 in the basis at no change in objective function value. This would produce tableau 2. Thus this problem has two basic optimal solutions.

The problem is solved graphically in Figure S.3. When the objective function is drawn at the optimal level, it coincides with constraint line (1). This means that all points on the line from point $(0, 6\frac{2}{5})$ to $(3, 4)$ represent optimal solutions. This situation can be stated as follows:

$$4x_1^* + 5x_2^* = 32, \qquad 0 \le x_1^* \le 3, \qquad 4 \le x_2^* \le 6\tfrac{2}{5}, \qquad x_0^* = 32.$$

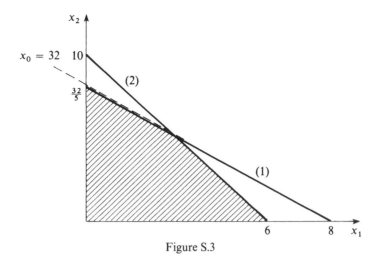

Figure S.3

5(a). In mathematical form the problem is

$$\begin{aligned}
\text{Maximize:} \quad & 3x_1 + 2x_2 + 4x_3 + x_4 \\
\text{subject to:} \quad & 8x_1 + 2x_2 + 5x_3 + 4x_4 \le 16 \\
& 6x_1 + 4x_2 + 3x_3 + 2x_4 \le 10 \\
& 3x_1 + 3x_2 + 2x_3 + x_4 \le 6\tfrac{5}{7}.
\end{aligned}$$

In standard form the problem becomes

$$\begin{aligned}
\text{Maximize:} \quad & 3x_1 + 2x_2 + 4x_3 + x_4 \\
\text{subject to:} \quad & 8x_1 + 2x_2 + 5x_3 + 4x_4 + x_5 && = 16 \\
& 6x_1 + 4x_2 + 3x_3 + 2x_4 + x_6 && = 10 \\
& 3x_1 + 3x_2 + 2x_3 + x_4 + x_7 && = 6\tfrac{5}{7} \\
& x_i \ge 0, \quad i = 1, 2, \ldots, 7.
\end{aligned}$$

The final two tableaux required to solve the problem are displayed:

2

	x_1	x_2	x_3	x_4	x_5	x_6	x_7	r.h.s.	Ratio
	$\frac{8}{5}$	$\frac{2}{5}$	1	$\frac{4}{5}$	$\frac{1}{5}$	0	0	$\frac{16}{5}$	8
	$\frac{6}{5}$	$\left(\frac{14}{5}\right)$	0	$-\frac{2}{5}$	$-\frac{3}{5}$	1	0	$\frac{2}{5}$	$\frac{1}{7}$
	$-\frac{1}{5}$	$\frac{11}{5}$	0	$-\frac{3}{5}$	$-\frac{2}{5}$	0	1	$\frac{11}{35}$	$\frac{1}{7}$
x_0	$\frac{17}{5}$	$-\frac{2}{5}$	0	$\frac{11}{5}$	$\frac{4}{5}$	0	0	$\frac{64}{5}$	

3

	x_1	x_2	x_3	x_4	x_5	x_6	x_7	r.h.s.
	$\frac{10}{7}$	0	1	$\frac{6}{7}$	$\frac{2}{7}$	$-\frac{1}{7}$	0	$\frac{22}{7}$
	$\frac{3}{7}$	1	0	$-\frac{1}{7}$	$-\frac{3}{14}$	$\frac{5}{14}$	0	$\frac{1}{7}$
	$-\frac{8}{7}$	0	0	$-\frac{2}{7}$	$\frac{1}{14}$	$-\frac{11}{14}$	1	0
x_0	$\frac{25}{7}$	0	0	$\frac{15}{7}$	$\frac{5}{7}$	$\frac{1}{7}$	0	$\frac{90}{7}$

It can be seen that x_2 should enter the basis in tableau 2 but a tie occurs on forming the ratios to decide which variable leaves the basis. In the next iteration one of the basic variables is $x_7 = 0$. This basic feasible solution is called a degenerate solution. The optimum is reached at the first stage of degeneracy.

The solution to the problem is that the farmer should cultivate $\frac{1}{7}$ acre of barley and $3\frac{1}{7}$ acres of wheat. His profit would be \$1285.71.

6(a). The problem can be expressed in mathematical terms as follows. Let

x_1 = the units of cheese produced

x_2 = the units of butter produced

x_3 = the units of milk powder produced

x_4 = the units of yoghurt produced.

The problem can now be restated:

$$\text{Maximize:} \quad 4x_1 + 3x_2 + 2x_3 + x_4$$
$$\text{subject to:} \quad 2x_1 + 3x_2 + 4x_3 + 2x_4 \leq 8$$
$$3x_1 + x_2 + 4x_3 + 2x_4 \leq 9$$
$$3x_1 + 2x_2 + 5x_3 + x_4 \leq 9,$$

which converted to standard form is

$$\text{Maximize:} \quad 4x_1 + 3x_2 + 2x_3 + x_4$$
$$\text{subject to:} \quad 2x_1 + 3x_2 + 4x_3 + 2x_4 + x_5 \qquad\qquad = 8$$
$$3x_1 + x_2 + 4x_3 + 2x_4 \qquad + x_6 \qquad = 9$$
$$3x_1 + 2x_2 + 5x_3 + x_4 \qquad\qquad + x_7 = 9.$$

The problem can now be solved using the simplex method.

1

	x_1	x_2	x_3	x_4	x_5	x_6	x_7	r.h.s.	Ratio
	2	3	4	2	1	0	0	8	$\frac{8}{2}$
	3	1	4	2	0	1	0	9	$\frac{9}{3}$
	③	2	5	1	0	0	1	9	$\frac{9}{3}$
x_0	-4	-3	-2	-1	0	0	0	0	

2

	x_1	x_2	x_3	x_4	x_5	x_6	x_7	r.h.s.	Ratio
	0	$\left(\frac{5}{3}\right)$	$\frac{2}{3}$	$\frac{4}{3}$	1	0	$-\frac{2}{3}$	2	$\frac{6}{5}$
	0	-1	-1	1	0	1	-1	0	—
	1	$\frac{2}{3}$	$\frac{5}{3}$	$\frac{1}{3}$	0	0	$\frac{1}{3}$	3	$\frac{9}{2}$
x_0	0	$-\frac{1}{3}$	$\frac{14}{3}$	$\frac{1}{3}$	0	0	$\frac{4}{3}$	12	

3

	x_1	x_2	x_3	x_4	x_5	x_6	x_7	r.h.s.
	0	1	$\frac{2}{5}$	$\frac{4}{5}$	$\frac{3}{5}$	0	$-\frac{2}{5}$	$\frac{6}{5}$
	0	0	$-\frac{3}{5}$	$\frac{9}{5}$	$\frac{3}{5}$	1	$-\frac{7}{5}$	$\frac{6}{5}$
	1	0	$\frac{7}{5}$	$-\frac{1}{5}$	$-\frac{2}{5}$	0	$\frac{3}{5}$	$\frac{11}{5}$
x_0	0	0	$\frac{24}{5}$	$\frac{3}{5}$	$\frac{1}{5}$	0	$\frac{6}{5}$	$12\frac{2}{5}$

The optimal solution can be found from tableau 3:

$$x_1^* = \tfrac{11}{5}, \qquad x_2^* = \tfrac{6}{5}, \qquad x_6^* = \tfrac{6}{5}, \qquad x_i^* = 0, \quad \text{otherwise} \qquad x_0^* = \$1{,}240.$$

Thus the dairy factory should produce $\frac{6}{5}$ ton of cheese and $\frac{11}{5}$ ton of butter daily to maximize profit at \$1,240.

Note that one of the basic feasible solutions produced by the simplex method was degenerate, as the variable x_6 had zero value. However, there is no degeneracy in the tableau of the next iteration. This is because the entering variable x_2 coefficient is negative in the x_6 row. Thus no ratio is formed.

7(a)

$$\text{Maximize:} \quad 4x_1 + 3x_2$$

subject to:	$3x_1 + 4x_2 \le 12$	(1)
	$5x_1 + 2x_2 \le 8$	(2)
	$x_1 + x_2 \ge 5$	(3)

$$x_1, x_2 \ge 0.$$

When this problem is expressed graphically (Figure S.4) it can be seen that there does not exist a point which will satisfy all constraints simultaneously. Hence the problem does not have a feasible solution.

Two-Phase Method

$$\text{Maximize:} \quad 4x_1 + 3x_2$$

$$\text{subject to:} \quad 3x_1 + 4x_2 + x_3 \qquad\qquad\quad = 12$$

$$5x_1 + 2x_2 \qquad + x_4 \qquad\quad = 8$$

$$x_1 + x_2 \qquad\qquad + x_5 - x_6 = 5.$$

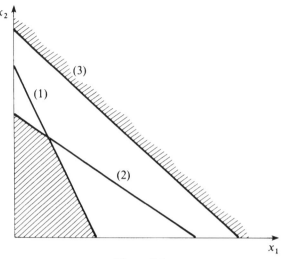

Figure S.4

Phase I

x_1	x_2	x_3	x_4	x_5	x_6	r.h.s.
3	4	1	0	0	0	12
5	2	0	1	0	0	8
1	1	0	0	1	-1	5
0	0	0	0	1	0	0
3	④	1	0	0	0	12
5	2	0	1	0	0	8
1	1	0	0	1	-1	5
-1	-1	0	0	0	-1	-5
$\frac{3}{4}$	1	$\frac{1}{4}$	0	0	0	3
$\frac{7}{2}$	0	$-\frac{1}{2}$	1	0	0	2
$\frac{1}{4}$	0	$-\frac{1}{4}$	0	1	-1	2
$-\frac{1}{4}$	0	$\frac{1}{4}$	0	0	-1	-2
0	1	$\frac{5}{14}$	$-\frac{3}{14}$	0	0	$\frac{18}{7}$
1	0	$-\frac{1}{7}$	$\frac{2}{7}$	0	0	$\frac{4}{7}$
0	0	$-\frac{3}{14}$	$-\frac{1}{14}$	1	-1	$\frac{13}{7}$
0	0	$\frac{3}{14}$	$\frac{1}{14}$	0	1	$-\frac{13}{7}$

$x_5^* > 0 \Rightarrow$ no feasible solution.

8(a). In mathematical form the problem is:
Primal

$$\text{Maximize:} \quad 5x_1 + 4x_2 = x_0$$
$$\text{subject to:} \quad 3x_1 + 4x_2 \le 14$$
$$4x_1 + 2x_2 \le 8$$
$$2x_1 + x_2 \le 6$$
$$x_1, x_2 \ge 0.$$

Since the number of constraints is greater than the number of variables, the problem is more easily solved when its dual is created. The problem can be written as follows.
Dual

$$\text{Minimize:} \quad 14y_1 + 8y_2 + 6y_3 = y_0'$$
$$\text{subject to:} \quad 3y_1 + 4y_2 + 2y_3 \ge 5$$
$$4y_1 + 2y_2 + y_3 \ge 4$$
$$y_1, y_2, y_3 \ge 0.$$

In standard form the problem is

$$\text{Maximize:} \quad -14y_1 - 8y_2 - 6y_3 - My_5 - My_7 = y_0$$
$$\text{subject to:} \quad 3y_1 + 4y_2 + 2y_3 - y_4 + y_5 = 5$$
$$4y_1 + 2y_2 + y_3 - y_6 + y_7 = 4$$
$$y_i \ge 0, \quad i = 1, 2, \ldots, 7.$$

The tableaux required to solve the problem are displayed next.

	y_1	y_2	y_3	y_4	y_5	y_6	y_7	r.h.s.
	3	4	2	-1	1	0	0	5
	4	2	1	0	0	-1	1	4
y_0	14	8	6	0	M	0	M	0

	y_1	y_2	y_3	y_4	y_5	y_6	y_7	r.h.s.	Ratio
	3	4	2	-1	1	0	0	5	$\frac{5}{3}$
	④	2	1	0	0	-1	1	4	$\frac{4}{4}$
y_0	$-(7M - 14)$	$-(6M - 8)$	$-(3M - 6)$	M	0	M	0	$-9M$	

	y_1	y_2	y_3	y_4	y_5	y_6	y_7	r.h.s.	Ratio
	0	$\boxed{\frac{5}{2}}$	$\frac{5}{4}$	-1	1	$\frac{3}{4}$	$-\frac{3}{4}$	2	$\frac{4}{5}$
	1	$\frac{1}{2}$	$\frac{1}{4}$	0	0	$-\frac{1}{4}$	$\frac{1}{4}$	1	2
y_0	0	$-(\frac{5}{2}M - 1)$	$-(\frac{5}{4}M - \frac{5}{2})$	M	0	$-(\frac{3}{4}M - \frac{7}{2})$	$\dfrac{7M - 14}{4}$	$-(2M + 14)$	

	y_1	y_2	y_3	y_4	y_5	y_6	y_7	r.h.s.
	0	1	$\frac{1}{2}$	$-\frac{2}{5}$	$\frac{2}{5}$	$\frac{3}{10}$	$-\frac{3}{10}$	$\frac{4}{5}$
	1	0	0	$\frac{1}{5}$	$-\frac{1}{5}$	$-\frac{2}{5}$	$\frac{2}{5}$	$\frac{3}{5}$
y_0	0	0	2	$\frac{2}{5}$	$\dfrac{10M - 4}{10}$	$\frac{16}{5}$	$\dfrac{5M - 8}{5}$	$-\frac{74}{5}$

Hence the solution to the original minimization problem is

$$y_1^* = \tfrac{3}{5}, \qquad y_2^* = \tfrac{4}{5},$$
$$y_i^* = 0, \qquad \text{otherwise}$$
$$y_0^* = \tfrac{74}{5} = 14\tfrac{4}{5}.$$

The solution to the primal problem can be found by observing the slack variables y_4 and y_6, in the objective function row. Thus x_1^* has value $\tfrac{2}{5}$ and x_2^* value $\tfrac{16}{5}$. The vintner should produce $\tfrac{2}{5}$ gallon of the medium white wine and $3\tfrac{1}{5}$ gallons of the dry white wine. He would then maximize his profit at $14.80.

9(a).

$$\begin{aligned}
\text{Maximize:} \quad & 5x_1 + 4x_2 \\
\text{subject to:} \quad & 3x_1 + 4x_2 + x_3 && = 14 \\
& 4x_1 + 2x_2 && + x_4 && = 8 \\
& 2x_1 + x_2 && && + x_5 = 6 \\
& x_i \geq 0, \quad i = 1, 2, \ldots, 5.
\end{aligned}$$

The problem is now solved using the simplex method.

	x_1	x_2	x_3	x_4	x_5	r.h.s.	Ratio
	3	4	1	0	0	14	$\frac{14}{3}$
	$\boxed{4}$	2	0	1	0	8	$\frac{8}{3}$
	2	1	0	0	1	6	
x_0	-5	-4	0	0	0	0	

	x_1	x_2	x_3	x_4	x_5	r.h.s.	Ratio
	0	$\frac{5}{2}$	1	$-\frac{3}{4}$	0	8	$\frac{5}{3}$
	1	$\frac{1}{2}$	0	$\frac{1}{4}$	0	2	4
	0	0	0	$-\frac{1}{2}$	1	2	
x_0	0	$-\frac{3}{2}$	0	$\frac{5}{4}$	0	10	

	x_1	x_2	x_3	x_4	x_5	r.h.s.
	0	1	$\frac{2}{5}$	$-\frac{3}{10}$	0	$\frac{16}{5}$
	1	0	$-\frac{1}{5}$	$\frac{2}{5}$	0	$\frac{2}{5}$
	0	0	0	$-\frac{1}{2}$	1	2
x_0	0	0	$\frac{3}{5}$	$\frac{4}{5}$	0	$14\frac{4}{5}$

Suppose c_2 is changed from 4 to $4 + p$. Then the initial simplex tableau for the problem becomes

	x_1	x_2	x_3	x_4	x_5	r.h.s.
	3	4	1	0	0	14
	4	2	0	1	0	8
	2	1	0	0	1	6
x_0	-5	$-(4 + p)$	0	0	0	0

The corresponding tableau from this table would be

	x_1	x_2	x_3	x_4	x_5	r.h.s.
	0	1	$\frac{2}{5}$	$-\frac{3}{10}$	0	$\frac{16}{5}$
	1	0	$-\frac{1}{5}$	$\frac{2}{5}$	0	$\frac{2}{5}$
	0	0	0	$-\frac{1}{2}$	1	2
x_0	0	$-p$	$\frac{3}{5}$	$\frac{4}{5}$	0	$14\frac{4}{5}$

In order for the present basis to remain optimal, x_2 must still be basic. Therefore the x_2 value in the x_0 row must have zero value. This results in the following tableau.

	x_1	x_2	x_3	x_4	x_5	r.h.s.
	0	1	$\frac{2}{5}$	$-\frac{3}{10}$	0	$\frac{16}{5}$
	1	0	$-\frac{1}{5}$	$\frac{2}{5}$	0	$\frac{2}{5}$
	0	0	0	$-\frac{1}{2}$	1	2
x_0	0	0	$\frac{3}{5} + \frac{2}{5}p$	$\frac{4}{5} - \frac{3}{10}p$	0	$\frac{74}{5} + \frac{16}{5}p$

For the present basis to remain optimal all x_0-row values must be non-negative. Thus

$$\tfrac{3}{5} + \tfrac{2}{5}p \geq 0$$

and

$$\tfrac{4}{5} - \tfrac{3}{10}p \geq 0.$$

This implies

$$-\tfrac{3}{2} \leq p \leq \tfrac{8}{3}.$$

Hence the range for c_2 is

$$(4 - \tfrac{3}{2}, 4 + \tfrac{8}{3}) = (\tfrac{5}{2}, \tfrac{20}{3}).$$

10(a). Consider the problem:

$$\text{Maximize:} \quad 4x_1 + 5x_2 = x_0$$
$$\text{subject to:} \quad 3x_1 + 4x_2 \leq 14$$
$$4x_1 + 2x_2 \leq 8$$
$$2x_1 + x_2 \leq 6.$$

The final tableau is

	x_1	x_2	x_3	x_4	x_5	r.h.s.
	0	1	$\tfrac{2}{5}$	$-\tfrac{3}{10}$	0	$\tfrac{16}{5}$
	1	0	$-\tfrac{1}{5}$	$\tfrac{2}{5}$	0	$\tfrac{2}{5}$
	0	0	0	$-\tfrac{1}{2}$	1	2
x_0	0	0	$\tfrac{3}{5}$	$\tfrac{4}{5}$	0	$14\tfrac{4}{5}$

Suppose we change the r.h.s. constant of the first constraint from 14 to $14 + y$. Since x_3 is the slack variable for this constraint, all the r.h.s. values in the final tableau will change to

$$\tfrac{16}{5} + \tfrac{2}{5}y$$
$$\tfrac{2}{5} - \tfrac{1}{5}y.$$

However, in order that the solution be feasible these values must be non-negative. Thus

$$\tfrac{16}{5} + \tfrac{2}{5}y \geq 0 \Rightarrow y \geq -8$$
$$\tfrac{2}{5} - \tfrac{1}{5}y \geq 0 \Rightarrow y \leq 2.$$

$y \geq -8$ implies that the r.h.s. constant must be greater than 6 and $y \leq 2$ implies that the r.h.s. constant must be smaller than 16 in order for the solution to be feasible. Thus the range is $-8 \leq y \leq 2$, with a r.h.s. constant range of 6 to 16. This means that for the problem to have an optimal and feasible solution the number of boxes of graphs can be no less than 6 or no greater than 16.

11(a). From 9(a), the tableau:

x_1	x_2	x_3	x_4	x_5	r.h.s.
3	4	1	0	0	14
4	2	0	1	0	8
2	1	0	0	1	6
-5	-4	0	0	0	0

becomes at optimality:

x_1	x_2	x_3	x_4	x_5	r.h.s.
0	1	$\frac{2}{5}$	$-\frac{3}{10}$	0	$\frac{16}{5}$
1	0	$-\frac{1}{5}$	$\frac{2}{5}$	0	$\frac{2}{5}$
0	0	0	$-\frac{1}{2}$	1	2
0	0	$\frac{3}{5}$	$\frac{4}{5}$	0	$\frac{74}{5}$

If a_{31} becomes $7\frac{1}{2}$ instead of 2, the same iterations produce:

x_1	x_2	x_3	x_4	x_5	r.h.s.
0	1	$\frac{2}{5}$	$-\frac{3}{10}$	0	$\frac{16}{5}$
1	0	$-\frac{1}{5}$	$\frac{2}{5}$	0	$\frac{2}{5}$
$\frac{11}{2}$	0	0	$-\frac{1}{2}$	1	2
0	0	$\frac{3}{5}$	$\frac{4}{5}$	0	$\frac{74}{5}$

which in canonical form is

x_1	x_2	x_3	x_4	x_5	r.h.s.
0	1	$\frac{2}{5}$	$-\frac{3}{10}$	0	$\frac{16}{5}$
1	0	$-\frac{1}{5}$	$\frac{2}{5}$	0	$\frac{2}{5}$
0	0	$\frac{11}{10}$	$-\frac{27}{10}$	1	$-\frac{1}{5}$
0	0	$\frac{3}{5}$	$\frac{4}{5}$	0	$\frac{74}{5}$

This is infeasible, as $x_5 < 0$. Using the dual simplex method, x_4 replaces x_5 in the basis (the only negative ratio):

x_1	x_2	x_3	x_4	x_5	r.h.s.
0	1	$\frac{5}{18}$	0	$-\frac{1}{9}$	$\frac{29}{9}$
1	0	$-\frac{7}{27}$	0	$\frac{4}{27}$	$\frac{10}{27}$
0	0	$-\frac{11}{27}$	1	$-\frac{10}{27}$	$\frac{2}{27}$
0	0	$\frac{25}{27}$	0	$-\frac{8}{27}$	$\frac{398}{27}$

Thus the new optimal solution is

$$x_1^* = \tfrac{10}{27},$$
$$x_2^* = \tfrac{29}{9},$$
$$x_0^* = \tfrac{398}{27}.$$

12(a). Solving 5(b), let

x_1 = the number of truckloads of A
x_2 = the number of truckloads of B
x_3 = the number of truckloads of C
x_4 = the number of truckloads of D.

Then the problem is to

maximize:	$2x_1 + \ 3x_2 + \ 4x_3 + \ 7x_4 = x_0$ (\$100)
subject to:	$16x_1 + 15x_2 + 20x_3 + 30x_4 \leq 150$ (Area)
	$x_1 + \ 9x_2 + \ \ x_3 + \ 2x_4 \leq 10$ (Manpower)
	$x_i \geq 0, \qquad i = 1, 2, 3, 4.$

(I) In standard form, this is

Maximize: $x_0 = 2x_1 + 3x_2 + 4x_3 + 7x_4$

subject to: $16x_1 + 15x_2 + 20x_3 + 30x_4 + x_5 \qquad\quad = 150$

                 $x_1 + \ 9x_2 + \ \ x_3 + \ 2x_4 \qquad\ + x_6 = 10$

                        $x_i \geq 0, \qquad i = 1, 2, 3, 4.$

Table a

x_1	x_2	x_3	x_4	x_5	x_6	r.h.s.
16	15	20	30	1	0	150
1	9	1	②	0	1	10
-2	-3	-4	-7	0	0	10
1	$\tfrac{3}{2}$	5	0	1	-15	0
$\tfrac{1}{2}$	$-\tfrac{9}{20}$	ⓛ$\tfrac{1}{2}$	1	0	$\tfrac{1}{2}$	5
$\tfrac{3}{2}$	$\tfrac{3}{20}$	$-\tfrac{1}{2}$	0	0	$\tfrac{7}{2}$	35
$\tfrac{1}{5}$	$\tfrac{3}{10}$	1	0	$\tfrac{1}{5}$	-3	0
$\tfrac{2}{5}$	$\tfrac{3}{10}$	0	1	$-\tfrac{1}{10}$	2	5
$\tfrac{8}{5}$	$\tfrac{3}{10}$	0	0	$\tfrac{1}{10}$	2	35

Table a shows the iterations to the optimal solution.

(II) Suppose a new variable x_7 is introduced which represents the amount of the new product to be stored. The problem then becomes:

Maximize: $2x_1 + 3x_2 + 4x_3 + 7x_4 + 5x_7 = x_0$

subject to: $16x_1 + 15x_2 + 20x_3 + 30x_4 + x_5 + 2x_7 = 150$

$$x_1 + 9x_2 + x_3 + 2x_4 + x_6 + x_7 = 10$$

$$x_i \geq 0, \qquad i = 1, 2, \ldots, 7.$$

The dual of this problem is

Minimize: $150y_1 + 10y_2 = y_0$

subject to: $16y_1 + y_2 \geq 2$

$$15y_1 + 9y_2 \geq 3$$
$$20y_1 + y_2 \geq 4$$
$$30y_1 + 2y_2 \geq 7$$
$$2y_1 + y_2 \geq 5$$
$$y_1, y_2 \geq 0.$$

The last constraint can be tested to see whether the present primal solution is optimal or not. Now

$$y_1 = x_5 = \tfrac{1}{10}, \qquad y_2 = x_6 = 2$$

and

$$2(\tfrac{1}{10}) + 2 < 5.$$

Hence the present primal solution is suboptimal. If the new primal (II) had the same primal iterations applied to it as had (I) to produce Table a, the final tableau would be

x_1	x_2	x_3	x_4	x_5	x_6	x_7	r.h.s.
$\frac{1}{5}$	$\frac{3}{10}$	1	0	$\frac{1}{5}$	-3	$[(\frac{1}{5})(2) + (-3)(1)]$	0
$\frac{2}{5}$	$\frac{3}{10}$	0	1	$-\frac{1}{10}$	2	$[(-\frac{1}{10})(2) + (2)(1)]$	5
$\frac{8}{5}$	$\frac{3}{10}$	0	0	$-\frac{1}{10}$	2	$[-5 + (\frac{1}{10})(2) + (2)(1)]$	35
$\frac{1}{5}$	$\frac{3}{10}$	1	0	$\frac{1}{5}$	-3	$-\frac{13}{5}$	0
$\frac{2}{5}$	$\frac{3}{10}$	0	1	$-\frac{1}{10}$	2	$\frac{9}{5}$	5
$\frac{8}{5}$	$\frac{3}{10}$	0	0	$\frac{1}{10}$	2	$-\frac{14}{5}$	35
$\frac{7}{9}$	$\frac{11}{5}$	1	$\frac{13}{9}$	$-\frac{1}{18}$	$-\frac{1}{9}$	0	$\frac{65}{9}$
$\frac{2}{9}$	$\frac{1}{6}$	0	$\frac{5}{9}$	$\frac{1}{18}$	$\frac{10}{9}$	1	$\frac{25}{9}$
$\frac{100}{45}$	$\frac{23}{30}$	0	$\frac{14}{9}$	$-\frac{1}{18}$	$\frac{46}{9}$	0	$\frac{385}{9}$
14	$\frac{66}{5}$	18	26	1	-2	0	130
1	$\frac{9}{10}$	1	2	0	1	1	10
3	$\frac{3}{2}$	1	3	0	5	0	50

So the new profit is \$5,000 and

$$x_7^* = 10,$$
$$x_i^* = 0, \qquad i = 1, 2, 3, 4.$$

13(a). Production constraints for breweries 1, 2, 3, 4 are

$$x_{11} + x_{12} + x_{13} + x_{14} \le 20$$
$$x_{21} + x_{22} + x_{23} + x_{24} \le 10$$
$$x_{31} + x_{32} + x_{33} + x_{34} \le 10$$
$$x_{41} + x_{42} + x_{43} + x_{44} \le 15.$$

Demand constraints for hotels 1, 2, 3, 4 are

$$x_{11} + x_{21} + x_{31} + x_{41} \ge 15$$
$$x_{12} + x_{22} + x_{32} + x_{42} \le 20$$
$$x_{13} + x_{23} + x_{33} + x_{43} \ge 10$$
$$x_{14} + x_{24} + x_{34} + x_{44} \ge 10.$$

All quantities transported must be nonnegative. Thus

$$x_{ij} \ge 0, \qquad i = 1, 2, 3, 4$$
$$j = 1, 2, 3, 4.$$

The objective was to find a supply schedule with minimum cost. The total cost is the sum of all costs from all breweries to all hotels. This cost x_0 can be expressed as

$$x_0 = 8x_{11} + 14x_{12} + 12x_{13} + 17x_{14} + 11x_{21} + 9x_{22} + 15x_{23} + 13x_{24}$$
$$+ 12x_{31} + 19x_{32} + 10x_{33} + 6x_{34} + 12x_{41} + 5x_{42} + 13x_{43} + 18x_{44}.$$

The problem can now be summarized in linear programming form:

Minimize: $x_0 = 8x_{11} + 14x_{12} + 12x_{13} + 17x_{14} + 11x_{21} + 9x_{22}$
$$+ 15x_{23} + 13x_{24} + 12x_{31} + 19x_{32} + 10x_{33} + 6x_{34}$$
$$+ 12x_{41} + 5x_{42} + 13x_{43} + 18x_{44}$$

subject to: $x_{11} + x_{12} + x_{13} + x_{14} \le 20$
$$x_{21} + x_{22} + x_{23} + x_{24} \le 10$$
$$x_{31} + x_{32} + x_{33} + x_{34} \le 10$$
$$x_{41} + x_{42} + x_{43} + x_{44} \le 15$$

$$x_{11} + x_{21} + x_{31} + x_{41} \ge 15$$
$$x_{12} + x_{22} + x_{32} + x_{42} \ge 20$$
$$x_{13} + x_{23} + x_{33} + x_{43} \ge 10$$
$$x_{14} + x_{24} + x_{34} + x_{44} \ge 10$$
$$x_{ij} \ge 0, \qquad i = 1, 2, 3, 4$$
$$j = 1, 2, 3, 4.$$

The tableau for the example problem is given below.

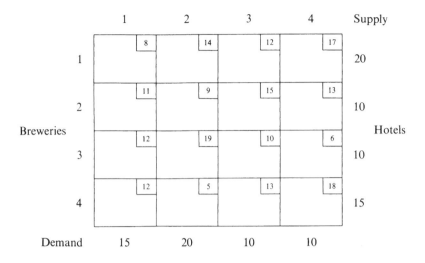

Identification of initial feasible solution by the four required methods follows.

(1) *The Northwest Corner Method.* The method starts by allocating as much as possible to the cell in the northwest corner of the tableau of the problem, cell 1, 1 or row 1, column 1. The maximum that can be allocated is 15 units, as the demand of hotel is 15 units. Column 1 is removed and cell 1, 2 becomes the new northwest corner. A maximum of 5 units is allocated to this cell, all that remains in brewery 1. Row 1 is removed and cell 2, 3

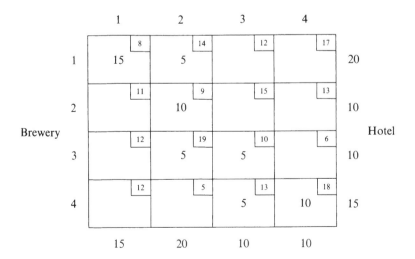

becomes the new northwest corner. This procedure continues until all demand is met. The tableau shows the feasible solution obtained.

Conclusion
  Brewery 1 supplies 15 units to hotel 1 and 5 units to hotel 2.
  Brewery 2 supplies 10 units to hotel 2.
  Brewery 3 supplies 5 units to hotel 2 and 5 units to hotel 3.
  Brewery 4 supplies 5 units to hotel 3 and 10 units to hotel 4.
Total cost: 670 units.

(2) *The Least Cost Method.* This method starts by allocating the largest possible amount to the cell in the tableau with the least unit cost. This means allocating 15 units to cell 4, 2, and row 4 is removed. The demand of hotel 2 is reduced to 5 units. The cell with the next smallest cost is identified, i.e., cell 3, 4 and 10 units are allocated to it removing row 3 and column 4. This procedure continues until all the demand is met. The following tableau illustrates the feasible solution which is obtained.

	1	2	3	4	
1	8 — 15	14	12 — 5	17	20
2	11	9 — 5	15 — 5	13	10
3	12	19	10	6 — 10	10
4	12	5 — 15	13	18	15
	15	20	10	10	

Conclusion
  Brewery 1 supplies 15 units to hotel 1 and 5 units to hotel 3.
  Brewery 2 supplies 5 units to hotel 2 and 5 units to hotel 3.
  Brewery 3 supplies 10 units to hotel 4.
  Brewery 4 supplies 15 units to hotel 2.
Total cost: 435 units.

(3) *The Vogel Approximation Method.* This method begins by first reducing the matrix of unit costs. This reduction is achieved by subtracting the minimum quantity in each row from all elements in that row. This results in the following tableau:

	1	2	3	4		
1	0	6	4	9	(-8)	20
2	2	0	6	4	(-9)	10
3	6	13	4	0	(-6)	10
4	7	0	8	13	(-5)	15
	15	20	10	10		

The costs are further reduced by carrying out this procedure on the columns of the new cost matrix:

	1	2	3	4
1	0	6	0	9
2	2	0	2	4
3	6	13	0	0
4	7	0	4	13
	(0)	(0)	(-4)	(0)

A penalty is then calculated for each cell which currently has zero unit cost. Each cell penalty is found by adding together the second smallest costs of the row and column of the cell:

	1	2	3	4	
1	0 (2)	6 (0)	0 (0)	9	(0)
2	2	0 (2)	2	4	(2)
3	6	13	0 (0)	0 (4)	(0)
4	7	0 (4)	4	13	(4)
	(2)	(0)	(0)	(4)	

The penalties are shown in the top right-hand corner of each appropriate cell and the cell with the largest penalty is identified. The maximum amount possible is then allocated to this cell. Cell 3, 4 will be arbitrarily chosen and 10 units are allocated to it. Row 3 and column 4 are removed from consideration. A further reduction in the cost matrix and a recalculation of some penalties is necessary. This results in the following tableau:

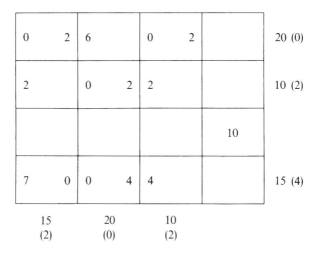

Cell 4, 2 is chosen and 15 units are allocated to it. Row 4 is then removed from consideration. This process is repeated until all demand is met.

0 2	6 0	0 2		20 (0)
2	0 8	2		10 (2)
			10	
	15			
15 (2)	20 (6)	10 (2)		

Cell 2, 2 is chosen and 5 units are allocated to it, removing column 2 from consideration.

0 2		0 2		20 (0)
2	5	2		10 (2)
			10	
	15			
15 (2)		10 (2)		

Cell 1, 1 is arbitrarily chosen and 15 units are allocated to it, removing column 1. Cell 1, 3 must be allocated 5 units and cell 2, 3 5 units in order that all demand shall be met.

The final allocation is shown in the following tableau:

	1	2	3	4	
1	8 · 15	14	12 · 5	17	20
2	11	9 · 5	15 · 5	13	10
3	12	19	10	6 · 10	10
4	12	5 · 15	13	18	15
	15	20	10	10	

Conclusion

Brewery 1 supplies 15 units to hotel 1 and 5 units to hotel 3.
Brewery 2 supplies 5 units to hotel 2 and 5 units to hotel 3.
Brewery 3 supplies 10 units to hotel 4.
Brewery 4 supplies 15 units to hotel 2.
Total cost: 435 units.

(4) *Stepping Stone Algorithm.* Consider the initial feasible solution found by the northwest corner method.

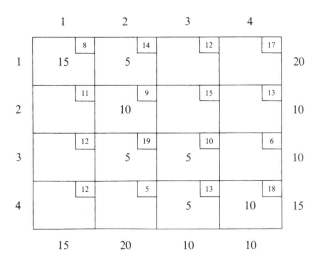

	1	2	3	4	
1	8 · 15	14 · 5	12	17	20
2	11	9 · 10	15	13	10
3	12	19 · 5	10 · 5	6	10
4	12	5	13 · 5	18 · 10	15
	15	20	10	10	

To determine whether this solution is optimal or not it is necessary to ask for each cell individually if the allocation of one unit to that cell would reduce the total cost. This is done for the cells which at present have no units assigned to them.

Cell 4, 2 has the greatest decrease (17 units) and as much as possible, (5 units) is allocated to this cell. This means a decrease in cost of $(17 \times 5) = \$85$.

The new solution is displayed in the following tableau.

	8		14		12		17
15		5					
	11		9		15		13
10		0					
	12		19		10		6
				10			
	12		5		13		18
		5			10		

The same procedure occurs—all empty cells in the new tableau are examined as before and the process is repeated. Since a basic feasible solution should contain $(m + n - 1)$ basic variables, one of the empty cells is assigned a zero. Cell 2, 4 has the greatest decrease (19 units) and as much as possible (10 units) is allocated to this cell. This means a decrease in cost of $(19 \times 5) = \$95$. The new solution is displayed.

	8		14		12		17
15		5					
	11		9		15		13
				10			
	12		19		10		6
0						10	
	12		5		13		18
		15		0			

The process is repeated and cell 2, 3 with a decrease of 8 units is allocated 5 units. This means a decrease in cost of $(8 × 5) = $40. The tableau is displayed next.

	1	2	3	4	
1	8 15	14	12 5	17	20
2	11	9 5	15 5	13	10
3	12	19	10	6 10	10
4	12	5 15	13	18 0	15
	15	20	10	10	

The process is repeated, but there is no allocation which will cause a cost reduction. Thus the optimal solution has been found.

Conclusion
Brewery 1 supplies 15 units to hotel 1 and 5 units to hotel 3
Brewery 2 supplies 5 units to hotel 2 and 5 units to hotel 3
Brewery 3 supplies 10 units to hotel 4.
Brewery 4 supplies 15 units to hotel 2.
Total cost: 435 units.

14(a)

		1	2	3	4	5	6	Tasks
	1	7	5	3	9	2	4	(-1)
	2	8	6	1	4	5	2	(-0)
	3	2	3	5	6	8	9	(-0)
Students	4	6	8	1	3	7	2	(-0)
	5	4	5	6	9	4	7	(-2)
	6	9	2	3	5	1	8	(-0)
		(-2)	(-2)	(-1)	(-3)	(-1)	(-2)	

(1) Subtract minimum quantity from each column and row of c_{ij} matrix to obtain:

$$
\begin{array}{cccccc}
4 & 2 & 1 & 5 & 0 & 1 \\
6 & 4 & 0 & 1 & 4 & 0 \\
0 & 2 & 4 & 3 & 7 & 7 \\
4 & 6 & 0 & 0 & 6 & 0 \\
0 & 1 & 3 & 4 & 1 & 3 \\
7 & 0 & 2 & 2 & 0 & 6
\end{array}
$$

(2) As the minimum number of lines is less than n, the minimum uncrossed number is subtracted from all the uncrossed numbers and added to all numbers with two lines passing through them to obtain:

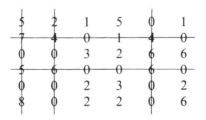

(3) The process of (1) and (2) is repeated to obtain

$$
\begin{array}{cccccc}
5 & 2 & 0 & 4 & 0 & 0 \\
8 & 5 & 0 & 1 & 5 & 0 \\
0 & 0 & 2 & 1 & 6 & 5 \\
6 & 7 & 0 & 0 & 7 & 0 \\
0 & 0 & 1 & 2 & 0 & 1 \\
8 & 0 & 1 & 1 & 0 & 5
\end{array}
$$

The solution for this problem is

		Student	Task
$x_{13} = 1$	i.e.	1	3
$x_{26} = 1$	i.e.	2	6
$x_{31} = 1$	i.e.	3	1
$x_{44} = 1$	i.e.	4	4
$x_{52} = 1$	i.e.	5	2
$x_{65} = 1$	i.e.	6	5.

The value of this solution is equal to the total of the numbers subtracted, i.e.,

$$x_0^* = 2 + 2 + 1 + 3 + 1 + 2 + 1 + 0 + 0 + 0 + 2 + 0 + 1 + 1 = 16.$$

By inspecting the original c_{ij} matrix we also obtain

$$x_0^* = 3 + 2 + 2 + 3 + 5 + 1 = 16.$$

Chapter 3

Section 3.7

1(a)

B_1	c_B	B_1^{-1}				π_1	b	c_1	Ratio	Entering	Leaving
x_5	0	1	0	0	0	0	9	3	3		
x_6	0	0	1	0	0	0	12	1	12		
x_7	0	0	0	1	0	0	8	2	4		
x_8	0	0	0	0	1	0	10	3	3		
$-\bar{c}_1$		-3	-2	-1	-2	0	0	0	0	x_1	x_5

B_2	c_B	B_2^{-1}				π_2	b	c_2	Ratio
x_1	-3	$\frac{1}{3}$	0	0	0	-1	3	$\frac{1}{3}$	9
x_6	0	$-\frac{1}{3}$	1	0	0	0	9	$\frac{5}{3}$	$\frac{27}{5}$
x_7	0	$-\frac{2}{3}$	0	1	0	0	2	$\frac{1}{3}$	6
x_8	0	-1	0	0	1	0	1	2	$\frac{1}{2}$
$-\bar{c}_2$		0	-1	0	0	1	0	0	0

B_3	c_B	B_3^{-1}				π_3	b	c_4	Ratio
x_1	-3	$\frac{1}{2}$	0	0	$-\frac{1}{6}$	$-\frac{1}{2}$	$\frac{17}{6}$	$\frac{5}{6}$	$\frac{17}{5}$
x_6	0	$\frac{1}{2}$	1	0	$-\frac{5}{6}$	0	$\frac{49}{6}$	$\frac{25}{6}$	$\frac{49}{25}$
x_7	0	$-\frac{1}{2}$	0	1	$-\frac{1}{6}$	0	$\frac{11}{6}$	$-\frac{1}{6}$	
x_2	-2	$-\frac{1}{2}$	0	0	$\frac{1}{2}$	$-\frac{1}{2}$	$\frac{1}{2}$	$-\frac{1}{2}$	
$-\bar{c}_3$		0	0	$\frac{1}{2}$	$-\frac{1}{2}$	$\frac{1}{2}$	0	0	$\frac{1}{2}$

B_4	c_b	B_4^{-1}				π_4	b	
x_1	-3	$\frac{2}{5}$	$-\frac{1}{5}$	0	0	$\frac{2}{5}$	$\frac{6}{5}$	
x_4	-2	$\frac{3}{25}$	$\frac{6}{25}$	0	$-\frac{1}{5}$	$-\frac{3}{25}$	$\frac{49}{5}$	
x_7	0	$-\frac{12}{25}$	$\frac{1}{25}$	1	$-\frac{1}{5}$	0	$\frac{54}{5}$	
x_2	-2	$-\frac{11}{25}$	$\frac{3}{25}$	0	$\frac{2}{5}$	$-\frac{2}{5}$	$\frac{37}{25}$	
$-\bar{c}_4$		0	0	$\frac{12}{25}$	0	$\frac{14}{25}$	$\frac{3}{25}$	$\frac{2}{5}$

$$x_1^* = \tfrac{6}{5}, \qquad x_4^* = \tfrac{49}{5}, \qquad x_7^* = \tfrac{54}{5}, \qquad x_2^* = \tfrac{37}{25}, \qquad x_3^*, x_5^*, x_6^*, x_8^* = 0,$$
$$x_0^* = c_B b = \tfrac{262}{25}.$$

2(a)

x_1	x_2	x_3	x_4	x_5	x_6	x_7	x_8	r.h.s.	Ratio
③	1	1	2	1	0	0	0	9	3
1	2	1	4	0	1	0	0	12	12
2	1	3	1	0	0	1	0	8	4
3	$\tfrac{1}{3}$	2	1	0	0	0	1	10	$\tfrac{10}{3}$
-3	-2	-1	-2	0	0	0	0	0	
1	$\tfrac{1}{3}$	$\tfrac{1}{3}$	$\tfrac{2}{3}$	$\tfrac{1}{3}$	0	0	0	3	9
0	$\tfrac{5}{3}$	$\tfrac{2}{3}$	$\tfrac{10}{3}$	$-\tfrac{1}{3}$	1	0	0	9	$\tfrac{27}{5}$
0	$\tfrac{1}{3}$	$\tfrac{7}{3}$	$-\tfrac{1}{3}$	$-\tfrac{2}{3}$	0	1	0	2	6
0	②	1	-1	-1	0	0	1	1	$\tfrac{1}{2}$
0	-1	0	0	1	0	0	0	9	
1	0	$\tfrac{1}{6}$	$\tfrac{5}{6}$	$\tfrac{1}{2}$	0	0	$-\tfrac{1}{6}$	$\tfrac{17}{6}$	$\tfrac{17}{5}$
0	0	$-\tfrac{1}{6}$	$\boxed{\tfrac{25}{6}}$	$\tfrac{1}{2}$	1	0	$-\tfrac{5}{6}$	$\tfrac{49}{6}$	$\tfrac{49}{5}$
0	0	$\tfrac{13}{6}$	$-\tfrac{1}{6}$	$-\tfrac{1}{2}$	0	1	$-\tfrac{1}{6}$	$\tfrac{11}{6}$	
0	1	$\tfrac{1}{2}$	$-\tfrac{1}{2}$	$-\tfrac{1}{2}$	0	0	$\tfrac{1}{2}$	$\tfrac{1}{2}$	
0	0	$\tfrac{1}{2}$	$-\tfrac{1}{2}$	$\tfrac{1}{2}$	0	0	$\tfrac{1}{2}$	$\tfrac{19}{2}$	
1	0	$\tfrac{1}{5}$	0	$\tfrac{2}{5}$	$-\tfrac{1}{5}$	0	0	$\tfrac{6}{5}$	
0	0	$-\tfrac{1}{25}$	1	$\tfrac{3}{25}$	$\tfrac{6}{25}$	0	$-\tfrac{1}{5}$	$\tfrac{49}{25}$	
0	0	$\tfrac{54}{25}$	0	$-\tfrac{12}{25}$	$\tfrac{1}{25}$	1	$-\tfrac{1}{5}$	$\tfrac{54}{25}$	
0	1	$\tfrac{12}{25}$	0	$-\tfrac{11}{25}$	$\tfrac{3}{25}$	0	$\tfrac{2}{5}$	$\tfrac{37}{25}$	
0	0	$\tfrac{12}{25}$	0	$\tfrac{14}{25}$	$\tfrac{3}{25}$	0	$\tfrac{2}{5}$	$\tfrac{265}{25}$	

The solution and its value are as in 1(a).

3(a). Forming the dual:

$$
\begin{aligned}
\text{Minimize:} \quad & 6y_1 + 8y_2 + 9y_3 + 12y_4 = y_0 \\
\text{subject to:} \quad & y_1 + 3y_2 + 2y_3 + 2y_4 \ge 3 \\
& 2y_1 + 4y_2 + 3y_3 + y_4 \ge 2 \\
& y_1 + 2y_2 + 3y_3 + 2y_4 \ge 1 \\
& 3y_1 + y_2 + y_3 + 2y_4 > 2 \\
& y_1, y_2, y_3, y_4 \ge 0.
\end{aligned}
$$

Introducing y_5, y_6, y_7, y_8 as slack variables:

	y_1	y_2	y_3	y_4	y_5	y_6	y_7	y_8	r.h.s.
	1	3	2	2	-1	0	0	0	3
	2	4	3	1	0	-1	0	0	2
	1	2	3	2	0	0	-1	0	1
	3	1	1	2	0	0	0	-1	2
y_0	6	8	9	12	0	0	0	0	0

Multiplying each constraint by (-1):

	y_1	y_2	y_3	y_4	y_5	y_6	y_7	y_8	r.h.s.
	-1	$\left(-3\right)$	-2	-2	1	0	0	0	-3
	-2	-4	-3	-1	0	1	0	0	-2
	-1	-2	-3	-2	0	0	1	0	-1
	-3	-1	-1	-2	0	0	0	1	-2
y_0	6	8	9	12	0	0	0	0	0
ratios	-6	$-\frac{8}{3}$	$-\frac{9}{2}$	-6					
	$\frac{1}{3}$	1	$\frac{2}{3}$	$\frac{2}{3}$	$-\frac{1}{3}$	0	0	0	1
	$-\frac{2}{3}$	0	$-\frac{1}{3}$	$\frac{5}{3}$	$-\frac{4}{3}$	1	0	0	2
	$-\frac{1}{3}$	0	$-\frac{5}{3}$	$-\frac{2}{3}$	$-\frac{2}{3}$	0	1	0	1
	$\left(-\frac{8}{3}\right)$	0	$-\frac{1}{3}$	$-\frac{4}{3}$	$-\frac{1}{3}$	0	0	1	-1
y_0	$\frac{10}{3}$	0	$\frac{11}{3}$	$\frac{20}{3}$	$\frac{8}{3}$	0	0	0	-8
Ratios	$-\frac{5}{4}$		-11	-5	-8	0	0	0	
	0	1	$\frac{5}{8}$	$\frac{1}{2}$	$-\frac{3}{8}$	0	0	$\frac{1}{8}$	$\frac{7}{8}$
	0	0	$-\frac{1}{4}$	2	$-\frac{5}{4}$	1	0	$-\frac{1}{4}$	$\frac{9}{4}$
	0	0	$-\frac{13}{8}$	$-\frac{1}{2}$	$-\frac{5}{8}$	0	1	$-\frac{1}{8}$	$\frac{9}{8}$
	1	0	$\frac{1}{8}$	$\frac{1}{2}$	$\frac{1}{8}$	0	0	$-\frac{3}{8}$	$\frac{3}{8}$
y_0	0	0	$\frac{13}{4}$	5	$\frac{9}{4}$	0	0	$\frac{5}{4}$	$-\frac{37}{4}$

Hence

$$y_1^* = \tfrac{3}{8}, \quad y_2^* = \tfrac{7}{8}, \quad y_6^* = \tfrac{9}{4}, \quad y_7^* = \tfrac{9}{8}, \quad y_0^* = \tfrac{37}{4},$$
$$x_1^* = \tfrac{9}{4}, \quad x_4^* = \tfrac{5}{4}, \quad x_2^*, x_3^* = 0, \quad x_0^* = \tfrac{37}{4}.$$

4(a) (*The Two-Phase Method*). Introducing artificial variables $y_9, y_{10}, y_{11}, y_{12}$:

Phase I

	y_1	y_2	y_3	y_4	y_5	y_9	y_6	y_{10}	y_7	y_{11}	y_8	y_{12}	r.h.s.
	1	3	2	2	-1	1	0	0	0	0	0	0	3
	2	4	3	1	0	0	-1	1	0	0	0	0	2
	1	2	3	2	0	0	0	0	-1	1	0	0	1
	3	1	1	2	0	0	0	0	0	0	-1	1	2
y_0'	0	0	0	0	0	1	0	1	0	1	0	1	0

In canonical form:

	y_1	y_2	y_3	y_4	y_5	y_9	y_6	y_{10}	y_7	y_{11}	y_8	y_{12}	r.h.s.	Ratio
	1	3	2	2	-1	1	0	0	0	0	0	0	3	1
	2	4	3	1	0	0	-1	1	0	0	0	0	2	$\frac{1}{2}$
	1	②	3	2	0	0	0	0	-1	1	0	0	1	$\frac{1}{2}$
	3	1	1	2	0	0	0	0	0	0	-1	1	2	2
y_0'	-7	-10	-9	-7	1	0	1	0	1	0	1	0	-8	
	$\frac{1}{2}$	0	$-\frac{5}{2}$	-1	-1	1	0	0	$\frac{3}{2}$	$-\frac{3}{2}$	0	0	$\frac{3}{2}$	1
	0	0	-3	-3	0	0	-1	1	②	-2	0	0	0	0
	$\frac{1}{2}$	1	$\frac{3}{2}$	1	0	0	0	0	$-\frac{1}{2}$	$\frac{1}{2}$	0	0	$\frac{1}{2}$	—
	$\frac{5}{2}$	0	$-\frac{1}{2}$	1	0	0	0	0	$\frac{1}{2}$	$-\frac{1}{2}$	-1	1	$\frac{3}{2}$	3
	-2	0	6	3	1	0	1	0	-4	5	1	0	-3	
	$-\frac{1}{2}$	0	$-\frac{1}{4}$	$\frac{5}{4}$	-1	1	$\frac{3}{4}$	$-\frac{3}{4}$	0	0	0	0	$\frac{3}{2}$	$\frac{6}{5}$
	0	0	$-\frac{3}{2}$	$-\frac{3}{2}$	0	0	$-\frac{1}{2}$	$\frac{1}{2}$	1	-1	0	0	0	—
	$\frac{1}{2}$	1	$\frac{3}{4}$	$\frac{1}{4}$	0	0	$-\frac{1}{4}$	$\frac{1}{4}$	0	0	0	0	$\frac{1}{2}$	2
	$\frac{5}{2}$	0	$\frac{1}{4}$	⑦⁄₄	0	0	$\frac{1}{4}$	$-\frac{1}{4}$	0	0	-1	1	$\frac{3}{2}$	$\frac{6}{7}$
y_0'	-2	0	0	-3	1	0	-1	2	0	1	1	0	-3	
	$-\frac{16}{7}$	0	$-\frac{3}{7}$	0	-1	1	$\frac{4}{7}$	$-\frac{4}{7}$	0	0	⑤⁄₇	$-\frac{5}{7}$	$\frac{3}{7}$	$\frac{3}{5}$
	$\frac{15}{7}$	0	$-\frac{9}{7}$	0	0	0	$-\frac{2}{7}$	$\frac{2}{7}$	1	-1	$-\frac{6}{7}$	$\frac{6}{7}$	$\frac{9}{7}$	—
	$\frac{1}{7}$	1	$\frac{5}{7}$	0	0	0	$-\frac{2}{7}$	$\frac{2}{7}$	0	0	$\frac{1}{7}$	$-\frac{1}{7}$	$\frac{2}{7}$	2
	$\frac{10}{7}$	0	$\frac{1}{7}$	1	0	0	$\frac{1}{7}$	$-\frac{1}{7}$	0	0	$-\frac{4}{7}$	$\frac{4}{7}$	$\frac{6}{7}$	—
y_0'	$\frac{16}{7}$	0	$\frac{3}{7}$	0	1	0	$-\frac{4}{7}$	$\frac{11}{7}$	0	1	$-\frac{5}{7}$	$\frac{12}{7}$	$-\frac{3}{7}$	
	$-\frac{16}{5}$	0	$-\frac{3}{5}$	0	$-\frac{7}{5}$	$\frac{7}{5}$	$\frac{4}{5}$	$-\frac{4}{5}$	0	0	1	-1	$\frac{3}{5}$	
	$-\frac{3}{5}$	0	$-\frac{9}{5}$	0	$-\frac{6}{5}$	$\frac{6}{5}$	$\frac{2}{5}$	$-\frac{2}{5}$	1	-1	0	0	$\frac{9}{5}$	
	$\frac{3}{5}$	1	$\frac{4}{5}$	0	$\frac{1}{5}$	$-\frac{1}{5}$	$-\frac{2}{5}$	$\frac{2}{5}$	0	0	0	0	$\frac{1}{5}$	
	$-\frac{2}{5}$	0	$-\frac{1}{5}$	1	$-\frac{4}{5}$	$\frac{4}{5}$	$\frac{3}{5}$	$-\frac{3}{5}$	0	0	0	0	$\frac{6}{5}$	
y_0'	0	0	0	0	0	1	0	1	0	1	0	1	0	

Phase II

$$\text{Minimize:} \quad 6y_1 + 8y_2 + 9y_3 + 12y_4 = y_0.$$

	y_1	y_2	y_3	y_4	y_5	y_6	y_7	y_8	r.h.s.
	$-\frac{16}{5}$	0	$-\frac{3}{5}$	0	$-\frac{7}{5}$	$\frac{4}{5}$	0	1	$\frac{3}{5}$
	$-\frac{3}{5}$	0	$-\frac{9}{5}$	0	$-\frac{6}{5}$	$\frac{2}{5}$	1	0	$\frac{9}{5}$
	$\frac{3}{5}$	1	$\frac{4}{5}$	0	$\frac{1}{5}$	$-\frac{2}{5}$	0	0	$\frac{1}{5}$
	$-\frac{2}{5}$	0	$-\frac{1}{5}$	1	$-\frac{4}{5}$	$\frac{3}{5}$	0	0	$\frac{6}{5}$
y_0	6	8	9	12	0	0	0	0	0

In canonical form:

	y_1	y_2	y_3	y_4	y_5	y_6	y_7	y_8	r.h.s.	Ratio
	$-\frac{16}{5}$	0	$-\frac{3}{5}$	0	$-\frac{7}{5}$	$\boxed{\frac{4}{5}}$	0	1	$\frac{3}{5}$	$\frac{3}{4}$
	$-\frac{3}{5}$	0	$-\frac{9}{5}$	0	$-\frac{6}{5}$	$\frac{2}{5}$	1	0	$\frac{9}{5}$	$\frac{9}{2}$
	$\frac{3}{5}$	1	$\frac{4}{5}$	0	$\frac{1}{5}$	$-\frac{2}{5}$	0	0	$\frac{1}{5}$	—
	$-\frac{2}{5}$	0	$-\frac{1}{5}$	1	$-\frac{4}{5}$	$\frac{3}{5}$	0	0	$\frac{6}{5}$	2
y_0	6	0	5	0	8	-4	0	0	-16	
	-4	0	$-\frac{3}{4}$	0	$-\frac{7}{4}$	1	0	$\frac{5}{4}$	$\frac{3}{4}$	—
	1	0	$-\frac{3}{2}$	0	$-\frac{1}{2}$	0	1	$-\frac{1}{2}$	$\frac{3}{2}$	$\frac{3}{2}$
	-1	1	$\frac{1}{2}$	0	$-\frac{1}{2}$	0	0	$\frac{1}{2}$	$\frac{1}{2}$	—
	$\boxed{2}$	0	$\frac{1}{4}$	1	$\frac{1}{4}$	0	0	$-\frac{3}{4}$	$\frac{3}{4}$	$\frac{3}{8}$
y_0	-10	0	2	0	1	0	0	5	-13	
	0	0	$-\frac{1}{4}$	2	$-\frac{5}{4}$	1	0	$-\frac{1}{4}$	$\frac{9}{4}$	
	0	0	$-\frac{13}{8}$	$-\frac{1}{2}$	$-\frac{5}{8}$	0	1	$-\frac{1}{8}$	$\frac{9}{8}$	
	0	1	$\frac{5}{8}$	$\frac{1}{2}$	$-\frac{3}{8}$	0	0	$\frac{1}{8}$	$\frac{7}{8}$	
	1	0	$\frac{1}{8}$	$\frac{1}{2}$	$\frac{1}{8}$	0	0	$-\frac{3}{8}$	$\frac{3}{8}$	
y_0	0	0	$\frac{13}{4}$	5	$\frac{9}{4}$	0	0	$\frac{5}{4}$	$-\frac{37}{4}$	

5(a)

x_1	x_2	x_3	x_4	x_5	x_6	r.h.s.	Ratio
2	1	4	1	0	0	10	5
1	2	1	0	1	0	4	4
$\boxed{3}$	-2	1	0	0	1	6	2
-3	-1	-2	0	0	0	0	

x_1	x_2	x_3	x_4	x_5	x_6	r.h.s.	Ratio
0	$\frac{7}{3}$	$\frac{10}{3}$	1	0	$-\frac{2}{3}$	6	$\frac{18}{7}$
0	$\left(\frac{8}{3}\right)$	$\frac{2}{3}$	0	1	$-\frac{1}{3}$	2	$\frac{3}{4}$
1	$-\frac{2}{3}$	$\frac{1}{3}$	0	0	$\frac{1}{3}$	2	—
0	-3	-1	0	0	1	6	
0	0	$\left(\frac{11}{4}\right)$	1	$-\frac{7}{8}$	$-\frac{3}{8}$	$\frac{17}{4}$	$\frac{17}{11}$
0	1	$\frac{1}{4}$	0	$\frac{3}{8}$	$-\frac{1}{8}$	$\frac{3}{4}$	3
1	0	$\frac{1}{2}$	0	$\frac{1}{4}$	$\frac{1}{4}$	$\frac{5}{2}$	5
0	0	$-\frac{1}{4}$	0	$\frac{9}{8}$	$\frac{5}{8}$	$\frac{33}{4}$	
0	0	1	$\frac{4}{11}$	$-\frac{7}{22}$	$-\frac{3}{22}$	$\frac{17}{11}$	
0	1	0	$-\frac{1}{11}$	$\frac{5}{11}$	$-\frac{1}{11}$	$\frac{4}{11}$	
1	0	0	$-\frac{2}{11}$	$\frac{9}{22}$	$\frac{7}{22}$	$\frac{19}{11}$	
0	0	0	$\frac{1}{11}$	$\frac{23}{22}$	$\frac{13}{22}$	$\frac{95}{11}$	

Hence

$$x_1^* = \tfrac{19}{11}, \qquad x_2^* = \tfrac{4}{11}, \qquad x_3^* = \tfrac{17}{11}, \qquad x_4^*, x_5^*, x_6^* = 0, \qquad x_0^* = \tfrac{95}{11}.$$

Let

θ = amount of elapsed time.

$$x_0 = (3 + \theta)x_1 + (1 + 2\theta)x_2 + (2 + 3\theta)x_3.$$

Then:

x_1	x_2	x_3	x_4	x_5	x_6	r.h.s.
0	0	1	$\frac{4}{11}$	$-\frac{7}{22}$	$-\frac{3}{22}$	$\frac{17}{11}$
0	1	0	$-\frac{1}{11}$	$\frac{5}{11}$	$-\frac{1}{11}$	$\frac{4}{11}$
1	0	0	$-\frac{2}{11}$	$\frac{9}{22}$	$\frac{7}{22}$	$\frac{19}{11}$
$-\theta$	-2θ	-3θ	$\frac{1}{11}$	$\frac{23}{22}$	$\frac{13}{22}$	$\frac{95}{11}$

In canonical form:

x_1	x_2	x_3	x_4	x_5	x_6	r.h.s.
0	0	1	$\frac{4}{11}$	$-\frac{7}{22}$	$-\frac{3}{22}$	$\frac{17}{11}$
0	1	0	$-\frac{1}{11}$	$\frac{5}{11}$	$-\frac{1}{11}$	$\frac{4}{11}$
1	0	0	$-\frac{2}{11}$	$\frac{9}{22}$	$\frac{7}{22}$	$\frac{19}{11}$
0	0	0	$\dfrac{1 + 8\theta}{11}$	$\dfrac{23 + \theta}{22}$	$\dfrac{13 - 6\theta}{22}$	$\dfrac{95 + 7\theta}{11}$

The first critical point is $\theta = \frac{13}{6}$.

x_1	x_2	x_3	x_4	x_5	x_6	r.h.s.
0	0	1	$\frac{4}{11}$	$-\frac{7}{22}$	$-\frac{3}{22}$	$\frac{17}{11}$
0	1	0	$-\frac{1}{11}$	$\frac{5}{11}$	$-\frac{1}{11}$	$\frac{4}{11}$
1	0	0	$-\frac{2}{11}$	$\frac{9}{22}$	$\frac{7}{22}$	$\frac{19}{11}$
0	0	0	$\frac{5}{3}$	$\frac{11}{6}$	0	24

Replacing x_1 by x_6:

x_1	x_2	x_3	x_4	x_5	x_6	r.h.s.
$\frac{3}{7}$	0	1	$\frac{2}{7}$	$-\frac{1}{7}$	0	$\frac{16}{7}$
$\frac{2}{7}$	1	0	$-\frac{1}{7}$	$\frac{4}{7}$	0	$\frac{6}{7}$
$\frac{22}{7}$	0	0	$-\frac{4}{7}$	$\frac{9}{7}$	1	$\frac{38}{7}$
0	0	0	$\frac{5}{3}$	$\frac{11}{6}$	0	24

Searching for further critical points:

x_1	x_2	x_3	x_4	x_5	x_6	r.h.s.
$\frac{3}{7}$	0	1	$\frac{2}{7}$	$-\frac{1}{7}$	0	$\frac{16}{7}$
$\frac{2}{7}$	1	0	$-\frac{1}{7}$	$\frac{4}{7}$	0	$\frac{6}{7}$
$\frac{22}{7}$	0	0	$-\frac{4}{7}$	$\frac{9}{7}$	1	$\frac{38}{7}$
$-\theta$	-2θ	-3θ	$\frac{5}{3}$	$\frac{11}{6}$	0	24

In canonical form:

x_1	x_2	x_3	x_4	x_5	x_6	r.h.s.
$\frac{3}{7}$	0	1	$\frac{2}{7}$	$-\frac{1}{7}$	0	$\frac{16}{7}$
$\frac{2}{7}$	1	0	$-\frac{1}{7}$	$\frac{4}{7}$	0	$\frac{6}{7}$
$\frac{22}{7}$	0	0	$-\frac{4}{7}$	$\frac{9}{7}$	1	$\frac{38}{7}$
$\frac{6}{7}\theta$	0	0	$\frac{5}{3} + \frac{4}{7}\theta$	$\frac{11}{6} + \frac{5}{7}\theta$	0	$24 + \frac{60}{7}\theta$

For $\theta > 0$ this basis is optimal. Thus

$$0 \leq \theta \leq \tfrac{13}{6}: \qquad x_1^* = \tfrac{19}{11}, \qquad x_2^* = \tfrac{4}{11}, \qquad x_3^* = \tfrac{17}{11}, \qquad x_0^* = \frac{95 + 7\theta}{11}$$

$$\theta > \tfrac{13}{6}: \qquad x_2^* = \tfrac{6}{7}, \qquad x_3^* = \tfrac{16}{7}, \qquad x_6^* = \tfrac{38}{7}, \qquad x_0^* = 24 + \tfrac{60}{7}\theta.$$

6(a). Dual:

Minimize: $\quad 10y_1 + 4y_2 + 6y_3$

subject to: $\quad 2y_1 + y_2 + 3y_3 \geq 3$

$\qquad\qquad\quad y_1 + 2y_2 - 2y_2 \geq 1$

$\qquad\qquad\quad 4y_1 + y_2 + y_3 \geq 2$

$\qquad\qquad\quad y_1, y_2, y_3 \geq 0.$

Phase I

Minimize: $\quad y_5 + y_7 + y_9$

y_1	y_2	y_3	y_4	y_5	y_6	y_7	y_8	y_9	r.h.s.	Ratios	
2	1	3	-1	1	0	0	0	0	3		θ
1	2	-2	0	0	-1	1	0	0	1		2θ
4	1	1	0	0	0	0	-1	1	2		3θ
0	0	0	0	1	0	1	0	1	0		0
2	1	3	-1	1	0	0	0	0	3	$\frac{3}{2}$	θ
1	2	-2	0	0	-1	1	0	0	1	1	2θ
④	1	1	0	0	0	0	-1	1	2	$\frac{1}{2}$	3θ
-7	-4	1	0	1	1	0	1	0	-6		-6θ
0	$\frac{1}{2}$	$\frac{5}{2}$	-1	1	0	0	$\frac{1}{2}$	$-\frac{1}{2}$	2	4	$-\frac{1}{2}\theta$
0	⑦⁄₄	$-\frac{9}{4}$	0	0	-1	1	$\frac{1}{4}$	$-\frac{1}{4}$	$\frac{1}{2}$	$\frac{2}{7}$	$\frac{5}{4}\theta$
1	$\frac{1}{4}$	$\frac{1}{4}$	0	0	0	0	$-\frac{1}{4}$	$\frac{1}{4}$	$\frac{1}{2}$	2	$\frac{3}{4}\theta$
0	$-\frac{9}{4}$	$-\frac{1}{4}$	1	0	1	0	$-\frac{3}{4}$	$\frac{7}{4}$	$-\frac{5}{2}$		$-\frac{3}{4}\theta$
0	0	㉒⁄₇	-1	1	$\frac{2}{7}$	$-\frac{2}{7}$	$\frac{3}{7}$	$-\frac{3}{7}$	$\frac{13}{7}$	$\frac{13}{22}$	$-\frac{6}{7}\theta$
0	1	$-\frac{9}{7}$	0	0	$-\frac{4}{7}$	$\frac{4}{7}$	$\frac{1}{7}$	$-\frac{1}{7}$	$\frac{2}{7}$	—	$\frac{5}{7}\theta$
1	0	$\frac{4}{7}$	0	0	$\frac{1}{7}$	$-\frac{1}{7}$	$-\frac{2}{7}$	$\frac{2}{7}$	$\frac{3}{7}$	$\frac{3}{4}$	$\frac{4}{7}\theta$
0	0	$-\frac{22}{7}$	1	0	$-\frac{2}{7}$	$\frac{9}{7}$	$-\frac{3}{7}$	$\frac{10}{7}$	$-\frac{13}{7}$		$\frac{6}{7}\theta$
0	0	1	$-\frac{7}{22}$	$\frac{7}{22}$	$\frac{1}{11}$	$-\frac{1}{11}$	$\frac{3}{22}$	$-\frac{3}{22}$	$\frac{13}{22}$		$-\frac{3}{11}\theta$
0	1	0	$-\frac{9}{22}$	$\frac{9}{22}$	$-\frac{5}{11}$	$\frac{5}{11}$	$\frac{7}{22}$	$-\frac{7}{22}$	$\frac{23}{22}$		$\frac{4}{11}\theta$
1	0	0	$\frac{2}{11}$	$-\frac{2}{11}$	$\frac{1}{11}$	$-\frac{1}{11}$	$-\frac{4}{22}$	$\frac{4}{11}$	$\frac{1}{11}$		$\frac{8}{11}\theta$
0	0	0	0	1	0	1	0	1	0		0

Phase II

$$\text{Minimize:} \quad 10y_1 + 4y_2 + 6y_3$$

y_1	y_2	y_3	y_4	y_5	y_6	y_7	y_8	y_9	r.h.s.	Ratios
0	0	1	$\left(-\frac{7}{22}\right)$		$\frac{1}{11}$		$\frac{3}{22}$		$\frac{13}{22}$	$-\frac{3}{11}\theta$
0	1	0	$-\frac{9}{22}$		$-\frac{5}{11}$		$\frac{7}{22}$		$\frac{23}{22}$	$\frac{4}{11}\theta$
1	0	0	$\frac{2}{11}$		$\frac{1}{11}$		$-\frac{4}{11}$		$\frac{1}{11}$	$\frac{8}{11}\theta$
10	4	6	0		0		0		0	

In canonical form:

y_1	y_2	y_3	y_4	y_5	y_6	y_7	y_8	y_9	r.h.s.	Ratios
0	0	1	$-\frac{7}{22}$		$\frac{1}{11}$		$\frac{3}{22}$		$\frac{13}{22}$	$-\frac{3}{11}\theta$
0	1	0	$-\frac{9}{22}$		$-\frac{5}{11}$		$\frac{7}{22}$		$\frac{23}{22}$	$\frac{4}{11}\theta$
1	0	0	$\frac{2}{11}$		$\frac{1}{11}$		$-\frac{4}{11}$		$\frac{1}{11}$	$\frac{8}{11}\theta$
0	0	0	$\frac{19}{11}$		$\frac{4}{11}$		$\frac{17}{11}$		$-\frac{95}{11}$	$-\frac{78}{11}\theta$

$$\left.
\begin{aligned}
y_1^* &= x_4^* = \tfrac{1}{11} & x_1^* &= \tfrac{19}{11} \\
y_2^* &= x_5^* = \tfrac{23}{22} & x_2^* &= \tfrac{4}{11} \\
y_3^* &= x_6^* = \tfrac{13}{22} & x_3^* &= \tfrac{17}{11} \\
y_4^* &= y_5^* = y_6^* = 0 & & \\
y_0^* &= \tfrac{95}{11} & &
\end{aligned}
\right\} \text{ by complementary slackness.}$$

y_1	y_2	y_3	y_4	y_6	y_8	r.h.s.
0	0	1	$-\frac{7}{22}$	$\frac{1}{11}$	$\frac{3}{22}$	$\dfrac{13 - 6\theta}{22}$
0	1	0	$-\frac{9}{22}$	$-\frac{5}{11}$	$\frac{7}{22}$	$\dfrac{23 + 8\theta}{22}$
1	0	0	$\frac{2}{11}$	$\frac{1}{11}$	$-\frac{4}{11}$	$\dfrac{1 + 8\theta}{11}$
0	0	0	$\frac{19}{11}$	$\frac{4}{11}$	$\frac{12}{11}$	$-\dfrac{95 - 78\theta}{11}$

For r.h.s. entries to be nonnegative, $0 \le \theta \le \frac{13}{6}$. y_4 enters the basis with $\theta = \frac{13}{6}$.

y_1	y_2	y_3	y_4	y_6	y_8	r.h.s.
0	0	1	$-\frac{7}{22}$	$\frac{1}{11}$	$\frac{3}{22}$	0
0	1	0	$-\frac{9}{22}$	$-\frac{5}{11}$	$\frac{7}{22}$	$\frac{11}{6}$
1	0	0	$\frac{2}{11}$	$\frac{1}{11}$	$-\frac{4}{11}$	$\frac{5}{3}$
0	0	0	$\frac{19}{11}$	$\frac{4}{11}$	$\frac{17}{11}$	-24

y_1	y_2	y_3	y_4	y_6	y_8		
0	0	$-\frac{22}{7}$	1	$-\frac{2}{7}$	$-\frac{3}{7}$	0	$\frac{6}{7}\theta$
0	1	$-\frac{9}{7}$	0	$-\frac{4}{7}$	$\frac{1}{7}$	$\frac{11}{6}$	$\frac{4}{11}\theta$
1	0	$\frac{4}{7}$	0	$\frac{1}{7}$	$-\frac{2}{7}$	$\frac{5}{3}$	$\frac{4}{7}\theta$
0	0	$\frac{38}{7}$	0	$\frac{6}{7}$	$\frac{16}{7}$	-24	$-\frac{60}{7}\theta$

$$y_1^* = x_4^* = \tfrac{5}{3} \qquad x_6^* = \tfrac{38}{7}$$
$$y_2^* = x_5^* = \tfrac{11}{6} \qquad x_2^* = \tfrac{6}{7}$$
$$y_4^* = x_1^* = 0 \qquad x_3^* = \tfrac{16}{7} \Big\} \text{ by complementary slackness.}$$
$$y_3^* = y_6^* = y_8^* = 0$$
$$y_0^* = 24$$

y_1	y_2	y_3	y_4	y_6	y_8	
0	0	$-\frac{22}{7}$	1	$-\frac{2}{7}$	$-\frac{3}{7}$	$0 + \frac{6}{7}\theta$
0	1	$-\frac{9}{7}$	0	$-\frac{4}{7}$	$\frac{1}{7}$	$\frac{11}{6} + \frac{4}{11}\theta$
1	0	$\frac{4}{7}$	0	$\frac{1}{7}$	$-\frac{2}{7}$	$\frac{5}{3} + \frac{4}{7}\theta$
0	0	$\frac{38}{7}$	0	$\frac{6}{7}$	$\frac{16}{7}$	$-24 - \frac{67}{7}\theta$

As all r.h.s. entries are nonnegative for $\theta \geq 0$, no further critical points can be found. Thus the solution is as in 5(a).

7(a). $\lambda = (2, 1, 3)$

y_1	y_2	y_3	y_4	y_6	y_8	r.h.s.
0	0	1	$-\frac{7}{22}$	$\frac{1}{11}$	$\frac{3}{22}$	$\frac{13}{22}$
0	1	0	$-\frac{9}{22}$	$-\frac{5}{11}$	$\frac{7}{22}$	$\frac{23}{22}$
1	0	0	$\frac{2}{11}$	$\frac{1}{11}$	$-\frac{4}{11}$	$\frac{1}{11}$
2λ	λ	3λ	$\frac{19}{11}$	$\frac{4}{11}$	$\frac{17}{11}$	$-\frac{95}{11}$
0	0	1	$-\frac{7}{22}$	$\frac{1}{11}$	$\frac{3}{22}$	$\frac{13}{22}$
0	1	0	$-\frac{9}{22}$	$-\frac{5}{11}$	$\frac{7}{22}$	$\frac{23}{22}$
1	0	0	$\frac{2}{11}$	$\frac{1}{11}$	$-\frac{4}{11}$	$\frac{1}{11}$
0	0	0	$\frac{19}{11} + \lambda$	$\frac{4}{11}$	$\frac{17}{11}$	$-\frac{95}{11} - 3\lambda$

Basis remains optimal for $\lambda \geq 0$. Therefore

$$x_1^* = \tfrac{19}{11} + \lambda, \qquad x_2^* = \tfrac{4}{11}, \qquad x_3^* = \tfrac{17}{11}, \qquad x_4^*, x_5^*, x_6^* = 0, \qquad x_0^* = \tfrac{95}{11} + 3\lambda.$$

Chapter 4

Section 4.6

1(a). The decision tree for this problem is shown in Figure S.5. The optimal solution is

$$x_1^*, x_3^* = 0, \qquad x_2^* = 1, \qquad x_0^* = 5.$$

To obtain the first node:

x_1	x_2	x_3	x_4	x_5	x_6	r.h.s.	Ratio
2	⑥	3	1	0	0	8	$\tfrac{4}{3}$
5	4	4	0	1	0	7	$\tfrac{7}{4}$
6	1	1	0	0	1	12	12
-3	-5	-4	0	0	0	0	
$\tfrac{1}{3}$	1	$\tfrac{1}{2}$	$\tfrac{1}{6}$	0	0	$\tfrac{4}{3}$	$\tfrac{8}{3}$
$\tfrac{11}{3}$	0	②	$\tfrac{2}{3}$	1	0	$\tfrac{5}{3}$	$\tfrac{5}{6}$
$\tfrac{17}{3}$	0	$\tfrac{1}{2}$	$-\tfrac{1}{6}$	0	1	$\tfrac{32}{3}$	$\tfrac{64}{3}$
$-\tfrac{4}{3}$	0	$-\tfrac{3}{2}$	$\tfrac{5}{6}$	0	0	$\tfrac{20}{3}$	
$-\tfrac{7}{12}$	1	0	$\tfrac{1}{3}$	$-\tfrac{1}{4}$	0	$\tfrac{11}{12}$	
$\tfrac{11}{6}$	0	1	$-\tfrac{1}{3}$	$\tfrac{1}{2}$	0	$\tfrac{5}{6}$	
$\tfrac{19}{4}$	0	0	0	$-\tfrac{1}{4}$	1	$\tfrac{41}{4}$	
$\tfrac{17}{4}$	0	0	$\tfrac{1}{3}$	$\tfrac{3}{4}$	0	$\tfrac{95}{12}$	

The problem has an optimal (noninteger) solution:

$$x_2^* = \tfrac{11}{12}, \qquad x_3^* = \tfrac{5}{6}, \qquad x_6^* = \tfrac{41}{4}, \qquad x_0^* = \tfrac{95}{12}.$$

To create nodes (I) and (II), examine $x_2^* = \tfrac{11}{12}$ and introduce

$$x_2 \leq 0, \quad \text{(I)}$$
$$x_2 \geq 1, \quad \text{(II)}.$$

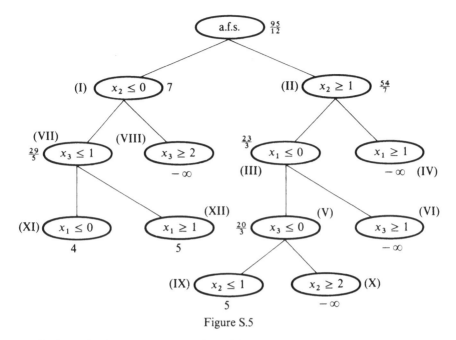

Figure S.5

(I) Introducing a new constraint with slack variable x_7:

$$x_2 + x_7 = 0.$$

x_1	x_2	x_3	x_4	x_5	x_6	x_7	r.h.s.
$-\frac{7}{12}$	1	0	$\frac{1}{3}$	$-\frac{1}{4}$	0	0	$\frac{11}{12}$
$\frac{11}{6}$	0	1	$-\frac{1}{3}$	$\frac{1}{2}$	0	0	$\frac{5}{6}$
$\frac{19}{4}$	0	0	0	$-\frac{1}{4}$	1	0	$\frac{41}{4}$
0	1	0	0	0	0	1	0
$\frac{17}{12}$	0	0	$\frac{1}{3}$	$\frac{3}{4}$	0	0	$\frac{95}{12}$

In canonical form:

x_1	x_2	x_3	x_4	x_5	x_6	x_7	r.h.s.
$-\frac{7}{12}$	1	0	$\frac{1}{3}$	$-\frac{1}{4}$	0	0	$\frac{11}{12}$
$\frac{11}{6}$	0	1	$-\frac{1}{3}$	$\frac{1}{2}$	0	0	$\frac{5}{6}$
$\frac{19}{4}$	0	0	0	$-\frac{1}{4}$	1	0	$\frac{41}{4}$
$\frac{7}{12}$	0	0	$\left(-\frac{1}{3}\right)$	$\frac{1}{4}$	0	1	$-\frac{11}{12}$
$\frac{17}{12}$	0	0	$\frac{1}{3}$	$\frac{3}{4}$	0	0	$\frac{95}{12}$

Applying the dual simplex method:

x_1	x_2	x_3	x_4	x_5	x_6	x_7	r.h.s.
0	1	0	0	0	0	0	0
$\frac{5}{4}$	0	1	0	$\frac{1}{4}$	0	0	$\frac{7}{4}$
$\frac{19}{4}$	0	0	0	$-\frac{1}{4}$	1	0	$\frac{41}{4}$
$-\frac{7}{4}$	0	0	1	$-\frac{3}{4}$	0	1	$\frac{11}{4}$
2	0	0	0	1	0	0	7

This has solution

$$x_2^* = 0$$

(as expected as $x_2 \leq 0$ and $x_2 \geq 0$),

$$x_3^* = \tfrac{7}{4}, \qquad x_4^* = \tfrac{11}{4}, \qquad x_0^* = 7.$$

(II) Introducing a new constraint with slack variable x_8:

$$x_2 - x_8 = 1.$$

x_1	x_2	x_3	x_4	x_5	x_6	x_8	r.h.s.
$-\frac{7}{12}$	1	0	$\frac{1}{3}$	$-\frac{1}{4}$	0	0	$\frac{11}{12}$
$\frac{11}{6}$	0	1	$-\frac{1}{3}$	$\frac{1}{2}$	0	0	$\frac{5}{6}$
$\frac{19}{4}$	0	0	0	$-\frac{1}{4}$	1	0	$\frac{41}{4}$
0	1	0	0	0	0	-1	1
$\frac{17}{12}$	0	0	$\frac{1}{3}$	$\frac{3}{4}$	0	0	$\frac{95}{12}$

In canonical form:

x_1	x_2	x_3	x_4	x_5	x_6	x_8	r.h.s.
$-\frac{7}{12}$	1	0	$\frac{1}{3}$	$-\frac{1}{4}$	0	0	$\frac{11}{12}$
$\frac{11}{6}$	0	1	$-\frac{1}{3}$	$\frac{1}{2}$	0	0	$\frac{5}{6}$
$\frac{19}{4}$	0	0	0	$-\frac{1}{4}$	1	0	$\frac{41}{4}$
$-\frac{7}{12}$	0	0	$\frac{1}{3}$	$-\frac{1}{4}$	0	1	$-\frac{1}{12}$
$\frac{17}{12}$	0	0	$\frac{1}{3}$	$\frac{3}{4}$	0	0	$\frac{95}{12}$

Applying the dual simplex method:

x_1	x_2	x_3	x_4	x_5	x_6	x_8	r.h.s.
0	1	0	0	0	0	-1	1
0	0	1	$\frac{5}{7}$	$-\frac{2}{7}$	0	$\frac{22}{7}$	$\frac{4}{7}$
0	0	0	$\frac{19}{7}$	$-\frac{16}{7}$	1	$\frac{57}{7}$	$\frac{67}{7}$
1	0	0	$-\frac{4}{7}$	$\frac{3}{7}$	0	$-\frac{12}{7}$	$\frac{1}{7}$
0	0	0	$\frac{8}{7}$	$\frac{1}{7}$	0	$\frac{17}{7}$	$\frac{54}{7}$

This has the solution:

$$x_1^* = \tfrac{1}{7}, \qquad x_2^* = 1, \qquad x_3^* = \tfrac{4}{7}, \qquad x_6^* = \tfrac{67}{7}, \qquad x_0^* = \tfrac{54}{7}.$$

(III) Introducing a new constraint with slack variable x_9:

$$x_1 + x_9 = 0.$$

x_1	x_2	x_3	x_4	x_5	x_6	x_8	x_9	r.h.s.
0	1	0	0	0	0	-1	0	1
0	0	1	$\frac{5}{7}$	$-\frac{2}{7}$	0	$\frac{22}{7}$	0	$\frac{4}{7}$
0	0	0	$\frac{19}{7}$	$-\frac{16}{7}$	1	$\frac{57}{7}$	0	$\frac{67}{7}$
1	0	0	$-\frac{4}{7}$	$\frac{3}{7}$	0	$-\frac{12}{7}$	0	$\frac{1}{7}$
1	0	0	0	0	0	0	1	0
0	0	0	$\frac{8}{7}$	$\frac{1}{7}$	0	$\frac{17}{7}$	0	$\frac{54}{7}$

In canonical form:

x_1	x_2	x_3	x_4	x_5	x_6	x_8	x_9	r.h.s.
0	1	0	0	0	0	-1	0	1
0	0	1	$\frac{5}{7}$	$-\frac{2}{7}$	0	$\frac{22}{7}$	0	$\frac{4}{7}$
0	0	0	$\frac{19}{7}$	$-\frac{16}{7}$	1	$\frac{57}{7}$	0	$\frac{67}{7}$
1	0	0	$-\frac{4}{7}$	$\frac{3}{7}$	0	$-\frac{12}{7}$	0	$\frac{1}{7}$
0	0	0	$\frac{4}{7}$	$\left(-\frac{3}{7}\right)$	0	$\frac{12}{7}$	1	$-\frac{1}{7}$
0	0	0	$\frac{8}{7}$	$\frac{1}{7}$	0	$\frac{17}{7}$	0	$\frac{54}{7}$

Applying the dual simplex method:

x_1	x_2	x_3	x_4	x_5	x_6	x_8	x_9	r.h.s.
0	1	0	0	0	0	-1	0	1
0	0	1	$\frac{1}{3}$	0	0	2	$-\frac{2}{3}$	$\frac{2}{3}$
0	0	0	$-\frac{1}{3}$	0	1	-1	$-\frac{16}{3}$	$\frac{31}{3}$
1	0	0	0	0	0	0	1	0
0	0	0	$-\frac{4}{3}$	1	0	-4	$-\frac{7}{3}$	$\frac{1}{3}$
0	0	0	$\frac{4}{3}$	0	0	3	$\frac{1}{3}$	$\frac{23}{3}$

which has the solution:

$$x_1^* = 0$$

(as expected, as $x_1 \geq 0$ and $x_1 \leq 0$),

$$x_2^* = 1, \qquad x_3^* = \tfrac{2}{3}, \qquad x_5^* = \tfrac{1}{3}, \qquad x_6^* = \tfrac{31}{3}, \qquad x_0^* = \tfrac{23}{3}.$$

(IV) Introducing a new constraint with slack variable x_{10}:

$$x_1 - x_{10} = 1.$$

x_1	x_2	x_3	x_4	x_5	x_6	x_8	x_{10}	r.h.s.
0	1	0	0	0	0	-1	0	1
0	0	1	$\frac{5}{7}$	$-\frac{2}{7}$	0	$\frac{22}{7}$	0	$\frac{4}{7}$
0	0	0	$\frac{19}{7}$	$-\frac{16}{7}$	1	$\frac{57}{7}$	0	$\frac{67}{7}$
1	0	0	$-\frac{4}{7}$	$\frac{3}{7}$	0	$-\frac{12}{7}$	0	$\frac{1}{7}$
1	0	0	0	0	0	0	-1	1
0	0	0	$\frac{8}{7}$	$\frac{1}{7}$	0	$\frac{17}{7}$	0	$\frac{54}{7}$

In canonical form:

x_1	x_2	x_3	x_4	x_5	x_6	x_8	x_{10}	r.h.s.
0	1	0	0	0	0	-1	0	1
0	0	1	$\frac{5}{7}$	$-\frac{2}{7}$	0	$\frac{22}{7}$	0	$\frac{4}{7}$
0	0	0	$\frac{19}{7}$	$-\frac{16}{7}$	1	$\frac{57}{7}$	0	$\frac{67}{7}$
1	0	0	$-\frac{4}{7}$	$\frac{3}{7}$	0	$-\frac{12}{7}$	0	$\frac{1}{7}$
0	0	0	$-\frac{4}{7}$	$\frac{3}{7}$	0	$\left(-\frac{12}{7}\right)$	1	$-\frac{6}{7}$
0	0	0	$\frac{8}{7}$	$\frac{1}{7}$	0	$\frac{17}{7}$	0	$\frac{54}{7}$

Applying the dual simplex method:

x_1	x_2	x_3	x_4	x_5	x_6	x_8	x_{10}	r.h.s.
0	1	0	$\frac{1}{3}$	$-\frac{1}{4}$	0	0	$-\frac{7}{12}$	$\frac{3}{2}$
0	0	1	$-\frac{1}{3}$	$\frac{1}{2}$	0	0	$\frac{11}{6}$	-1
0	0	0	0	$-\frac{1}{4}$	1	0	$\frac{57}{12}$	$\frac{11}{12}$
1	0	0	0	0	0	0	-1	1
0	0	0	$\frac{1}{3}$	$-\frac{1}{4}$	0	1	$-\frac{7}{12}$	$\frac{1}{2}$
0	0	0	$\frac{1}{3}$	$\frac{3}{4}$	0	0	$\frac{17}{12}$	$\frac{13}{2}$
0	1	1	0	$\frac{1}{4}$	0	0	$\frac{5}{4}$	$\frac{1}{2}$
0	0	-3	1	$-\frac{3}{2}$	0	0	$-\frac{11}{2}$	3
0	0	0	0	$-\frac{1}{4}$	1	0	$\frac{57}{12}$	$\frac{11}{2}$
1	0	0	0	0	0	0	-1	1
0	0	1	0	$\frac{1}{4}$	0	1	$\frac{5}{4}$	$-\frac{1}{2}$
0	0	1	0	$\frac{5}{4}$	0	0	$\frac{13}{4}$	$\frac{11}{2}$

Note that this subproblem required two iterations. The final tableau indicates that it does not have a feasible solution.

(V) Introducing a new constraint with slack variable x_{11}:

$$x_3 + x_{11} = 0.$$

x_1	x_2	x_3	x_4	x_5	x_6	x_8	x_9	x_{11}	r.h.s.
0	1	0	0	0	0	-1	0	0	1
0	0	1	$\frac{1}{3}$	0	0	2	$-\frac{2}{3}$	0	$\frac{2}{3}$
0	0	0	$-\frac{1}{3}$	0	1	-1	$-\frac{16}{3}$	0	$\frac{31}{3}$
1	0	0	0	0	0	0	1	0	0
0	0	0	$-\frac{4}{3}$	1	0	-4	$-\frac{7}{3}$	0	$\frac{1}{3}$
0	0	1	0	0	0	0	0	1	0
0	0	0	$\frac{4}{3}$	0	0	3	$\frac{1}{3}$	0	$\frac{23}{3}$

In canonical form:

x_1	x_2	x_3	x_4	x_5	x_6	x_8	x_9	x_{11}	r.h.s.
0	1	0	0	0	0	-1	0	0	1
0	0	1	$\frac{1}{3}$	0	0	2	$-\frac{2}{3}$	0	$\frac{2}{3}$
0	0	0	$-\frac{1}{3}$	0	1	-1	$-\frac{16}{3}$	0	$\frac{31}{3}$
1	0	0	0	0	0	0	1	0	0
0	0	0	$-\frac{4}{3}$	1	0	-4	$-\frac{7}{3}$	0	$\frac{1}{3}$
0	0	0	$-\frac{1}{3}$	0	0	$\boxed{-2}$	$\frac{2}{3}$	1	$\frac{2}{3}$
0	0	0	$\frac{4}{3}$	0	0	3	$\frac{1}{3}$	0	$\frac{23}{3}$

Applying the dual simplex method:

x_1	x_2	x_3	x_4	x_5	x_6	x_8	x_9	x_{11}	r.h.s.
0	1	0	$\frac{1}{6}$	0	0	0	$-\frac{1}{3}$	$-\frac{1}{2}$	$\frac{4}{3}$
0	0	1	0	0	0	0	0	1	0
0	0	0	$-\frac{1}{6}$	0	1	0	$-\frac{17}{3}$	$-\frac{1}{2}$	$\frac{32}{3}$
1	0	0	0	0	0	0	1	0	0
0	0	0	$\frac{2}{3}$	1	0	0	$-\frac{11}{3}$	-2	$\frac{5}{3}$
0	0	0	$\frac{1}{6}$	0	0	1	$-\frac{1}{3}$	$-\frac{1}{2}$	$\frac{1}{3}$
0	0	0	$\frac{5}{6}$	0	0	0	$\frac{4}{3}$	$\frac{3}{2}$	$\frac{20}{3}$

which has solution:

$$x_1^*, x_3^* = 0, \quad \text{(as expected)}, \quad x_2^* = \tfrac{4}{3}, \quad x_5^* = \tfrac{5}{3},$$
$$x_6^* = \tfrac{32}{3}, \quad x_8^* = \tfrac{1}{3}, \quad x_0^* = \tfrac{20}{3}.$$

(VI) Introducing a new constraint with slack variable x_{12}:

$$x_3 - x_{12} = 1.$$

x_1	x_2	x_3	x_4	x_5	x_6	x_8	x_9	x_{12}	r.h.s.
0	1	0	0	0	0	-1	0	0	1
0	0	1	$\frac{1}{3}$	0	0	2	$-\frac{2}{3}$	0	$\frac{2}{3}$
0	0	0	$-\frac{1}{3}$	0	1	-1	$-\frac{16}{3}$	0	$\frac{31}{3}$
1	0	0	0	0	0	0	1	0	0
0	0	0	$-\frac{4}{3}$	1	0	-4	$-\frac{7}{3}$	0	$\frac{1}{3}$
0	0	1	0	0	0	0	0	-1	1
0	0	0	$\frac{4}{3}$	0	0	3	$\frac{1}{3}$	0	$\frac{23}{3}$

In canonical form:

x_1	x_2	x_3	x_4	x_5	x_6	x_8	x_9	x_{12}	r.h.s.
0	1	0	0	0	0	-1	0	0	1
0	0	1	$\frac{1}{3}$	0	0	2	$-\frac{2}{3}$	0	$\frac{2}{3}$
0	0	0	$-\frac{1}{3}$	0	1	-1	$-\frac{16}{3}$	0	$\frac{31}{3}$
1	0	0	0	0	0	0	1	0	0
0	0	0	$-\frac{4}{3}$	1	0	-4	$-\frac{7}{3}$	0	$\frac{1}{3}$
0	0	0	$\frac{1}{3}$	0	0	2	$\left(-\frac{2}{3}\right)$	1	$-\frac{1}{3}$
0	0	0	$\frac{4}{3}$	0	0	3	$\frac{1}{3}$	0	$\frac{23}{3}$

Applying the dual simplex method:

x_1	x_2	x_3	x_4	x_5	x_6	x_8	x_9	x_{12}	r.h.s.
0	1	0	0	0	0	-1	0	0	1
0	0	1	0	0	0	0	0	-1	1
0	0	0	-3	0	1	-17	0	-8	13
1	0	0	$\frac{1}{2}$	0	0	3	0	$\frac{3}{2}$	$-\frac{1}{2}$
0	0	0	$-\frac{5}{2}$	1	0	11	0	$-\frac{7}{2}$	$\frac{3}{2}$
0	0	0	$-\frac{1}{2}$	0	0	-3	1	$-\frac{3}{2}$	$\frac{1}{2}$
0	0	0	$\frac{3}{2}$	0	0	4	0	$\frac{1}{2}$	15

This does not have a feasible solution.

(VII) Introducing a new constraint with slack variable x_{13}:

$$x_3 + x_{13} = 1.$$

x_1	x_2	x_3	x_4	x_5	x_6	x_7	x_{13}	r.h.s.
0	1	0	0	0	0	1	0	0
$\frac{5}{4}$	0	1	0	$\frac{1}{4}$	0	-1	0	$\frac{7}{4}$
$\frac{19}{4}$	0	0	0	$-\frac{1}{4}$	1	0	0	$\frac{41}{4}$
$-\frac{7}{4}$	0	0	1	$-\frac{3}{4}$	0	3	0	$\frac{11}{4}$
0	0	1	0	0	0	0	1	1
2	0	0	0	1	0	1	0	7

In canonical form:

x_1	x_2	x_3	x_4	x_5	x_6	x_7	x_{13}	r.h.s.
0	1	0	0	0	0	1	0	0
$\boxed{\frac{5}{4}}$	0	1	0	$\frac{1}{4}$	0	-1	0	$\frac{7}{4}$
$\frac{19}{4}$	0	0	0	$-\frac{1}{4}$	1	0	0	$\frac{41}{4}$
$-\frac{7}{4}$	0	0	1	$-\frac{3}{4}$	0	3	0	$\frac{11}{4}$
$-\frac{5}{4}$	0	0	0	$-\frac{1}{4}$	0	1	1	$-\frac{3}{4}$
2	0	0	0	1	0	1	0	7

Applying the dual simplex method:

x_1	x_2	x_3	x_4	x_5	x_6	x_7	x_{13}	r.h.s.
0	1	0	0	0	0	1	0	0
0	0	1	0	0	0	0	1	1
0	0	0	0	$-\frac{6}{5}$	1	$\frac{19}{5}$	$\frac{19}{5}$	$\frac{37}{5}$
0	0	0	1	$-\frac{2}{5}$	0	$-\frac{22}{5}$	$-\frac{27}{5}$	$\frac{19}{5}$
1	0	0	0	$\frac{1}{5}$	0	$-\frac{4}{5}$	$-\frac{4}{5}$	$\frac{3}{5}$
0	0	0	0	$\frac{3}{5}$	0	$\frac{13}{5}$	$\frac{8}{5}$	$\frac{29}{5}$

This has solution:

$$x_1^* = \tfrac{3}{5}, \qquad x_2^* = 0 \quad \text{(as expected)}, \qquad x_3^* = 1,$$
$$x_4^* = \tfrac{19}{5}, \qquad x_6^* = \tfrac{37}{5}, \qquad x_0^* = \tfrac{29}{5}.$$

(VIII) Introducing a new constraint with slack variable x_{14}:

$$x_3 - x_{14} = 2.$$

x_1	x_2	x_3	x_4	x_5	x_6	x_7	x_{14}	r.h.s.
0	1	0	0	0	0	1	0	0
$\frac{5}{4}$	0	1	0	$\frac{1}{4}$	0	-1	0	$\frac{7}{4}$
$\frac{19}{4}$	0	0	0	$-\frac{1}{4}$	1	0	0	$\frac{41}{4}$
$-\frac{7}{4}$	0	0	1	$-\frac{3}{4}$	0	-3	0	$\frac{11}{4}$
0	0	1	0	0	0	0	-1	20
2	0	0	0	1	0	1	0	7

In canonical form:

x_1	x_2	x_3	x_4	x_5	x_6	x_7	x_{14}	r.h.s.
0	1	0	0	0	0	1	0	0
$\frac{5}{4}$	0	1	0	$\frac{1}{4}$	0	-1	0	$\frac{7}{4}$
$\frac{19}{4}$	0	0	0	$-\frac{1}{4}$	1	0	0	$\frac{41}{4}$
$-\frac{7}{4}$	0	0	1	$-\frac{3}{4}$	0	-3	0	$\frac{11}{4}$
$\frac{5}{4}$	0	0	0	$\frac{1}{4}$	0	$\boxed{-1}$	1	$-\frac{1}{4}$
2	0	0	0	1	0	1	0	7

Applying the dual simplex method:

x_1	x_2	x_3	x_4	x_5	x_6	x_7	x_{14}	r.h.s.
$\frac{5}{4}$	1	0	0	$\frac{1}{4}$	0	0	1	$-\frac{1}{4}$
0	0	1	0	0	0	0	-1	2
$\frac{19}{4}$	0	0	0	$-\frac{1}{4}$	1	0	0	$\frac{41}{4}$
$-\frac{11}{2}$	0	0	1	$-\frac{3}{2}$	0	0	-3	$\frac{7}{2}$
$-\frac{5}{4}$	0	0	0	$-\frac{1}{4}$	0	1	-1	$\frac{1}{4}$
$\frac{13}{4}$	0	0	0	$\frac{5}{4}$	0	0	1	$\frac{27}{4}$

This does not have a feasible solution.

(IX) Introducing a new constraint with slack variable x_{15}:

$$x_2 + x_{15} = 1.$$

x_1	x_2	x_3	x_4	x_5	x_6	x_8	x_9	x_{11}	x_{15}	r.h.s.
0	1	0	$\frac{1}{6}$	0	0	0	$-\frac{1}{3}$	$-\frac{1}{2}$	0	$\frac{4}{3}$
0	0	1	0	0	0	0	0	1	0	0
0	0	0	$-\frac{1}{6}$	0	1	0	$-\frac{17}{3}$	$-\frac{1}{2}$	0	$\frac{32}{3}$
1	0	0	0	0	0	0	1	0	0	0
0	0	0	$\frac{2}{3}$	1	0	0	$-\frac{11}{3}$	-2	0	$\frac{5}{3}$
0	0	0	$\frac{1}{6}$	0	0	1	$-\frac{1}{3}$	$-\frac{1}{2}$	0	$\frac{1}{3}$
0	1	0	0	0	0	0	0	0	1	1
0	0	0	$\frac{5}{6}$	0	0	0	$\frac{4}{3}$	$\frac{3}{2}$	0	$\frac{20}{3}$

In canonical form:

x_1	x_2	x_3	x_4	x_5	x_6	x_8	x_9	x_{11}	x_{15}	r.h.s.
0	1	0	$\frac{1}{6}$	0	0	0	$-\frac{1}{3}$	$-\frac{1}{2}$	0	$\frac{4}{3}$
0	0	1	0	0	0	0	0	1	0	0
0	0	0	$-\frac{1}{6}$	0	1	0	$-\frac{17}{3}$	$-\frac{1}{2}$	0	$\frac{32}{3}$
1	0	0	0	0	0	0	1	0	0	0
0	0	0	$\frac{2}{3}$	1	0	0	$-\frac{11}{3}$	-2	0	$\frac{5}{3}$
0	0	0	$\frac{1}{6}$	0	0	1	$-\frac{1}{3}$	$-\frac{1}{2}$	0	$\frac{1}{3}$
0	0	0	$\left(-\frac{1}{6}\right)$	0	0	0	$\frac{1}{3}$	$\frac{1}{2}$	1	$-\frac{4}{3}$
0	0	0	$\frac{5}{6}$	0	0	0	$\frac{4}{3}$	$\frac{3}{2}$	0	$\frac{20}{3}$

Applying the dual simplex method:

x_1	x_2	x_3	x_4	x_5	x_6	x_8	x_9	x_{11}	x_{15}	r.h.s.
0	1	0	0	0	0	0	0	0	1	1
0	0	1	0	0	0	0	0	1	0	0
0	0	0	0	0	1	0	-6	-1	-1	11
1	0	0	0	0	0	0	1	0	0	0
0	0	0	0	1	0	0	$-\frac{7}{3}$	0	4	$\frac{1}{3}$
0	0	0	0	0	0	1	0	0	1	0
0	0	0	1	0	0	0	-2	-3	-6	2
0	0	0	0	0	0	0	3	4	5	5

This has a feasible *integral* solution:

$$x_1^*, x_3^* = 0, \qquad x_2^* = 1, \qquad x_0^* = 5.$$

This is the first incumbent.

(X) Introducing a new constraint with slack variable x_{16}:

$$x_2 - x_{16} = 2.$$

x_1	x_2	x_3	x_4	x_5	x_6	x_8	x_9	x_{11}	x_{16}	r.h.s.
0	1	0	$\frac{1}{6}$	0	0	0	$-\frac{1}{3}$	$-\frac{1}{2}$	0	$\frac{4}{3}$
0	0	1	0	0	0	0	0	1	0	0
0	0	0	$-\frac{1}{6}$	0	1	0	$-\frac{17}{3}$	$-\frac{1}{2}$	0	$\frac{32}{3}$
1	0	0	0	0	0	0	1	0	0	0
0	0	0	$\frac{2}{3}$	1	0	0	$-\frac{11}{3}$	-2	0	$\frac{5}{3}$
0	0	0	$\frac{1}{6}$	0	0	1	$-\frac{1}{3}$	$-\frac{1}{2}$	0	$\frac{1}{3}$
0	1	0	0	0	0	0	0	0	-1	2
0	0	0	$\frac{5}{6}$	0	0	0	$\frac{4}{3}$	$\frac{3}{2}$	0	$\frac{20}{3}$

In canonical form:

x_1	x_2	x_3	x_4	x_5	x_6	x_8	x_9	x_{11}	x_{16}	r.h.s.
0	1	0	$\frac{1}{6}$	0	0	0	$-\frac{1}{3}$	$-\frac{1}{2}$	0	$\frac{4}{3}$
0	0	1	0	0	0	0	0	1	0	0
0	0	0	$-\frac{1}{6}$	0	1	0	$-\frac{17}{3}$	$-\frac{1}{2}$	0	$\frac{32}{3}$
1	0	0	0	0	0	0	1	0	0	0
0	0	0	$\frac{2}{3}$	1	0	0	$\frac{11}{3}$	-2	0	$\frac{5}{3}$
0	0	0	$\frac{1}{6}$	0	0	1	$-\frac{1}{3}$	$-\frac{1}{2}$	0	$\frac{1}{3}$
0	0	0	$\frac{1}{6}$	0	0	0	$-\frac{1}{3}$	$\left(-\frac{1}{2}\right)$	1	$-\frac{2}{3}$
0	0	0	$\frac{5}{6}$	0	0	0	$\frac{4}{3}$	$\frac{3}{2}$	0	$\frac{20}{3}$

Applying the dual simplex method:

x_1	x_2	x_3	x_4	x_5	x_6	x_8	x_9	x_{11}	x_{16}	r.h.s.
0	1	0	0	0	0	0	0	0	-1	2
0	0	1	$\frac{1}{3}$	0	0	0	$\boxed{-\frac{2}{3}}$	0	2	$-\frac{4}{3}$
0	0	0	$-\frac{1}{3}$	0	1	0	$\frac{16}{3}$	0	-1	10
1	0	0	0	0	0	0	1	0	0	0
0	0	0	0	1	0	0	$-\frac{7}{3}$	0	-4	$\frac{12}{3}$
0	0	0	0	0	0	1	0	0	-1	1
0	0	0	$-\frac{1}{3}$	0	0	0	$\frac{2}{3}$	1	-2	$\frac{4}{3}$
0	0	0	$\frac{4}{3}$	0	0	0	$\frac{1}{3}$	0	3	$\frac{14}{3}$

x_1	x_2	x_3	x_4	x_5	x_6	x_8	x_9	x_{11}	x_{16}	r.h.s.
0	1	0	0	0	0	0	0	0	-1	2
0	0	$-\frac{3}{2}$	$-\frac{1}{2}$	0	0	0	1	0	-3	2
0	0	8	$\frac{7}{3}$	0	1	0	0	0	15	$-\frac{2}{3}$
1	0	$\frac{3}{2}$	$\frac{1}{2}$	0	0	0	0	0	3	-2
0	0	$-\frac{7}{2}$	$-\frac{7}{6}$	1	0	0	0	0	-11	9
0	0	0	0	0	0	1	0	0	-1	1
0	0	1	0	0	0	0	0	1	0	0
0	0	$-\frac{1}{2}$	3	0	0	0	0	0	4	4

This does not have a feasible solution.

(XI) Introducing a new constraint with slack variable x_{17}:

$$x_1 + x_{17} = 0.$$

x_1	x_2	x_3	x_4	x_5	x_6	x_7	x_{13}	x_{17}	r.h.s.
0	1	0	0	0	0	1	0	0	0
0	0	1	0	0	0	0	1	0	1
0	0	0	0	$-\frac{6}{5}$	1	$\frac{19}{5}$	$\frac{19}{5}$	0	$\frac{37}{5}$
0	0	0	1	$-\frac{2}{5}$	0	$-\frac{22}{5}$	$-\frac{27}{5}$	0	$\frac{19}{5}$
1	0	0	0	$\frac{1}{5}$	0	$-\frac{4}{5}$	$-\frac{4}{5}$	0	$\frac{3}{5}$
1	0	0	0	0	0	0	0	1	0
0	0	0	0	$\frac{3}{5}$	0	$\frac{13}{5}$	$\frac{8}{5}$	0	$\frac{29}{5}$

In canonical form:

x_1	x_2	x_3	x_4	x_5	x_6	x_7	x_{13}	x_{17}	r.h.s.
0	1	0	0	0	0	1	0	0	0
0	0	1	0	0	0	0	1	0	1
0	0	0	0	$-\frac{6}{5}$	1	$\frac{19}{5}$	$\frac{19}{5}$	0	$\frac{37}{5}$
0	0	0	1	$-\frac{2}{5}$	0	$-\frac{22}{5}$	$-\frac{27}{5}$	0	$\frac{19}{5}$
1	0	0	0	$\frac{1}{5}$	0	$-\frac{4}{5}$	$-\frac{4}{5}$	0	$\frac{3}{5}$
0	0	0	0	$\left(-\frac{1}{5}\right)$	0	$\frac{4}{5}$	$\frac{4}{5}$	1	$-\frac{3}{5}$
0	0	0	0	$\frac{3}{5}$	0	$\frac{13}{5}$	$\frac{8}{5}$	0	$\frac{29}{5}$

x_1	x_2	x_3	x_4	x_5	x_6	x_7	x_{13}	x_{17}	r.h.s.
0	1	0	0	0	0	1	0	0	0
0	0	1	0	0	0	0	1	0	1
0	0	0	0	0	1	-1	-1	-6	11
0	0	0	1	0	0	-6	-7	-2	5
1	0	0	0	0	0	0	0	1	0
0	0	0	0	1	0	-4	-4	-5	3
0	0	0	0	0	0	5	4	3	4

This has a feasible *integral* solution:

$$x_1^*, x_2^* = 0, \qquad x_3^* = 1, \qquad x_0^* = 4.$$

This is less than the value of the incumbent, so it can be discarded as sub-optimal.

(XII) Introducing a new constraint with slack variable x_{18}:

x_1	x_2	x_3	x_4	x_5	x_6	x_7	x_{13}	x_{18}	r.h.s.
0	1	0	0	0	0	1	0	0	0
0	0	1	0	0	0	0	1	0	1
0	0	0	0	$-\frac{6}{5}$	1	$\frac{19}{5}$	$\frac{19}{5}$	0	$\frac{37}{5}$
0	0	0	1	$-\frac{2}{5}$	0	$-\frac{22}{5}$	$-\frac{7}{5}$	0	$\frac{19}{5}$
1	0	0	0	$\frac{1}{5}$	0	$-\frac{4}{5}$	$-\frac{4}{5}$	0	$\frac{3}{5}$
1	0	0	0	0	0	0	0	-1	1
0	0	0	0	$\frac{3}{5}$	0	$\frac{13}{5}$	$\frac{8}{5}$	0	$\frac{29}{5}$

Solutions to Selected Exercises

In canonical form:

x_1	x_2	x_3	x_4	x_5	x_6	x_7	x_{13}	x_{18}	r.h.s.
0	1	0	0	0	0	1	0	0	0
0	0	1	0	0	0	0	1	0	1
0	0	0	0	$-\frac{6}{5}$	1	$\frac{19}{5}$	$\frac{19}{5}$	0	$\frac{37}{5}$
0	0	0	1	$-\frac{2}{5}$	0	$-\frac{22}{5}$	$-\frac{7}{5}$	0	$\frac{19}{5}$
1	0	0	0	$\frac{1}{5}$	0	$-\frac{4}{5}$	$-\frac{4}{5}$	0	$\frac{3}{5}$
0	0	0	0	$\frac{1}{5}$	0	$-\frac{4}{5}$	$\left(-\frac{4}{5}\right)$	1	$-\frac{2}{5}$
0	0	0	0	$\frac{3}{5}$	0	$\frac{13}{5}$	$\frac{8}{5}$	0	$\frac{29}{5}$

Applying the dual simplex method:

x_1	x_2	x_3	x_4	x_5	x_6	x_7	x_{13}	x_{18}	r.h.s.
0	1	0	0	0	0	1	0	0	0
0	0	1	0	$\frac{1}{4}$	0	-1	0	$\frac{5}{4}$	$\frac{1}{2}$
0	0	0	0	$-\frac{1}{4}$	1	0	0	$\frac{19}{4}$	$\frac{11}{2}$
0	0	0	1	$-\frac{3}{4}$	0	3	0	$-\frac{7}{4}$	$\frac{9}{2}$
1	0	0	0	0	0	0	0	-1	1
0	0	0	0	$-\frac{1}{4}$	0	1	1	$-\frac{5}{4}$	$\frac{1}{2}$
0	0	0	0	1	0	1	0	2	5

This has the solution:

$$x_1^* = 1, \qquad x_2^* = 0, \qquad x_3^* = \tfrac{1}{2}, \qquad x_0^* = 5,$$

which is no better than the incumbent and can be ignored.

Thus there is no node in the decision tree with bound greater than that of node (IV). As this represents a feasible solution it is optimal. Hence

$$x_1^* = x_3^* = 0, \qquad x_2^* = 1, \qquad x_0^* = 5.$$

2(a). The decision tree is shown in Figure S.6. The optimal solution is

$$x_1^*, x_3^* = 0, \qquad x_2^* = 1, \qquad x_0^* = 5.$$

3(a). This has the same solution as 2(a).

3(b)

$$x_1 = y_0^1 + 2y_1^1 \le 2$$
$$x_2 = y_0^2 + 2y_1^2 \le 3$$
$$x_3 = y_0^3 + 2y_1^3 \le 2.$$

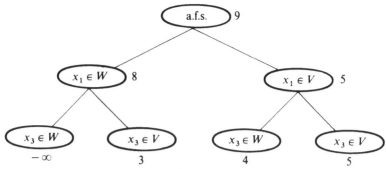

Figure S.6

Hence the problem becomes

$$\text{Maximize:} \quad 4y_0^1 + 8y_1^1 + 3y_0^2 + 6y_1^2 + 3y_0^3 + 6y_1^3$$
$$\text{subject to:} \quad 3y_0^1 + 6y_1^1 + 4y_0^2 + 8y_1^2 + 2y_0^3 + 4y_1^3 \le 14$$
$$4y_0^1 + 8y_1^1 + 2y_0^2 + 4y_1^2 + y_0^3 + 2y_1^3 \le 10$$
$$2y_0^1 + 4y_1^1 + y_0^2 + 2y_1^2 + 3y_0^3 + 6y_1^3 \le 7$$
$$y_i^j = 0 \text{ or } 1, \quad i = 0, 1$$
$$j = 1, 2, 3.$$

The optimal solution is

$$y_0^1 = y_1^2 = y_0^3 = 1$$
$$y_1^1 = y_0^2 = y_1^3 = 0,$$

with value 13.

4(a). From 1(a) the final tableau is

x_1	x_2	x_3	x_4	x_5	x_6	r.h.s.
$-\frac{7}{12}$	1	0	$\frac{1}{3}$	$-\frac{1}{4}$	0	$\frac{11}{12}$
$\frac{11}{6}$	0	1	$-\frac{1}{3}$	$\frac{1}{2}$	0	$\frac{5}{6}$
$\frac{19}{4}$	0	0	0	$-\frac{1}{4}$	1	$\frac{41}{4}$
$\frac{17}{12}$	0	0	$\frac{1}{3}$	$\frac{3}{4}$	0	$\frac{95}{12}$

Now

$$x_2 - \tfrac{7}{12} + \tfrac{1}{3}x_4 - \tfrac{1}{4}x_5 = \tfrac{11}{12}$$
$$x_2 = \tfrac{11}{12} - (-1 + \tfrac{5}{12})x_1 - (0 + \tfrac{1}{3}x_4) - (-1 + \tfrac{3}{4})x_5.$$

Adding in slack variable x_7, the constraint is

$$-\tfrac{5}{12}x_1 - \tfrac{1}{3}x_4 - \tfrac{3}{4}x_5 + x_7 = -\tfrac{11}{12}.$$

Introducing this into the above tableau:

x_1	x_2	x_3	x_4	x_5	x_6	x_7	r.h.s.
$-\frac{7}{12}$	1	0	$\frac{1}{3}$	$-\frac{1}{4}$	0	0	$\frac{11}{12}$
$\frac{11}{6}$	0	1	$-\frac{1}{3}$	$\frac{1}{2}$	0	0	$\frac{5}{6}$
$\frac{19}{4}$	0	0	0	$-\frac{1}{4}$	1	0	$\frac{41}{4}$
$-\frac{5}{12}$	0	0	$\left(-\frac{1}{3}\right)$	$-\frac{3}{4}$	0	1	$-\frac{11}{12}$
$\frac{17}{12}$	0	0	$\frac{1}{3}$	$\frac{3}{4}$	0	0	$\frac{95}{12}$

Applying the dual simplex method:

x_1	x_2	x_3	x_4	x_5	x_6	x_7	r.h.s.
-1	1	0	0	-1	0	1	0
$\frac{9}{4}$	0	1	0	$\frac{5}{4}$	0	-1	$\frac{7}{4}$
$\frac{19}{4}$	0	0	0	$-\frac{1}{4}$	1	0	$\frac{41}{4}$
$\frac{5}{4}$	0	0	1	$\frac{9}{4}$	0	-3	$\frac{11}{4}$
1	0	0	0	0	0	1	7

$$x_3 + \tfrac{9}{4}x_1 + \tfrac{5}{4}x_5 - x_7 = \tfrac{7}{4}.$$

Therefore

$$x_3 = (1 + \tfrac{3}{4}) - (2 + \tfrac{1}{4})x_1 - (1 + \tfrac{1}{4})x_5 - (-1 + 0)x_7.$$

Adding in slack variable x_8, the constraint is:

$$-\tfrac{1}{4}x_1 - \tfrac{1}{4}x_5 + x_8 = -\tfrac{3}{4}.$$

Introducing this into the preceding tableau:

x_1	x_2	x_3	x_4	x_5	x_6	x_7	x_8	r.h.s.
-1	1	0	0	-1	0	1	0	0
$\frac{9}{4}$	0	1	0	$\frac{5}{4}$	0	-1	0	$\frac{7}{4}$
$\frac{19}{4}$	0	0	0	$-\frac{1}{4}$	1	0	0	$\frac{41}{4}$
$\frac{5}{4}$	0	0	1	$\frac{9}{4}$	0	-3	0	$\frac{11}{4}$
$-\frac{1}{4}$	0	0	0	$\left(-\frac{1}{4}\right)$	0	0	1	$-\frac{3}{4}$
1	0	0	0	0	0	1	0	7

x_1	x_2	x_3	x_4	x_5	x_6	x_7	x_8	r.h.s.
0	1	0	0	0	0	1	-4	3
1	0	1	0	0	0	-1	5	-2
5	0	0	0	0	1	0	-1	11
-1	0	0	1	0	0	$\boxed{-3}$	9	-4
1	0	0	0	1	0	0	-4	3
1	0	0	0	0	0	1	0	7
$-\frac{1}{3}$	1	0	$\frac{1}{3}$	0	0	0	-1	$\frac{5}{3}$
$\frac{4}{3}$	0	1	$\boxed{-\frac{1}{3}}$	0	0	0	2	$-\frac{2}{3}$
5	0	0	0	0	1	0	-1	11
$\frac{1}{3}$	0	0	$-\frac{1}{3}$	0	0	1	-3	$\frac{4}{3}$
1	0	0	0	1	0	0	-4	3
$\frac{2}{3}$	0	0	$\frac{1}{3}$	0	0	0	3	$\frac{17}{3}$
1	1	1	0	0	0	0	1	1
-4	0	-3	1	0	0	0	-6	2
5	0	0	0	0	1	0	-1	11
-1	0	-1	0	0	0	1	-5	2
1	0	0	0	1	0	0	-4	3
2	0	1	0	0	0	0	5	5

The last tableau represents the optimal solution:

$$x_1^*, x_3^* = 0, \qquad x_2^* = 1, \qquad x_0^* = 5.$$

5(a). From 1(a) the final tableau is

x_1	x_2	x_3	x_4	x_5	x_6	r.h.s.
$-\frac{7}{12}$	1	0	$\frac{1}{3}$	$-\frac{1}{4}$	0	$\frac{11}{12}$
$\frac{11}{6}$	0	1	$-\frac{1}{3}$	$\frac{1}{2}$	0	$\frac{5}{6}$
$\frac{19}{4}$	0	0	0	$-\frac{1}{4}$	1	$\frac{41}{4}$
$\frac{17}{12}$	0	0	$\frac{1}{3}$	$\frac{3}{4}$	0	$\frac{95}{12}$

Now substituting in:

$$\bar{b}_j' = \bar{b}_j'(\bar{b}_j' - 1)^{-1} \sum_{k \in S_-} \bar{a}_{jk} y_k + \sum_{k \in S_+} \bar{a}_{jk} y_k - x_r$$

$$0 + \tfrac{11}{12} - x_2 = -\tfrac{7}{12}x_1 + \tfrac{1}{3}x_4 - \tfrac{1}{4}x_5$$

$$\tfrac{11}{12} = \{\tfrac{11}{12}(\tfrac{11}{12} - 1)^{-1}[-\tfrac{7}{12}x_1 - \tfrac{1}{4}x_5]\} + \tfrac{1}{3}x_4 - x_7.$$

Therefore

$$\tfrac{11}{12} = \tfrac{77}{12}x_1 + \tfrac{11}{4}x_5 + \tfrac{1}{3}x_4 - x_7.$$

Introducing this into the tableau:

x_1	x_2	x_3	x_4	x_5	x_6	x_7	r.h.s.
$-\frac{7}{12}$	1	0	$\frac{1}{3}$	$-\frac{1}{4}$	0	0	$\frac{11}{12}$
$\frac{11}{6}$	0	1	$-\frac{1}{3}$	$\frac{1}{2}$	0	0	$\frac{5}{6}$
$\frac{19}{4}$	0	0	0	$-\frac{1}{4}$	1	0	$\frac{41}{4}$
$\left(-\frac{77}{12}\right)$	0	0	$-\frac{1}{3}$	$-\frac{11}{4}$	0	1	$-\frac{11}{12}$
$\frac{17}{12}$	0	0	$\frac{1}{3}$	$\frac{3}{4}$	0	0	$\frac{95}{12}$
0	1	0	$\frac{4}{11}$	0	0	$-\frac{1}{11}$	1
0	0	1	$-\frac{3}{7}$	$-\frac{2}{7}$	0	$\frac{2}{7}$	$\frac{4}{7}$
0	0	0	$-\frac{19}{77}$	$\frac{16}{7}$	1	$\frac{57}{77}$	$\frac{67}{7}$
1	0	0	$-\frac{4}{77}$	$\frac{3}{7}$	0	$-\frac{12}{77}$	$\frac{1}{7}$
0	0	0	$\frac{20}{77}$	$\frac{2}{7}$	0	$\frac{17}{77}$	$\frac{54}{7}$

Hence the optimal solution is

$$x_1^* = \tfrac{1}{7},$$
$$x_2^* = 1 \quad \text{(which was constrained to be integral)}$$
$$x_3^* = \tfrac{4}{7}, \qquad x_6^* = \tfrac{67}{7}, \qquad x_0^* = \tfrac{54}{7}.$$

Chapter 5

Section 5.6

1(a). The graph is shown in Figure S.7. The shortest path is $\langle p_1, p_2, p_5, p_8, p_{11} \rangle$ with a length of 40.

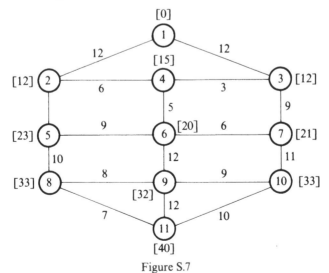

Figure S.7

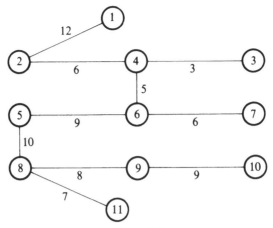

Figure S.8

2(a) and 3(a). The graph is shown in Figure S.8. The order of inclusion of the lines for Kruskal's method is

$$\{p_3, p_4\}, \{p_4, p_6\}, \{p_6, p_7\}, \{p_4, p_2\}, \{p_8, p_{11}\}, \{p_8, p_9\}, \{p_6, p_5\},$$
$$\{p_9, p_{10}\} \text{ [reject: } \{p_3, p_7\}, \{p_8, p_5\}] \{p_8, p_5\} \text{ [reject: } \{p_{10}, p_{11}\},$$
$$\{p_2, p_5\}, \{p_7, p_{10}\}, \{p_6, p_9\}] \{p_1, p_2\}.$$

The weight of the minimal spanning tree is 75.

4(a). The minimum cut is $\{(p_1, p_2), (p_6, p_2), (p_6, p_8), (p_7, p_8)\}$, with a capacity of 18. The arc capacities are shown in Figure S.9.

5(a). The optimal flow assignment is

$$\begin{array}{lllll}
f_{12} = 7, & f_{15} = 11, & f_{24} = 4, & f_{23} = 5, & f_{62} = 2, \\
f_{56} = 6, & f_{57} = 5, & f_{34} = 2, & f_{38} = 3, & f_{48} = 6, \\
f_{68} = 4, & f_{67} = 0, & f_{78} = 5.
\end{array}$$

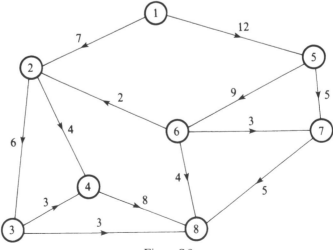

Figure S.9

6(a). The optimal solution is shown in Figure S.10. The cost of this flow is 325. Note that it is different from the solution to 5(a).

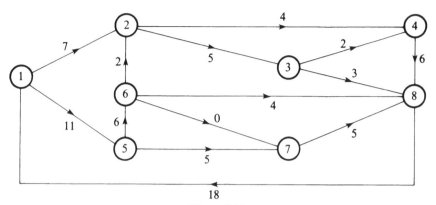

Figure S.10

7(b)	i	t_i	es_i	ls_i	ef_i	lf_i	tf_i	ff_i	
	α	0	0	0	0	0	0	0	critical
	1	5	0	0	5	5	0	0	critical
	2	10	0	1	10	11	1	1	
	3	8	0	6	8	14	6	0	
	4	5	5	5	11	11	0	0	critical
	5	12	5	6	17	18	1	1	
	6	7	11	11	18	18	0	0	critical
	7	4	8	14	12	18	6	6	
	8	6	18	18	24	24	0	0	critical
	9	10	8	14	18	24	6	6	
	w	0	24	24	24	24	0	0	critical

7(c)	i	t_i	es_i	ls_i	ef_i	lf_i	tf_i	ff_i	
	0	0	0	0	0	0	0	0	critical
	1	6	0	30	6	36	30	30	
	2	4	0	0	4	4	0	0	critical
	3	5	4	24	9	29	20	0	
	4	6	4	4	10	10	0	0	critical
	5	4	4	6	8	10	2	2	
	6	3	9	29	12	32	20	20	
	7	10	10	10	20	20	0	0	critical
	8	12	20	20	32	32	0	0	critical
	9	4	32	32	36	36	0	0	critical
	w	0	36	36	36	36	0	0	critical

8. Activity 4 ceases to be critical. The new critical path is

$$\langle \alpha, 2, 5, 7, 8, 9, w \rangle.$$

The earliest completion time is now 34.

Chapter 6

Section 6.9

1(a). The solution is shown in Figure S.11. The shortest path is

$$\langle p_1, p_2, p_5, p_8, p_{11} \rangle$$

with a length of 40.

2(a). Let $f_n(s)$ be the return when s has been allocated to $x_1, x_2, x_n, n = 1, 2, 3, 4$. Let

$$s_n = x_1 + x_2 + \cdots + x_n.$$

Then

$$f_n(s_n) = \max_{\substack{x_n \\ 0 \le x_n \le s_n}} \{ f_{n-1}(s_n - x_n) + 7x_n - nx_n^2 \}, \qquad n = 2, 3, 4 \ldots,$$

$$s_1 = x_1$$

$$f_1(s_1) = [7x_1 - x_1^2]_{x_1 = s_1}$$

$$= 7s_1 - s_1^2,$$

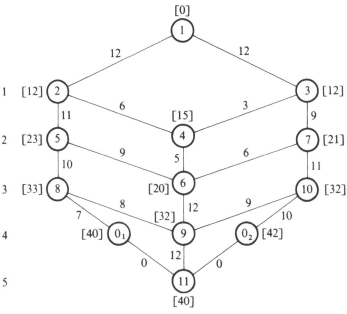

Figure S.11

$$s_2 = x_2 + s_1$$

$$f_2(s_2) = \max_{\substack{x_2 \\ 0 \le x_2 \le s_2}} \{f_1(s_2 - x_2) + 7x_2 - x_2^2\}$$

$$= \max_{\substack{x_2 \\ 0 \le x_2 \le s_2}} \{7(s_2 - x_2) - (s_2 - x_2)^2 + 7x_2 - 2x_2^2\}$$

$$f_3(s_3) = 7s_3 + \tfrac{2}{3}s_3^2 + \tfrac{8}{33}s_3^2 - \tfrac{4}{33}s_3^2$$

$$= 7s_3 - \tfrac{6}{11}s_3^2,$$

$$s_4 = x_4 + s_3 = 8$$

$$f_4(8) = \max_{\substack{x_4 \\ 0 \le x_4 \le 8}} \{f_3(8 - x_4) + 7x_4 - 4x_4^2\}$$

$$= \max_{\substack{x_4 \\ 0 \le x_4 \le 8}} \{7(8 - x_4) - \tfrac{6}{11}(8 - x_4)^2 + 7x_4 - 4x_4^2\}$$

$$= \max_{\substack{x_4 \\ 0 \le x_4 \le 8}} \{\tfrac{232}{11} + \tfrac{96}{11}x_4 - \tfrac{50}{11}x_4^2\}.$$

Let

$$F_4(x_4) = \tfrac{232}{11} + \tfrac{96}{11}x_4 - \tfrac{50}{11}x_4^2.$$

Then

$$\frac{dF(x_4^*)}{dx_4} = \tfrac{96}{11} - \tfrac{100}{11}x_4^* = 0.$$

Therefore

$$x_4^* = \tfrac{24}{25} \in [0, 8].$$

Also,

$$\frac{d^2 F(x_4^*)}{dx_4^2} = -\tfrac{100}{11} < 0,$$

hence x_4^* is a maximum point. Therefore

$$f_4(8) = \tfrac{232}{11} + (\tfrac{96}{11})(\tfrac{24}{25})^2 = \tfrac{632}{25}.$$

Recapitulating:

$$s_4 = 8 \qquad\qquad \Rightarrow x_4^* = \tfrac{24}{25}$$

$$s_3 = 8 - \tfrac{24}{25} = \tfrac{176}{125} \quad \Rightarrow x_3 = (\tfrac{2}{11})(\tfrac{176}{125}) = \tfrac{32}{25}$$

$$s_2 = \tfrac{176}{25} - \tfrac{32}{25} = \tfrac{144}{25} \Rightarrow x_2^* = (\tfrac{1}{3})(\tfrac{144}{25}) = \tfrac{48}{25}$$

$$s_1 = \tfrac{144}{25} - \tfrac{48}{25} = \tfrac{96}{25} \Rightarrow x_1^* = \tfrac{96}{25}.$$

The optimum is $\tfrac{632}{25}$.

2(b). Let $f_n(s)$ be the return when s has been allocated to $x_4, x_3, \ldots, x_n$.
Let

$$s_n = x_4 + x_3 + \cdots + x_n, \qquad n = 1, 2, 3, 4.$$

Then

$$f_n(s_n) = \max_{\substack{x_n \\ 0 \le x_n \le s_n}} \{f_{n+1}(s_n - x_n) + 7x_n - nx_n^2\},$$

$$s_4 = x_4$$

$$f_4(s_4) = [7x_4 - 4x_4^2]_{x_4 = s_4}$$

$$= 7s_4 - 4s_4^2,$$

$$s_3 = s_4 + x_3$$

$$f_3(s_3) = \max_{\substack{x_3 \\ 0 \le x_3 \le s_3}} \{f_4(s_3 - x_3) + 7x_3 - 3x_3^2\}$$

$$= \max_{\substack{x_3 \\ 0 \le x_3 \le s_3}} \{7(s_3 - x_3) - 4(s_3 - x_3)^2 + 7x_3 - 3x_3^2\}$$

$$= \max_{\substack{x_3 \\ 0 \le x_3 \le s_3}} \{7s_3 - 4s_3^2 + 8s_3 x_3 - 7x_3^2\}.$$

Let

$$F_3(x_3) = 7s_3 - 4s_3^2 + 8s_3 x_3 - 7x_3^2.$$

Then

$$\frac{\partial F_3(x_3^*)}{\partial x_3} = 8s_3 - 14x_3^* = 0.$$

Therefore

$$x_3^* = \tfrac{4}{7}s_3 \in [0, s_3]$$

and

$$\frac{\partial^2 F_3(x_3^*)}{\partial x_3^*} = -14 < 0,$$

hence x_3^* is a maximum point. Hence

$$f_3(s_3) = 7s_3 - 4s_3^2 + \tfrac{32}{7}s_3^2 - \tfrac{16}{7}s_3^2$$

$$= 7s_3 - \tfrac{12}{7}s_3^2,$$

$$s_2 = s_2 - x_2$$

$$f_2(s_2) = \max_{\substack{x_2 \\ 0 \le x_2 \le s_2}} \{f_3(s_2 - x_2) + 7x_2 - 2x_2^2\}$$

$$= \max_{\substack{x_2 \\ 0 \le x_2 \le s_2}} \{7s_2 - \tfrac{12}{7}s_2^2 + \tfrac{24}{7}s_2 x_2 - \tfrac{26}{7}x_2^2\}.$$

Let

$$F_2(x_2) = 7s_2 - \tfrac{12}{7}s_2^2 + \tfrac{24}{7}s_2 x_2 - \tfrac{26}{7}x_2^2.$$

Hence

$$\frac{\partial F_2(x_2^*)}{\partial x_2} = \tfrac{24}{7}s_2 - \tfrac{52}{7}x_2^* = 0$$

and

$$x_2^* = \tfrac{6}{13}s_2 \in [0, s_2].$$

Hence

$$\frac{\partial^2 F_2(x_2^*)}{\partial x_2^2} = -\tfrac{52}{7} < 0,$$

and x_2^* is a maximum point. Therefore

$$f_2(s_2) = 7s_2 - \tfrac{12}{7}s_2^2 + \tfrac{144}{91}s_2^2 - \tfrac{72}{91}s_2^2$$
$$= 7s_2 - \tfrac{84}{91}s_2^2.$$

Now

$$s_2 = s_1 - x_1 = 8 - x_1$$

$$f_1(s_1) = \max_{\substack{x_1 \\ 0 \le x_1 \le 8}} \{f_2(8 - x_1) + 7x_1 - x_1^2\}$$

$$= \max_{\substack{x_1 \\ 0 \le x_1 \le 8}} \{7(8 - x_1) - \tfrac{84}{91}(8 - x_1)^2 + 7x_1 - x_1^2\}.$$

Let

$$F_1(x_1) = 7(8 - x_1) - \tfrac{84}{91}(8 - x_1)^2 + 7x_1 - x_1^2.$$

Hence

$$\frac{\partial F_1(x_1^*)}{\partial x_1} = \tfrac{1344}{91} - \tfrac{350}{91}x_1^* = 0$$

and

$$x_1^* = \tfrac{96}{25} \in [0, 8]$$

$$\frac{\partial^2 F_1(x_1^*)}{\partial x_1^2} = -\tfrac{350}{91} < 0,$$

hence x_1^* is a maximum point. Therefore

$$f_1(s_1) = 7(8 - \tfrac{96}{25}) - \tfrac{84}{91}(8 - \tfrac{96}{25})^2 + 7(\tfrac{96}{25}) - (\tfrac{96}{25})^2$$
$$= \tfrac{632}{25}.$$

Recapitulating:

$$s_1 = 8 \qquad\qquad \Rightarrow x_1^* = \tfrac{96}{25}$$
$$s_2 = 8 - \tfrac{96}{25} = \tfrac{104}{25} \Rightarrow x_2^* = (\tfrac{6}{13})(\tfrac{104}{25}) = \tfrac{48}{25}$$
$$s_3 = \tfrac{104}{25} - \tfrac{48}{25} = \tfrac{56}{25} \Rightarrow x_3^* = (\tfrac{4}{7})(\tfrac{56}{25}) = \tfrac{32}{25}$$
$$s_4 = \tfrac{56}{25} - \tfrac{32}{25} = \tfrac{24}{25} \Rightarrow x_4^* = \tfrac{24}{25}.$$

The optimum is $\tfrac{632}{25}$.

There is slightly less effort involved in forward recursion.

3. Let $f_n(s)$ be the return when s has been allocated to $x_1, x_2, \ldots, x_n$, $n = 1, 2, 3, 4$. Let

$$s_n = x_1 + x_2 + \cdots + x_n.$$

Then

$$f_n(s_n) = \max_{\substack{x_n \\ 0 \le x_n \le s_n}} \{f_{n-1}(s_n - x_n) + 7x_n - nx_n^2\}.$$

Now

$$s_1 = x_1$$
$$f_1(s_1) = \max_{\substack{x_1 \\ x_1 = s_1}} \{7x_1 - x_1^2\}.$$

s_1	0	1	2	3	4	5	6	7	8
$f_1(s_1)$	0	6	10	12	12	10	6	0	-8

Now

$$s_2 = x_2 + s_1$$
$$f_2(s_2) = \max_{\substack{x_2 \\ 0 \le x_2 \le s_2}} \{f_1(s_1) + 7x_2 - 2x_2^2\}.$$

				x_2							
s_2	0	1	2	3	4	5	6	7	8	x_2^*	$f_2(s_2)$
0	0	—	—	—	—	—	—	—	—	0	0
1	6	5	—	—	—	—	—	—	—	1	6
2	10	11	6	—	—	—	—	—	—	1	11
3	12	15	12	3	—	—	—	—	—	1	15
4	12	17	16	9	-4	—	—	—	—	1	17
5	10	17	18	13	2	-15	—	—	—	2	18
6	6	15	18	15	6	-9	-30	—	—	2	18
7	0	11	16	15	8	-5	-24	-49	—	2	16
8	-8	5	12	13	8	-3	-20	-43	-72	3	13

Now

$$s_3 = x_3 + s_2$$

$$f_3(s_3) = \max_{\substack{x_2 \\ 0 \le x_2 \le s_2}} \{f_2(s_2) + 7x_3 - 3x_3^2\}.$$

				x_3							
s_3	0	1	2	3	4	5	6	7	8	x_3^*	$f_2(s_3)$
0	0	—	—	—	—	—	—	—	—	0	0
1	1	4	—	—	—	—	—	—	—	0	6
2	11	10	2	—	—	—	—	—	—	0	11
3	15	15	8	−6	—	—	—	—	—	0, 1	15
4	17	19	13	0	−20	—	—	—	—	1	19
5	18	21	17	5	−14	−40	—	—	—	1	21
6	18	22	19	9	−10	−34	−66	—	—	1	22
7	16	22	20	11	−8	−30	−60	−98	—	1	22
8	13	20	20	12	−8	−28	−56	−92	−136	1, 2	20

Now

$$s_4 = s_3 + x_4 = 8$$

$$f_4(s_4) = \max_{\substack{x_4 \\ 0 \le x_4 \le 8}} \{f_3(s_3) + 7x_4 - 4x_4^2\}.$$

				x_4						
s_4	0	1	2	3	4	5	6	7	8	x_4^*	$f_4(s_4)$
8	20	25	20	0	−17	−44	−80	−125	−180	1	25

$$s_4 = 8 \qquad \Rightarrow x_4^* = 1$$
$$s_3 = 8 - 1 = 7 \Rightarrow x_3^* = 1$$
$$s_2 = 7 - 1 = 6 \Rightarrow x_2^* = 2$$
$$s_1 = 6 - 2 = 4 \Rightarrow x_1^* = 4$$

with an optimum of 25.

4. Let

x_n = the number of units produced at stage.

$f_n(s)$ = return when a total of s units have been produced by the end of stage n.

$s_n = x_1 + x_2 + \cdots + x_n$.

$c_{x,j}$ = production cost of x_i units in period j.

$$f_n(s_n) = \min_{\substack{x_4 \\ 0 \le x_n \le s_n}} \{f_{n-1}(s_n - x_n) + c_{x_n n} + (4 - n)x_n\}, \qquad n = 1, 2, 3, 4.$$

Now

$$s_1 = x_1.$$

Therefore

$$f_1(s_1) = \min_{\substack{x_1 \\ 0 \le x_1 \le s_1}} \{c_{x_1 1} + 3x_1\}.$$

$s_1 (= x_1)$	0	1	2	3	4	5	6
$f_1(s_1)$	2	7	14	18	23	27	32

Now

$$s_2 = s_1 + x_2$$

$$f_2(s_2) = \min_{\substack{x_2 \le 6 \\ 1 \le x_2 \le s_2 \\ 5 \le s_2 \le 12}} \{f_1(s_2 - x_2) + c_{x_2 2} + 2x_2\}.$$

The restrictions $1 \le x_2$ and $5 \le s_2 \le 12$ arise from the facts that 19 units must be produced but that no more than 6 can be produced per period.

	x_2							
s_2	1	2	3	4	5	6	x_2^*	$f_2(s_2)$
7	41	40	40	41	43	39	2, 3	39
8	—	45	44	46	47	46	3	44
9	—	—	49	50	52	50	3	49
10	—	—	—	55	56	55	4, 6	55
11	—	—	—	—	61	59	6	59
12	—	—	—	—	—	64	6	64

$$s_3 = s_2 + x_3$$

$$f_3(s_3) = \min_{\substack{1 \le x_3 \le 6 \\ 13 \le s_3 \le 18}} \{f_2(s_3 - x_3) + c_{x_3 3} + x_3\}.$$

	x_2							
s_2	1	2	3	4	5	6	x_3^*	$f_3(s_3)$
13	73	72	73	69	66	66	5,6	65
14		77	77	75	71	70	6	70
15			82	79	77	75	6	75
16				84	81	81	5,6	81
17					86	85	6	85
18						90	6	90

$$s_4 = s_3 + x_3 = 19$$

$$f_4(s_4) = \min_{1 \le x_4 \le 6} \{f_3(19 - x_4) + c_{x_4 4}\}.$$

	x_4							
s_4	1	2	3	4	5	6	x_4^*	$f_4(s_4)$
19	95	94	94	90	87	87	5	87

The minimum cost is $87. Backtracking:

$$x_4^* = 5 \quad \text{or} \quad x_4^* = 6$$
$$x_3^* = 6 \qquad\qquad x_3^* = 6$$
$$x_2^* = 3 \qquad\qquad x_2^* = 6$$
$$x_1^* = 5 \qquad\qquad x_1^* = 1.$$

5. Let

$$f_n(s) = \text{return when } s \text{ has been allocated to } x_1, x_2, \ldots, x_n$$
$$s_n = x_1 + x_2 + \cdots + x_n.$$

Then

$$f_n(s_n) = \max_{\substack{x_n \\ 0 < x_n < s_n}} \{f_{n-1}(s_n - x_n)x_n\}$$

$$s_1 = x_1$$

$$f_1(s_1) = \max_{\substack{x_1 \\ x_1 = s_1}} \{x_1\} = s_1$$

$$s_2 = x_1 + s_1$$

$$f_2(s_2) = \max_{\substack{x_2 \\ 0 < x_2 < s_2}} \{f(s_2 - x_2)x_2\}$$

$$\max_{\substack{x_2 \\ 0 < x_2 < s_2}} \{(s_2 - x_2)x_2\}.$$

Let

$$F_2(x_2) = (s_2 - x_2)x_2.$$

Hence

$$\frac{\partial F_2(x_1^*)}{\partial x_2} = s_2 - sx_2^* = 0$$

and

$$x_2^* = s_2/2 \in [0, s_2].$$

Also

$$\frac{\partial^2 F_2(x_2^*)}{\partial x_2^2} = -2 < 0,$$

hence x_2^* is a maximum point.

$$s_3 = x_3 + s_2$$

$$f_3(s_3) = \max_{\substack{x_3 \\ 0 < x_3 < s_3}} \{f_2(s_3 - x_3)x_3\}$$

$$= \max_{\substack{x_3 \\ 0 < x_3 < s_3}} \left\{ \frac{(s_3 - x_3)^2}{4} x_3 \right\}.$$

Let

$$F_3(x_3) = \tfrac{1}{4}(s_3 - x_3)^2 x_3.$$

Then

$$\frac{\partial F_3(x_3^*)}{\partial x_3} = \tfrac{1}{4}[s_3^2 - 4s_3 x_3^* + 3(x_3^*)^2] = 0$$

and

$$x_3^* = s_3/3 \text{ or } s_3, \quad \text{both in } [0, s_3].$$

$$\frac{\partial^2 F_3(s_3/3)}{\partial x_3^2} = -2s_3 < 0,$$

hence $s_3/3$ a maximum point, and

$$\frac{\partial^2 F_3(s_3)}{\partial x_3^2} = 2s_3 > 0,$$

hence s_3 is a minimum point. Therefore

$$x_3^* = s_3/3$$
$$f_3(s_3) = \tfrac{1}{4}(s_3/3)(s_3 - (s_3/3))^2 = (s_3/3)^3,$$

$$s_4 = x_4 + s_3 = 9.$$

Hence

$$f_4(s_4) = \max_{\substack{x_4 \\ 0 < x_4 < 9}} \{f_3(9 - x_4)x_4\}$$

$$= \max_{\substack{x_4 \\ 0 < x_4 < 9}} \left\{ \left(\frac{9 - x_4}{3}\right)^3 x_4 \right\}.$$

Let

$$F_4(x_4) = \left(\frac{9 - x_4}{3}\right)^3 x_4.$$

Then

$$\frac{\partial F_4(x_4^*)}{\partial x_4} = (9 - x_4^*)^2 \frac{x_4^*}{9} = 0$$

and

$$x_4^* = 0, \tfrac{9}{4}, \text{ or } 9.$$

But as $0 < x_4^* < 9$,

$$x_4^* = \tfrac{9}{4}$$

$$f_4(s_4) = \frac{9}{4}\left(\frac{9 - \tfrac{9}{4}}{3}\right)^3 = \left(\frac{9}{4}\right)^4.$$

Backtracking:

$$s_4 = 9 \qquad\qquad \Rightarrow \qquad x_4^* = \tfrac{9}{4}$$
$$s_3 = 9 - x_4^* = \tfrac{27}{4} \Rightarrow x_3^* = s_3/3 = \tfrac{9}{4}$$
$$s_2 = \tfrac{27}{4} - x_3^* = \tfrac{18}{4} \Rightarrow x_2^* = s_2/2 = \tfrac{9}{4}$$
$$s_1 = \tfrac{18}{4} - x_1^* = \tfrac{9}{4} \Rightarrow x_1^* = s_1 = \tfrac{9}{4}.$$

The value of the optimal solution is $(\tfrac{9}{4})^4$.

6. Let

$$f_n(s) = \text{return when } s \text{ has been allocated to } x_1, x_2, \ldots, x_n$$
$$s_n = x_1 + x_2 + \cdots + x_n.$$

Then

$$f_n(s_n) = \min_{\substack{x_n \\ 0 < x_n}} \{f_{n-1}(s_n/x_n) + x_n^2\},$$

$$s_1 = x_1.$$

Hence

$$f_1(s_1) = \min_{\substack{x_1 \\ x_1 = s_1}} \{x_1^2\} = s_1^2$$

$$s_2 = s_1 x_2$$

$$f_2(s_2) = \min_{\substack{x_2 \\ 0 < x_2}} \{f_1(s_2/x_2) + x_2^2\}$$

$$= \min_{\substack{x_2 \\ 0 < x_2}} \{(s_2/x_2)^2 + x_2^2\}.$$

Let

$$F_2(x_2) = (s_2/x_2)^2 + x_2^2.$$

Then

$$\frac{\partial F_2(x_2^*)}{\partial x_2} = -(2s_2^2/x_2^{*3}) + 2x_2^* = 0.$$

Therefore

$$x_2^* = \sqrt{s_2}.$$

Also

$$\frac{\partial^2 F_2(x_2^*)}{\partial x_2^2} = \frac{6s_2^2}{(\sqrt{s_2})^4} + 2 > 0,$$

hence x_2^* is a minimum point. Therefore

$$f_2(s_2) = 2s_2,$$

$$s_2 = s_3/x_3.$$

Hence

$$f_3(s_3) = \min_{\substack{x_3 \\ 0 < x_3}} \{f_2(s_3/x_3) + x_3^2\}$$

$$= \min_{\substack{x_3 \\ 0 < x_3}} \{2(s_3/x_3) + x_3^2\}.$$

Let

$$F_3(x_3) = 2(s_3/x_3) + x_3^*.$$

Then

$$\frac{\partial F_3(x_3^*)}{\partial x_3} = -2(s_3/(x_3^*)^2) + 2x_3^* = 0.$$

Therefore

$$x_3^* = 3\sqrt{s_3}.$$

Also

$$\frac{\partial^2 F_3(x_3^*)}{\partial x_3^2} = \frac{4s_3}{(3\sqrt{s_3})^3} + 2 > 0,$$

hence x_3^* is a minimum point. Therefore

$$f_3(s_3) = 3s_3^{2/3},$$

$$s_3 = s_4/x_4.$$

Hence

$$f_4(s_4) = \min_{\substack{x_4 \\ 0 < x_4}} \{f_3(s_4/x_4) + x_4^2\}$$

$$= \min_{\substack{x_4 \\ 0 < x_4}} \{3(s_4/x_4)^{2/3} + x_4^2\}.$$

Let

$$F_4(x_4) = 3(s_4/x_4)^{2/3} + x_4^2.$$

Then

$$\frac{\partial F_4(x_4^*)}{\partial x_4} = -\frac{2s_4^{2/3}}{(x_4^*)^{5/3}} + 2x_4^* = 0.$$

Therefore

$$x_4^* = s_4^{1/4}.$$

Also

$$\frac{\partial^2 F_4(x_4^*)}{\partial x_4^*} = \frac{10s_4^{2/3}}{3s_4^{2/3}} + 2 > 0,$$

hence x_4^* is a minimum point. Therefore

$$f_4(s_4) = 3s_4^{2/3}s_4^{-1/6} + s_4^{1/2} = 4s_4^{1/2},$$

$$s_4 = s_5/x_5 = 11/x_5.$$

Hence

$$f_5(s_5) = f_5(11) = \min_{\substack{x_5 \\ 0 < x_5}} \{f_4(11/x_5) + x_5^2\}$$

$$= \min_{\substack{x_5 \\ 0 < x_5}} \{4(11/x_5)^{1/2} + x_5^2\}.$$

Let

$$F_5(x_5) = 4(11/x_5)^{1/2} + x_5^2.$$

Then

$$\frac{\partial F_5(x_5^*)}{\partial x_5} = (-2.11^{1/2})(x_5^*)^{-3/2} + 2x_5^* = 0$$

and

$$x_5^* = 11^{1/5}.$$

Hence

$$f_s(11) = \frac{4(11)^{1/2}}{11^{1/10}} + 11^{2/5} = 5(11)^{2/5}.$$

Backtracking:

$$s_5 = 11 \qquad \Rightarrow \qquad x_5^* = 11^{1/5}$$

$$s_4 = \frac{11}{11^{1/5}} = 11^{4/5} \Rightarrow x_4^* = (11^{4/5})^{1/4} = 11^{1/5}$$

$$s_3 = \frac{11^{4/5}}{11^{1/5}} = 11^{3/5} \Rightarrow x_3^* = (11^{3/5})^{1/3} = 11^{1/5}$$

$$s_2 = \frac{11^{3/5}}{11^{1/5}} = 11^{2/5} \Rightarrow x_2^* = (11^{2/5})^{1/2} = 11^{1/5}$$

$$s_1 = \frac{11^{2/5}}{11^{1/5}} = 11^{1/5} \Rightarrow \qquad x_1^* = 11^{1/5}.$$

7. Let $f_n(s)$ be the return when s has been allocated to $v_1 x_1 + \cdots + v_n x_n$, $n = 1, 2$, where

$$v_1 = 3, \qquad v_2 = 4.$$

Then

$$s_n = \sum_{i=1}^{n} v_i x_i$$

$$s_n = s_{n-1} + v_n x_n, \qquad n = 2,$$

$$s_2 \geq 12,$$

hence

$$f_n(s_n) = \min_{\substack{x_n \\ 0 \leq x_n \leq s_n/v_n}} \{f_{n-1}(s_{n-1}) + x_n\},$$

$$3x_1 = s_1.$$

Hence

$$f_1(s_1) = \min_{\substack{x_1 \\ 0 \leq x_1 \leq s_1/3}} \{x_1\} = s_1/3.$$

Therefore

$$x_1^* = s_1/3,$$

$$s_2 = s_1 + 4x_2.$$

Hence

$$f_2(s_2) = \min_{\substack{x_2 \\ 0 \leq x_2 \leq s_2/4}} \{f_1(s_2 - 4x_2) + x_2\}$$

$$= \min_{\substack{x_2 \\ 0 \leq x_2 \leq s_2/4}} \left\{\frac{(s_2 - 4x_2)}{3} + x_2\right\}.$$

Hence

$$x_2^* = s_2/4$$

and

$$f_2(s_2) = s_2/4,$$

$$s_2 \geq 12.$$

Therefore

$$f_2(s_2) = 3.$$

Recapitulating:

$$s_2 = 12, \qquad x_2^* = 3, \qquad x_1^* = 0.$$

8(a). Let $f_n(s, t)$ be the return when

$$\sum_{i=1}^{n} v_i x_i = s_n, \qquad n = 1, 2$$

and

$$\sum_{i=1}^{n} w_i x_i = t_n, \qquad n = 1, 2$$

where

$$(v_1, v_2) = (3, 4)$$

$$(w_1, w_2) = (4, -5)$$

$$s_2 \leq 24$$

$$t_2 \leq 20.$$

From the constraints,

$$x_1 \leq 5, \qquad x_2 \leq 4.$$

Therefore

$$f_n(s_n, t_n) = \max_{x_n} \{ f_{n-1}(s_{n-1}, t_{n-1}) + d_n x_n^2 + e_n x_n \}, \qquad n = 2$$

where

$$(d_1, d_2) = (8, 4)$$

$$(e_1, e_2) = (-3, -4).$$

s_1	t_1	x_1	$f_1(s_1, t_1)$
0	0	0	0
3	4	1	5
6	8	2	26
9	12	3	63
12	16	4	116
15	20	5	185

$$f_2(s_2, t_2) = \max_{\substack{x_2 \\ 0 \leq x_2 \leq 6}} \{ f_1(s_1, t_1) + 4x_2^2 - 4x_2 \}$$

$$s_2 = s_1 + 4x_2$$

$$t_2 = t_1 + 5x_2.$$

Therefore

$$f_2(s_2, t_2) = \max_{\substack{x_2 \\ 0 \le x_2 \le 6}} \{f_1(s_2 - 4x_2, t_2 - 5x_2) + 4x_2^2 - 4x_2\}.$$

s_2	t_2	x_2	$f_2(s_2, t_2)$
0	0	0	0
3	4	0	5
4	5	1	0
6	8	0	26
7	9	1	5
8	10	2	8
9	12	0	63
10	13	1	26
11	14	2	13
12	16	0	116
13	17	1	63
14	18	2	34
15	20	0	185
16	20	4	116

Hence the maximum value occurs when $s_2 = 15$, $t_2 = 20$, $x_2 = 0$. Backtracking:

$$x_1^* = 5, \qquad x_2^* = 0, \qquad x_0^* = 185.$$

Let $f_n(s, t)$ be the return when

$$\sum_{i=n}^{2} v_i x_i = s_n, \qquad n = 1, 2$$

and

$$\sum_{i-n}^{2} w_i x_i = t_n, \qquad n = 1, 2.$$

Then

$$f_1(s_1, t_1) = \max_{x_1} \{f_2(s_2, t_2) + 8x_1^2 - 3x_1\}.$$

s_2	t_2	x_2	$f_2(s_2, t_2)$
0	0	0	0
4	5	1	0
8	10	2	8
12	15	3	24
16	20	4	48

s_1	t_1	x_1	$f_1(s_1, t_1)$
0	0	0	0
3	4	1	5
4	5	0	0
6	8	0	26
7	9	1	5
8	10	0	8
9	12	3	63
10	13	2	26
11	14	1	13
12	15	0	24
12	16	4	116
13	17	3	63
14	18	2	34
15	20	5	185
16	20	4	116

Hence, as before,

$$x_1^* = 5, \qquad x_2^* = 0, \qquad x_0^* = 185.$$

9. Number the objects $1, 2, \ldots, 7$ in nonincreasing order of weight. Let

$$x_i = \begin{cases} 1, & \text{if object } i \text{ is taken} \\ 0, & \text{otherwise} \end{cases}$$

w_i = weight of object i

v_i = value of object i, $\qquad i = 1, 2, \ldots, 7$

$$S_n = \sum_{i=1}^{n} w_i x_i, \qquad n = 1, 2, \ldots, 7$$

$f_n(s)$ = return when the weight of the objects selected after the first n objects have been considered is s.

Then

$$f_n(s_n) = \max_{\substack{x_n \\ \sum_{i=1}^{n} w_i x_i \le s_n \\ s_7 \le 100}} \{f_{n-1}(s_{n-1}) + v_n x_n\}$$

s_1	0	50
x_1	0	1
$f_1(s_1)$	0	60

s_2	0	40	50	90
x_2	0	1	0	1
$f_2(s_2)$	0	40	60	100

s_3	0	40	50	80	90
x_3	0	0	0	1	0
$f_1(s_3)$	0	40	60	60	100

s_4	0	30	40	50	70	80	90
x_4	0	1	0	0	1	1	0
$f_4(s_4)$	0	10	40	60	50	70	100

s_5	0	30	40	50	70	80	90
x_5	0	1	0	0	1	1	0
$f_5(s_5)$	0	60	40	60	100	120	100

s_6	0	10	30	40	50	60	70	80	90	100
x_6	0	1	0	1	0	0,1	0	1	1	1
$f_6(s_6)$	0	10	60	70	60	70	100	110	130	110

s_7	0	10	30	40	50	60	70	80	90	100
x_7	0	0	0	0	0	0	0	0	0	1
$f_7(s_7)$	0	10	60	70	60	70	100	110	130	133

Backtracking:

$$x_7^* = 1, \quad x_6^* = 1, \quad x_5^* = 1, \quad x_4^* = 0,$$
$$x_3^* = 0, \quad x_2^* = 0, \quad x_1^* = 1, \quad x_0^* = 133.$$

10. Let

u_i = volume of object i

$$t_n = \sum_{i=1}^{n} u_i x_i, \qquad n = 1, 2, \dots, 7$$

$t_7 \leq 100$

$g_n(s, t)$ = return when the weight and volume of the objects selected after the first n objects have been considered in s and t, respectively.

s_1	0	50
t_1	0	50
x_1	0	1
$g_1(s_1, t_1)$	0	60

s_2	0	40	50	90
t_2	0	25	50	75
x_2	0	1	0	1
$g_2(s_2, t_2)$	0	40	60	100

s_3	0	40	40	50	80	90	90
t_3	0	0	25	50	25	75	50
x_3	0	1	0	0	1	0	1
$g_3(s_3,t_3)$	0	20	40	60	60	100	80

s_4	0	30	40	40	50	70	70	80	80	90	90
t_4	0	25	0	25	50	25	50	75	25	75	50
x_4	0	1	0	0	0	1	1	1	0	0	0
$g_4(s_4,t_4)$	0	10	20	40	60	30	50	70	60	100	80

s_5	0	30	30	40	40	50	60	70	70	70	70	80	80	90	90
t_5	0	25	75	0	25	50	100	25	50	100	75	75	25	75	90
x_5	0	0	1	0	0	0	1	0	0	1	1	0	0	0	0
$g_5(s_5,t_5)$	0	10	60	20	40	60	70	30	50	100	80	70	60	100	80

s_6	0	10	30	30	40	40	50	50	60	60	70	70	70	70	80	80	80	80
t_6	0	25	25	75	0	25	50	25	75	100	25	100	75	50	75	25	50	100
x_6	0	1	0	0	0	0	0	1	1	0	0	0	0	0	0	0	1	1
g_6	0	10	10	60	20	40	50	30	70	70	30	100	80	50	70	60	40	90

s_6	90	90	90	100	100
t_6	75	50	100	100	75
x_6	0	0	1	1	1
g_6	100	80	80	110	90

s_7	0	10	10	20	30	30	40	40	40	50	50	50	50	60	60
t_7	0	24	50	75	25	75	0	25	75	50	25	75	100	75	100
x_7	0	0	1	1	0	0	0	0	1	0	0	1	1	0	0
g_7	0	10	3	13	10	60	20	40	13	60	30	43	63	70	70

s_7	60	70	70	70	70	80	80	80	80	90	90	90	100	100
t_7	50	25	100	75	50	75	25	50	100	75	50	100	100	75
x_7	1	0	0	0	0	0	0	0	0	0	0	0	0	0
g_7	63	30	100	80	70	70	60	40	90	100	80	80	110	90

Backtracking:

$$x_7^* = 0, \quad x_6^* = 1, \quad x_5^* = 0, \quad x_4^* = 0,$$
$$x_3^* = 0, \quad x_2^* = 1, \quad x_1^* = 1, \quad x_0^* = 110.$$

11. First divide all constants by 10.

s_1	0	10	20	30	40	50
t_1	0	100	200	300	400	500
x_1	0	1	2	3	4	5
$f_1(s_1,t_1)$	0	9	18	27	36	45

s_2	0	10	20	20	30	30	40	40	40	50
t_2	0	100	50	200	150	300	100	250	400	200
x_2	0	0	1	0	1	0	2	1	0	2
$f_2(s_2, t_2)$	0	9	10	18	19	27	20	28	36	29

s_2	50	50	60	60	60	70	70	70	80
t_2	350	500	150	300	450	250	400	550	200
x_2	1	0	3	2	1	3	2	1	4
$f_2(s_2, t_2)$	37	45	30	38	46	39	47	55	40

s_2	80	80	90	90	100	100	100
t_2	350	500	300	450	250	400	550
x_2	3	2	4	3	5	4	3
$f_2(s_2, t_2)$	48	56	49	57	50	58	66

s_3	0	5	10	10	15	15	20	20	25	25	25	30
t_3	0	150	100	300	250	450	50	400	200	350	550	150
x_3	0	1	0	2	1	3	0	2	1	1	3	0
$f_3(s_3, t_3)$	0	15	9	30	24	45	10	39	25	33	54	19

s_3	30	30	35	35	35	40	40	40	40	45	45	45
t_3	300	500	300	450	500	100	250	400	450	250	400	550
x_3	0	2	1	1	3	0	0	0	2	1	1	1
$f_3(s_3, t_3)$	27	48	34	42	55	20	28	36	49	35	43	51

s_3	50	50	50	50	50	55	55	55	60	60	60	60
t_3	200	350	400	500	550	350	500	550	150	300	450	500
x_3	0	0	2	0	2	1	1	3	0	0	0	2
$f_3(s_3, t_3)$	29	37	50	45	58	44	52	65	30	38	46	59

s_3	65	65	70	70	70	70	75	75	80	80	80	80	85
t_3	300	450	250	400	450	550	400	550	200	350	500	550	350
x_3	1	1	0	0	2	0	1	1	0	0	0	2	1
$f_3(s_3, t_3)$	45	53	39	47	60	55	54	62	40	48	56	69	55

s_3	85	90	90	90	95	100	100	100
t_3	500	300	450	500	450	250	400	550
x_3	1	0	0	2	1	0	0	0
$f_3(s_3, t_3)$	63	49	57	70	64	50	58	66

Backtracking:

$$x_3^* = 2, \qquad x_2^* = 4, \qquad x_1^* = 0$$

$$x_0^* = 70, \quad \text{or 700 in terms of the original data.}$$

12. Let

s_i = number of tons available at the end of year i

x_i = number of tons sold at the end of year i, $i = 1, 2, \ldots, 5$

$s_n = 3(s_{n-1} - x_{n-1})$, $n = 2, 3, 4, 5.$

Using backward recursion: Let

$f_n(s)$ = the return that can be accrued from years n, $n + 1, \ldots, 5$
given s tons are available at the end of year n.

Then

$$f_n(s_n) = \max_{\substack{x_n \\ 0 \le x_n \le s_n}} \{f_{n+1}(3(s_n - x_n)) + d_n x_n\}, \qquad n = 1, 2, 3, 4$$

where

$$(d_1, d_2, d_3, d_4, d_5) = (400, 330, 44, 15, 5),$$

$$s_1 = 20.$$

Hence

$$f_5(s_5) = \max_{x_5 = s_5} \{5x_5\} = 5s_5, \quad \text{and} \quad x_5^* = s_5,$$

$$f_4(s_4) = \max_{\substack{x_4 \\ 0 \le x_4 \le s_4}} \{f_5(3(s_4 - x_4)) + 15x_4\}$$

$$= 15s_4, \quad \text{and} \quad x_4^* = 0,$$

$$f_3(s_3) = \max_{\substack{x_3 \\ 0 \le x_3 \le s_3}} \{f_4(3(s_3 - x_3)) + 44x_3\}$$

$$= \max_{\substack{x_3 \\ 0 \le x_3 \le s_3}} \{45s_3 - x_3\}$$

$$= 45s_3, \quad \text{and} \quad x_3^* = 0,$$

$$f_2(s_2) = \max_{\substack{x_2 \\ 0 \le x_2 \le s_2}} \{f_3(3(s_2 - x_2)) + 330x_2\}$$

$$= \max_{\substack{x_2 \\ 0 \le x_2 \le s_2}} \{135s_2 + 195x_2\}$$

$$= 330s_2, \quad \text{and} \quad x_2^* = s_2,$$

$$f_1(s_1) = \max_{\substack{x_1 \\ 0 \le x_1 \le 20}} \{f_2(3(20 - x_1)) + 400x_1\}$$

$$= \max_{\substack{x_1 \\ 0 \le x_1 \le 20}} = \{990(20 - x_1) + 400x_1\}$$

$$= 19{,}800, \quad \text{and} \quad x_1^* = 0.$$

Backtracking:

$$x_1^* = 0, \qquad s_1 = 20$$
$$s_2 = 3(s_1 - 0) = 60$$
$$x_2^* = 60$$
$$s_3 = 3(60 - 60) = 0$$
$$s_4 = 0, \qquad s_5 = 0 \qquad x_3^*, x_4^*, x_5^* = 0.$$

Hence all potatoes should be sold at the end of the second year for a total profit of 19,800 units.

Chapter 7

Section 7.6

1(a)

$$f(x) = x^3 + \tfrac{3}{2}x^2 - 18x + 19$$
$$f'(x) = 3x^2 + 3x - 18$$
$$f''(x) = 6x + 3.$$

If

$$f'(x^*) = 0,$$

then

$$x^* = -3 \text{ or } 2,$$

$$f''(-3) < 0,$$

hence $x^* = -3$ is the global maximum; also,

$$f''(2) > 0,$$

hence $x^* = 2$ is the global minimum.

2(a). Referring to Exercise 1(a), as $-3 \notin [-\tfrac{5}{2}, \tfrac{5}{2}]$ the global maximum of f occurs at one of the endpoints of $[-\tfrac{5}{2}, \tfrac{5}{2}]$. As $f(-\tfrac{5}{2}) > f(\tfrac{5}{2})$,

$$x^* = -\tfrac{5}{2}$$

is the global maximum, and $2 \notin [-\tfrac{5}{2}, \tfrac{5}{2}]$. Therefore $x^* = 2$ is still the global minimum.

8(a)

$$f(x) = \sin x$$
$$f'(x) = \cos x.$$

If

$$f'(x^*) = 0,$$

then

$$\cos x^* = 0.$$

Therefore

$$x^* = \pi/2, \qquad x^* \in [0, \pi].$$

Hence as f is concave, by Theorem 7.7,

$$x^* = \pi/2$$

is a global maximum. f has global minima at $x = 0$ and $x = \pi$.

10(a)

$$f(x_1, x_2) = x_1^2 - x_1 + 3x_2^2 + 18x_2 + 14$$

$$\frac{\partial f}{\partial x_1} = 2x_1 - 1$$

$$\frac{\partial^2 f}{\partial x_1^2} = 2$$

$$\frac{\partial f}{\partial x_2} = 6x_2 + 18$$

$$\frac{\partial^2 f}{\partial x_2^2} = 6$$

$$\frac{\partial^2 f}{\partial x_1 x_2} = 0$$

$$\frac{\partial^2 f}{\partial x_2 \, \partial x_1} = 0$$

$$\frac{\partial f}{\partial x_1} = 0 \quad \text{at} \quad x_1 = x_1^*.$$

Hence

$$2x_1^* - 1 = 0.$$

Therefore

$$x_1^* = \tfrac{1}{2},$$

$$\frac{\partial f}{\partial x_2} = 0 \quad \text{at} \quad x_2 = x_2^*.$$

Hence

$$6x_2^* + 18 = 0.$$

Therefore

$$x_2^* = -3.$$

Therefore

$$X_0 = (\tfrac{1}{2}, -3)$$

is the only candidate for an extreme point. As

$$H(X) = \begin{pmatrix} 2 & 0 \\ 0 & 6 \end{pmatrix}$$

is positive definite, X_0 is a global minimum.

$$f(X_0) = \tfrac{51}{4}.$$

11(a). From Exercise 10(a),

$$x_1^* = \tfrac{1}{2}, \qquad x_2^* = -3.$$

But

$$x_2^* \notin [-2, 6].$$

Hence the global minimum lies on the boundary of

$$S = \{(x_1, x_2): -1 \le x_1 \le 5, \ -2 \le x_2 \le 6\}.$$

The boundaries of S are

$$B_1 = \{\alpha(-1, -2) + (1 - \alpha)(5, -2): 0 \le \alpha \le 1\}$$
$$B_2 = \{\alpha(5, -2) + (1 - \alpha)(5, 6): 0 \le \alpha \le 1\}$$
$$B_3 = \{\alpha(-1, 6) + (1 - \alpha)(5, 6): 0 \le \alpha \le 1\}$$
$$B_4 = \{\alpha(-1, -2) + (1 - \alpha)(-1, 6): 0 \le \alpha \le 1\}.$$

For B_1. Let

$$\begin{aligned}
g(\alpha) &= f(\alpha(-1, -2) + (1 - \alpha)(5, -2)), \qquad 0 \le \alpha \le 1 \\
&= f(5 - 6\alpha, -2) \\
&= (5 - 6\alpha)^2 - (5 - 6\alpha) + 3(-2)^2 + 18(-2) + 14.
\end{aligned}$$

Then

$$g'(\alpha) = 2(5 - 6\alpha)(-6) + 6,$$

which is zero at α^*. Therefore

$$\alpha^* = \tfrac{3}{4} \in [0, 1] \quad \text{and} \quad g''(\alpha^*) > 0,$$

indicating a minimum. Therefore

$$\begin{aligned}
(x_1^*, x_2^*) &= \tfrac{3}{4}(-1, -2) + \tfrac{1}{4}(5, -2) \\
&= (\tfrac{1}{2}, -2) \\
f(x_1^*, x_2^*) &= -\tfrac{41}{4}.
\end{aligned}$$

For B_2. Let

$$\begin{aligned}
g(\alpha) &= f(\alpha(5, -2) + (1 - \alpha)(5, 6)), \qquad 0 \le \alpha \le 1 \\
&= f(5, 6 - 8\alpha) \\
&= 5^2 - 5 + 3(6 - 8\alpha)^2 + 18(6 - 8\alpha) + 14.
\end{aligned}$$

Then

$$g'(\alpha) = 6(6 - 8\alpha)(-8) - 144,$$

which is zero at α^*. Therefore

$$\alpha^* = \tfrac{9}{8} \notin [0, 1].$$

Hence the global minimum for f cannot lie on B_2.

For B_3. Let

$$\begin{aligned}
g(\alpha) &= f(\alpha(-1, 6) + (1 - \alpha)(5, 6)) \\
&= f(5 - 6\alpha, 6) \\
&= (5 - 6\alpha)^2 - (5 - 6\alpha) + (3)6^2 + 18(6) + 14.
\end{aligned}$$

Then
$$g'(\alpha) = 2(5 - 6\alpha)(-6) + 6,$$

which is zero at α^*. Therefore

$$\alpha^* = \tfrac{3}{4} \in [0,1] \quad \text{and} \quad g''(\alpha^*) > 0,$$

indicating a minimum. Therefore

$$(x_1^*, x_2^*) = \tfrac{3}{4}(-1,6) + \tfrac{1}{4}(5,6).$$
$$= (\tfrac{1}{2}, 6)$$

and

$$f(x_1^*, x_2^*) = 229\tfrac{3}{4}.$$

For B_4. Let

$$\begin{aligned}
g(\alpha) &= f(\alpha(-1,-2) + (1-\alpha)(-1,6)) \\
&= f(-1, 6 - 8\alpha) \\
&= (-1)^2 - (-1) + 3(6 - 8\alpha)^2 + 18(6 - 8\alpha) + 14.
\end{aligned}$$

Then

$$g'(\alpha) = 6(6 - 8\alpha)(-8) - 8(18),$$

which is zero at α^*. Therefore

$$\alpha^* = \tfrac{9}{8} \notin [0,1].$$

Hence the global minimum for f cannot lie on B_4.

As the lowest value for f occurs on the boundary of B_1, we have

$$X^* = (\tfrac{1}{2}, -2)$$
$$f(X^*) = -\tfrac{41}{4}.$$

14(a)

$$h_1(X) = x_1 + x_2 + x_3 + x_4 - 2 = 0$$
$$h_2(X) = 3x_1 + 2x_2 + 4x_3 + x_4 - 3 = 0$$
$$h_3(X) = x_1 + 4x_2 + 3x_3 + x_4 - 1 = 0.$$

Let

$$X = (x_1, x_2, x_3, x_4) = (w_1, w_2, w_3, y_1)$$
$$W = (w_1, w_2, w_3)$$
$$Y = y_1$$

$$J = \begin{pmatrix} 1 & 1 & 1 \\ 3 & 2 & 4 \\ 1 & 4 & 3 \end{pmatrix}$$

$$J^{-1} = \begin{pmatrix} 2 & -\tfrac{1}{5} & -\tfrac{2}{5} \\ 1 & -\tfrac{2}{5} & \tfrac{1}{5} \\ -2 & \tfrac{3}{5} & \tfrac{1}{5} \end{pmatrix}$$

$$K = \begin{pmatrix} 1 \\ 1 \\ 1 \end{pmatrix}.$$

Then

$$\frac{\partial f(W, Y)}{\partial y} = Vf(x_4) - Vf(x_1, x_2, x_3)J^{-1}K$$

$$= -2x_4 + (8x_1, 6x_2, 12x_3)J^{-1}K$$

$$= -2x_4 + (8x_1, 6x_2, 12x_3)\begin{pmatrix} 2 & -\frac{1}{5} & -\frac{2}{5} \\ 1 & -\frac{2}{5} & \frac{1}{5} \\ -2 & \frac{3}{5} & \frac{1}{5} \end{pmatrix}\begin{pmatrix} 1 \\ 1 \\ 1 \end{pmatrix}$$

$$= \frac{56}{5}x_1 + \frac{24}{5}x_2 - \frac{72}{5}x_3 - 2x_4$$

$$= 0, \quad \text{for a stationary point.}$$

Combining this equation with the previous three, we get

$$\frac{56}{5}x_1 + \frac{24}{5}x_2 - \frac{72}{5}x_3 - 2x_4 = 0$$
$$x_1 + x_2 + x_3 + x_4 = 2$$
$$3x_1 + 2x_2 + 4x_3 + x_4 = 3$$
$$x_1 + 4x_2 + 3x_3 + x_4 = 1,$$

which can be solved to yield

$$X^* = (0.58, -0.39, 0.08, 1.73).$$

This point is a maximum and

$$f(X^*) = -4.86.$$

15(a)

Maximize: $4x_1 + 3x_2 \qquad = x_0$
subject to: $3x_1 + 4x_2 + \bar{x}_3 = 12$
$3x_1 + 3x_2 + \bar{x}_4 = 10$
$4x_1 + 2x_2 + \bar{x}_5 = 8$
$x_1, x_2, \bar{x}_3, \bar{x}_4, \bar{x}_5 \geq 0, \qquad i = 1, 2, \ldots, 5.$

In order to guarantee that the nonnegativity conditions are satisfied we introduce squared slack variables:

Maximize: $4x_1 + 3x_2 \qquad = x_0$
subject to: $3x_1 + 4x_2 + x_3^2 = 12$
$3x_1 + 3x_2 + x_4^2 = 10$
$4x_1 + 2x_2 + x_5^2 = 8$
$x_1, x_2 \geq 0.$

We must still guarantee that x_1 and x_2 are nonnegative. Hence we introduce the following constraints:

$$x_1 \geq 0$$
$$x_2 \geq 0$$

with squared slack variables:

$$x_1 - x_6^2 = 0$$
$$x_2 - x_7^2 = 0.$$

Substituting for x_1 and x_2, we get

Maximize: $4x_6^2 + 3x_7^2 \qquad = x_0$

subject to: $3x_6^2 + 4x_7^2 + x_3^2 = 12$ (A)

$3x_6^2 + 3x_7^2 + x_4^2 = 10$ (B)

$4x_6^2 + 2x_7^2 + x_5^2 = 8.$ (C)

There is now no need for nonnegativity conditions and the problem is in a form which is amenable to solution by the Jacobian method. Hence

$$m = 3, n = 5.$$

We must choose which variables are assigned to W and which to Y. In order to do this we call on Theorem 7.12. If the Hessian matrix is negative definite at X^*, then X^* is a candidate for a local maximum for x_0.
Let

$$W = (w_1, w_2, w_3) = (x_3, x_4, x_5)$$
$$Y = (y_1, y_2) = (x_6, x_7).$$

Then

$$x_0 = 4x_6^2 + 3x_7^2$$

and

$$\frac{\partial x_0}{\partial x_6} = 8x_6$$

$$\frac{\partial^2 x_0}{\partial x_6^2} = 8$$

$$\frac{\partial x_0}{\partial x_7} = 6x_7$$

$$\frac{\partial^2 x_0}{\partial^2 x_7} = 6.$$

Therefore

$$H_y = \begin{pmatrix} 8 & 0 \\ 0 & 6 \end{pmatrix},$$

which is positive definite, not negative definite. This choice will not lead to a local maximum.

Now let

$$W = (w_1, w_2, w_3) = (x_4, x_6, x_7)$$
$$Y = (y_1, y_2) = (x_3, x_5).$$

By (A) and (C), we have

$$5x_6^2 - x_3^2 + 2x_5^2 = 4$$

and

$$10x_7^2 + 4x_3^2 - 3x_5^2 = 24.$$

Therefore

$$x_6^2 = \tfrac{1}{5}(x_3^2 - 2x_5^2 + 4)$$
$$x_7^2 = \tfrac{1}{10}(3x_5^2 - 4x_3^2 + 24)$$
$$x_0 = \tfrac{4}{5}(x_3^2 - 2x_5^2 + 4) + \tfrac{3}{10}(3x_5^2 - 4x_3^2 + 24).$$

Therefore

$$\frac{\partial x_0}{\partial x_3} = -\frac{4}{5}x_3$$

$$\frac{\partial^2 x_0}{\partial x_3^2} = -\frac{4}{5}$$

$$\frac{\partial x_0}{\partial x_5} = -\frac{7}{5}x_5$$

$$\frac{\partial^2 x_0}{\partial x_5^2} = -\frac{7}{5}.$$

Therefore

$$H_y = \begin{pmatrix} -\frac{4}{5} & 0 \\ 0 & -\frac{7}{5} \end{pmatrix},$$

which is negative definite. Proceeding with the Jacobian method:

$$\nabla_w x_0 = (0, 8x_6, 6x_7)$$
$$\nabla_y^c x_0 = (-\tfrac{4}{5}x_3, -\tfrac{7}{5}x_5)$$

$$J = \begin{pmatrix} 0 & 6x_6 & 8x_7 \\ 2x_4 & 6x_6 & 6x_7 \\ 0 & 8x_6 & 4x_7 \end{pmatrix}.$$

Therefore

$$J^{-1} = \begin{pmatrix} -\dfrac{3}{10x_4} & \dfrac{1}{2x_4} & -\dfrac{3}{20x_4} \\ -\dfrac{1}{10x_6} & 0 & \dfrac{1}{5x_6} \\ \dfrac{1}{5x_7} & 0 & -\dfrac{3}{20x_7} \end{pmatrix}$$

$$C = \begin{pmatrix} 2x_3 & 0 \\ 0 & 0 \\ 0 & 2x_5 \end{pmatrix}.$$

Now

$$\nabla_y^c x_0 = \nabla_y x_0 - \nabla_w x_0 J^{-1} C$$

$$\nabla_y^c x_0 = (0,0) - (0, 8x_6, 6x_7) \begin{pmatrix} -\dfrac{3}{10x_4} & \dfrac{1}{2x_4} & -\dfrac{3}{20x_4} \\ -\dfrac{1}{10x_6} & 0 & \dfrac{1}{5x_6} \\ \dfrac{1}{5x_7} & 0 & -\dfrac{3}{20x_7} \end{pmatrix} \begin{pmatrix} 2x_3 & 0 \\ 0 & 0 \\ 0 & 2x_5 \end{pmatrix}$$

$$= \left(\frac{-16x_3}{10} + \frac{12x_3}{5}, \frac{16x_5}{5} - \frac{18x_5}{10} \right)$$

$$= (-\tfrac{4}{5}x_3, -\tfrac{7}{5}x_5),$$

which is what we expect from D. Therefore

$$\nabla_y^c x_0 = 0 \Rightarrow x_3^* = 0, \qquad x_5^* = 0.$$

Hence the original constraints become

$$\begin{aligned}
3x_6^2 + 4x_7^2 &= 12 \\
3x_6^2 + 3x_7^2 + x_4^2 &= 10 \\
4x_6^2 + 2x_9^2 &= 8.
\end{aligned}$$

This means that

$$\begin{aligned}
x_6^2 &= \tfrac{4}{5} \, (= x_1^*) \\
x_7^2 &= \tfrac{12}{5} \, (= x_2^*) \\
x_4^2 &= \tfrac{2}{5} \\
x_0 &= \tfrac{52}{5}.
\end{aligned}$$

17(a). Let

$$F(X, \lambda) = f(X) - \sum_{i=1}^{m} \lambda_i h_i(X),$$

where

$$\begin{aligned}
f(X) &= -4x_1^2 - 3x_2^2 - 6x_3^2 - x_4^2 \\
h_1(X) &= x_1 + x_2 + x_3 + x_4 - 2 \\
h_2(X) &= 3x_1 + 2x_2 + 4x_3 + x_4 - 3 \\
h_3(X) &= x_1 + 4x_2 + 3x_3 + x_4 - 1.
\end{aligned}$$

Then

$$\frac{\partial F}{\partial x_1} = -8x_1 - \lambda_1 - 3\lambda_2 - \lambda_3 = 0$$

$$\frac{\partial F}{\partial x_2} = -6x_2 - \lambda_1 - 2\lambda_2 - 4\lambda_3 = 0$$

$$\frac{\partial F}{\partial x_3} = -12x_3 - \lambda_1 - 4\lambda_2 - 3\lambda_3 = 0$$

$$\frac{\partial F}{\partial x_4} = -2x_4 - \lambda_1 - \lambda_2 - \lambda_3 = 0$$

$$\frac{\partial F}{\partial \lambda_1} = -(x_1 + x_2 + x_3 + x_4 - 2) = 0$$

$$\frac{\partial F}{\partial \lambda_2} = -(3x_1 + 2x_2 + 4x_3 + x_4 - 3) = 0$$

$$\frac{\partial F}{\partial \lambda_3} = -(x_1 + 4x_2 + 3x_3 + x_4 - 1) = 0,$$

which can be solved to yield:

$$x_1^* = 0.58, \quad x_2^* = -0.39, \quad x_3^* = 0.08, \quad x_4^* = 1.74$$
$$\lambda_1^* = -5.02, \quad \lambda_2^* = -0.57, \quad \lambda_3^* = 2.12.$$

Therefore

$$f(X^*) = -4.86.$$

Chapter 8

Section 8.4

The number of evaluations n must be such that

$$r \geq \frac{b_m - a_m}{b_1 - a_1} = \frac{\varepsilon}{b_1 - a_1} + \frac{1}{3A_{n-2} + 2A_{n-3}}.$$

Therefore

$$\frac{1}{10} \geq \frac{\frac{1}{9}}{10} + \frac{1}{3A_{n-2} + 2A_{n-3}}.$$

Therefore

$$n = 6.$$

Thus 6 evaluations will be necessary. The first two points are placed at

$$S_1 = -5 + (5 - (-5))(A_5/A_7)$$
$$\overline{S}_1 = -5 + (5 - (-5))(A_6/A_7).$$

Therefore

$$\overline{S}_1 = -5 + 10(\tfrac{5}{13}) = -\tfrac{15}{13}$$
$$\overline{S}_1 = -5 + 10(\tfrac{8}{13}) = \tfrac{15}{13},$$

$$f(S_1) < f(\overline{S}_1).$$

Hence the new interval becomes $[a_1 b_1] = [-\frac{15}{13}, 5]$:

$$S_2 = \bar{S}_1 = \tfrac{15}{13}$$
$$\bar{S}_2 = -\tfrac{15}{13} + (5 - (-\tfrac{15}{13}))(A_5/A_6)$$
$$= -\tfrac{15}{13} + (\tfrac{80}{13})(\tfrac{5}{8}) = \tfrac{35}{13},$$

$$f(S_2) = -2(\tfrac{15}{13})^2 + (\tfrac{15}{13}) + 4 = -2.66 + 1.15 + 4 = 2.49$$
$$f(\bar{S}_2) = -2(\tfrac{35}{13})^2 + (\tfrac{35}{13}) + 4 = 14.4 + 2.69 + 4 = -7.8,$$

$$f(S_2) > f(\bar{S}_2).$$

Hence the new interval becomes $[a_2 b_2] = [-\frac{15}{13}, \frac{35}{13}]$:

$$S_3 = -\tfrac{15}{13} + (\tfrac{35}{13} - (-\tfrac{15}{13}))(A_3/A_5)$$
$$= -\tfrac{15}{13} + \tfrac{50}{13}(\tfrac{2}{5}) = \tfrac{5}{13}$$
$$\bar{S}_3 = S_2 = \tfrac{15}{13},$$

$$f(S_3) = -2(\tfrac{5}{13})^2 + (\tfrac{5}{13}) + 4 = 0.29 + 0.38 + 4 = 4.09$$
$$f(\bar{S}_3) = -2(\tfrac{15}{13})^2 + 4 = -2.66 + 1.15 + 4 = 2.49,$$

$$f(S_3) > f(\bar{S}_3).$$

Hence the new interval becomes $[a_3, b_3] = [-\frac{15}{13}, \frac{15}{13}]$:

$$S_4 = -\tfrac{15}{13} + (\tfrac{15}{13} - (-\tfrac{15}{13}))(A_2/A_4)$$
$$= -\tfrac{15}{13} + (\tfrac{30}{13})(\tfrac{1}{3}) = -\tfrac{5}{13}$$
$$\bar{S}_4 = S_3 = \tfrac{5}{13},$$

$$f(S_4) < f(\bar{S}_4).$$

Hence the new interval becomes $[a_4, b_4] = [-\frac{5}{13}, \frac{15}{13}]$. It can be seen that the point remaining in the interval $[a_4, b_4]$, namely, $\bar{S}_4 = \frac{5}{13}$, is exactly at the centre of $[a_4, b_4]$. Also $S_5 = \bar{S}_4$. Thus in order to place $\bar{S}_5$ symmetrically it must coincide with S_5, which is of no advantage. Hence $\bar{S}_5$ is placed ε to the right of S_5. Thus

$$\bar{S}_5 = \tfrac{1}{2}(-\tfrac{5}{13} + \tfrac{15}{13}) + \tfrac{1}{9} = \tfrac{58}{117},$$

$$f(S_5) = -2(\tfrac{5}{13})^2 + (\tfrac{5}{13}) + 4 = -0.29 + 0.38 + 4 = 4.09$$
$$f(\bar{S}_5) = -2(\tfrac{58}{117})^2 + (\tfrac{58}{117}) + 4 = -0.49 + 0.49 + 4 = 4,$$

$$f(S_5) > f(\bar{S}_5).$$

Hence the final interval is $\left[-\frac{5}{13}, \frac{58}{117}\right]$. The length of the interval is $\frac{103}{117}$, which is only 8.8% as long as the original interval.

The first two points are placed at

$$S_0 = -5 + (5 - (-5))\left(\frac{3 - \sqrt{5}}{2}\right)$$

$$= 10 - 5\sqrt{5} = 5(\sqrt{2} - \sqrt{5})$$

$$\bar{S}_0 = -5 + (5 - (-5))\frac{(\sqrt{5} - 1)}{2}$$

$$= 5\sqrt{5} - 10 = 5(\sqrt{5} - 2),$$

$$f(S_1) > f(\bar{S}_1).$$

Hence the new interval becomes $[a_2, b_2] = [10 - 5\sqrt{5}, 25 - 10\sqrt{5}]$:

$$S_2 = 5(2 - \sqrt{5}) + [5(5 - 2\sqrt{5}) - 5(2 - \sqrt{5})]\left(\frac{3 - \sqrt{5}}{2}\right)$$

$$= 45 - 20\sqrt{5} = 5(9 - 4\sqrt{5})$$
$$\bar{S}_2 = S_1 = 5\sqrt{5} - 10,$$

$$f(S_2) > f(\bar{S}_2).$$

Hence the new interval becomes $[a_3, b_3] = [10 - 5\sqrt{5}, 5\sqrt{5} - 10]$:

$$S_3 = 5(2 - \sqrt{5}) + [5(\sqrt{5} - 2) - 5(2 - \sqrt{5})]\left(\frac{3 - \sqrt{5}}{2}\right)$$

$$= -45 + 20\sqrt{5} = 5(4\sqrt{5} - 9)$$
$$\bar{S}_3 = S_2 = 5(9 - 4\sqrt{5}) = 45 - 20\sqrt{5},$$

$$f(S_3) < f(\bar{S}_3).$$

Hence the new interval becomes $[a_4, b_4] = [5(4\sqrt{5} - 9), 5(\sqrt{5} - 2)]$.

$$S_5 = \bar{S}_4 = 45 - 20\sqrt{5}$$

$$\bar{S}_5 = 5(4\sqrt{5} - 9) + [5(\sqrt{5} - 2) - 5(4\sqrt{5} - 9)]\left[\frac{\sqrt{5} - 1}{2}\right]$$

$$= 45\sqrt{5} - 100 = 5(9\sqrt{5} - 20),$$

$$f(S_4) > f(\bar{S}_4).$$

Hence the final interval is $[a_5, b_5] = [5(4\sqrt{5} - 9), 5(9\sqrt{5} - 20)]$. This interval has length

$$5(9\sqrt{5} - 20) - 5(4\sqrt{5} - 9) = 25\sqrt{5} - 55$$
$$= 5(5\sqrt{5} - 11).$$

The original interval is 10; this interval is a little over 9% of the original interval in length.

3(a)

$$f(S) = -2S^2 + 5 + 4 \quad \text{and} \quad [a_0, b_0] = [-5, 5].$$

Therefore

$$f'(S) = -4S + 1.$$

Let

$$S_i = \frac{a_i + b_i}{2}.$$

Then

$$f'(S_0) > 0, \quad \text{so that} \quad [a_1, b_1] = [0, 5],$$

and

$$f'(S_1) = f'(2.5) < 0, \quad \text{so that} \quad [a_2, b_2] = [0, 2.5].$$
$$f'(S_2) = f'(1.25) < 0 \quad \text{so that} \quad [a_3, b_3] = [0, 1.25].$$
$$f'(S_3) = f'(0.625) < 0 \quad \text{so that} \quad [a_4, b_4] = [0, 0.625].$$
$$f'(S_4) = f'(0.3125) < 0 \quad \text{so that} \quad [a_5, b_5] = [0, 0.3125].$$
$$f'(S_5) = f'(0.15625) > 0 \quad \text{so that} \quad [a_6, b_6] = [0.15625, 0.3125].$$

Hence, after only 6 iterations the interval has been reduced to one of length $0.3125 - 0.15626$. This is equal to $0.15625/10$ or 1.5625% of the original length

4(a). Refer to the solution to Exercise 3.

5(a). $X^* = (0, 0)$.

(b). $X^* = (2, 4)$.

(c). $X^* = (\frac{3}{2}, 1)$.

(d). $X^* = (-3, -\frac{3}{2})$.

(e). $X^* = (0, 0)$.

(f). $X^* = (0, 0)$.

(g). $X^* = (1, -2)$.

(h). $X^* = (2, 3)$.

6. Refer to Exercise 5.

7. Refer to Exercise 5.

8. Refer to Exercise 5.

9(a). Let

$$X_0 = (0, 0, 0)$$

then

$$f(X_0) = -98$$

and

$$\frac{\partial f}{\partial x_1} = -6(x_1 - 2)$$

$$\frac{\partial f}{\partial x_2} = -8(x_2 - 3)$$

$$\frac{\partial f}{\partial x_3} = -4(x_3 + 5).$$

Therefore

$$Vf = (-6(x_1 - 2), -8(x_2 - 3), -4(x_3 + 5))$$
$$Vf(X_0) = (12, 24, -20)$$
$$X_1 = X_0 + sD$$
$$= (0, 0, 0) + s(12, 24, -20).$$

Hence

$$f(X_1) = -3(12s - 2)^2 - 4(24s - 3)^2 - 2(-20s + 5)^2$$

and

$$\frac{df(X_1)}{ds} = -6(12s - 2)(12) - 8(24s - 3)(24) - 4(-20s + 5)(-20)$$

which is zero at s^*. Therefore

$$s^* = 0.158 \quad \text{and} \quad \frac{d^2 f(X_1)}{ds^2} < 0$$

indicating a maximum. Therefore

$$X_1 = (1.9, 3.8, -3.2)$$
$$f(X_1) = -9.07 > f(X_0)$$
$$Vf(X_1) = (0.6, -6.4, -7.2),$$
$$X_2 = (1.9, 3.8, -3.2) + 5(0.6, -6.4, -7.2).$$

Therefore

$$f(X_2) = -9.07 - 2.68.6s^2 + 96.16s$$

and

$$\frac{df(X_2)}{ds} = -537.2s + 93.16,$$

which is zero at s^*. Therefore

$$s^* = 0.173 \quad \text{and} \quad \frac{d^2 f(X_2)}{ds^2} < 0,$$

indicating a maximum. Therefore

$$X_2 = (2, 2.7, -4.4).$$
$$f(X_2) = -1.08 > f(X_1)$$
$$Vf(X_2) = (0, 2.4, -2.4),$$
$$X_3 = (2, 2.7 + 2.4s, -4.4 - 2.4s).$$

Therefore

$$f(X_3) = -1.08 + 11.52s - 34.56s^2$$

and

$$\frac{df(X_3)}{ds} = 11.52 - 69.12s,$$

which is zero at s^*. Therefore

$$s^* = 0.1666 \quad \text{and} \quad \frac{d^2f(x_3)}{ds^2} < 0,$$

indicating a maximum. Therefore

$$X_3 = (2.0, 3.0, -4.8), \qquad f(X_3) > f(X_2)$$
$$Vf(X_3) = (0, 0, -0.8),$$
$$X_4 = (2, 3, -4.8) + s(0, 0, -0.8)$$
$$= (2, 3, -4.8 - 0.8s).$$

Therefore

$$f(X_4) = -1.28s^2 + 0.64s - 0.08$$

and

$$\frac{df(X_4)}{ds} = -2.56s + 0.64,$$

which is zero at s^* Therefore

$$s^* = 0.25, \quad \text{and} \quad \frac{d^2f(X_4)}{ds^2} < 0,$$

indicating a maximum. Therefore

$$X_4 = (2, 3, -4.8) + (0, 0, -0.2)$$
$$= (2, 3, -5)$$
$$f(X_4) = 0 > f(X_3)$$
$$Vf(X_4) = 0.$$

Therefore

$$X^* = (2, 3, -5)$$
$$f(X^*) = 0.$$

10(a). From 9(a),

$$X_0 = (0,0,0)$$

$$X_2 = (2, 2.7, -4.4).$$

The search direction is

$$X_2 - X_0 = (-2, -2.7, 4.4).$$

Therefore

$$X_3 = X_0 + s(X_2 - X_0)$$
$$= (0,0,0) + s(-2, -2.7, 4.4)$$

so that

$$f(X_3) = -3(2s - 2)^2 - 4(2.7s - 3)^2 - 2(-4.4s + 5)^2$$

and

$$\frac{df(X_3)}{ds} = -24s + 24 - 58.32s + 64.8 - 77.44s + 88,$$

which is zero at s^*. Therefore

$$s^* = 1.106 \quad \text{and} \quad \frac{d^2 f(X_3)}{ds^2} < 0,$$

indicating a maximum. Therefore

$$X_3 = (2.2, 3.0, -4.9)$$

$$X_4 = X_3 + sD$$

$$\frac{df}{dx_1} = -6(2.2 - 2) = -1.2$$

$$\frac{df}{dx_2} = -8(3 - 3) = 0$$

$$\frac{df}{dx_3} = -4(-4.9 + 5) = -0.4.$$

Therefore

$$X_4 = (2.2, 3.0, -4.9) + s(-1.2, 0, -0.4),$$

so that

$$f(X_4) = -3(2.2 - 1.2s - 2)^2 - 4(3 - 3)^2 - 2(-4.9 - 0.4s + s)^2$$

and

$$\frac{df(X_4)}{ds} = -6(-1.2s + .2)(-1.2) - 4(0.1 - 0.4s)(-0.4),$$

which is zero at s^*. Therefore

$$s^* = 0.1724, \quad \text{and} \quad \frac{d^2f(X_4)}{ds} < 0,$$

indicating a maximum. Therefore

$$X_4 = (2.0, 3.0, -5.0),$$

which is the maximum point of f from 9(a).

11(a)

$$f(X) = (12, 24, -20)\begin{pmatrix} x_1 \\ x_2 \\ x_3 \end{pmatrix} + \tfrac{1}{2}(x_1, x_2, x_3)\begin{pmatrix} -6 & 0 & 0 \\ 0 & -8 & 0 \\ 0 & 0 & -4 \end{pmatrix}\begin{pmatrix} x_1 \\ x_2 \\ x_3 \end{pmatrix} - 98$$

$$X_0 = 0$$

$$\nabla f(X_0) = (12, 24, -20)$$

$$\frac{df}{\partial x_1} = -6(x_1 - 2) = 12 \text{ at } x_1 = 0$$

$$\frac{\partial f}{\partial x_2} = -8(x_2 - 3) = 24 \text{ at } x_2 = 0$$

$$\frac{\partial f}{\partial x_3} = -4(x_3 + 5) = -20 \text{ at } x_3 = 0.$$

From 9(a),

$$X_1 = (1.9, 3.8, -3.2)$$

$$a_0 = \frac{(-0.6, 6.4, 7.2)\begin{pmatrix} -6 & 0 & 0 \\ 0 & -8 & 0 \\ 0 & 0 & -4 \end{pmatrix}\begin{pmatrix} 12 \\ 24 \\ -20 \end{pmatrix}}{(12, 24, -20)\begin{pmatrix} -6 & 0 & 0 \\ 0 & -8 & 0 \\ 0 & 0 & -4 \end{pmatrix}\begin{pmatrix} 12 \\ 24 \\ -20 \end{pmatrix}}$$

$$= -0.2843$$

$$D_1 = \nabla f(X_1) + a_0 D_0$$
$$= (0.6, -6.4, 7.2) - 0.2843(12, 24, -20)$$
$$= (-2.8166, -13.2232, -1.514)$$

$$X_2 = X_1 + s_1 D_1$$
$$= (1.9, 3.8, -3.2) + s(-2.8, -13.2, -1.5)$$

$$f(X_2) = -3(1.9 - 2.8s - 2)^2 - 4(3.8 - 13.2s - 3)^2 - 2(-3.2 - 1.5s + 5)^2$$

$$\frac{df(X_2)}{ds} = -6(-2.8s - 0.1)(-2.8) - 8(0.8 - 13.2s)(-13.2) - 4(1.8 - 1.5s)(-1.5),$$

which is zero at s^*. Also

$$\frac{d^2 f(X_2)}{ds^2} < 0,$$

indicating a maximum point. Therefore

$$s^* = 0.0645$$

$$X_2 = (1.9, 3.8, -3.2) + 0.0645(-2.8, -13.2, -1.5)$$
$$= (1.7, 2.9, -3.3)$$

$$Vf(X_2) = (-6(1.7 - 2), -8(2.9 - 3), -4(3.3 + 5))$$
$$= (1.8, 0.8, -6.06)$$

$$a_1 = \frac{(-1.8, -0.8, 6.06)\begin{pmatrix} -6 & 0 & 0 \\ 0 & -8 & 0 \\ 0 & 0 & -4 \end{pmatrix}\begin{pmatrix} -2.81 \\ -13.22 \\ -1.514 \end{pmatrix}}{(-2.81, -13.22, -1.51)\begin{pmatrix} -6 & 0 & 0 \\ 0 & -8 & 0 \\ 0 & 0 & -4 \end{pmatrix}\begin{pmatrix} -2.81 \\ -13.22 \\ -1.51 \end{pmatrix}}$$

$$= 0.538257$$

$$D_2 = Vf(X_2) + a_1 D_1$$
$$= (1.8, 0.8, -6.06) + 0.0538257(-2.8166, -13.2232, -1.514)$$
$$= (1.648, 0.088, -6.141)$$

$$X_3 = X_2 + sD_2$$
$$= (1.7, 2.9, -3.3) + s(1.648, 0.088, -6.141)$$

$$f(X_3) = -3(-0.3 + 1.648s)^2 - 4(-0.1 + 0.088s)^2 - 2(1.7 - 6.141s)^2$$

$$\frac{df(X_3)}{ds} = -6(-0.3 + 1.648s)(1.648) - 8(-0.1 + 0.088s)(0.088)$$
$$- 4(1.7 - 6.141s)(-6.141),$$

which is zero at s^*. Therefore

$$s^* = 0.4275 \quad \text{and} \quad \frac{d^2 f(X_3)}{ds^2} < 0,$$

indicating a maximum point. Therefore

$$X_3 = (1.7, 2.9, -3.3) + 0.4275(1.648, 0.088, -6.141)$$
$$= (2.4, 3.0, -5.9).$$

At the next iteration

$$X_4 = (2.0, 3.0, -5.0),$$

which is the maximum point of f.

12(a)

$$\frac{\partial f}{\partial x_1} = -3(x_1 - 3)^2$$

$$\frac{\partial^2 f}{\partial x_1^2} = -6(x_1 - 3)$$

$$\frac{\partial f}{\partial x_2} = -2(x_2 - 4)$$

$$\frac{\partial^2 f}{\partial x_2^2} = -2$$

$$\frac{\partial^2 f}{\partial x_1 \, \partial x_2} = \frac{\partial^2 f}{\partial x_2 \, \partial x_1} = 0.$$

Therefore

$$H(X) = \begin{pmatrix} -6(x_1 - 3) & 0 \\ 0 & -2 \end{pmatrix}.$$

Assume

$$X_0 = (0, 0).$$

Then

$$X_1 = X_0 - H^{-1}(X_0)\nabla f(X_0)$$

$$= (0, 0) - \begin{pmatrix} \frac{1}{18} & 0 \\ 0 & -\frac{1}{2} \end{pmatrix} \begin{pmatrix} -27 \\ 8 \end{pmatrix}$$

$$= (\tfrac{27}{18}, 4)$$

$$H(X_1) = \begin{pmatrix} \frac{117}{4} & 0 \\ 0 & -2 \end{pmatrix}$$

$$\nabla f(X_1) = \begin{pmatrix} -\frac{27}{4} \\ 0 \end{pmatrix}.$$

Therefore

$$X_2 = X_1 - H^{-1}(X_1)\nabla f(X_1)$$

$$= (\tfrac{27}{18}, 4) - \begin{pmatrix} \frac{4}{117} & \\ & -\frac{1}{2} \end{pmatrix} \begin{pmatrix} -\frac{27}{4} \\ 0 \end{pmatrix}$$

$$= (1.73, 4)$$

$$\nabla f(X_2) = \begin{pmatrix} -4.83 \\ 0 \end{pmatrix}$$

$$H(X_2) = \begin{pmatrix} 7.62 & 0 \\ 0 & -2 \end{pmatrix}.$$

Therefore

$$X_3 = X_2 - H^{-1}(X_2)\nabla f(X_2)$$

$$= (1.73, 4) - \begin{pmatrix} 0.13 & 0 \\ 0 & -0.5 \end{pmatrix}\begin{pmatrix} -4.83 \\ 0 \end{pmatrix}$$

$$= (2.36, 4)$$

$$\nabla f(X_3) = \begin{pmatrix} -1.23 \\ 0 \end{pmatrix}$$

$$H(X_3) = \begin{pmatrix} 3.84 & 0 \\ 0 & -2 \end{pmatrix}$$

$$X_4 = X_3 - H^{-1}(X_3)\nabla f(X_3)$$

$$= (2.34, 4) - \begin{pmatrix} 0.26 & 0 \\ 0 & -0.5 \end{pmatrix}\begin{pmatrix} -1.23 \\ 0 \end{pmatrix}$$

$$= (2.75, 4)$$

$$\nabla f(X_4) = \begin{pmatrix} -18 \\ 0 \end{pmatrix}$$

$$H(X_4) = \begin{pmatrix} 1.5 & 0 \\ 0 & -2 \end{pmatrix}.$$

Therefore

$$X_5 = X_4 - H^{-1}(X_4)\nabla f(X_4)$$

$$= (2.75, 4) - \begin{pmatrix} 0.67 & 0 \\ 0 & -0.5 \end{pmatrix}\begin{pmatrix} -0.18 \\ 0 \end{pmatrix}$$

$$= (2.87, 4)$$

$$\nabla f(X_5) = \begin{pmatrix} -0.06 \\ 0 \end{pmatrix}$$

$$H(X_5) = \begin{pmatrix} 0.78 & 0 \\ 0 & -2 \end{pmatrix}.$$

Therefore

$$X_6 = X_5 - H^{-1}(X_5)\nabla f(X_5)$$

$$= (2.87, 4) - \begin{pmatrix} 1.28 & 0 \\ 0 & -0.5 \end{pmatrix}\begin{pmatrix} -0.06 \\ 0 \end{pmatrix}$$

$$= (2.95, 4)$$

Eventually

$$X^* = (3, 4), \qquad f(X^*) = 1.$$

13(a)

$$X^* = (3, 4), \qquad f(X^*) = 1.$$

13(b)
$$X^* = (-2, 1).$$

13(c)
$$X^* = (0, 2, 0).$$

14(b). $X^* = (\frac{8}{3}, 0.4, 0.0182)$.

   (c). $X^* = (1.094, 1.077, 1.041)$.

   (d). $X^* = (0.9073, 1.0561, 0.8211)$.

   (e). $X^* = (2(\frac{1}{4})^{1/4}, (\frac{1}{4})^{1/14}, (\frac{1}{4})^{-1/28})$.

   (f). $X^* = (1.82, 0.234, \sqrt{2}/2)$.

   (g). $X^* = (1.5004, 0.4992, 3.9936)$.

   (h). $X^* = (0.2878, 2.3438, 1.8420)$.

   (i). $X^* = (1.03, 0.103, 0.872, -1.05)$.

Index

In the case of duplicated page numbers the major reference is given in boldface. Authors not referred to here can be found in the references.

Undergraduate Texts in Mathematics